Collected Papers

Volume V

Alice Silverberg · Hiroyuki Yoshida
Editors

Goro Shimura

Collected Papers

Volume V

2002–2012

Springer

Editors
Alice Silverberg
Department of Mathematics
University of California
Irvine, CA
USA

Hiroyuki Yoshida
Department of Mathematics
Kyoto University
Kyoto
Japan

ISBN 978-3-319-32547-7 ISBN 978-3-319-32548-4 (eBook)
DOI 10.1007/978-3-319-32548-4

Library of Congress Control Number: 2016941602

Printed on acid-free paper

This Springer imprint is published by Springer Nature
The registered company is Springer International Publishing AG Switzerland

Contents

Preface

This book is Volume V of Goro Shimura's Collected Papers, containing the articles from the year 2002 onwards, excluding books, and including a note [06e] on the "Woods Hole formula" that was not published but was circulated around 2006. The following List of Articles contains a list of the eleven articles that appear in this volume, along with six books that appeared in English and eight books that appeared in Japanese, in addition to the lists that appeared in Volumes I through IV.

February 2016

Alice Silverberg
Hiroyuki Yoshida

List of Articles

The numbers in parentheses are the articles not included in this collection.

Volume 1

[(52)] On a certain ideal of the center of a Frobeniusean algebra, Scientific Papers of the College of General Education, University of Tokyo, 2 (1952), 117–124.

[54] A note on the normalization-theorem of an integral domain, Scientific Papers of the College of General Education, University of Tokyo, 4 (1954), 1–8.

[55] Reduction of algebraic varieties with respect to a discrete valuation of the basic field, American Journal of Mathematics, 77 (1955), 134–176.

[56] On complex multiplications, Proceedings of the International Symposium on Algebraic Number Theory, Tokyo-Nikko, 1955, Science Council of Japan, Tokyo (1956), 23–30.

[57a] La fonction ζ du corps des fonctions modulaires elliptiques, Comptes rendus des séances de l' Académie des Sciences, 244 (1957), 2127–2130.

[(57b)] Kindai-teki Seisu-ron (Modern Number Theory, in Japanese, with Yutaka Taniyama), Kyoritsu Shuppan, 1957.

[58a] Correspondances modulaires et les fonctions zeta de courbes algébriques, Journal of the Mathematical Society of Japan, 10 (1958), 1–28.

[58b] Modules des variétés abéliennes polarisées et fonctions modulaires, Séminaire Henri Cartan, École Normale Supérieure, 1957/58, *Fonctions Auomorphes*, Exposé 18–20 (1958).

[(58c)] Fonctions automorphes et variétés abéliennes, Séminaire Bourbaki, 1957/58, Exposé 167 (1958).

[59a] Fonctions automorphes et correspondances modulaires, Proceedings of the International Congress of Mathematicians, Edinburgh, 1958 (1959), 330–338.

[59b] On the theory of automorphic functions, Annals of Mathematics, 70 (1959), 101–144.

[59c] Sur les intégrales attachées aux formes automorphes, Journal of the Mathematical Society of Japan, 11 (1959), 291–311.

[59d] On specializations of abelian varieties (with Shoji Koizumi), Scientific Papers of the College of General Education, University of Tokyo, 9 (1959), 187–211.

[60a] On vector differential forms attached to automorphic forms (with Michio Kuga), Journal of the Mathematical Society of Japan, 12 (1960), 258–270.

[(60b)] Automorphic functions and number theory: I (in Japanese), Sugaku, 11 (1960), 193–205.

[(61a)] Complex multiplication of abelian varieties and its applications to number theory (with Yutaka Taniyama), Publications of the Mathematical Society of Japan, No. 6 (1961).

[61b] On the zeta functions of the algebraic curves uniformized by certain automorphic functions, Journal of the Mathematical Society of Japan, 13 (1961), 275–331.

[(61c)] Automorphic functions and number theory: II (in Japanese), Sugaku, 13 (1961), 65–80.

[62a] On Dirichlet series and abelian varieties attached to automorphic forms, Annals of Mathematics, 76 (1962), 237–294.

[62b] On the class-fields obtained by complex multiplication of abelian varieties, Osaka Mathematical Journal, 14 (1962), 33–44.

[63a] Arithmetic of alternating forms and quaternion hermitian forms, Journal of the Mathematical Society of Japan, 15 (1963), 33–65.

[63b] On analytic families of polarized abelian varieties and automorphic functions, Annals of Mathematics, 78 (1963), 149–192.

[63c] On the cohomology groups attached to certain vector valued differential forms on the product of the upper half planes (with Yozo Matsushima), Annals of Mathematics, 78 (1963), 417–449.

[63d] On modular correspondences for $Sp(n, \mathbf{Z})$ and their congruence relations, Proceedings of the National Academy of Sciences, 49 (1963), 824–828.

[63e] On the fields of definition for fields of automorphic functions, Proceedings of the 1963 Number Theory Conference, Boulder, Colorado, August 5–24, 1963, 26–32.

[64a] Arithmetic of unitary groups, Annals of Mathematics, 79 (1964), 369–409.

[64b] On the field of definition for a field of automorphic functions, Annals of Mathematics, 80 (1964), 160–189.

[64c] Class-fields and automorphic functions, Annals of Mathematics, 80 (1964), 444–463.

[64d] On purely transcendental fields of automorphic functions of several variables, Osaka Journal of Mathematics, 1 (1964), 1–14.

[64e] The zeta function of an algebraic variety and automorphic functions, Summer Research Institute on Algebraic Geometry, Woods Hole, Massachusetts, July 6–31, 1964 (mimeographed notes of lectures).

[65a] On the field of definition for a field of automorphic functions: II, Annals of Mathematics, 81 (1965), 124–165.

[65b] On the zeta function of a fibre variety whose fibres are abelian varieties (with Michio Kuga), Annals of Mathematics, 82 (1965), 478–539.

[66a] A reciprocity law in non-solvable extensions, Journal für die reine und angewandte Mathematik, 221 (1966), 209–220.

[66b] Moduli and fibre systems of abelian varieties, Annals of Mathematics, 83 (1966), 294–338.

[66c] On the field of definition for a field of automorphic functions: III, Annals of Mathematics, 83 (1966), 377–385.

[66d] Moduli of abelian varieties and number theory, Proceedings of the Symposium on Pure Mathematics, 9, *Algebraic groups and discontinuous subgroups*, 1965, American Mathematical Society, (1966), 312–332.

Volume II

[67a] Discontinuous groups and abelian varieties, Mathematische Annalen, 168 (1967), 171–199.

[67b] Construction of class fields and zeta functions of algebraic curves, Annals of Mathematics, 85 (1967), 58–159.

[67c] Number fields and zeta functions associated with discontinuous groups and algebraic varieties, Proceedings of the International Congress of Mathematicians, Moscow, 1966, 100–107 (1967).

[67d] Algebraic number fields and symplectic discontinuous groups, Annals of Mathematics, 86 (1967), 503–592.

[68a] Algebraic varieties without deformation and the Chow variety, Journal of the Mathematical Society of Japan, 20 (1968), 336–341.

[(68b)] Automorphic functions and number theory, Lecture notes in mathematics 54, Springer, 1968.

[68c] An *l*-adic method in the theory of automorphic forms, the text of a lecture at the conference *Automorphic functions for arithmetically defined groups*, Oberwolfach, Germany, July 28–August 3, 1968.

[69] Local representations of Galois groups, Annals of Mathematics, 89 (1969), 99–124.

[70a] On canonical models of arithmetic quotients of bounded symmetric domains, Annals of Mathematics, 91 (1970), 144–222.

[70b] On canonical models of arithmetic quotients of bounded symmetric domains: II, Annals of Mathematics, 92 (1970), 528–549.

[71a] On arithmetic automorphic functions, Proceedings of the International Congress of Mathematicians, Nice, 1970, vol. 2, 343–348 (1971).

[71b] On the zeta-function of an abelian variety with complex multiplication, Annals of Mathematics, 94 (1971), 504–533.

[71c] Class fields over real quadratic fields in the theory of modular functions, in *Several Complex Variables* II, Maryland 1970, Lecture notes in mathematics 185 (1971), 169–188.

[(71d)] Introduction to the arithmetic theory of automorphic functions, Publications of the Mathematical Society of Japan, No. 11, Iwanami Shoten and Princeton University Press, 1971.

[71e] On elliptic curves with complex multiplication as factors of the Jacobians of modular function fields, Nagoya Mathematical Journal, 43 (1971), 199–208.

[72a] On the field of rationality for an abelian variety, Nagoya Mathematical Journal, 45 (1972), 167–178.

[72b] Class fields over real quadratic fields and Hecke operators, Annals of Mathematics, 95 (1972), 130–190.

[73a] On modular forms of half integral weight, Annals of Mathematics, 97 (1973), 440–481.

[(73b)] Complex multiplication, Proceedings of the International Summer School of Modular Functions of One Variable, Antwerp, 1972, Lecure notes in mathematics, 320 (1973), 37–56.

[(73c)] Modular forms of half integral weight, Proceedings of the International Summer School of Modular Functions of One Variable, Antwerp, 1972, Lecure notes in mathematics, 320 (1973), 57–74.

[73d] On the factors of the jacobian variety of a modular function field, Journal of the Mathematical Society of Japan, 25 (1973), 523–544.

[74] On the trace formula for Hecke operators, Acta mathematica, 132 (1974), 245–281.

[75a] On the holomorphy of certain Dirichlet series, Proceedings of the London Mathematical Society, 3rd ser. 31 (1975), 79–98.

[75b] On the real points of an arithmetic quotient of a bounded symmetric domain, Mathematische Annalen, 215 (1975), 135–164.

[75c] On some arithmetic properties of modular forms of one and several variables, Annals of Mathematics, 102 (1975), 491–515.

[75d] On the Fourier coefficients of modular forms of several variables, Nachrichten der Akademie der Wissenschaften in Göttingen, Mathematisch-Physikalische Klasse, 1975, 261–268.

[76a] Theta functions with complex multiplication, Duke Mathematical Journal, 43 (1976), 673–696.

[76b] The special values of the zeta functions associated with cusp forms, Communications on pure and applied Mathematics, 29 (1976), 783–804.

[77a] On abelian varieties with complex multiplication, Proceedings of the London Mathematical Society, 3rd ser. 34 (1977), 65–86.

[77b] Unitary groups and theta functions, Proceedings of the International Symposium on Algebraic Number Theory, Kyoto, 1976 (1977), 195–200.

[77c] On the derivatives of theta functions and modular forms, Duke Mathematical Journal, 44 (1977), 365–387.

[77d] On the periods of modular forms, Mathematische Annalen, 229 (1977), 211–221.

Volume III

[78a] On certain reciprocity-laws for theta functions and modular forms, Acta mathematica, 141 (1978), 35–71.

[78b] The arithmetic of automorphic forms with respect to a unitary group, Annals of Mathematics, 107 (1978), 569–605.

[78c] The special values of the zeta functions associated with Hilbert modular forms, Duke Mathematical Journal, 45 (1978), 637–679.

[79a] Automorphic forms and the periods of abelian varieties, Journal of the Mathematical Society of Japan, 31 (1979), 561–592.

[79b] On some problems of algebraicity, Proceedings of the International Congress of Mathematicians, Helsinki, 1978 (1979), 373–379.

[80] The arithmetic of certain zeta functions and automorphic forms on orthogonal groups, Annals of Mathematics, 111 (1980), 313–375.

[81a] The critical values of certain zeta functions associated with modular forms of half-integral weight, Journal of the Mathematical Society of Japan, 33 (1981), 649–672.

[81b] On certain zeta functions attached to two Hilbert modular forms: I. The case of Hecke characters, Annals of Mathematics, 114 (1981), 127–164.

[81c] On certain zeta functions attached to two Hilbert modular forms: II. The case of automorphic forms on a quaternion algebra, Annals of Mathematics, 114 (1981), 569–607.

[81d] Arithmetic of differential operators on symmetric domains, Duke Mathematical Journal, 48 (1981), 813–843.

[(81e)] Corrections to "The special values of the zeta functions associated with Hilbert modular forms," vol 45 (1978), 637–679, Duke Mathematical Journal, 48 (1981), 697.

[82a] Models of an abelian variety with complex multiplication over small fields, Journal of Number Theory, 15 (1982), 25–35.

[82b] The periods of certain automorphic forms of arithmetic type, Journal of the Faculty of Science, University of Tokyo, Sec. IA, 28 (1982), 605–632.

[82c] Confluent hypergeometric functions on tube domains, Mathematische Annalen, 260 (1982), 269–302.

[83a] Algebraic relations between critical values of zeta functions and inner products, American Journal of Mathematics, 105 (1983), 253–285.

[83b] On Eisenstein series, Duke Mathematical Journal, 50 (1983), 417–476.

[84a] Differential operators and the singular values of Eisenstein series, Duke Mathematical Journal, 51 (1984), 261–329.

[84b] On differential operators attached to certain representations of classical groups, Inventiones mathematicae, 77 (1984), 463–488.

[85a] On Eisenstein series of half-integral weight, Duke Mathematical Journal, 52 (1985), 281–314.

[85b] On the Eisenstein series of Hilbert modular groups, Revista Matemática Iberoamericana, 1 (1985), 1–42.

[86] On a class of nearly holomorphic automorphic forms, Annals of Mathematics, 123 (1986), 347–406.
[87a] Nearly holomorphic functions on hermitian symmetric spaces, Mathematische Annalen, 278 (1987), 1–28.
[87b] On Hilbert modular forms of half-integral weight, Duke Mathematical Journal, 55 (1987), 765–838.
[88] On the critical values of certain Dirichlet series and the periods of automorphic forms, Inventiones mathematicae, 94 (1988), 245–305.

Volume IV

[89a] Yutaka Taniyama and his time, Bulletin of the London Mathematical Society, 21 (1989), 186–196.
[89b] *L*-functions and eigenvalue problems, *Algebraic analysis, geometry, and number theory*, Proceedings of the JAMI Inaugural Conference 1988, Supplement to the American Journal of Mathematics, 1989, 341–396.
[90a] Invariant differential operators on hermitian symmetric spaces, Annals of Mathematics, 132 (1990), 237–272.
[90b] On the fundamental periods of automorphic forms of arithmetic type, Inventiones mathematicae, 102 (1990), 399–428.
[(90c)] Some old and recent arithmetical results concerning modular forms and related zeta functions, University of Istanbul, Faculty of Science, Journal of Mathematics, 49 (1990), 45–56.
[91] The critical values of certain Dirichlet series attached to Hilbert modular forms, Duke Mathematical Journal, 63 (1991), 557–613.
[(93a)] Arithmeticity of the special values of various zeta functions and the periods of abelian integrals (in Japanese), Sugaku, 45 (1993), 111–127; English translation by Toshitsune Miyake, Sugaku expositions, 8 (1995), 17–38.
[93b] On the transformation formulas of theta series, American Journal of Mathematics, 115(1993), 1011–1052.
[93c] On the Fourier coefficients of Hilbert modular forms of half-integral weight, Duke Mathematical Journal, 71 (1993), 501–557.
[94a] Fractional and trigonometric expressions for matrices, American Mathematical Monthly, 101 (1994), 744–758.
[94b] Euler products and Fourier coefficients of automorphic forms on symplectic groups, Inventiones mathematicae, 116 (1994), 531–576.
[94c] Differential operators, holomorphic projection, and singular forms, Duke Mathematical Journal, 76 (1994), 141–173.
[95a] Eisenstein series and zeta functions on symplectic groups, Inventiones mathematicae, 119 (1995), 539–584.
[95b] Zeta functions and Eisenstein series on metaplectic groups, Inventiones mathematicae, 121 (1995), 21–60.
[96a] Convergence of zeta functions on symplectic and metaplectic groups, Duke Mathematical Journal, 82 (1996), 327–347.

[96b] Response, Notices of the American Mathematical Society, vol. 43, No. 11 (November 1996), 1344–1347.

[(97a)] Euler Products and Eisenstein series, CBMS Regional Conference Series in Mathematics, No. 93, American Mathematical Society, 1997.

[97b] Zeta functions and Eisenstein series on classical groups, Proceedings of the National Academy of Sciences, 94,11133–11137 (1997).

[(98)] Abelian varieties with complex multiplication and modular functions, Princeton University Press, 1998.

[99a] An exact mass formula for orthogonal groups, Duke Mathematical Journal, 97 (1999), 1–66.

[99b] The number of representations of an integer by a quadratic form, Duke Mathematical Journal, 100 (1999), 59–92.

[99c] Generalized Bessel functions on symmetric spaces, Journal für die reine und angewandte Mathematik, 509 (1999), 35–66.

[99d] Some exact formulas on quaternion unitary groups, Journal für die reine und angewandte Mathematik, 509 (1999), 67–102.

[99e] André Weil as I knew him, Notices of the American Mathematical Society, vol. 46, No. 4 (April 1999), 428-433.

([00)] Arithmeticity in the theory of automorphic forms, Mathematical Surveys and Monographs, vol. 82, American Mathematical Society, 2000.

[(01a)] Letter to the editor, Notices of the American Mathematical Society, vol. 48, No. 7 (August 2001), 678.

[01b] Arithmeticity of Dirichlet series and automorphic forms on unitary groups, unpublished.

[01c] The relative regulator of an algebraic number field, unpublished.

Volume V

List of papers 2002–2010

[02] The representation of integers as sums of squares, Amer. J. Math. 124 (2002), 1059–1081.

[04] Inhomogeneous quadratic forms and triangular numbers, Amer. J. Math. 126 (2004), 191–214.

[06a] Quadratic Diophantine equations and orders in quaternion algebras, Amer. J. Math. 128 (2006), 481–518.

[06b] Quadratic Diophantine equations, the class number, and the mass formula, Bull. Amer. Math. Soc. (N.S.) 43 (2006), 285–304.

[06c] Integer-valued quadratic forms and quadratic Diophantine equations, Doc. Math. 11 (2006), 333–367.

[06d] Classification, construction, and similitudes of quadratic forms, Amer. J. Math. 128 (2006), 1521–1552.

[08a] Classification of integer-valued symmetric forms, Amer. J. Math. 130 (2008), 685–711.

[08b] Arithmetic of Hermitian forms. Doc. Math. 13 (2008), 739–774.

[08c] The critical values of certain Dirichlet series, Doc. Math. 13 (2008), 775–794.

[10] The critical values of generalizations of the Hurwitz zeta function, Doc. Math. 15 (2010), 489–506.

[06e] Some remarks and open problems on the Woods Hole fixed point formula, unpublished, 2 pp.

List of books 2004–2012
(* indicates that the book treats non-mathematical topics exclusively.)

[(B1)] Arithmetic and analytic theories of quadratic forms and Clifford groups, Mathematical Surveys and Monographs, 109. American Mathematical Society, Providence, RI, 2004. x+275 pp. ISBN: 0-8218-3573-4.

[(B2)] Elementary Dirichlet series and modular forms, Springer Monographs in Mathematics. Springer, New York, 2007. viii+147 pp. ISBN: 978-0-387-72473-7.

[(B3)] The map of my life, Springer, New York, 2008. vi+211 pp. ISBN: 978-0-387-79714-4.

[(B4)] Arithmetic of quadratic forms, Springer Monographs in Mathematics. Springer, New York, 2010. xii+237 pp. ISBN: 978-1-4419-1731-7.

[(B5)] Modular forms: basics and beyond, Springer Monographs in Mathematics. Springer, New York, 2012. x+175 pp. ISBN: 978-1-4614-2124-5; 978-1-4614-2125-2.

[(*B6)] The story of Imari, Ten speed press, 2008,. ISBN-13: 978-1-58-8-896-1. ISBN-10: 1-58008-896-1.

List of books written in Japanese 2006–2014

- *中国説話文学とその背景 (Chugoku setsuwabungaku to sono haikei), ちくま学芸文庫, 筑摩書房, 2006. ISBN: 4-480-09007-X.
- 記憶の切繪図 (Kioku no kiriezu), 筑摩書房, 2008. ISBN: 978-4-480-86069-9.
- *中国古典文学私選ー凡人と非凡人の物語 (Chugoku kotenbungaku shisen—bonjin to hibonjin no aida), 明倫出版社, 2008. ISBN: 978-4-89619-191-2.
- *鳥のように (tori no yôni), 筑摩書房, 2010. ISBN: 978-4-480-86071-2.
- 数学をいかに使うか (Sugaku wo ikani tsukauka), ちくま学芸文庫, 筑摩書房, 2010. ISBN: 978-4-480-09325-7. (How to use mathematics)
- 数学の好きな人のために (Sugaku no sukina hito no tameni), ちくま学芸文庫, 筑摩書房, 2012. ISBN: 978-4-480-09439-1. (For those who love mathematics)
- 数学で何が重要か (Sugaku de naniga jyuuyouka), ちくま学芸文庫, 筑摩書房, 2013. ISBN: 978-4-480-09534-3. (What are important in mathematics)
- 数学をいかに教えるか (Sugaku wo ikani oshieruka), ちくま学芸文庫, 筑摩書房, 2014. ISBN: 978-4-480-09630-2. (How to teach mathematics)

THE REPRESENTATION OF INTEGERS AS SUMS OF SQUARES

By Goro Shimura

Abstract. We present a uniform method by which we obtain an explicit formula for the number of representations of an integer as the sum of n squares for each n in the range $2 \le n \le 8$. The final formulas for odd n are new, though those for even n are not. We also present a formula for the dominant term for the number for an arbitrary n.

Introduction. In this paper we are interested in the sum of squares $\Phi[x] = \sum_{i=1}^{n} x_i^2$, viewed as a quadratic form on an n-dimensional vector space V of row vectors over $\mathbf{Q}$. For a $\mathbf{Z}$-lattice M in V and a positive integer h we denote by $N(M, h)$ the number of elements $x \in M$ such that $\Phi[x] = h$. Here we assume that $\Phi[x] \in \mathbf{Z}$ for every $x \in M$. If L denotes the standard lattice consisting of all row vectors with components in $\mathbf{Z}$, then $N(L, h)$ is the number of representations of h as the sum of n squares, and the determination of this number is one of the most frequently discussed classical problems in number theory. If n is even and ≤ 8, then this was settled in a clear-cut way. However, though the case of odd $n \le 7$ has been treated, the previous results are far from decisive. Indeed, referring to the odd-dimensional case, Hardy wrote, "The results do not seem to have been worked out so systematically as those for even k, and it is not possible to refer to a comprehensive statement of them" (see [H2, p. 157]). That was in 1940, but the situation after more than 60 years seems to be the same. The main aim of this paper is to present a systematic treatment and a comprehensive, and in fact clear-cut, statement. We will do so by treating both odd and even cases by the same methods, though we have to state the final formulas separately. Our formulas for a certain quantity $\mathfrak{n}(L, h)$ introduced below will be first given for an arbitrary n, from which the formulas for $N(L, h)$ for $n \le 8$ will be derived as immediate consequences, as $\mathfrak{n}(L, h) = N(L, h)$ for any such n.

To explain our methods and results, we first take a set of lattices $\{M_i\}_{i=1}^{k}$ which are representatives for the classes belonging to the genus of M with respect to the orthogonal group $G = \{\gamma \in GL_n(\mathbf{Q}) \mid \gamma \cdot {}^t\gamma = 1_n\}$. Then we put

$$\mathfrak{m}(M) = \sum_{i=1}^{k} [\Gamma_i : 1]^{-1}, \tag{0.1}$$

Manuscript received January 14, 2002.
Shimura, Goro. "The representation of integers as sums of squares." American Journal of Mathematics 124:5 (2002), 1059–1081.

$$(0.2)\qquad R(M,h)=\sum_{i=1}^{k}[\Gamma_i:1]^{-1}N(M_i,h),$$

$$(0.3)\qquad \mathfrak{n}(M,h)=R(M,h)/\mathfrak{m}(M),$$

where $\Gamma_i=\{\gamma\in G \mid M_i\gamma=M_i\}$.

In the previous paper [S3] we gave exact formulas for $R(M,h)$ and $\mathfrak{n}(M,h)$ when M is *maximal* in the sense that it is maximal among the lattices on which Φ takes values in $\mathbf{Z}$. (In fact, our formulas in [S3] cover the case of an arbitrary totally definite quadratic form Φ over an arbitrary totally real algebraic number field.) This result does not apply to the standard lattice L if $n>3$, since such an L is not maximal. However, in this paper we shall determine the quotient $\mathfrak{n}(L,h)/\mathfrak{n}(M,h)$, which leads to explicit formulas for $\mathfrak{n}(L,h)$ (Theorem 2.2, Propositions 3.6 and 3.7). If the genus of L consists of a single class, which is the case if $n\le 8$, we obtain an explicit formula for $N(L,h)$ for each such n; see §3.9.

Since the case of even n is relatively easy, we state here the formula for $\mathfrak{n}(L,h)$ only for odd n. Let $h=2^\nu g$ with a nonnegative integer ν and an odd positive integer g; put $g=r^2t$ with a positive integer r and a squarefree integer t; take the prime decomposition $r=\prod_p p^{\lambda_p}$. Then, for $n=2m+1>3$ we have

$$\mathfrak{n}(L,h)=\frac{2^n m(m-1)!(2\pi)^{-m}L(m,\psi_h)(2^m-\psi_h(2))}{|B_{n-1}|(2^{2m}-1)}\cdot g^{(n-2)/2}$$
$$\cdot\,\tilde{b}(h)\prod_{p|r}\left\{\sum_{k=0}^{\lambda_p}p^{k(2-n)}-\psi_h(p)\sum_{i=1}^{\lambda_p}p^{i(2-n)+(n-3)/2}\right\}.$$

Here B_{n-1} is the Bernoulli number, ψ_h is the primitive Dirichlet character corresponding to $\mathbf{Q}((-1)^{m/2}h^{1/2})$, $L(s,\psi_h)$ is the L-function of ψ_h, and $\tilde{b}(h)$ is the quantity given as

$$q^\nu+\varphi_1(n)2^{-1/2}\cdot\frac{q^{\nu+1}-2^{n-1}+1}{2^{n-2}-1}\qquad\text{or}$$
$$q^\nu+\varphi_1(n)2^{m-1}\cdot\frac{q^\nu-1}{q^2-1}+\varphi_1(n)\varphi(ng)2^{-m}+\varphi((g+2-n)/2)2^{2-n},$$

according as ν is odd or even, where $q=2^{(n-2)/2}, \varphi_1(n)=(-1)^{(n^2-1)/8}$, and φ is the primitive Dirichlet character modulo 4. If $n=5$ or 7, then $\mathfrak{n}(L,h)$ is the number of representations of h as the sum of n squares. All previous results concern either the case of squarefree h, or somewhat more generally, the number of primitive representations of h, that is, representations of h as the sum of n squares that have no nontrivial common divisors. The paper [H1] of Hardy was his attempt at the determination of $N(L, h)$ for $n = 5$ or 7, which he called "one of the most interesting and difficult problems in the Theory of Numbers," but he gave only an answer to the case of primitive representations, without presenting a decisive result.

In our proof we first show that $\mathfrak{n}(L, h)/\mathfrak{n}(M, h)$ can be expressed as a certain product whose essential factor is an infinite series involving Gauss sums over $\mathbf{Z}/2^k\mathbf{Z}$. The main technical point is to determine this series as a rational function of $2^{(n-2)/2}$ in an explicit form, which can be done in a few pages, and so the whole proof is short, in fact, much shorter than what the reader might expect from the above expression.

Though most readers will principally be interested in the case of sums of squares, we will also state the results about $\mathfrak{n}(L_0, h)$ for maximal L_0 (Theorem 1.5), as we believe they are as basic as those for sums of squares. They include the formula for the number of representations by three squares, which seems to be new.

1. Preliminaries and the theorem for a maximal lattice.

1.1. For an associative ring R with identity element we denote by $R^\times$ the group of all its invertible elements and by R_n^m the R-module of all $m \times n$-matrices with entries in R. For a finite set X we denote by $\#(X)$ or simply by $\#X$ the number of elements in X. For a Dirichlet character χ we denote by $L(s, \chi)$ the L-function of χ, that is, $L(s, \chi) = \sum_{k=1}^{\infty} \chi(k)k^{-s}$. Our Dirichlet character χ will always be primitive, and we adopt the usual convention that $\chi(k) = 0$ for every $k \in \mathbf{Z}$ that is not prime to the conductor of χ. The most frequently used nontrivial character is the primitive character modulo 4, which we denote by φ. For a prime number p we denote by $\mathbf{Z}_p$ and $\mathbf{Q}_p$ the ring of p-adic integers and the field of p-adic numbers. We put $\mathbf{e}(z) = \exp(2\pi i z)$ for $z \in \mathbf{C}$, and define a (standard) charater $\mathbf{e}_p$ of the additive group $\mathbf{Q}_p$ so that $\mathbf{e}_p(p^{-m}) = \mathbf{e}(-p^{-m})$ for $0 \leq m \in \mathbf{Z}$ and $\mathbf{e}_p(\mathbf{Z}_p) = 1$.

Throughout the paper, we put $V = \mathbf{Q}_n^1$ and $L = \mathbf{Z}_n^1$, and denote by L_0 a maximal lattice containing L. Given any $\mathbf{Z}$-lattice M in V, we put

$$\tilde{M} = \{y \in V \mid 2x \cdot {}^t y \in \mathbf{Z} \text{ for every } x \in M\}. \tag{1.1}$$

For each p we denote by r_p the dimension of a maximal totally isotropic subspace of $V_p = V \otimes_{\mathbf{Q}} \mathbf{Q}_p$, and put $t_p = n - 2r_p$. Then our quadratic form is equivalent, over $\mathbf{Q}_p$, to the form represented by a matrix

$$\begin{bmatrix} 0 & 0 & 1_{r_p} \\ 0 & c_p 1_{t_p} & 0 \\ 1_{r_p} & 0 & 0 \end{bmatrix} \tag{1.2}$$

with $c_p = \pm 1$. For $p = 2$, the numbers r_p and t_p can be determined as in [S2, §5.16]. If $p \equiv 1 \pmod 4$, then 1_2 is isotropic, and the matter is easy. If $p \equiv 3 \pmod 4$, then 1_2 is anisotropic and equivalent to the norm form of the unramified quadratic extension K_p of Q_p. Since -1 is the norm of an element of K_p, 1_2 is equivalent to -1_2 over $\mathbf{Q}_p$. Therefore it is easy to determine r_p and t_p. The results are as follows:

(I) $p \neq 2$:

$$t_p = \begin{cases} 1 & \text{if } n \text{ is odd,} \\ 2 & \text{if } n \equiv 2 \pmod 4 \text{ and } p \equiv 3 \pmod 4, \\ 0 & \text{otherwise.} \end{cases} \tag{1.3}$$

(II) $p = 2$:

$$t_p = a \qquad \text{if } n \equiv \pm a \pmod 8,\ 0 \leq a \leq 4. \tag{1.4}$$

We need c_p only when $p = 2$ and $n \equiv 2 \pmod 4$, in which case

$$c_2 = (-1)^{(n-2)/4}. \tag{1.5}$$

We also note (see [S2, (3.2.1)]) that

$$[\tilde{L}_0 : L_0] = \begin{cases} 2^a & \text{if } n \equiv \pm a \pmod 8,\ 0 \leq a \leq 3, \\ 4 & \text{if } n \equiv 4 \pmod 8. \end{cases} \tag{1.6}$$

We first recall [S3, Theorem 1.5] specialized to the present situation.

PROPOSITION 1.2. *Let* $0 < h \in \mathbf{Z}$. *Define* $\gamma_p(s)$ *for each prime* p *by the formulas given in §1.3 below.*

(I) *Suppose* n *is odd; put* $n = 2m+1$ *and* $K = \mathbf{Q}(\kappa^{1/2})$ *with* $\kappa = (-1)^m h$; *denote by* ψ_h *the primitive Dirichlet character corresponding to* K. *Then we have*

$$\mathfrak{n}(L_0, h) = \frac{2^{(n-2)}(m-1)!(2\pi)^{-m}L(m, \psi_h)}{f(n-2)!(2\pi)^{1-n}\zeta(n-1)} \cdot h^{(n-2)/2} \prod_{p|2h} \gamma_p(n/2),$$

where $f = 2$ *if* $n \equiv \pm 3 \pmod 8$ *and* $f = 1$ *otherwise.*

(II) *Suppose* n *is even; put* $n = 2m$ *and denote by* ψ *the trivial character or* φ *according as* m *is even or odd. Then*

$$\mathfrak{n}(L_0, h) = \frac{a_n h^{m-1}}{[\tilde{L}_0 : L_0]^{1/2}(m-1)!(2\pi)^{-m}L(m, \psi)} \prod_{p|h^*} \gamma_p(m).$$

Here $a_n = 1$ *if* $n > 2$, *and* $a_2 = 1/2$; $h^* = h$ *if* $8|n$ *and* $h^* = 2h$ *if* $8 \nmid n$.

The formula for $\mathfrak{n}(L_0, h)$ in [S3, Theorem 1.5] for even n involves a factor $N(\mathfrak{d})^{1/2}N(\mathfrak{f})$, which equals $[\tilde{L}_0 : L_0]^{1/2}$ by [S3, (5.4.3)].

1.3. To give the formulas for $\gamma_p(s)$, determine $\nu \in \mathbf{Z}$ so that p^ν is the highest power of p that divides h, and put $\lambda = [\nu/2]$. If t_p is odd, we put $\psi_h(p) = \xi$. Now $\gamma_p(s)$ can be given as follows:

$$\gamma_p(s) = (1 - p^{(\nu+1)(1-s)})/(1 - p^{1-s}) \quad \textit{if} \quad t_p = 0,$$

$$\gamma_p(s) = \frac{1 - p^{(\lambda+1)(2-2s)} - \xi p^{(1/2)-s}(1 - p^{\lambda(2-2s)})}{1 - p^{2-2s}}$$

if $t_p = 1$ *except when* $p = 2$ *and* $\nu \in 2\mathbf{Z}$,

$$\gamma_p(s) = \frac{1 - p^{(\nu+4)(1-s)} - \xi p^{(1/2)-s}(1 - p^{(\nu+2)(1-s)})}{1 - p^{2-2s}}$$

if $t_p = 1,\ p = 2, \nu \in 2\mathbf{Z},$ *and* $\xi \neq 0$,

$$\gamma_p(s) = \frac{1 - p^{(\nu+2)(1-s)}}{1 - p^{2-2s}} \quad \textit{if} \quad t_p = 1,\ p = 2, \nu \in 2\mathbf{Z},\ \textit{and}\ \xi = 0,$$

$$\gamma_p(s) = (1 - (-p^{1-s})^{\nu+1})/(1 + p^{1-s}) \quad \textit{if} \quad t_p = 2 \ \textit{and}\ p \neq 2,$$

$$\gamma_p(s) = 1 + \varphi(2^{-\nu}mh)p^{(\nu+2)(1-s)} \quad \textit{if} \quad t_p = 2 \ \textit{and}\ p = 2,$$

$$\gamma_p(s) = \frac{1 - p^{3-2s} - p^{(\lambda+1)(2-2s)}(1 - \xi p^{(3/2)-s})(p^{2-2s} + \xi p^{(3/2)-s})}{(1 + \xi p^{(1/2)-s})(1 - p^{2-2s})}$$

if $t_p = 3$ *and* $\xi \neq 0$,

$$\gamma_p(s) = \frac{1 - p^{3-2s} + p \cdot p^{(\nu+\varepsilon+1)(1-s)} - p^{(\nu+\varepsilon+3)(1-s)}}{(1 - p^{1-2s})(1 - p^{2-2s})} \quad \textit{if} \quad t_p = 3 \ \textit{and}\ \xi = 0,$$

where $\varepsilon = 1$ *if* $\nu \in 2\mathbf{Z}$ *and* $\varepsilon = 0$ *if* $\nu \notin 2\mathbf{Z}$,

$$\gamma_p(s) = \frac{1 - p^{(\nu+2)(1-s)} - p^{2-s}(1 - p^{\nu(1-s)})}{(1 - p^{-s})(1 - p^{1-s})} \quad \textit{if} \quad t_p = 4.$$

These formulas are special cases of those given in [S3, §1.6]. If $p = 2$ and $t_2 = 2$, then the number c_2 of (1.5) equals $\varphi(m)$, and hence the form $1 + \psi_v(c_v h)(|h|_v^{-1}N(\mathfrak{d}_v))^{1-s}$ in [S3, p. 66] becomes $1 + \varphi(2^{-\nu}mh)p^{(\nu+2)(1-s)}$. We also needed integers κ, e, and ε in [S3, §1.6]. Clearly $\kappa = 1$ and $e = 3$, since $\mathbf{Q}$ is the basic field. Also, as noted on the last line of [S3, p. 66], we have $\varepsilon = 1$ if $\nu \in 2\mathbf{Z}$ and $\psi_h(2) = 0$. Therefore the formulas for γ_p there take the forms as stated above.

1.4. The above formulas for $\mathfrak{n}(L_0, h)$ involve special values of the Riemann zeta function and Dirichlet L-functions. There are well-known expressions for these values as algebraic numbers up to powers of π. For the reader's convenience, we note the formulas here.

Given a nontrivial primitive Dirichlet character χ of conductor d, we put

$$\tau(\chi) = \sum_{a=1}^{d} \chi(a)\mathbf{e}(a/d),$$

and define the Bernoulli polynomial $B_k(x)$ for $0 \le k \in \mathbf{Z}$ by

$$\frac{te^{xt}}{e^t - 1} = \sum_{k=0}^{\infty} \frac{B_k(x)t^k}{k!},$$

where t and x are independent variables. Then, for every positive integer k such that $\chi(-1) = (-1)^k$ we have

$$2 \cdot k! \cdot \tau(\bar{\chi})(2\pi i)^{-k}L(k, \chi) = -\sum_{a=1}^{d} \bar{\chi}(a)B_k(a/d). \tag{1.7}$$

We also note two well-known facts:

$$2 \cdot m! \cdot (2\pi)^{-m}\zeta(m) = |B_m| \quad \text{if } 0 < m \in 2\mathbf{Z}, \tag{1.8}$$

$$2^{2m+1}(2k)! \cdot (2\pi)^{-m}L(m, \varphi) = E_k \quad \text{if } m = 2k+1 \text{ with } 0 \le k \in \mathbf{Z}, \tag{1.9}$$

where B_m is the mth Bernoulli number and E_k the kth Euler number.

The product $h^{(n-2)/2}\prod \gamma_p(n/2)$ can be simplified considerably depending on the nature of n and h. Indeed, employing (1.8) and (1.9), we can state the formulas of Proposition 1.2 in simpler forms as follows:

THEOREM 1.5. *The notation being as in Proposition 1.2, let $h = 2^\nu g$ with a nonnegative integer ν and an odd positive integer g; let $g = r^2 t$ with a positive integer r and a squarefree integer t. Then $\mathfrak{n}(L_0, h)$ can be given as follows:*

$$\frac{n}{|B_m|}\sum_{d|h} d^{m-1} \qquad \text{if } n \equiv 0 \pmod 8,$$

$$\frac{m}{|B_m|}2^{\nu(m-1)}\gamma_2(m)\sum_{d|g} d^{m-1} \qquad \text{if } n \equiv 4 \pmod 8,$$

$$\frac{a_n 2^n}{E_{(m-1)/2}}\{\varphi(g)2^{\nu(m-1)} + \varphi(m)2^{2-n}\}\sum_{d|g}\varphi(d)d^{m-1} \qquad \text{if } n \equiv 2 \pmod 4,$$

$$\frac{2^n m \cdot (m-1)!(2\pi)^{-m}L(m, \psi_h)}{f|B_{n-1}|} \cdot h^{(n-2)/2}\gamma_2(n/2)$$
$$\cdot \prod_{p|r}\left\{\sum_{k=0}^{\lambda_p} p^{k(2-n)} - \psi_h(p)\sum_{i=1}^{\lambda_p} p^{i(2-n)+(n-3)/2}\right\} \qquad \text{if } n \text{ is odd.}$$

Here $\sum_{d|h}$ *or* $\sum_{d|g}$ *denotes the sum over all the positive divisors d of h or g;* λ_p *is determined so that* $r = \prod_p p^{\lambda_p}$ *is the prime decomposition of r.*

Proof. For even n and odd p we have $t_p = 0$ except when $n \equiv 2 \pmod 4$ and $\varphi(p) = -1$, in which case $t_p = 2$. Therefore, from the formulas for γ_p in §1.3 we obtain

$$g^{m-1}\prod_{p|g}\gamma_p(m) = \begin{cases} \sum_{d|g} d^{m-1} & \text{if } n \equiv 0 \pmod 4, \\ \sum_{d|g} \varphi(g/d)d^{m-1} & \text{if } n \equiv 2 \pmod 4. \end{cases} \tag{1.10}$$

Combining this with the formula for γ_2, we easily obtain

$$h^{m-1}\prod_{p|h^*}\gamma_p(m) = \begin{cases} \sum_{d|h} d^{m-1} & \text{if } n \equiv 0 \pmod 8, \\ 2^{\nu(m-1)}\gamma_2(m)\sum_{d|g} d^{m-1} & \text{if } n \equiv 4 \pmod 8, \end{cases} \tag{1.11}$$

$$h^{m-1}\prod_{p|2h}\gamma_p(m) = \{2^{\nu(m-1)} + \varphi(mg)2^{2-n}\}\sum_{d|g}\varphi(g/d)d^{m-1} \quad \text{if } n \equiv 2 \pmod 4. \tag{1.12}$$

Substituting these into the formulas of Proposition 1.2 and employing (1.6), (1.8), and (1.9), we obtain the formulas for even n as above. If n is odd, we have $t_p = 1$ for all odd p, and $\gamma_p(s) = 1$ for $p \nmid r$, and hence

$$h^{(n-2)/2}\prod_{p|2h}\gamma_p(n/2) = h^{(n-2)/2}\prod_{p|2r}\gamma_p(n/2). \tag{1.13}$$

In view of the formula for γ_p when p is odd and $t_p = 1$, we obtain the formulas for $\mathfrak{n}(L_0, h)$ as stated in our theorem.

The quantity $\gamma_2(m)$ for $n \equiv 4 \pmod 8$ and $\gamma_p(n/2)$ for odd n are determined by the formulas for γ_p in §1.3. Though they are not so complicated, they can be simplified considerably in some special cases. For example, the notation being the same as in the above theorem, for $n = 3$ we have

$$\gamma_p(3/2) = 1 \qquad \text{if } p \neq 2 \text{ and } \psi_h(p) = 1, \tag{1.14a}$$

$$2^{\nu/2}\gamma_2(3/2) = \begin{cases} 0 & \text{if } \psi_h(2) = 1, \\ 2 & \text{if } \nu \in 2\mathbf{Z} \text{ and } \psi_h(2) \neq 1, \\ 2^{3/2} & \text{if } \nu \notin 2\mathbf{Z}. \end{cases} \tag{1.14b}$$

For $n = 4$ we have

(1.15) $$\gamma_2(2) = 2^{1-\nu}.$$

In the general case, however, it is best to keep them as given in §1.3, without further trying to put them in some other forms.

LEMMA 1.6. *The genus of L_0 consists of a single class if $n \leq 9$.*

Proof. This is well known for $n \leq 4$. The result for $n = 8$ is also classical; see [S2, p. 34], in which we also showed that the class number is 1 for $n = 9$. To treat the remaining cases, put $\Gamma = \Gamma_n = \{\gamma \in G \mid L_0\gamma = L_0\}$ and denote by $\{e_i\}_{i=1}^n$ the standard basis of $\mathbf{Q}_n^1$. Let $n = 7$ and $M = \sum_{i=1}^4 \mathbf{Z}e_i + \sum_{i=1}^3 \mathbf{Z}f_i$, where $f_i = 2^{-1}\sum_{\nu=0}^2 e_{i+\nu} + 2^{-1}e_{i+4}$. Then we easily see that our quadratic form takes integral values on M and $[\tilde{M} : M] = 2$, and hence we can put $L_0 = M$. Now this lattice is what is usually written E_7, and it is well known that Γ_7 has order $2^{10} \cdot 3^4 \cdot 5 \cdot 7$ (see [C, p. 422]). On the other hand, from the formulas for $\mathfrak{m}(L_0)$ in [S2, p. 34] we easily find that $\mathfrak{m}(L_0)$ is exactly the inverse of that order. Therefore the class number is 1. Next let $n = 5$. Put $V' = \sum_{i=1}^4 \mathbf{Q}e_i$ and consider it a subspace of $V = \sum_{i=1}^5 \mathbf{Q}e_i$. Take a maximal lattice Λ' in V' and put $\Lambda = \Lambda' + \mathbf{Z}e_5$. Since $[\tilde{\Lambda}' : \Lambda'] = 4$, we obtain $[\tilde{\Lambda} : \Lambda] = 8$. Comparing this with (1.6), we can put $L_0 = \Lambda$. Now it can easily be verified that $\pm e_5$ are the only vectors in L_0 whose norm is 1 and whose orthogonal complement in L_0 is a maximal lattice in a four-dimensional space. Thus every element of Γ_5 sends e_5 to $\pm e_5$. From this we can conclude that $[\Gamma_5 : 1] = 2[\Gamma_4 : 1] = 2^8 \cdot 3^2$, since it is well known that $[\Gamma_4 : 1] = 2^7 \cdot 3^2$. Now the formula for $\mathfrak{m}(L_0)$ in [S2, p. 34] for $n = 5$ shows that $\mathfrak{m}(L_0)$ is exactly the inverse of that number, and hence the class number is 1. Finally let $n = 6$. Denote by G_p and $(L_0)_p$ the localizations of G and L_0 at a prime p; put

$$C_p = \{\gamma \in G_p \mid (L_0)_p\gamma = (L_0)_p\},$$
$$C_p^* = \{\gamma \in C_p \mid (\tilde{L}_0)_p(\gamma - 1) \subset 2p(L_0)_p\}.$$

If $p \neq 2$, then $(\tilde{L}_0)_p = (L_0)_p$, and a well-known principle shows that Γ can be injectively mapped into C_p/C_p^*, so that $[\Gamma : 1]$ divides $[C_p : C_p^*]$. For $p = 2$ we have $C_2^* \subset \{\gamma \in C_2 \mid (L_0)_2(\gamma - 1) \subset 4(L_0)_2\}$, and so the same principle shows that $[\Gamma : 1]$ divides $[C_2 : C_2^*]$. Now $[C_p : C_p^*]$ was determined in [S2, Proposition 3.9]. The formulas given there (in the case $n = 6$) show that the largest odd divisor of $[C_2 : C_2^*]$ is $3^2 \cdot 5$ and the highest power of 2 dividing $[C_3 : C_3^*]$ is 2^{10}. Therefore $[\Gamma : 1]$ divides $2^{10} \cdot 3^2 \cdot 5$. On the other hand, the formula for $\mathfrak{m}(L_0)$ in [S2, p. 34] for $n = 6$ shows that $\mathfrak{m}(L_0)$ is exactly the inverse of the last number, which proves the case $n = 6$.

1.7. By the above lemma we have $N(L_0, h) = \mathfrak{n}(L_0, h)$ if $n \leq 9$. For simplicity put $R_n(h) = N(L_0, h)$. Specializing the formulas of Theorem 1.5 to such cases and employing (1.14a, b) and (1.15), we thus obtain the following formulas:

$$\begin{aligned} R_2(h) &= 2(1+\varphi(g))\sum_{d|g}\varphi(d), \\ R_4(h) &= 24\sum_{d|g} d, \\ R_6(h) &= \{\varphi(g)2^{2\nu+6}-4\}\sum_{d|g}\varphi(d)d^2, \\ R_8(h) &= 240\sum_{d|h} d^3, \end{aligned}$$

$$R_3(h) = \begin{cases} 0 & \text{if } \nu \in 2\mathbf{Z} \text{ and } g \equiv 7 \pmod 8, \\ A\pi^{-1}L(1,\psi_h)g^{1/2}\prod_p^* \left\{ \sum_{k=0}^{\lambda_p} p^{-k} - \psi_h(p)\sum_{i=1}^{\lambda_p} p^{-i} \right\} & \text{otherwise}, \end{cases}$$

where A is 24 or $24\sqrt{2}$ according as $\nu \in 2\mathbf{Z}$ or $\nu \notin 2\mathbf{Z}$, the symbol $\prod_p^*$ denotes the product over all the prime factors p of r such that $\psi_h(p) \neq 1$, and λ_p is determined for each p as in Theorem 1.5.

Since $L = L_0$ for $n \leq 3$, $R_2(h)$ resp. $R_3(h)$ is the number of representations of h as the sum of 2 resp. 3 squares. It can easily be seen that $R_2(h) = 4\sum_{d|g}\varphi(d)$, which is due to Jacobi. Formulas for the number of representations of h as the sum of three squares that have no common divisors were given by Dirichlet and later researchers; see [H2, p. 158]. The above formula for $R_3(h)$ is perhaps new, except that a similar but more complicated formula was given by Bateman [B]. Since $L(1, \psi_h)$ can be expressed in terms of the class number of $\mathbf{Q}(\sqrt{-h})$, we have a formula for the class number in terms of $R_3(h)$, a fact observed by Gauss and Dirichlet.

As for $R_4(h)$, we can derive the formula more directly from the fact that $(1-2^{1-s})\zeta(s)\zeta(s-1)$ is the zeta function of a quaternion algebra over $\mathbf{Q}$ whose norm form is the sum of four squares.

For odd $n > 3$, it does not seem worth stating the formula for $R_n(h)$ beyond what is given in Theorem 1.5. We merely state here a relatively simple special case when $n = 9$ under the assumption that h is squarefree, including the case of even h:

$$R_9(h) = 2^6 3^2 5 \cdot (2\pi)^{-4} L(4,\psi_h)h^{7/2} \cdot \begin{cases} \left\{129 - 8\left(\dfrac{h}{2}\right)\right\} & \text{if } h \equiv 1 \pmod 4, \\ 2^7 & \text{otherwise}, \end{cases}$$

where $\left(\dfrac{h}{2}\right)$ is 1 or -1 according as $h \equiv 1 \pmod 8$ or $h \equiv 5 \pmod 8$.

1.8. In the general setting of (0.2), put $f(z, M) = \sum_{x\in M} \mathbf{e}(\Phi[x]z)$, where z is a variable on the upper half plane $\{z \in \mathbf{C} \mid \mathrm{Im}(z) > 0\}$. This is a modular form of weight $n/2$. Clearly $f(z, M) = 1 + \sum_{h=1}^{\infty} N(M, h)\mathbf{e}(hz)$. Put also $p(z, M) = 1 + \sum_{h=1}^{\infty} \mathfrak{n}(M, h)\mathbf{e}(hz)$. Then $f(z, M) - p(z, M)$ is a cusp form. This was shown by Siegel for an arbitrary positive definite quadratic form Φ over $\mathbf{Q}$ (see [Si, p. 376]); for a simple proof of the fact for the forms over an arbitrary totally real algebraic number field, see [S4, Theorem A4.4]. Siegel also showed in [Si, p. 373, Satz 3] that $p(z, M)$ is an Eisenstein series if $n > 4$. Actually the fact is true for $n \geq 2$; see [S3, Section 5].

Thus $\mathfrak{n}(M, h)$ gives the dominant term for $N(M, h)$ in the sense that the difference $N(M, h) - \mathfrak{n}(M, h)$ is the h-th coefficient of a cusp form of weight $n/2$.

2. Main theorem on $\mathfrak{n}(L, h)$.

2.1. We now turn to $\mathfrak{n}(L, h)$ with $L = \mathbf{Z}_n^1$. If $n = 2$ or 3, then $L = L_0$, and so the problem is already settled in Theorem 1.5 and §1.7. Thus in this section we assume that $n > 3$. To obtain $\mathfrak{n}(L, h)$, we first note

$$\mathfrak{n}(L, h) = [\tilde{L} : L]^{-1/2}(2\pi)^{n/2}\Gamma(n/2)^{-1}h^{(n-2)/2}\left\{\prod_p b_p(h, L; s)\right\}_{s=0}. \tag{2.1}$$

This is a special case of [S3, (5.3.2)], and

$$b_p(h, L; s) = \int_{\mathbf{Q}_p} \nu(\sigma)^{-s}\mathbf{e}_p(-\sigma h)\int_{L_p} \mathbf{e}_p(\sigma x \cdot {}^t x)\, dx\, d\sigma. \tag{2.2}$$

Here $\nu(\sigma)$ is the denominator of σ, that is, the smallest power $p^e \geq 1$ of p such that $p^e\sigma \in \mathbf{Z}_p$; the measures on $\mathbf{Q}_p$ and V_p are taken so that $\int_{\mathbf{Z}_p} d\sigma = \int_{L_p} dx = 1$. (Notice that the proof of [S3, (5.3.2)] requires the formula for $c(h, s)$ in [S3, page 73, line 8], where L is assumed to be maximal. But it can easily be seen that the formulas are valid for an arbitrary lattice L.)

Take L_0 in place of L in (2.1) and divide the formula for $\mathfrak{n}(L, h)$ by that for $\mathfrak{n}(L_0, h)$. We have $b_p(h, L; s) = b_p(h, L_0; s)$ for $p \neq 2$, since $L_p = (L_0)_p$ for such p. Therefore we obtain a fundamental formula

$$\frac{\mathfrak{n}(L, h)}{\mathfrak{n}(L_0, h)} = \left\{\frac{[\tilde{L}_0 : L_0]}{[\tilde{L} : L]}\right\}^{1/2}\left\{\frac{b_2(h, L; s)}{b_2(h, L_0; s)}\right\}_{s=0}. \tag{2.3}$$

This is valid only if $n > 3$, since $\mathfrak{n}(L_0, h)$ can be 0 otherwise. Since $[\tilde{L} : L] = 2^n$, from (1.6) we obtain

$$\{[\tilde{L}_0 : L_0]/[\tilde{L} : L]\}^{1/2} = 2^{(a-n)/2} \quad \text{if } n \not\equiv 4 \pmod 8, \tag{2.4}$$

where a is as in (1.6). If $n = 2m + 1$ and f is as in Proposition 1.2 (I), then $2^{(a-n)/2}/f = 2^{-m}$. By [S3, (3.1.2) and (5.4.1)] we have

$$(2.5) \qquad b_2(h, L_0; s) = \gamma_2(s + n/2) \cdot \begin{cases} 1 - \psi(2)2^{-s-n/2} & \text{if } n \in 2\mathbf{Z}, \\ \dfrac{1 - 2^{1-n-2s}}{1 - \psi_h(2)2^{(1-n)/2-s}} & \text{if } n \notin 2\mathbf{Z}, \end{cases}$$

where γ_2 is the function of §1.3. Therefore from the formulas of Proposition 1.2 combined with (1.8), (1.10), and (2.5) we obtain the following theorem.

THEOREM 2.2. *Let ψ and ψ_h be as in Proposition 1.2. Let $h = 2^\nu g$ with a nonnegative integer ν and an odd positive integer g; let $g = r^2 t$ with a positive integer r and a squarefree integer t. Put*

$$(2.6) \qquad \tilde{b}(h) = 2^{\nu(n-2)/2} b_2(h, L; 0).$$

Then, for the standard lattice $L = \mathbf{Z}_n^1$ the quantity $\mathfrak{n}(L, h)$ can be given as follows:

(I) $n = 2m + 1 > 3$:

$$\mathfrak{n}(L, h) = \frac{2^n m(m-1)!(2\pi)^{-m} L(m, \psi_h)(2^m - \psi_h(2))}{|B_{n-1}|(2^{2m} - 1)} \cdot g^{(n-2)/2}$$
$$\cdot \tilde{b}(h) \prod_{p|r} \left\{ \sum_{k=0}^{\lambda_p} p^{k(2-n)} - \psi_h(p) \sum_{i=1}^{\lambda_p} p^{i(2-n)+(n-3)/2} \right\}.$$

Here λ_p is determined as in Theorem 1.5.

(II) $n = 2m > 2$:

$$\mathfrak{n}(L, h) = \begin{cases} \dfrac{2^{m+1}}{E_{(m-1)/2}} \cdot \tilde{b}(h) \displaystyle\sum_{d|g} \varphi(g/d) d^{m-1} & \textit{if } m \notin 2\mathbf{Z}, \\ \dfrac{n}{(2^m - 1)|B_m|} \cdot \tilde{b}(h) \displaystyle\sum_{d|g} d^{m-1} & \textit{if } m \in 2\mathbf{Z}. \end{cases}$$

We will determine $\tilde{b}(h)$ explicitly in Propositions 3.6 and 3.7. Combining those results with the above formulas, we have completely explicit formulas for $\mathfrak{n}(L, h)$, which we will state in §3.8. One of the differences of the above formulas from those of Theorem 1.5 is that instead of $L(m, \psi_h)$ there, we have $L(m, \psi_h)(2^m - \psi_h(2))$ here, which can be written $2^m \sum_{2 \nmid k} \psi_h(k) k^{-m}$.

3. Gauss sums and $b_2(h, L; s)$.

3.1. Temporarily we denote by F, $\mathfrak{g}$, and $\mathfrak{p}$ a finite algebraic extension of $\mathbf{Q}_2$, the integral closure of $\mathbf{Z}_2$ in F, and the maximal ideal of $\mathfrak{g}$; we put $q = [\mathfrak{g} : \mathfrak{p}]$ and let π denote an arbitrarily fixed prime element of F. We define characters $\mathbf{e}_\mathfrak{p}$ and χ of the additive group F by $\mathbf{e}_\mathfrak{p}(x) = \mathbf{e}_2(\mathrm{Tr}_{F/\mathbf{Q}_2}(x))$ and $\chi(x) = \mathbf{e}_\mathfrak{p}(\delta^{-1}x)$, where δ is an arbitrarily fixed element of $F^\times$ such that $\delta^{-1}\mathfrak{g} = \{y \in F \mid \mathrm{Tr}_{F/\mathbf{Q}_2}(y\mathfrak{g}) \subset \mathbf{Z}_2\}$. Then $\mathfrak{g} = \{y \in F \mid \chi(y\mathfrak{g}) = 1\}$. For $\sigma \in F$ we define a Gauss sum $\tau(\sigma)$ by

$$\tau(\sigma) = \int_\mathfrak{g} \chi(\sigma x^2)\, dx, \tag{3.1}$$

where the integral is with respect to the Haar measure of F such that $\mathfrak{g}$ has measure 1. Clearly $\tau(\sigma)$ depends only on σ modulo $\mathfrak{g}$, and $\tau(0) = 1$. We easily see that

$$\tau(\pi^{-m}a) = q^{-m} \sum_{x\in\mathfrak{g}/\mathfrak{p}^m} \chi(\pi^{-m}ax^2) \text{ if } a \in \mathfrak{g} \text{ and } 0 \le m \in \mathbf{Z}. \tag{3.2}$$

In particular,

$$\tau(\pi^{-1}a) = 0 \text{ if } a \in \mathfrak{g}^\times. \tag{3.3}$$

Indeed, since $x \mapsto x^2$ is an automorphism of $\mathfrak{g}/\mathfrak{p}$, we have $\sum_{x\in\mathfrak{g}/\mathfrak{p}} \chi(\pi^{-1}ax^2) = \sum_{y\in\mathfrak{g}/\mathfrak{p}} \chi(\pi^{-1}ay) = 0$ if $a \in \mathfrak{g}^\times$.

LEMMA 3.2. (i) $1 + 4\mathfrak{p} \subset \{x^2 \mid x \in \mathfrak{g}^\times\}$.

(ii) *For every nonnegative integer k, the map $x \mapsto x^2$ gives a homomorphism of $(\mathfrak{g}/2\mathfrak{p}^k)^\times$ into $(\mathfrak{g}/4\mathfrak{p}^k)^\times$, whose kernel consists of ± 1.*

Proof. Assertion (i) is included in [S3, Lemma 3.2(1)]. The only nontrivial point of (ii) is the statement about the kernel. Suppose $x \in \mathfrak{g}^\times$ and $x^2 - 1 \in 4\mathfrak{p}^k$. If both $x+1$ and $x-1$ belong to $2\mathfrak{p}$, then $2x \in 2\mathfrak{p}$, a contradiction. Thus $x+1 \notin 2\mathfrak{p}$ or $x - 1 \notin 2\mathfrak{p}$. Since $(x+1)(x-1) \in 4\mathfrak{p}^k$, we see that $x - 1 \in 2\mathfrak{p}^k$ or $x + 1 \in 2\mathfrak{p}^k$, which proves (ii).

LEMMA 3.3. *If κ is the positive integer such that $2\mathfrak{g} = \mathfrak{p}^\kappa$, then $\tau(\pi^{-m}a) = q^{-1}\tau(\pi^{2-m}a)$ for every $a \in \mathfrak{g}^\times$ and every $m > 2\kappa + 1$.*

Proof. Put $k = m - 2\kappa$. Then $\mathfrak{p}^m = 4\mathfrak{p}^k$ and

$$\begin{aligned}\sum_{x\in\mathfrak{g}/\mathfrak{p}^m} \chi(\pi^{-m}ax^2) &= \sum_{x\in(\mathfrak{g}/\mathfrak{p}^m)^\times} \chi(\pi^{-m}ax^2) + \sum_{y\in\mathfrak{p}/\mathfrak{p}^m} \chi(\pi^{-m}ay^2) \\ &= q^\kappa \sum_{x\in(\mathfrak{g}/2\mathfrak{p}^k)^\times} \chi(\pi^{-m}ax^2) + \sum_{z\in\mathfrak{g}/\mathfrak{p}^{m-1}} \chi(\pi^{2-m}az^2).\end{aligned}$$

Applying Lemma 3.2 (ii) to the sum over $(\mathfrak{g}/2\mathfrak{p}^k)^\times$, we find that the sum equals $2\sum_{\xi\in X} \chi(\pi^{-m}a\xi)$, where $X = \{x^2 \mid x \in (\mathfrak{g}/\mathfrak{p}^m)^\times\}$. Let Y denote $1 + 4\mathfrak{p}$ modulo

$\mathfrak{p}^m$. Then we have $\{1\} \neq Y \subset X$, and hence

$$\begin{aligned}\sum_{\xi \in X} \chi(\pi^{-m} a\xi) &= \sum_{\zeta \in X/Y} \sum_{\eta \in Y} \chi(\pi^{-m} a\zeta\eta) = \sum_{\zeta \in X/Y} \sum_{\varepsilon \in \mathfrak{p}/\mathfrak{p}^k} \chi(\pi^{-m} a\zeta(1 + \pi^{2\kappa}\varepsilon)) \\ &= \sum_{\zeta \in X/Y} \chi(\pi^{-m} a\zeta) \sum_{\varepsilon \in \mathfrak{p}/\mathfrak{p}^k} \chi(\pi^{-k} a\zeta\varepsilon) = 0,\end{aligned}$$

since $k > 1$. Thus $\tau(\pi^{-m}a) = q^{-m} \sum_{z \in \mathfrak{g}/\mathfrak{p}^{m-1}} \chi(\pi^{2-m} az^2) = q^{-1}\tau(\pi^{2-m}a)$, which proves our lemma.

The following lemma, if easy, is the main vehicle that smooths our calculations.

LEMMA 3.4. *Put* $\varepsilon_w = \varepsilon(w) = 2^{-1}\{\mathbf{e}(w/4) + \mathbf{e}(-w/4)\}$ *for* $w \in \mathbf{Z}$. *Then* $\varepsilon_w = \varphi(w + 1)$, *and for* $r \geq 2$ *we have*

$$\sum_{a \in (\mathbf{Z}/2^r\mathbf{Z})^\times} \mathbf{e}(2^{-r} aw) = \begin{cases} 2^{r-1}\varepsilon(2^{2-r}w) & \textit{if } w \in 2^{r-2}\mathbf{Z}, \\ 0 & \textit{if } w \notin 2^{r-2}\mathbf{Z}. \end{cases}$$

Proof. This is obvious if $w \in 2^{r-2}\mathbf{Z}$. Suppose $w = 2^s u$ with $0 \leq s < r - 2$ and an odd u; put $t = r - s$. Then $r \geq t > 2$ and the sum equals

$$\sum_{a \in \mathbf{Z}/2^r\mathbf{Z}} \mathbf{e}(2^{-t}au) - \sum_{a \in 2\mathbf{Z}/2^r\mathbf{Z}} \mathbf{e}(2^{-t}au).$$

Since both sums are 0, we obtain the desired equality for $w \notin 2^{r-2}\mathbf{Z}$.

3.5. We now take $F = \mathbf{Q}_2, \mathfrak{g} = \mathbf{Z}_2$, $\pi = 2$, and $\delta = 1$, so that $\chi(x) = \mathbf{e}_2(x)$. We have $\tau(0) = 1$ and $\tau(a/2) = 0$ for every odd integer a. Also direct calculations show that $\tau(a/4) = 2^{-1}(1 + i^{-a})$ and $\tau(a/8) = 2^{-1}\mathbf{e}(-a/8)$ for every odd integer a, where $i = \mathbf{e}(1/4)$. Therefore, from Lemma 3.3 we obtain

$$\tau(2^{-2k-1}a) = 2^{-k}\mathbf{e}(-a/8), \tag{3.4}$$

$$\tau(2^{-2k}a) = 2^{-k}(1 + i^{-a}), \tag{3.5}$$

for every odd integer a and every positive integer k.

For $0 < n \in \mathbf{Z}$ and a fixed $h \in \mathbf{Z}$ we have, as a special case of (2.2),

$$b_2(h, L; s) = \int_{\mathbf{Q}_2} \nu(\sigma)^{-s}\mathbf{e}_2(-\sigma h) \int_{L_2} \mathbf{e}_2(\sigma x \cdot {}^t x)\, dx\, d\sigma.$$

Observing that the last integral over L_2 equals $\tau(\sigma)^n$, we find that

$$b_2(h, L; s) = \sum_{\sigma \in \mathbf{Q}_2/\mathbf{Z}_2} \nu(\sigma)^{-s}\mathbf{e}_2(-\sigma h)\tau(\sigma)^n$$
$$= 1 + \sum_{r=2}^{\infty} 2^{-rs} \sum_{a \in (\mathbf{Z}/2^r\mathbf{Z})^\times} \mathbf{e}(2^{-r}ah)\tau(2^{-r}a)^n.$$

Decomposing the last sum into the sum over even r and the sum over odd r, we have, in view of (3.4) and (3.5),

$$b_2(h, L; s) = 1 + A + B, \tag{3.6}$$

$$A = \sum_{k=1}^{\infty} 2^{-k(2s+n)} \sum_{a \in (\mathbf{Z}/2^{2k}\mathbf{Z})^\times} \mathbf{e}(2^{-2k}ah)(1 + i^{-a})^n, \tag{3.6a}$$

$$B = \sum_{k=1}^{\infty} 2^{-k(2s+n)-s} \sum_{a \in (\mathbf{Z}/2^{2k+1}\mathbf{Z})^\times} \mathbf{e}(2^{-2k-1}a(h - 2^{2k-2}n)). \tag{3.6b}$$

If $n = 2m \in 2\mathbf{Z}$, then $(1 + i^{-a})^n = (2i^{-a})^m = 2^m\mathbf{e}(-am/4)$, and hence

$$A = 2^m \sum_{k=1}^{\infty} 2^{-k(2s+n)} \sum_{a \in (\mathbf{Z}/2^{2k}\mathbf{Z})^\times} \mathbf{e}(2^{-2k}a(h - 2^{2k-2}m)). \tag{3.7}$$

Similarly, if $n = 2m + 1$, we have $(1 + i^{-a})^n = 2^m\{\mathbf{e}(-am/4) + \mathbf{e}(-a(m+1)/4)\}$, and hence

$$A = 2^m \sum_{k=1}^{\infty} 2^{-k(2s+n)} \cdot \sum_{a \in (\mathbf{Z}/2^{2k}\mathbf{Z})^\times} \{\mathbf{e}(2^{-2k}a(h - 2^{2k-2}m)) + \mathbf{e}(2^{-2k}a(h - 2^{2k-2}(m+1)))\}. \tag{3.8}$$

PROPOSITION 3.6. (The case of even h.) *Let φ denote the primitive Dirichlet character modulo 4 and let $\varphi_1(n) = (-1)^{(n^2-1)/8}$ for odd n; let $h = 2^\nu g$ with an odd positive integer g and a positive integer ν; put $q = 2^{(n-2)/2}$ and $m = [n/2]$. Then the quantity $q^\nu b_2(h, L; 0)$, which we called $\tilde{b}(h)$ in (2.6), can be given for each n as follows:*

(Ia) *$n = 2m + 1$ and ν odd:*

$$q^\nu + \varphi_1(n)2^{-1/2} \cdot \frac{q^{\nu+1} - 2^{n-1} + 1}{2^{n-2} - 1}.$$

(Ib) $n = 2m + 1$ *and* ν *even:*

$$q^\nu + \varphi_1(n)2^{m-1} \cdot \frac{q^\nu - 1}{q^2 - 1} + \varphi_1(n)\varphi(ng)2^{-m} + \varphi((g+2-n)/2)2^{2-n}.$$

(II) $n = 2m$:

$$\begin{aligned} &2^{\nu(m-1)} + \varphi(mg)2^{1-m} && \text{if } n \equiv 2 \pmod 4, \\ &\sum_{k=0}^{\nu} 2^{k(m-1)} - 2 && \text{if } n \equiv 0 \pmod 8, \\ &\sum_{k=0}^{\nu} 2^{k(m-1)} - 2\sum_{i=1}^{\nu-1} 2^{i(m-1)} && \text{if } n \equiv 4 \pmod 8. \end{aligned}$$

The quantity of (Ia) can be written

$$q^\nu + \varphi_1(n)2^{-1/2}\sum_{k=0}^{\mu} 2^{k(n-2)} - \varphi_1(n)2^{1/2} \tag{3.9}$$

if $\nu = 2\mu + 1$, which becomes $2^{-1/2}\{2^m - \varphi_1(n)\}$ if $\nu = 1$.

Proof. Suppose $n = 2m + 1$; then the sum over $(\mathbf{Z}/2^{2k}\mathbf{Z})^\times$ in (3.8) is

$$\sum_{a \in (\mathbf{Z}/2^{2k}\mathbf{Z})^\times} \{\mathbf{e}(2^{-2k}a(2^\nu g - 2^{2k-2}m)) + \mathbf{e}(2^{-2k}a(2^\nu g - 2^{2k-2}(m+1)))\}.$$

By Lemma 3.4 this is 0 for $\nu < 2k - 2$, and $2^{\nu+1}(\varepsilon_{g-m} + \varepsilon_{g-m-1})$ for $\nu = 2k - 2$. Similarly it is $2^\nu(\varepsilon_{2g-m} + \varepsilon_{2g-m-1})$ for $\nu = 2k - 1$, and $2^{2k-1}(\varepsilon_m + \varepsilon_{m+1})$ for $\nu \geq 2k$. Since $\varepsilon_k = \varepsilon_{-k}$ and $2g - 2 \in 4\mathbf{Z}$, we have $\varepsilon_{2g-m} = \varepsilon_{m-2}$ and $\varepsilon_{2g-m-1} = \varepsilon_{m-1}$. Taking the sum over all $k \geq 1$, and putting $t = 2^{2-n}$, we find that the quantity A at $s = 0$ is

$$\sum_{k=1}^{\mu} 2^{m-1}t^k(\varepsilon_m + \varepsilon_{m+1}) + 2^{m-1}t^{\mu+1} \cdot \begin{cases} (\varepsilon_{g-m} + \varepsilon_{g-m-1}) & (\nu = 2\mu), \\ (\varepsilon_{m-1} + \varepsilon_{m-2}) & (\nu = 2\mu + 1), \end{cases}$$

where $\mu = [\nu/2]$. Similarly the terms of (3.6b) for $2k \neq \nu + 2$ vanish, and we find that the value of B at $s = 0$ is $t^{\mu+1}\varepsilon((g-n)/2)$ if $\nu = 2\mu$ and 0 if $\nu = 2\mu + 1$. Since $\varepsilon_k = \varphi(k+1)$, we have $\varepsilon((g-n)/2) = \varphi((g+2-n)/2)$. Also, we easily see that $\varepsilon_m + \varepsilon_{m+1} = -\varepsilon_{m-1} - \varepsilon_{m-2} = \varphi_1(n)$ for $n = 2m + 1$. Thus $\varepsilon_{g-m-1} + \varepsilon_{g-m} = \varphi_1(2g - 2m - 2 + 1) = \varphi_1(2g - n) = \varphi_1(n)\varphi_1(2ng - 1)$, as $n^2 - 1 \in 8\mathbf{Z}$. Now $\varphi_1(2k - 1) = \varphi(k)$ for odd k, and hence $\varphi_1(2ng - 1) = \varphi(ng)$. Combining all these, we obtain our formulas for odd n.

Next suppose $n = 2m$; then by Lemma 3.4 the sum over $(\mathbf{Z}/2^{2k}\mathbf{Z})^\times$ in (3.7) is $0, 2^{\nu+1}\varepsilon_{g-m}, 2^\nu\varepsilon_{2g-m}$, or $2^{2k-1}\varepsilon_m$ accordingly as $\nu < 2k-2, \nu = 2k-2, \nu = 2k-1$, or $\nu \geq 2k$. Similarly the sum over $(\mathbf{Z}/2^{2k+1}\mathbf{Z})^{\times}$ in (3.6b) is $0, 2^{\nu+1}\varepsilon_{g-m}, 2^{\nu}\varepsilon_{2g-m}$, or $2^{2k}\varepsilon_m$ accordingly as $\nu < 2k-1, \nu = 2k-1, \nu = 2k$, or $\nu > 2k$. Adding all the terms, we find that $q^{\nu}b_2(h, L; 0)$ equals

$$q^{\nu} + \varepsilon_m(q^{\nu} - q)/(q-1) + \varepsilon_{2-m} + \varepsilon_{g-m}q^{-1} \tag{3.10}$$

for both even and odd ν. We easily see that $\varepsilon_{g-m} = \varphi(mg)$ for every $m \in \mathbf{Z}$ and every odd g. Therefore we obtain the formulas for even n as stated in our theorem. Notice that the formulas are valid even for $\nu = 1$.

PROPOSITION 3.7. (The case of odd h.) *Let φ and φ_1 be as in Proposition 3.6. Then for an odd positive integer g the value of $b_2(g, L; 0)$, which equals $\tilde{b}(h)$ if $h = g$, can be given as follows:*

$$\begin{array}{ll} 1 + \varphi(gm)2^{1-m} & \textit{if } n \textit{ is even,} \\ 1 + \varphi_1(n)\varphi(ng)2^{-m} + \varphi((g+2-n)/2)2^{2-n} & \textit{if } n \textit{ is odd.} \end{array}$$

In particular,

$$b_2(1, L; 0) = \begin{cases} 1 + \varphi(m)2^{1-m} & \textit{if } n \textit{ is even,} \\ 1 + \varphi_1(n)\varphi(n)2^{-m} - \varphi((n-3)/2)2^{2-n} & \textit{if } n \textit{ is odd.} \end{cases}$$

Proof. Take $\nu = 0$ in the proof of Proposition 3.6. We have to ignore the terms for $\nu \geq 2k$, since k must be positive; otherwise our computation is the same. For $n = 2m+1$, we find that the formula of (Ib) with $\mu = 0$ is valid, and obtain $b_2(g, L; 0)$ as stated above. If $n = 2m$, however, the second and third terms of (3.10) must be eliminated, so that $b_2(g, L; 0) = 1 + \varepsilon_{g-m}q^{-1}$, which equals $1 + \varphi(gm)2^{1-m}$. This completes the proof.

3.8. Substituting the values of $\tilde{b}(h)$ determined in Propositions 3.6 and 3.7 into the formulas of Theorem 2.2, we can make them for even n more explicit as follows. The notation is the same as in that theorem and those propositions.

(A) $2 < n \equiv 2 \pmod 4$. For $\nu \geq 0$ we have

$$\mathfrak{n}(L, h) = \frac{4}{E_{(m-1)/2}}\{\varphi(g)2^{(\nu+1)(m-1)} + \varphi(m)\}\sum_{d|g}\varphi(d)d^{m-1}.$$

(B) $n \equiv 4 \pmod 8$. The case of odd h must be separated from that of even h; thus

$$\mathfrak{n}(L, h) = \frac{n}{(2^m - 1)|B_m|} \cdot \sum_{d|g} d^{m-1} \cdot \begin{cases} 1 & \text{if } \nu = 0, \\ \sum_{k=0}^{\nu} 2^{k(m-1)} - 2\sum_{i=1}^{\nu-1} 2^{i(m-1)} & \text{if } \nu > 0. \end{cases}$$

(C) $n \equiv 0 \pmod 8$. Similarly we have

$$\mathfrak{n}(L,h) = \frac{n}{(2^m-1)|B_m|}\cdot\begin{cases}\sum_{d|h} d^{m-1} & \text{if } \nu = 0,\\ \sum_{d|h} d^{m-1} - 2\sum_{d|g} d^{m-1} & \text{if } \nu > 0.\end{cases}$$

3.9. For simplicity put $S_n(h) = N(L,h)$. This is the number of representations of h as the sum of n squares. We have $S_n(h) = R_n(h)$ for $n \le 3$, and $R_2(h)$ and $R_3(h)$ have been determined in Theorem 1.5 and §1.7. It is well known that the genus of L consists of a single class if $n \le 8$. For the reader's convenience, we prove this fact in §5.8. Therefore, specializing the formulas of §3.8, we obtain the formulas for $S_n(h)$ for $4 \le n \le 8$ as follows:

$$S_4(h) = \begin{cases} 8\sum_{d|g} d & \text{if } h \text{ is odd},\\ 24\sum_{d|g} d & \text{if } h \text{ is even},\end{cases}$$

$$S_6(h) = \{\varphi(g)2^{2\nu+4} - 4\}\sum_{d|g}\varphi(d)d^2,$$

$$S_8(h) = 16\cdot\begin{cases}\sum_{d|h} d^3 & \text{if } h \text{ is odd},\\ \sum_{d|h} d^3 - 2\sum_{d|g} d^3 & \text{if } h \text{ is even},\end{cases}$$

$$S_5(h) = 2^7(2\pi)^{-2}L(2,\psi_h)\{4-\psi_h(2)\}g^{3/2} \cdot \tilde{b}(h)\prod_{p|r}\left\{\sum_{k=0}^{\lambda_p} p^{-3k} - \psi_h(p)\sum_{i=1}^{\lambda_p} p^{1-3i}\right\},$$

$$S_7(h) = 2^9(2\pi)^{-3}L(3,\psi_h)\{8-\psi_h(2)\}g^{5/2} \cdot \tilde{b}(h)\prod_{p|r}\left\{\sum_{k=0}^{\lambda_p} p^{-5k} - \psi_h(p)\sum_{i=1}^{\lambda_p} p^{2-5i}\right\}.$$

Here r and λ_p for each p are determined as in Theorem 1.5. The values of $\tilde{b}(h)$ for even h given in Proposition 3.6 cannot be much simplified in the most general case. We content ourselves here with noting only the case in which $4 \nmid h$. Thus, if $n = 5$ and $4 \nmid h$, then

$$\{4 - \psi_h(2)\}\tilde{b}(h) = \begin{cases} 2^{3/2} \cdot 5 & \text{if } h \equiv 2 \pmod 4, \\ 5 & \text{if } h \equiv 3 \pmod 4, \\ 2^{-3} \cdot 3 \cdot 5 & \text{if } h \equiv 1 \pmod 8, \\ 2^{-3} \cdot 5 \cdot 7 & \text{if } h \equiv 5 \pmod 8. \end{cases}$$

Similarly if $n = 7$ and $4 \nmid h$, then

$$\{8 - \psi_h(2)\}\tilde{b}(h) = \begin{cases} 2^{5/2} \cdot 7 & \text{if } h \equiv 2 \pmod 4, \\ 7 & \text{if } h \equiv 1 \pmod 4, \\ 2^{-5} \cdot 3^2 \cdot 5 \cdot 7 & \text{if } h \equiv 3 \pmod 8, \\ 2^{-5} \cdot 7 \cdot 37 & \text{if } h \equiv 7 \pmod 8. \end{cases}$$

The formulas for $S_4(h), S_6(h)$, and $S_8(h)$ are classical, except that they have been given in various forms equivalent to the above ones. For instance, $S_6(h)$ is traditionally given as $16\sum_{d|g}\varphi(d)(h/d)^2 - 4\sum_{d|g}\varphi(d)d^2$. The history of the representation of integers by sums of squares is long. A good account can be found in Hardy [H1]. A somewhat shorter version is also given in Lecture IX of [H2]; see also [Si, p. 530].

As for $S_5(h)$ and $S_7(h)$, all previous results concern the number of primitive representations, as we said in the introduction. As remarked in §1.8, $\sum_{h=1}^{\infty}\{N(L,h) - \mathfrak{n}(L,h)\}\mathbf{e}(hz)$ is a cusp form of weight $n/2$, whose determination does not seem particularly interesting.

4. Reduction of $\mathfrak{m}(L)$ to $\mathfrak{m}(L_0)$. Our next topic is $\mathfrak{m}(L)$. Since its exact value is known, we cannot add anything substantially new to the question of determining it. However, we treat it here for the following reasons: (i) our methods are applicable to $\mathfrak{m}(M)$ for other lattices M; (ii) some of the auxiliary results stated in the lemmas below may be of independent interest; (iii) the results in this section can make this paper self-contained.

The principal idea is to compare $\mathfrak{m}(L)$ with the quantity $\mathfrak{m}(L_0)$ which was determined in [S2]. Put $G = \{\alpha \in GL_n(\mathbf{Q}) \mid \alpha \cdot {}^t\alpha = 1_n\}$ and define the adelization $G_{\mathbf{A}}$ and the localization G_p of G as usual; then put

$$C = \{\gamma \in G_{\mathbf{A}} \mid L_0\gamma = L_0\}, \quad D = \{\gamma \in G_{\mathbf{A}} \mid L\gamma = L\},$$
$$C_p = C \cap G_p, \qquad D_p = D \cap G_p,$$

where $L\gamma$ is the $\mathbf{Z}$-lattice such that $(L\gamma)_p = L_p\gamma_p$ for every p. Since $(L_0)_p = L_p$ for every $p \neq 2$, we have $C_p = D_p$ for every $p \neq 2$. Therefore $[C : C \cap D] = [C_2 : C_2 \cap D_2]$ and $[D : C \cap D] = [D_2 : C_2 \cap D_2]$. By [S1, Lemma 24.2] we have

$[C : C \cap D]\mathfrak{m}(L_0) = \mathfrak{m}(C \cap D) = [D : C \cap D]\mathfrak{m}(L)$, and hence

$$\mathfrak{m}(L) = \mathfrak{m}(L_0)[C_2 : C_2 \cap D_2]/[D_2 : C_2 \cap D_2]. \tag{4.1}$$

As observed in [S2, §3.1], there exists a $\mathbf{Z}_2$-basis $\{f_i\}_{i=1}^n$ of $(L_0)_2$ such that our quadratic form $x \mapsto x \cdot {}^t x$ can be represented, with repect to $\{f_i\}_{i=1}^n$, by a matrix of the form

$$\psi = \begin{bmatrix} 0 & 0 & 2^{-1}1_r \\ 0 & \theta & 0 \\ 2^{-1}1_r & 0 & 0 \end{bmatrix}, \tag{4.2}$$

with a suitable $\theta \in GL_t(\mathbf{Q}_2)$, where $r = r_2$ and $t = t_2$. Clearly $\psi = (f_i \cdot {}^t f_j)_{i,j=1}^n$. Let α be the matrix whose i-th row is f_i. Then $\psi = \alpha \cdot {}^t\alpha$ and $(L_0)_2 = L_2\alpha$. Put

$$G^\psi = \{\xi \in GL_n(\mathbf{Q}_2) \mid \xi\psi \cdot {}^t\xi = \psi\}, \qquad C^\psi = G^\psi \cap GL_n(\mathbf{Z}_2),$$
$$C_2' = \{\gamma \in C_2 \mid (\tilde{L}_0)_2(\gamma - 1) \subset 2(L_0)_2\}, \quad D_2' = \{\gamma \in D_2 \mid \tilde{L}_2(\gamma - 1) \subset 2L_2\}.$$

Then $G^\psi = \alpha G_2 \alpha^{-1}$, $C^\psi = \alpha C_2 \alpha^{-1}$, $\tilde{L}_2 = 2^{-1}L_2$, and $(\tilde{L}_0)_2 = L_2(2 \cdot {}^t\alpha)^{-1}$, and hence

$$\alpha C_2' \alpha^{-1} = \{\gamma \in C^\psi \mid L_2\psi^{-1}(\gamma - 1) \subset 4L_2\} = \{\gamma \in C^\psi \mid \gamma - 1 \in 4\psi(\mathbf{Z}_2)_n^n\}.$$

Take our setting to be that of [S2, Section 8] with $\mathbf{Q}_2$ as the field F there. Define the representation density $g(\psi)$ by [S2, (8.1.7)]. Now the last group in the above equality is written D_1^ψ in [S2, Lemma 8.2 (3)], and hence, by [S2, Theorem 8.6 (1)] we have $g(\psi) = 2^{-k}[C_2 : C_2']$, where $k = n(n-1)/2$. We have similarly $g(1_n) = 2^{-k}[D_2 : D_2']$.

Next, in [S2, Lemma 8.7] take (ψ, φ, a) there to be $(\psi, 1_n, \alpha)$ here; then C^φ there is D_2. Since $\alpha^{-1}C^\psi\alpha = C_2$, from that lemma we obtain

$$g(1_n)[C_2 : C_2 \cap D_2] = |\det(\alpha)|^{n-1} g(\psi)[D_2 : C_2 \cap D_2],$$

and thus

$$\frac{\mathfrak{m}(L)}{\mathfrak{m}(L_0)} = \frac{[C_2 : C_2 \cap D_2]}{[D_2 : C_2 \cap D_2]} = |\det(\alpha)|^{n-1}\frac{g(\psi)}{g(1_n)} = |\det(\alpha)|^{n-1}\frac{[C_2 : C_2']}{[D_2 : D_2']}.$$

We have $|\det(\alpha)^2| = |\det(\psi)| = 2^{2r}|\det(\theta)|$, and $|\det(\theta)| = 4$ or 1 accordingly as $t_2 = 4$ or $t_2 < 4$ (see [S2, (3.1.9) and (3.2.1)]). Thus $|\det(\alpha)| = 2^r$ if $t_2 < 4$ and

$|\det(\alpha)| = 2^{r+1}$ if $t_2 = 4$, where $r = r_2$. Therefore

$$\frac{\mathfrak{m}(L)}{\mathfrak{m}(L_0)} = \frac{[C_2 : C_2']}{[D_2 : D_2']} \cdot \begin{cases} 2^{r_2(n-1)} & \text{if } t_2 < 4, \\ 2^{(r_2+1)(n-1)} & \text{if } t_2 = 4. \end{cases} \tag{4.3}$$

We already determined $\mathfrak{m}(L_0)$ in [S2, p. 34]. The index $[C_2 : C_2']$ is given in [S2, Proposition 3.9]; in fact, it is $[C^\varphi : D_1^\varphi]$ in that proposition, and therefore, taking $q = 2$ and $\nu = 1$, and putting $m = [n/2]$, we find that $[C_2 : C_2']$ equals

$$\begin{array}{ll} 2^{m^2-m+1}(2^m - 1)\prod_{i=1}^{m-1}(2^{2i} - 1) & \text{if } n \equiv 0 \pmod 8, \\ 2^{m^2+1}\prod_{i=1}^{m}(2^{2i} - 1) & \text{if } n \equiv \pm 1 \pmod 8, \\ 2^{m^2+2}\prod_{i=1}^{m-1}(2^{2i} - 1) & \text{if } n \equiv 2 \pmod 4, \\ 2^{m^2+2m+2}\cdot 3\prod_{i=1}^{m-1}(2^{2i} - 1) & \text{if } n \equiv \pm 3 \pmod 8, \\ 2^{m^2+m+1}\cdot 3(2^{m-1} - 1)^{-1}\prod_{i=1}^{m-1}(2^{2i} - 1) & \text{if } n \equiv 4 \pmod 8. \end{array}$$

In the next section we will determine $[D_2 : D_2']$. Combining all these results, we will obtain $\mathfrak{m}(L)$.

5. Calculation of $[D_2 : D_2']$.

5.1. Put

$$\begin{aligned} T_n &= \{\alpha \in (\mathbf{Z}_2)_n^n \mid {}^t\alpha = \alpha, \alpha_{ii} \in \mathbf{Z}_2, 2\alpha_{ij} \in \mathbf{Z}_2\}, \\ Z_n &= \{\alpha \in (\mathbf{Z}_2)_n^n \mid {}^t\alpha\alpha - 1 \in 8T_n\}, \\ I_n &= \{x \in (\mathbf{Z}_2/4\mathbf{Z}_2)_n^1 \mid x \cdot {}^t x - 1 \in 8\mathbf{Z}_2\}, \\ K_n &= \text{the set of the first rows of the elements of } Z_n \text{ modulo } 4(\mathbf{Z}_2)_n^1, \\ J_n &= \{x \in I_n \mid x \notin K_n\}. \end{aligned}$$

Notice that we can replace ${}^t\alpha\alpha$ in the definition of Z_n by $\alpha \cdot {}^t\alpha$; also, Z_n is a subgroup of $GL_n(\mathbf{Z}_2)$, and the sets I_n, K_n, and J_n are stable under right multiplication by the elements of Z_n.

LEMMA 5.2. (i) *There exists an element α of Z_3 such that* $(1, 2, 2)\alpha \equiv (1, 0, 0) \pmod 8$.

(ii) *There exists an element* β *of* Z_6 *such that* $(2,1,1,1,1,1)\beta \equiv (1,0,0,0,0,0) \pmod 8$.

(iii) *Given* $x \in I_n$, *there exists an element* γ *of* Z_n *such that the components of* $x\gamma$ *are either* 0 *or* 1 *(as elements of* $\mathbf{Z}_2/4\mathbf{Z}_2$*).*

Proof. The desired element α of (i) (resp. β of (ii)) can be given as the element of $(\mathbf{Z}_2)_3^3$ (resp $(\mathbf{Z}_2)_6^6$) whose diagonal elements are 1 (resp. 2) and nondiagonal elements are 2 (resp. 1). To prove (iii), we first note that $\mathbf{Z}_2/4\mathbf{Z}_2$ consists of $\pm1, 2$, and 0. Given $x \in I_n$, let $a(x), b(x)$, and $c(x)$ denote the number of the components of x which are equal to $\pm1, 2$, and 0, respectively. Clearly $a(x) > 0$ and $a(x) + 4b(x) - 1 \in 8\mathbf{Z}$. Since Z_n contains all the diagonal matrices whose diagonal entries are ±1, we can find an element δ of Z_n such that the odd components of $x\delta$ are all equal to 1. Put $y = x\delta$ and suppose $b(y) > 1$. Then the components of y include $(1,2,2)$. By (i) we can find an element ξ of Z_n such that $a(y\xi) = a(y)$ and $b(y\xi) = b(y) - 2$. Repeating this procedure, we find, for any $x \in I_n$, an element η of Z_n such that $b(x\eta) \le 1$. Put $w = x\eta$; suppose $b(w) = 1$; then $a(w) + 3 \in 8\mathbf{Z}$. Thus $a(w) \ge 5$. Then by (ii) we can find an element ζ of Z_n such that $b(w\zeta) = 0$. This proves (iii)

LEMMA 5.3. *If* $x \in I_n$ *and* $b(x) + c(x) > 0$, *then* $x \in K_n$.

Proof. The proof of Lemma 5.2 shows that if $b(x) > 0$, then there exists an element γ of Z_n such that $b(x\gamma) = 0$ and $c(x\gamma) > 0$. Therefore, to prove our lemma, we may assume that $b(x) = 0$ and $c(x) > 0$. Then $a(x) - 1 \in 8\mathbf{Z}$. Let $a = a(x)$ and let z be the $(a + 1)$-dimensional vector whose first component is 0 and the remaining components are all equal to 1. Let ε be the matrix of size $a + 1$ whose diagnal entries are 0 and nondiagonal entries are 1. Then $\varepsilon \in Z_{a+1}$ and $z\varepsilon \equiv (1,0,\dots,0) \pmod 8$. Thus $z \in K_{a+1}$. Since x consists of z and a zero vector, we have $x \in K_n$. This proves our lemma.

LEMMA 5.4. $J_n \ne \emptyset$ *only if* $0 < n - 1 \in 8\mathbf{Z}$, *in which case* $\#(J_n) = 2^n$.

Proof. Let $x \in J_n$. By Lemma 5.3 every component of x is ±1. Then $n = a(x) \in 1 + 8\mathbf{Z}$. Since $I_n = K_n$ if $n = 1$, we have $n > 1$. Suppose conversely that $\xi \in Z_n, 0 < n - 1 \in 8\mathbf{Z}$, and the entries of ξ on the first row are all equal to 1. Let p, q, r, and s be the numbers of the entries on the second row of ξ which are equal to $1, -1, 2$, and 0, respectively. Then $p - q + 2r \in 4\mathbf{Z}$ and $p + q + 4r - 1 \in 8\mathbf{Z}$, and hence $2p + 6r - 1 \in 4\mathbf{Z}$, a contradiction. Therefore such a ξ does not exist. This shows that J_n consists of the vectors whose components are all ±1. This proves our lemma.

LEMMA 5.5. $[D_2 : D_2'] = \prod_{i=1}^n \#(K_i)$.

Proof. We prove this by induction on n. The case $n = 1$ is clear. For any matrix or vector X with components in $\mathbf{Z}_2$ let $\mu(X)$ denote the object obtained by taking X modulo $4\mathbf{Z}_2$. Put $A_n = [D_2 : D_2']$. By [S2, Lemma 8.2 (3)], A_n equals $\#\mu(Z_n)$. Let $\{e_i\}_{i=1}^n$ be the standard $\mathbf{Z}$-basis of L. By the same lemma the map $\zeta \mapsto \mu(e_1\zeta)$ gives a surjective map of Z_n onto K_n. Given $\xi \in K_n$ let $H(\xi)$ denote the set of elements $\zeta \in Z_n$ such that $\xi = \mu(e_1\zeta)$. To complete our induction, it is sufficient to prove that $\#\mu(H(\xi)) = A_{n-1}$. Given $\xi \in K_n$, fix an element $\zeta_0 \in H(\xi)$; let $\zeta \in Z_n$. Then $\zeta \in H(\xi)$ if and only if $\mu(e_1\zeta\zeta_0^{-1}) = \mu(e_1)$. For such a ζ, put $\zeta\zeta_0^{-1} = \begin{bmatrix} c & u \\ x & y \end{bmatrix}$ with $c \in \mathbf{Z}_2$, a square matrix y of size $n-1$, and matrices u and x of appropriate shape. Then $\mu(c) = 1, \mu(u) = 0, \mu(x) = 0$, and $y \in Z_{n-1}$. Thus $\mu(\zeta) = \mathrm{diag}[1, \mu(y)]\mu(\zeta_0)$. Conversely such a ζ with $y \in Z_{n-1}$ belongs to $H(\xi)$, and hence $\#\mu(H(\xi)) = \#\mu(Z_{n-1})$, which proves our lemma.

LEMMA 5.6. $\#(I_n) = 2^{2n-3}b_2(1, L; 0)$.

Proof. Put $W_n^k = \{x \in L_2/2^kL_2 \mid x \cdot {}^tx - 1 \in 2^k\mathbf{Z}_2\}$. By [S1, Proposition 14.3], $2^{k(1-n)}\#(W_n^k)$ is constant for $k > 2$, and by [S3, Lemma 2.5], $b_2(1, L; 0)$ equals this constant; thus $b_2(1, L; 0) = 2^{3-3n}\#(W_n^3)$. Clearly we have $\#(W_n^3) = 2^n\#(I_n)$, and hence we obtain our lemma.

LEMMA 5.7. *Let φ and m be as in Proposition 3.6. Then $\#(K_n)$ can be given as follows: $\#(K_1) = 2$; $\#(K_n)$ for $n > 1$ equals*

$$\begin{array}{ll} 2^{2n-3} + \varphi(m)2^{n+m-2} & \text{if } n - 1 \notin 4\mathbf{Z}, \\ 2^{2n-3} + 2^{n+m-2} - 2^n + 2^{n-1} & \text{if } n \equiv 1 \pmod 8, \\ 2^{2n-3} - 2^{n+m-2} - 2^{n-1} & \text{if } n \equiv 5 \pmod 8. \end{array}$$

Proof. Combining Lemma 5.6 with Proposition 3.7, we obtain the explicit forms for $\#(I_n)$, which equal $\#(K_n)$ unless $0 < n - 1 \in 8\mathbf{Z}$, in which case $\#(K_n) = \#(I_n) - 2^n$ by Lemma 5.4. Thus we obtain our formulas.

5.8. We can now use (4.3) to compute $\mathfrak{m}(L)$. As already mentioned, $\mathfrak{m}(L_0)$ was given in [S2, p. 34], and $[C_2 : C_2']$ at the end of Section 4. Now $[D_2 : D_2']$ can be determined by Lemmas 5.5 and 5.7. Therefore, by means of (4.3) we easily obtain the numerical values of $\mathfrak{m}(L)$. In particular, we have $\mathfrak{m}(L) = (2^n \cdot n!)^{-1}$ for $2 \le n \le 8$. We easily see that $\{\gamma \in G \mid L\gamma = L\}$ has order $2^n \cdot n!$, and hence we see that the genus of L for $2 \le n \le 8$ consists of a single class, so that $\mathfrak{n}(L, h) = N(L, h) = S_n(h)$ for any such n. These facts are classical.

DEPARTMENT OF MATHEMATICS, PRINCETON UNIVERSITY, PRINCETON, NJ 08544

REFERENCES

[B] P. T. Bateman, On the representaions of a number as the sum of three squares, *Trans. Amer. Math. Soc.* **71** (1951), 70–101.
[C] H. S. M. Coxeter, Extreme forms, *Canad. J. Math.* **3** (1951), 391–441.
[H1] G. H. Hardy, On the representation of a number as the sum of any number of squares, and in particular of five, *Trans. Amer. Math. Soc.* **21** (1920), 255–284.
[H2] ———, *Ramanujan*, Cambridge Univ. Press, 1940.
[S1] G. Shimura, *Euler Products and Eisenstein Series, CBMS Reg. Conf. Ser. Math.*, vol. 93, Amer. Math. Soc., Providence, RI, 1997.
[S2] ———, An exact mass formula for orthogonal groups, *Duke Math. J.* **97** (1999), 1–66.
[S3] ———, The number of representations of an integer by a quadratic form, *Duke Math. J.* **100** (1999), 59–92.
[S4] ———, *Arithmeticity in the Theory of Automorphic Forms, Math. Surveys Monogr.*, vol. 82, Amer. Math. Soc., Providence, RI, 2000.
[Si] C. L. Siegel, Über die analytische Theorie der quadratischen Formen I, *Ann. of Math.* **36** (1935), 527–606.

INHOMOGENEOUS QUADRATIC FORMS AND TRIANGULAR NUMBERS

By GORO SHIMURA

Abstract. We prove an explicit formula for the number of representations of an integer as the sum of n triangular numbers for each n in the range $2 \leq n \leq 8$ as special cases of a more general formula applicable to an inhomogeneous quadratic form over a totally real number field. The formula can be derived by calculating explicitly the Fourier coefficients of a certain Eisenstein series that appears in the Siegel-Weil formula for an inhomogeneous quadratic form.

Introduction. It is a famous statement of Fermat that every natural number is the sum of at most three triangular numbers, and also of four squares, of five pentagonal numbers, etc. In spite of his repeated claims, he presented no proof, and the solutions were given by Lagrange, Gauss, Cauchy, and Legendre many years later. Since then mathematicians seem to have lost interest in those polygonal numbers, except that the number of representations of an integer as sums of squares have been investigated by many researchers, often with obsessive zeal. They could have asked about the number of representations of integers as sums of three triangular numbers, but no decisive answer was given to this problem, though there is an easy but unsatisfactory answer, and the cases of two and four triangular numbers were settled long ago. As for the sums of pentagonal numbers, no meaningful investigations seem to have been made.

One of the principal aims of this paper is to give a clear-cut formula for the number of representations of an integer as sums of n triangular numbers for each n in the range $2 \leq n \leq 8$. In fact, we shall give a formula for the dominant term for the number of such representations even for $n > 8$, in the same sense as in the case of sums of n squares, for which we gave explicit formulas in our recent paper [S8]. We shall also do the same for the sums of pentagonal numbers.

Aside from the point that nobody is expected to be unhappy with such formulas, there is a methodological reason for presenting these results: that the product formula of Siegel for the representation densities of a number by a quadratic form is applicable even to an inhomogeneous quadratic form. This is a consequence of the fact that a certain integral of a theta series is an Eisenstein series, which was first proved by Siegel and then generalized by Weil. It should be noted that when Siegel treated inhomogeneous forms, he excluded definite cases, and so his

Manuscript received May 16, 2003.
Shimura, Goro. "Inhomogeneous quadratic forms and triangular numbers." American Journal of Mathematics 126:1 (2004), 191–214.

results are not sufficient for our purposes. However, the formula, nowadays called the Siegel-Weil formula, in Weil's formulation is applicable to an inhomogeneous totally definite form, which is a key point of our methods.

Let us now explain how the formula can be employed for the purpose of answering the question about triangular numbers. Let H denote the upper half complex plane; take a positive definite quadratic form $\varphi[x]$ with rational coefficients defined for x in the space V of n-dimensional row vectors over $\mathbf{Q}$. Let M be a coset in V modulo a lattice in V. Then we can define the genus and class of M with respect to the orthogonal group G of φ in exactly the same fashion as in the case where M is a lattice. Let $\{M_i\}_{i=1}^{\nu}$ be a set of representatives for the classes in the genus of M. We then put

$$\theta_i(z) = \sum_{\xi\in M_i} \exp\bigl(2\pi i\varphi[\xi]z\bigr) \qquad (z \in H),$$

$$\Gamma_i = \{\,\gamma \in G \mid M_i\gamma = M_i\,\}, \qquad N(h, M_i) = \#\{\,\xi \in M_i \mid \varphi[\xi] = h\,\},$$

$$f(z) = \left\{\sum_{i=1}^{\nu} [\Gamma_i : 1]^{-1}\right\}^{-1} \sum_{i=1}^{\nu} [\Gamma_i : 1]^{-1}\theta_i(z),$$

where $\#X$ denotes the number of elements in X. Clearly

$$\theta_i(z) = \sum_{h\in\mathbf{Q}} N(h, M_i)\exp(2\pi ihz).$$

Now, as will be explained in the text, the Siegel-Weil formula shows that $f(z)$ can be given as an Eisenstein series of weight $n/2$ that is determined by φ and M, and belongs to a congruence subgroup of $SL_2(\mathbf{Q})$. This means that for a number $h \in \mathbf{Q}$, > 0, the average

$$(*) \qquad \left\{\sum_{i=1}^{\nu} [\Gamma_i : 1]^{-1}\right\}^{-1} \sum_{i=1}^{\nu} [\Gamma_i : 1]^{-1}N(h, M_i)$$

can be given as a Fourier coefficient of the Eisenstein series f. Also, $\theta_i - f$ is a cusp form. Take, for example, $\varphi[\xi] = \sum_{i=1}^{n} \xi_i^2$ and $M = \{\xi \in V \mid \xi_i - (1/2) \in \mathbf{Z}\,\}$. Given an integer k, put $h = 2k + (n/4)$. Then we easily see that $N(h, M)$ is exactly the number of representations of k as the sum of n triangular numbers. In this case we can show that $\theta_1 = f$ for each n in the range $2 \le n \le 8$, and therefore the desired number is a Fourier coefficient of the Eisenstein series f. For $n > 8$, the theta series θ_1 is the sum of f and a cusp form, so that $N(h, M_1)$ is the number of $(*)$ plus a quantity of order at most $O(h^{n/4})$.

Thus the question can be reduced to the determination of such Fourier coefficients. By means of the methods developed in our earlier papers, we can express each coefficient as the product of local integrals. Each nonarchimedean local integral is a local Dirichlet series with a parameter s, whose value at $s = 0$ is essentially equal to the local representation density in Siegel's sense, and also to a local version of Hardy's singular series, if the problem concerns sums of squares. However, our integral expression makes the calculation more manageable. In fact, the values of the integrals for almost all primes are the same as those we already determined in [S6], and so our task here is to show how the present problem can be understood in the framework of that paper, and then to calculate the values for exceptional primes. In the case of triangular numbers, 2 is the only exceptional prime, and our calculation is simple. We shall eventually give an explicit formula for the number of (∗) for each $n > 1$ in the cases of triangular and pentagonal numbers (Theorems 4.4 and 5.3).

To show that $\theta_1 = f$, it is sufficient to prove that there is no nontrivial cusp form satisfying the same transformation formulas as θ_1, which can be done without introducing any drastically new ideas, though the case of odd n is technically somewhat involved.

Though our applications are restricted to the cases of triangular and pentagonal numbers, in the first part of the paper we treat the general case of totally definite quadratic forms over a totally real algebraic number field, in order to show that the basic principle is general enough to be applicable to that case. Indeed, we merely specialize general principles to obtain the desired results, without recourse to any fortuitous identities like those employed by early researchers.

1. Inhomogeneous forms over a totally real algebraic number field.

1.1. Notation. For a finite set X we denote by $\#X$ or $\#(X)$ the number of elements in X, as we did above. Also, H will denote the upper half complex plane throughout the paper. For an associative ring R with identity element we denote by $R^\times$ the group of all its invertible elements. For a prime number p we denote by $\mathbf{Q}_p$ the field of p-adic numbers, and by $\mathbf{Z}_p$ the ring of p-adic integers.

We fix a totally real algebraic number field F of finite degree and denote by $\mathbf{a}$ and $\mathbf{h}$ the sets of archimedean and nonarchimedean primes of F; we put $\mathbf{v} = \mathbf{a} \cup \mathbf{h}$; further we denote by $\mathfrak{g}$, $\mathfrak{d}_F$, and D_F the maximal order of F, the different of F relative to $\mathbf{Q}$, and the discriminant of F, respectively. For each $v \in \mathbf{h}$ we denote by $\mathfrak{p}_v$ the prime ideal at v, and put $q_v = \#(\mathfrak{g}/\mathfrak{p}_v)$. Given a set S, we denote by $S^{\mathbf{a}}$ the product of $\mathbf{a}$ copies of S, that is, the set of all indexed elements $(x_v)_{v\in\mathbf{a}}$ with x_v in S. For a Hecke character χ of F we denote by $L(s, \chi)$ the L-function of χ and by $\zeta_F(s)$ the Dedekind zeta function of F.

Given an algebraic group X defined over F, we define X_v for each $v \in \mathbf{v}$ and the adelization $X_{\mathbf{A}}$ as usual, and view X as a subgroup of $X_{\mathbf{A}}$. We then denote by $X_{\mathbf{a}}$ and $X_{\mathbf{h}}$ the archimedean and nonarchimedean factors of $X_{\mathbf{A}}$, respectively. For $y \in X$ we denote by y_v, $y_{\mathbf{a}}$, and $y_{\mathbf{h}}$ its projections to X_v, $X_{\mathbf{a}}$, and $X_{\mathbf{h}}$. For a fractional ideal $\mathfrak{x}$ in F we denote by $N(\mathfrak{x})$ the norm of $\mathfrak{x}$. For $t \in F_{\mathbf{A}}^\times$ we put $|t|_v = |t_v|_v$ with the normalized valuation $|\ \ |_v$ at v and $|t|_{\mathbf{A}} = \prod_{v\in\mathbf{v}} |t|_v$. We put also

$$\mathbf{T} = \{\zeta \in \mathbf{C} \mid |\zeta| = 1\},$$
$$\mathbf{e}(z) = e^{2\pi iz} \qquad (z \in \mathbf{C}),$$

and define characters $\mathbf{e_A} : F_\mathbf{A} \to \mathbf{T}$ and $\mathbf{e}_v : F_v \to \mathbf{T}$ for each $v \in \mathbf{v}$ as follows: if $v \in \mathbf{a}$, then $\mathbf{e}_v(x) = \mathbf{e}(x)$; if $v \in \mathbf{h}$ and p is the rational prime divisible by v, then $\mathbf{e}_v(x) = \mathbf{e}_p(\mathrm{Tr}_{F_v/\mathbf{Q}_p}(x))$, where $\mathbf{e}_p(z) = \mathbf{e}(-y)$ with $y \in \bigcup_{m=1}^{\infty} p^{-m}\mathbf{Z}$ such that $z - y \in \mathbf{Z}_p$. We then put $\mathbf{e_A}(x) = \prod_{v \in \mathbf{v}} \mathbf{e}_v(x_v)$, $\mathbf{e_h}(x) = \mathbf{e_A}(x_\mathbf{h})$, and $\mathbf{e_a}(x) = \mathbf{e_A}(x_\mathbf{a})$ for $x \in F_\mathbf{A}$. We extend the last symbol $\mathbf{e_a}$ by putting $\mathbf{e_a}(z) = \mathbf{e}\left(\sum_{v \in \mathbf{a}} z_v\right)$ for $z \in \mathbf{C}^\mathbf{a}$. Thus, for $\xi \in F$ and $z \in \mathbf{C}^\mathbf{a}$ we understand that $\mathbf{e_a}(\xi z) = \mathbf{e}\left(\sum_{v \in \mathbf{a}} \xi_v z_v\right)$.

1.2. Given a finite-dimensional vector space V over F of dimension > 1 and a nondegenerate symmetric F-bilinear form $\varphi : V \times V \to F$, we put

$$G = \{\alpha \in GL(V) \mid \varphi(x\alpha, y\alpha) = \varphi(x, y)\},$$
$$\varphi[x] = \varphi(x, x) \qquad (x \in V),$$

where $GL(V)$ denotes the group of all F-linear automorphisms of V; we denote by $x\alpha$ the image of x under α.

Hereafter we assume the following two conditions:

(1.1a) *φ is positive definite at every archimedean prime of F;*
(1.1b) *If n is odd, then* $\det(\varphi)\mathfrak{g}$ *is a square of a fractional ideal in F.*

The latter is not a strong condition, since if we take $\det(\varphi)\varphi$ in place of φ, then $\det(\varphi)\varphi$ is still totally positive and $\det(\det(\varphi)\varphi) = \det(\varphi)^{n+1}$, which is a square if n is odd. If φ is replaced by $c\varphi$, then the equation $\varphi[x] = h$ can be reduced to $c\varphi[x] = ch$, and so the solution of the problem concerning the latter equation with *no condition* on ch can give a completely satisfactory solution for that concerning $\varphi[x] = h$. The group G does not change by this replacement.

By a $\mathfrak{g}$-*lattice* in V we understand a finitely generated $\mathfrak{g}$-submodule of V that spans V over F. Given a $\mathfrak{g}$-lattice L in V and $\alpha \in GL(V)_\mathbf{A}$, we denote by $L\alpha$ the $\mathfrak{g}$-lattice in V such that $(L\alpha)_v = L_v\alpha_v$ for every $v \in \mathbf{h}$. This can be generalized to the case of a coset of V/L. Namely, for $y \in V$ let $L + y$ denote the set $\{x \in V \mid x - y \in L\}$. Its localization $(L + y)_v = L_v + y$ is a coset of V_v/L_v. Then for $\alpha \in GL(V)_\mathbf{A}$ we denote by $(L+y)\alpha$ the coset of $V/L\alpha$ such that $((L+y)\alpha)_v = (L_v + y)\alpha_v$ for every $v \in \mathbf{h}$. Such a coset $(L+y)\alpha$ can be found in the following way: Since $V_\mathbf{A} = V + V_\mathbf{a} + \prod_{v \in \mathbf{h}} L_v\alpha_v$, given y and α, there exists an element z of V such that $y\alpha_v - z \in L_v\alpha_v$ for every $v \in \mathbf{h}$. Then $L\alpha + z$ is the desired coset.

We call a $\mathfrak{g}$-lattice L in V *maximal* if $\varphi[x] \in \mathfrak{g}$ for every $x \in L$ and L is maximal among the $\mathfrak{g}$-lattices on which φ has values in $\mathfrak{g}$.

1.3. We consider a nonzero **C**-valued function λ on $V_{\mathbf{h}}$ of the type

$$\lambda(x) = \prod_{v\in\mathbf{h}} \lambda_v(x_v) \quad \text{for} \quad x \in V_{\mathbf{h}} \tag{1.2}$$

with **C**-valued functions λ_v on V_v, for which there exist a $\mathfrak{g}$-lattice L and another $\mathfrak{g}$-lattice M contained in L such that: (i) λ_v is the characteristic function of L_v for almost all $v \in \mathbf{h}$; (ii) $\lambda_v(y) = 0$ for $y \notin L_v$, and $\lambda_v(y)$ for $y \in L_v$ depends only on $M_v + y$ for every $v \in \mathbf{h}$.

Given such a λ, we put

$$C = \{\, \alpha \in G_{\mathbf{A}} \mid \lambda(x\alpha_{\mathbf{h}}) = \lambda(x) \textit{ for every } x \in V_{\mathbf{h}} \,\}. \tag{1.3}$$

This is an open subgroup of $G_{\mathbf{A}}$ containing $G_{\mathbf{a}}$. Clearly $C = (C \cap G_{\mathbf{h}})G_{\mathbf{a}}$, and $G_{\mathbf{a}}$ is compact in view of (1.1a). Since we are assuming that λ is nontrivial, we easily see that $C \cap G_{\mathbf{h}}$ is compact, and hence C is compact.

By a well-known principle we can take a finite subset $\mathscr{B}$ of $G_{\mathbf{h}}$ so that

$$G_{\mathbf{A}} = \bigsqcup_{a\in\mathscr{B}} GaC. \tag{1.4}$$

We then define the mass $\mathfrak{m}(C)$ of C, which we write also $\mathfrak{m}(\lambda)$, by

$$\mathfrak{m}(C) = \mathfrak{m}(\lambda) = \sum_{a\in\mathscr{B}} [\Gamma^a : 1]^{-1}, \qquad \Gamma^a = G \cap aCa^{-1}. \tag{1.5}$$

This is a special case of [S4, (24.1.2)]. For each $h \in F$ and $g \in G_{\mathbf{A}}$ put

$$N(h;\, \lambda,\, gC) = \sum_{\xi\in V,\varphi[\xi]=h} \lambda(\xi g_{\mathbf{h}}), \tag{1.6a}$$

$$R(h;\, \lambda) = \sum_{a\in\mathscr{B}} [\Gamma^a : 1]^{-1} N(h;\, \lambda,\, aC). \tag{1.6b}$$

Since φ is totally positive definite, $N(h;\, \lambda,\, gC)$ is finite, and it is not zero only if h is totally positive or $h = 0$. Notice that $N(h;\, \lambda,\, gC)$ indeed depends only on gC, once φ, h, and λ are fixed. We do not assume that $h \in \mathfrak{g}$.

As a typical example, we take a coset $X = L+y$ with a $\mathfrak{g}$-lattice L and $y \in V$, and take λ_v to be the characteristic function of $X_v = L_v + y$ for each $v \in \mathbf{h}$. For $h \in F$ put

$$N(h,\, X) = \#\{\, \xi \in X \mid \varphi[\xi] = h \,\}. \tag{1.7}$$

Then we have $C = \{\alpha \in G_{\mathbf{A}} \mid X\alpha = X\}$, $\Gamma^a = \{\, \gamma \in G \mid Xa^{-1}\gamma = Xa^{-1}\}$, and

$$N(h;\, \lambda,\, aC) = N\big(h,\, Xa^{-1}\big). \tag{1.8}$$

The set of the cosets $X\alpha$ for $\alpha \in G_{\mathbf{A}}$ may be called the *genus* of X, and the cosets Xa^{-1} for $a \in \mathscr{B}$ represent the *classes* in this genus of X. Thus $\mathfrak{m}(C)$ may be called the *mass* of this genus of cosets. If $X = L$, then $R(h, \lambda)/\mathfrak{m}(C)$ equals $\mathfrak{m}(C)^{-1}\sum_{a\in\mathscr{B}}[\Gamma^a : 1]^{-1}N(h, La^{-1})$, which is the quantity for which Siegel gave his product formula.

1.4. To state our main theorem, we need to introduce some more symbols. First, if n is even, we define an extension K of F by

$$K = F(\kappa^{1/2}) \quad \text{with} \quad \kappa = (-1)^{n/2}\det(\varphi), \tag{1.9}$$

and call it *the discriminant field* of φ. This is either a quadratic extension of F or F itself. Next, for each $v \in \mathbf{h}$ we consider the localization $(V, \varphi)_v = (V_v, \varphi_v)$, and denote by r_v the dimension of a maximal totally isotropic subspace of V_v with respect to φ_v. We put $t_v = n - 2r_v$, and call t_v the *core dimension* of $(V, \varphi)_v$. Then φ_v is equivalent, over F_v, to a form represented by a matrix

$$\begin{bmatrix} 0 & 0 & 1_{r_v} \\ 0 & \theta_v & 0 \\ 1_{r_v} & 0 & 0 \end{bmatrix}, \tag{1.10}$$

where 1_r denotes the identity matrix of size r and $\theta_v = {}^t\theta_v \in GL_{t_v}(F_v)$; we ignore θ_v if $t_v = 0$. If $t_v > 0$, then θ_v is anisotropic. To describe the nature of θ_v when $t_v = 2$, put $K_v = K \otimes_F F_v$. Then K_v is a quadratic extension of F_v, and the quadratic form represented by θ_v is equivalent to the form $x \mapsto c_v N_{K_v/F_v}(x)$ defined for $x \in K_v$ with $c_v \in F_v^\times$. We can normalize c_v as follows: it is a prime element of F_v or an element of $\mathfrak{g}_v^\times$; it is always an element of $\mathfrak{g}_v^\times$ if K_v is ramified over F_v. For these, see [S5, §3.2 and Proposition 5.2].

Given a $\mathfrak{g}$-lattice L in V, we put

$$\widetilde{L} = \{\, x \in V \mid 2\varphi(x, L) \subset \mathfrak{g} \,\}. \tag{1.11}$$

This is also a $\mathfrak{g}$-lattice in V. We have $L \subset \widetilde{L}$ if $\varphi[x] \in \mathfrak{g}$ for every $x \in L$.

Fixing an arbitrary $\mathfrak{g}$-lattice L in V, we put, for each $v \in \mathbf{h}$ and $h \in F_v$,

$$b_v(h, \lambda, s) = \int_{F_v} \nu(\sigma)^{-s} \int_{V_v} \mathbf{e}_v\big(\delta_v^{-1}(\varphi[x_v] - h)\sigma\big)\lambda_v(x_v)dx_v^* d\sigma. \tag{1.12}$$

Here the measures dx_v^* and $d\sigma$ are determined so that $\int_{L_v} dx_v^* = \int_{\mathfrak{g}_v} d\sigma = 1$; $\nu(\sigma) = 1$ if $\sigma \in \mathfrak{g}_v$ and $\nu(\sigma) = |\sigma|_v^{-1}$ otherwise; δ_v is an element of $F_v^\times$ such that $\delta_v\mathfrak{g}_v = (\mathfrak{d}_F)_v$. Then b_v is convergent for sufficiently large $\mathrm{Re}(s)$, and in fact, it is a rational expression of q_v^{-s}. Indeed, the problem can be reduced to the case where λ_v is the characteristic function of a $\mathfrak{g}_v$-lattice in V_v. In that special case the fact can be seen from the formula on the last three lines of the proof of [S6, Lemma 2.5].

THEOREM 1.5. *Suppose that conditions (1.1a, b) are satisfied. Let* L *be a* $\mathfrak{g}$*-lattice in* V *such that* $\varphi[L] \subset \mathfrak{g}$, *and* h *a totally positive element of* F *(not necessarily in* $\mathfrak{g}$*); let* $\mathfrak{e}'$ *be the product of all primes* $v \in \mathbf{h}$ *such that* $\widetilde{L}_v \neq L_v$, *and* $\mathfrak{e}_1$ *be the squarefree integral ideal characterized by the property that* $v \nmid \mathfrak{e}_1$ *if and only if* $h \in \mathfrak{g}_v$, L_v *is maximal, and* λ_v *is the characteristic function of* L_v. *Put* $d = [F : \mathbf{Q}]$. *Define* $\gamma_v(s)$ *by the formulas given in* [S6, §1.6] *(see also §1.6 below); define also* $b_v(h, \lambda, s)$ *by (1.12) with respect to* L.

(I) *Suppose* n *is odd; put* $n = 2\mu + 1$ *and* $K_h = F(\kappa^{1/2})$ *with* $\kappa = (-1)^{(n-1)/2}\det(\varphi)h$; *denote by* ψ_h *the Hecke ideal character of* F *corresponding to* K_h/F. *Then*

$$\frac{R(h, \lambda)}{\mathfrak{m}(\lambda)} = \frac{(2\pi)^{nd/2} N_{F/\mathbf{Q}}(h)^{(n-2)/2} L(\mu, \psi_h)}{D_F^{(n-1)/2}[\widetilde{L} : L]^{1/2}\Gamma(n/2)^d \zeta_F(2\mu)} \cdot \prod_{v|\mathfrak{e}_1} \frac{b_v(h, \lambda, 0)(1 - \psi_h(\mathfrak{p}_v)q_v^{-\mu})}{(1 - q_v^{-2\mu})} \cdot \prod_{v|h\mathfrak{e}',\, v\nmid\mathfrak{e}_1} \gamma_v(n/2).$$

(II) *Suppose* n *is even; put* $n = 2\mu$; *let* ψ *be the Hecke ideal character of* F *corresponding to the extension* K *of* F *given by (1.9). Then*

$$\frac{R(h, \lambda)}{\mathfrak{m}(\lambda)} = \frac{c_n (2\pi)^{nd/2} N_{F/\mathbf{Q}}(h)^{(n-2)/2}}{D_F^{(n-1)/2}[\widetilde{L} : L]^{1/2}\Gamma(\mu)^d L(\mu, \psi)} \cdot \prod_{v|\mathfrak{e}_1} \frac{b_v(h, \lambda, 0)}{(1 - \psi(\mathfrak{p}_v)q_v^{-\mu})} \cdot \prod_{v|h\mathfrak{e}',\, v\nmid\mathfrak{e}_1} \gamma_v(\mu),$$

where $c_n = 1$ *if* $n > 2$, *and* $c_2 = 1/2$.

Proof. We normalize the Haar measure on $G_{\mathbf{A}}$ so that the measure of $G\backslash G_{\mathbf{A}}$ is 1. With C and $\mathscr{B}$ as in (1.3) and (1.4) we easily see that $G\backslash G_{\mathbf{A}}$ can be identified with $\bigsqcup_{a\in\mathscr{B}} (\Gamma^a\backslash aC)$, and hence we have $1 = \sum_{a\in\mathscr{B}} [\Gamma^a : 1]^{-1}\mathrm{vol}(C)$. Thus

$$\mathfrak{m}(\lambda)^{-1} = \mathfrak{m}(C)^{-1} = \mathrm{vol}(C) = \int_C dx. \tag{1.13}$$

Next we define a theta series $\theta(z; \lambda, g)$ by

$$\theta(z; \lambda, g) = \sum_{\xi\in V} \lambda(\xi g_{\mathbf{h}})\mathbf{e}_{\mathbf{a}}(\varphi[\xi]z) \qquad (g \in G_{\mathbf{A}},\ z \in H^{\mathbf{a}}). \tag{1.14}$$

Then we have

$$\begin{aligned}\int_{G\backslash G_{\mathbf{A}}}\theta(z;\ \lambda,\ g)dg &= \sum_{a\in\mathscr{B}}\int_{\Gamma^a\backslash aC}\sum_{\xi\in V}\lambda(\xi g_{\mathbf{h}})\mathbf{e}_{\mathbf{a}}(\varphi[\xi]z)dg \\ &= \mathrm{vol}(C)\sum_{a\in\mathscr{B}}[\Gamma^a:1]^{-1}\sum_{\xi\in V}\lambda(\xi a)\mathbf{e}_{\mathbf{a}}(\varphi[\xi]z) \\ &= \mathrm{vol}(C)\sum_{a\in\mathscr{B}}[\Gamma^a:1]^{-1}\sum_{h\in F}N(h;\ \lambda,\ aC)\mathbf{e}_{\mathbf{a}}(hz) \\ &= \mathfrak{m}(\lambda)^{-1}\sum_{h\in F}R(h,\ \lambda)\mathbf{e}_{\mathbf{a}}(hz).\end{aligned}$$

In other words, $\mathfrak{m}(\lambda)^{-1}R(h,\ \lambda)$ is a Fourier coefficient of $\int_{G\backslash G_{\mathbf{A}}}\theta(z;\ \lambda,\ g)\,dg$.

Now, as a consequence of the Siegel-Weil formula in [W], we know that our integral over $G\backslash G_{\mathbf{A}}$ is an Eisenstein series. In order to apply that formula to our setting, we have to clarify various points; also, if $n\le 4$, we have to invoke the results of Rallis and Kudla; for details of all these the reader is referred to [S6, §§5.1 and 5.3]. Once we know that the integral equals an Eisenstein series defined in an explicit way, we can compute its Fourier coefficients by means of the technique developed in our previous papers, and we eventually obtain

$$(1.15)\qquad \frac{R(h,\ \lambda)}{\mathfrak{m}(\lambda)}=\frac{c_n(2\pi)^{nd/2}N_{F/\mathbf{Q}}(h)^{(n-2)/2}}{D_F^{(n-1)/2}[\widetilde{L}:L]^{1/2}\Gamma(n/2)^d}\cdot\left\{\prod_{v\in\mathbf{h}}b_v(h,\ \lambda,\ s)\right\}_{s=0}.$$

The details of calculation are given in [S6, §§2.1–2.4 and 5.3]; in particular, (1.15) is [S6, (5.3.2)]. In [S6, §2.6] we assumed that L is maximal, but it can easily be seen that formula (1.15) holds for a more general L, provided $L\subset\widetilde{L}$.

Suppose $v\nmid\mathfrak{e}_1$; then $h\in\mathfrak{g}_v$, L_v is maximal, and λ_v is the characteristic function of L_v. In [S6, (5.4.1)] for such a v we determined the explicit form of $b_v(h,\ \lambda,\ s)$ as follows:

$$(1.16)\quad b_v(h,\ \lambda,\ s)=\gamma_v(s+n/2)\cdot\begin{cases}1-\psi(\mathfrak{p}_v)q_v^{-(n/2)-s} & \text{if}\quad n\in 2\mathbf{Z},\\ \dfrac{1-q_v^{1-n-2s}}{1-\psi_h(\mathfrak{p}_v)q_v^{(1-n)/2-s}} & \text{if}\quad n\notin 2\mathbf{Z},\end{cases}$$

where γ_v is the rational expression of q_v^{-s} given in [S6, p. 66]. Therefore, expressing $\prod_{v\in\mathbf{h}}$ as the product $\prod_{v|\mathfrak{e}_1}\prod_{v\nmid\mathfrak{e}_1}$, we obtain the formulas of our theorem, since $\gamma_v(s)=1$ for $v\nmid h\mathfrak{e}'$. Indeed, let M be a maximal lattice containing L, and $\mathfrak{e}$ the product of all $v\in\mathbf{h}$ such that $\widetilde{M}_v\neq M_v$. This is the ideal $\mathfrak{e}$ in [S6], and $\gamma_v(s)=1$ for $v\nmid h\mathfrak{e}$ as noted in [S6, p. 88, line 4]. If $v\nmid\mathfrak{e}_1$, then $L_v=M_v$, and hence, for $v\nmid\mathfrak{e}_1$, we have $v|h\mathfrak{e}$ if and only if $v|h\mathfrak{e}'$. Therefore we obtain the results as stated in our theorem.

Remark. (A) If λ is the characteristic function of a lattice, then $b_v(h, \lambda, 0)$ is a local version of Hardy's singular series, and essentially equal to the representation density in Siegel's sense. Though in general its determination is highly nontrivial, the cases of triangular and pentagonal numbers are relatively easy, as will be shown in Sections 4 and 5. The problem is more complex in the case of sums of squares, but still the value can be determined; see [S8].

(B) In [Si1, I] Siegel proved that $\sum_{h\in F} R(h, \lambda)\mathbf{e_a}(hz)$ is an Eisenstein series when $F = \mathbf{Q}$ and λ is the characteristic function of a lattice, which is the first case of what we call the Siegel-Weil formula. He treated the case of indefinite quadratic forms in [Si1, II], and sixteen years later in [Si2] proved that a certain integral of a theta series for an inhomogeneous indefinite quadratic form is an Eisenstein series, but he excluded the totally definite case; see [Si2, II, p. 378]. The formula given by Weil in [W] is applicable to inhomogeneous forms both in definite and indefinite cases.

(C) Formula (1.15) involves L, but actually it is independent of L. Indeed, if we change L, then the measure dx_v^* in (1.12) is changed accordingly, but that is compensated by the factor $[\widetilde{L} : L]^{1/2}$.

1.6. Here we give the formulas for γ_v only when $F = \mathbf{Q}$. The reader is referred to [S6, §1.6] for those in the general case. Taking $F = \mathbf{Q}$, we write p, t_p, and γ_p instead of v, t_v, and γ_v. Also, let c_p be an element of $\mathbf{Q}_p^\times$, determined as in §1.4 when $t_p = 2$; this is a prime element of $\mathbf{Q}_p$ or an element of $\mathbf{Z}_p^\times$. We denote by the same symbols ψ and ψ_h the primitive Dirichlet characters corresponding to the ideal characters defined in the above theorem.

Given $h \in \mathbf{Z}_p$, $\neq 0$, determine $\nu \in \mathbf{Z}$ so that p^ν is the largest power of p that divides h, and put $\lambda = [\nu/2]$. If t_p is odd, we put $\psi_h(p) = \xi$. Now $\gamma_p(s)$ can be given as follows:

$$\gamma_p(s) = (1 - p^{(\nu+1)(1-s)})/(1 - p^{1-s}) \quad \textit{if} \ \ t_p = 0,$$

$$\gamma_p(s) = \frac{1 - p^{(\lambda+1)(2-2s)} - \xi p^{(1/2)-s}(1 - p^{\lambda(2-2s)})}{1 - p^{2-2s}}$$
$$\textit{if} \ \ t_p = 1 \ \textit{except when} \ p = 2 \ \textit{and} \ \nu \in 2\mathbf{Z},$$

$$\gamma_p(s) = \frac{1 - p^{(\nu+4)(1-s)} - \xi p^{(1/2)-s}(1 - p^{(\nu+2)(1-s)})}{1 - p^{2-2s}}$$
$$\textit{if} \ \ t_p = 1, \ p = 2, \ \nu \in 2\mathbf{Z}, \ \textit{and} \ \xi \neq 0,$$

$$\gamma_p(s) = \frac{1 - p^{(\nu+2)(1-s)}}{1 - p^{2-2s}} \quad \textit{if} \ \ t_p = 1, \ p = 2, \ \nu \in 2\mathbf{Z}, \ \textit{and} \ \xi = 0,$$

$$\gamma_p(s) = (1 - (-p^{1-s})^{\nu+1})/(1 + p^{1-s}) \quad \textit{if} \ \ t_p = 2, \ \psi(p) \neq 0, \ \textit{and} \ c_p \in \mathbf{Z}_p^\times,$$

$$\gamma_p(s) = \frac{1 - (-p^{1-s})^{\nu+2} + p^{2-s}(1 - (-p^{1-s})^{\nu})}{(1 + p^{-s})(1 + p^{1-s})}$$

$$\textit{if}\ \ t_p = 2,\ \psi(p) \neq 0,\ \textit{and}\ c_p \notin \mathbf{Z}_p^\times,$$

$$\gamma_p(s) \ = \ 1 + \xi_p(c_p h)p^{(\nu+\sigma)(1-s)} \qquad \textit{if}\ \ t_p = 2\ \textit{and}\ \psi(p) = 0,$$

where p^σ *is the largest power of* p *that divides the discriminant of* K *and* ξ_p *is the isomorphism* $\mathbf{Q}_p^\times / N_{K/\mathbf{Q}}(K_p^\times) \to \{\pm 1\}$,

$$\gamma_p(s) \ = \ \frac{1 - p^{3-2s} - p^{(\lambda+\kappa)(2-2s)}(1 - \xi p^{(3/2)-s})(p^{2-2s} + \xi p^{(3/2)-s})}{(1 + \xi p^{(1/2)-s})(1 - p^{2-2s})}$$

if $t_p = 3$ *and* $\xi \neq 0$, where $\kappa = 1$ if $p = 2$ and $\kappa = 0$ otherwise,

$$\gamma_p(s) \ = \ \frac{1 - p^{3-2s} + p \cdot p^{(\nu+\varepsilon+1)(1-s)} - p^{(\nu+\varepsilon+3)(1-s)}}{(1 - p^{1-2s})(1 - p^{2-2s})} \qquad \textit{if}\ \ t_p = 3\ \textit{and}\ \xi = 0,$$

where $\varepsilon = 0$ *if* $\nu \notin 2\mathbf{Z}$ *or* $p \neq 2$, *and* $\varepsilon = 1$ *otherwise,*

$$\gamma_p(s) \ = \ \frac{1 - p^{(\nu+2)(1-s)} - p^{2-s}(1 - p^{\nu(1-s)})}{(1 - p^{-s})(1 - p^{1-s})} \qquad \textit{if}\ \ t_p = 4.$$

These formulas are special cases of those given in [S6, §1.6]. If $t_p = 2$ and $\psi(p) = 0$, then the form $1 + \psi_v(c_v h)\big(|h|_v^{-1}N(\mathfrak{d}_v)\big)^{1-s}$ in [S6, p. 66] becomes $1+\xi_p(c_p h)p^{(\nu+\sigma)(1-s)}$. We also needed integers κ, e, and ε in [S6, §1.6]. Clearly $\kappa = 1$ and $e = 3$ if $p = 2$, since $\mathbf{Q}$ is the basic field. Also, as noted on the last line of [S6, p. 66], we have $\varepsilon = 1$ if $\nu \in 2\mathbf{Z}$, $p = 2$, and $\psi_h(2) = 0$. Therefore the formulas for γ_p there take the forms as stated above.

1.7. The above formulas for $R(h, \lambda)/\mathfrak{m}(\lambda)$ involve special values of the Dedekind zeta function and Hecke L-functions, which become the Riemann zeta function and Dirichlet L-functions if $F = \mathbf{Q}$. There are well-known expressions for these values as algebraic numbers up to powers of π. For the reader's convenience, we note the formulas here.

Given a nontrivial primitive Dirichlet character χ of conductor d, we put

$$\tau(\chi) = \sum_{a=1}^{d} \chi(a)\mathbf{e}(a/d),$$

and define the Bernoulli polynomial $B_k(x)$ for $0 \leq k \in \mathbf{Z}$ by

$$\frac{te^{xt}}{e^t - 1} = \sum_{k=0}^{\infty} \frac{B_k(x)t^k}{k!},$$

where t and x are independent variables. Then, for every positive integer k such that $\chi(-1) = (-1)^k$ we have

$$(1.17) \qquad 2 \cdot k! \cdot \tau(\overline{\chi})(2\pi i)^{-k}L(k, \chi) = -\sum_{a=1}^{d} \overline{\chi}(a)B_k(a/d).$$

Here are two well-known facts:

$$(1.18)\qquad 2 \cdot m! \cdot (2\pi)^{-m}\zeta(m) = |B_m| \ \text{ if } \ 0 < m \in 2\mathbf{Z},$$

$$(1.19)\qquad 2^{2\mu+1}(\mu-1)! \cdot (2\pi)^{-\mu}L(\mu, \eta) = E_{(\mu-1)/2} \ \text{ if } \ 0 \leq \mu - 1 \in 2\mathbf{Z},$$

where B_m denotes the m-th Bernoulli number, E_k the k-th Euler number, and η the Dirichlet character of conductor 4. We note also

$$(1.20)\qquad \Gamma(n/2) = \begin{cases} \dfrac{(n-1)! \cdot \pi^{1/2}}{2^{n-1} \cdot \mu!} & \text{if } \ n = 2\mu + 1 \notin 2\mathbf{Z}, \ 0 \leq \mu \in \mathbf{Z}, \\ (\mu - 1)! & \text{if } \ n = 2\mu \in 2\mathbf{Z}. \end{cases}$$

2. Theta series as Hilbert modular forms.

2.1. Clearly we have

$$(2.1)\qquad \theta(z, \lambda, g) = \textstyle\sum_{h\in F} N(h, \lambda, gC)\mathbf{e}_{\mathbf{a}}(hz).$$

This is a Hilbert modular form of weight $n/2$. Moreover,

$$(2.2)\qquad \textstyle\sum_{h\in F} \left\{ N(h, \lambda, gC) - \dfrac{R(h, \lambda)}{\mathfrak{m}(\lambda)} \right\} \mathbf{e}_{\mathbf{a}}(hz)$$

is a cusp form for every $g \in G_{\mathbf{A}}$. To explain about these points, let us first recall some basic facts of such theta series and Hilbert modular forms in general.

For a matrix $\gamma = \begin{bmatrix} a & b \\ c & d \end{bmatrix} \in SL_2(F)$ and $z = (z_v)_{v\in\mathbf{a}} \in H^{\mathbf{a}}$ we define $\gamma(z) \in H^{\mathbf{a}}$ by $\gamma(z) = \left((a_v z_v + b_v)(c_v z_v + d_v)^{-1}\right)_{v\in\mathbf{a}}$, where x_v for $x \in F$ is the image of x under the injection $F \to F_v = \mathbf{R}$. We then define Hilbert modular forms as follows. In general, a weight of a Hilbert modular form is an element of $2^{-1}\mathbf{Z}^{\mathbf{a}}$, but in this paper we consider only those forms belonging to the factors of automorphy $j_k(\gamma, z)$ of the following two types:

$$(2.3\text{a})\qquad j_k(\gamma, z) = \prod_{v\in\mathbf{a}} (c_v z_v + d_v)^k \qquad (0 < k \in \mathbf{Z}),$$

$$(2.3\text{b})\qquad j_k(\gamma, z) = h(\gamma, z) \prod_{v\in\mathbf{a}} (c_v z_v + d_v)^m \qquad (0 \leq k - (1/2) = m \in \mathbf{Z}).$$

Here j_k of (2.3a) is meaningful for every $\gamma \in SL_2(F)$; however, $h(\gamma, z)$ is a factor of automorphy defined for γ in a certain congruence subgroup of $SL_2(F)$, such that $h(\gamma, z)^2 = \zeta_\gamma j_1(\gamma, z)$ with a root of unity ζ_γ. Either type of j_k is called the canonical factor of automorphy of weight k. For details the reader is referred to [S3]. By a Hilbert modular form of weight k with respect to a congruence subgroup Γ of $SL_2(F)$ we understand a holomorphic function f on $H^{\mathbf{a}}$ such that $f(\gamma(z)) = j_k(\gamma, z)f(z)$ for every $\gamma \in \Gamma$, and that f is finite at every cusp; the last condition at the cusps is automatically satisfied if $F \neq \mathbf{Q}$. If f is a Hilbert modular form of weight k with $k \in 2^{-1}\mathbf{Z}$, then f^2 is a Hilbert modular form of weight $2k$ in the classical sense. We then call f a *cusp form* if f^2 is a cusp form.

To state a basic lemma, let $\mathscr{S}(V_{\mathbf{h}})$ denote the set of all finite $\mathbf{C}$-linear combinations of functions of type (1.2), and define $\theta(z;\, \mu,\, g)$ for $\mu \in \mathscr{S}(V_{\mathbf{h}})$ by (1.14) with μ in place of λ.

LEMMA 2.2. (1) *For every fixed* $\mu \in \mathscr{S}(V_{\mathbf{h}})$ *and* $g \in G_{\mathbf{A}}$, $\theta(z;\, \mu,\, g)$ *as a function of* $z \in H^{\mathbf{a}}$ *is a Hilbert modular form of weight* $n/2$ *belonging to a congruence subgroup of* $SL_2(F)$.

(2) *Let* $\mu \in \mathscr{S}(V_{\mathbf{h}})$ *and* $\gamma \in SL_2(F)$; *let* $t(z)$ *be a holomorphic function on* $H^{\mathbf{a}}$ *such that* $t(z)^2 = \zeta j_n(\gamma,\, z)$ *with* $\zeta \in \mathbf{T}$. *Suppose the equality*

$$t(z)^{-1}\theta\big(\gamma(z);\, \mu,\, g\big) = \theta(z;\, \mu,\, g) \quad \textit{for every} \quad z \in H^{\mathbf{a}}$$

holds for $g = 1$; *then it holds for every* $g \in G_{\mathbf{A}}$.

(3) $\theta(z;\, \mu,\, 1) - \theta(z;\, \mu,\, g)$ *is a cusp form for every* $g \in G_{\mathbf{A}}$.

(4) *Given* λ *as in (1.2), let* $f(z) = \theta(z;\, \lambda,\, g) - \mathfrak{m}(\lambda)^{-1}\sum_{h \in F} R(h,\, \lambda)\mathbf{e}_{\mathbf{a}}(hz)$ *with any* $g \in G_{\mathbf{A}}$. *Then* f *is a cusp form; moreover, if* $t(z)^{-1}\theta\big(\gamma(z);\, \lambda,\, 1\big) = \theta(z;\, \lambda,\, 1)$ *with* $\big(\gamma,\, t(z)\big)$ *as in (2), then* $t(z)^{-1}f\big(\gamma(z)\big) = f(z)$.

Assertion (1) is included in [S3, Proposition 11.8] and [S7, Theorem A3.8]; (2) is a special case of [S7, Theorem A4.2]; (3) is [S7, Theorem A4.3 (2)]; (4) follows from (2) and (3), since $f(z) = \mathfrak{m}(\lambda)^{-1}\sum_{a \in \mathscr{B}} [\Gamma^a : 1]^{-1}\{\theta(z;\, \lambda,\, g) - \theta(z;\, \lambda,\, a)\}$. If $F = \mathbf{Q}$ and μ is the characteristic function of a lattice, then (3) and the first half of (4) were given in [Si1, I].

Now the function of (2.2) is exactly f of (4), so that it is a cusp form, and hence as an easiest estimate we have

$$\left| N(h,\, \lambda,\, gC) - \frac{R(h,\, \lambda)}{\mathfrak{m}(\lambda)} \right| \leq M_0 N_{F/\mathbf{Q}}(h)^{n/4} \tag{2.4}$$

for every $h \in F$ with a positive constant M_0 independent of h; see [S7, Proposition A6.4]. This means that the quantity $R(h,\, \lambda)/\mathfrak{m}(\lambda)$ gives the dominant term for $N(h,\, \lambda,\, gC)$. If n is even, then there is a better estimate for the Fourier coefficients of a cusp form. We will not go into details here, as the topic is well understood.

3. Known results obtained by elementary methods. Given $2 \leq n \in \mathbf{Z}$ and $k \in \mathbf{Z}$, let us denote by $T_n(k)$, $S_n(k)$, and $P_n(k)$ respectively the number of $x \in \mathbf{Z}^n$ such that $k = \sum_{i=1}^{n} x_i(x_i + 1)/2$, $k = \sum_{i=1}^{n} x_i^2$, and $k = \sum_{i=1}^{n} x_i(3x_i - 1)/2$. These are the number of representations of k as the sum of n triangular numbers, n squares, and n pentagonal numbers, respectively, but there is one noticeable point that distinguishes one of these three types from the others. Namely, in the case of triangular numbers, given $a \in \mathbf{Z}$, the number of $x \in \mathbf{Z}$ such that $x(x+1)/2 = a$, if nonzero, is always 2; the number of $x \in \mathbf{Z}$ such that $x^2 = a$, if nonzero, is 1 if $a = 0$, and 2 if $a \neq 0$; finally the number of $x \in \mathbf{Z}$ such that $x(3x-1)/2 = a$, if nonzero, is always 1. Therefore, from the expression $1 = 0 + 0 + 1$ we find that $T_3(1) = 24$, $S_3(1) = 6$, and $P_3(1) = 3$. The pentagonal numbers in the traditional sense are $x(3x-1)/2$ for $0 < x \in \mathbf{Z}$, but in this paper we include $x(3x-1)/2$ for all $x \in \mathbf{Z}$.

There is one more type: we denote by $T_n^0(k)$ the number of $x \in \mathbf{Z}^n$ such that $k = \sum_{i=1}^n (2x_i^2 - x_i)$. We easily see that if y runs over $\mathbf{Z}$, then $2y^2 - y$ runs over the set of all triangular numbers without repetition. Therefore $T_n^0(k)$ is the number of representations of k as the sum of n triangular numbers in the "exact sense," and so $T_n(k) = 2^n T_n^0(k)$. Notice that the numbers $2y^2 - y$ for $0 < y \in \mathbf{Z}$ are exactly the hexagonal numbers.

Let us now state here several easy formulas:

$$T_2(m) = S_2(8m+2) = 4 \sum_{d|4m+1} (-1)^{(d-1)/2}, \tag{3.1}$$

$$T_3(m) = S_3(8m+3), \tag{3.2}$$

$$T_4(m) = 2S_4(2m+1) = 16 \sum_{d|2m+1} d, \tag{3.3}$$

$$4P_2(m) = S_2(24m+2) = 4 \sum_{d|12m+1} (-1)^{(d-1)/2}, \tag{3.4}$$

$$8P_3(m) = S_3(24m+3) \quad \text{if} \quad m - 1 \notin 3\mathbf{Z}. \tag{3.5}$$

The first three can be found in Dickson's History [D, Chapter I, pp. 6-19]; in particular, (3.3) is due to Legendre. The last two are probably known, but we have been unable to find any literature on those. In any case the proofs of these are easy. For example, let $m = \sum_{i=1}^3 (3x_i^2 - x_i)/2$. Then $24m+3 = \sum_{i=1}^3 (6x_i - 1)^2$. Conversely, if $24m+3 = \sum_{i=1}^3 y_i^2$ with $y_i \in \mathbf{Z}$ and $m-1 \notin 3\mathbf{Z}$, then $24m+3 \notin 9\mathbf{Z}$, and we easily see that $y_i \notin 3\mathbf{Z}$ for every i. Thus we can put $y_i^2 = (6x_i - 1)^2$ with a unique $x_i \in \mathbf{Z}$. Then we obtain (3.5). We can show in a similar manner that $T_2(m) = S_2(8m+2)$, $T_3(m) = S_3(8m+3)$, and $4P_2(m) = S_2(24m+2)$. Now the formulas for $S_k(n)$ are classically known for even $k \leq 8$, and so all formulas are established, except (3.3), which can be proved as follows.

Take the quaternion algebra $B = \mathbf{Q} + \mathbf{Q}i + \mathbf{Q}j + \mathbf{Q}k$ with the standard quaternion units i, j, k. Let $R_0 = \mathbf{Z} + \mathbf{Z}i + \mathbf{Z}j + \mathbf{Z}k$ and $R = R_0 + \mathbf{Z}\ell$ with $\ell = 2^{-1}(1+i+j+k)$. Then R is a maximal order in B. Now $m = \sum_{\nu=1}^4 (a_\nu^2 - a_\nu)/2$ if and only if $2m + 1 = \sum_{\nu=1}^4 (a_\nu - 1/2)^2$. This means that $T_4(m)$ is the number of elements $x \in R$ not contained in R_0 such that the norm of x is $2m+1$. Now the zeta function of B is $(1 - 2^{1-s})\zeta(s)\zeta(s-1)$. Since R is a principal ideal ring and $R^\times$ has 24 elements, it can easily be verified that the number of $x \in R$ whose norm is $2m+1$ is $24\sum_{d|2m+1} d$. On the other hand $S_4(2m+1) = 8\sum_{d|2m+1} d$, and hence we obtain (3.3).

The remaining point is how to deal with $S_3(8m+3)$. Now a formula for $S_3(k)$ is classically known for a squarefree k; more generally the number of primitive representations of k, that is, representations of k as sums of n squares with no nontrivial common divisors, is known for $n = 3$, 5, or 7. For a brief history of sums of squares, the reader is referred to [H, Lecture IX]. But clearly this is a completely inadequate answer to the question about $T_3(m)$.

In our recent paper [S8] we gave explicit formulas for $S_n(k)$ for $n = 3$, 5, and 7, with no condition on k, and therefore we can state a clear-cut formula for $T_3(m)$. We shall not do it here, since our results stated in §4.5 below will include such as a special case.

4. Triangular numbers.

LEMMA 4.1. (i) *For $0 < m \in \mathbf{Z}$ and an odd $b \in \mathbf{Z}$ the map $x \mapsto bx^2 - x$ is a surjection of $\mathbf{Z}/2^m\mathbf{Z}$ onto $2\mathbf{Z}/2^m\mathbf{Z}$, and every element of $2\mathbf{Z}/2^m\mathbf{Z}$ corresponds exactly to two elements of $\mathbf{Z}/2^m\mathbf{Z}$.*

(ii) *For $0 < m \in \mathbf{Z}$ and a prime number p, the map $x \mapsto px^2 - x$ is a bijection of $\mathbf{Z}/p^m\mathbf{Z}$ onto itself.*

Proof. Clearly the image of the map of (i) is contained in $2\mathbf{Z}/2^m\mathbf{Z}$. Suppose $bx^2 - x - (by^2 - y) \in 2^m\mathbf{Z}$ with $x, y \in \mathbf{Z}$. Then $(x-y)(bx+by-1) \in 2^m\mathbf{Z}$. If $x - y \in 2\mathbf{Z}$, then $bx + by - 1 \notin 2\mathbf{Z}$, so that $x - y \in 2^m\mathbf{Z}$. If $x - y \notin 2\mathbf{Z}$, then $bx + by - 1 \in 2^m\mathbf{Z}$. Thus we obtain (i). Assertion (ii) can be proved in a similar and simpler way.

4.2. For $\sigma \in \mathbf{Q}_2$ and an odd $b \in \mathbf{Z}$ we define $\tau_b(\sigma)$ by

$$\tau_b(\sigma) = \int_{\mathbf{Z}_2} \mathbf{e}_2\big(\sigma(bx^2 - x)\big)dx, \tag{4.1}$$

where the measure on $\mathbf{Z}_2$ is such that $\mathbf{Z}_2$ has measure 1. Clearly we can change $bx^2 - x$ for $bx^2 + x$ without changing the value of the integral. Let us now prove

$$\tau_b(\sigma) = \begin{cases} 1 & \textit{if } \ \sigma \in 2^{-1}\mathbf{Z}_2, \\ 0 & \textit{if } \ \sigma \notin 2^{-1}\mathbf{Z}_2. \end{cases} \tag{4.2}$$

Indeed, for $0 < m \in \mathbf{Z}$ and $a \in \mathbf{Z}_2$ we have

$$\tau_b(2^{-m}a) = 2^{-m} \sum_{x \in \mathbf{Z}_2/2^m\mathbf{Z}_2} \mathbf{e}_2\big(2^{-m}a(bx^2 - x)\big).$$

By Lemma 4.1, the last sum can be written

$$2\sum_{y\in 2\mathbf{Z}_2/2^m\mathbf{Z}_2}\mathbf{e}_2(2^{-m}ay)=2\sum_{z\in\mathbf{Z}_2/2^{m-1}\mathbf{Z}_2}\mathbf{e}_2(2^{1-m}az),$$

which is 0 if $m>1$ and $a\notin 2\mathbf{Z}_2$. This proves the part of (4.2) for which $\tau_b(\sigma)=0$. The other part can easily be seen by putting $m=1$.

4.3. We now consider the sum of n squares $\varphi[x]=\sum_{i=1}^n x_i^2$, and view it as a quadratic form on $\mathbf{Q}^n$. The core dimension t_p of the localization of φ at a prime p can easily be determined. For $p\neq 2$ we have

$$t_p=\begin{cases}1 & \text{if } n \text{ is odd},\\ 2 & \text{if } n\equiv 2 \pmod 4 \text{ and } p\equiv 3 \pmod 4,\\ 0 & \text{otherwise.}\end{cases}\tag{4.3}$$

The number c_p in §1.6 is always ± 1. For these and the value of t_2, the reader is referred to [S8, (1.3), (1.4)] and [S5, §5.16]; we do not need t_2 in the present paper.

Take $L=\mathbf{Z}^n$ and define λ by $\lambda(x)=\prod_p\lambda_p(x_p)$ for $x\in V_\mathbf{h}$ with the characteristic function of $L_p+2^{-1}u$ as λ_p, where u is the element of L whose components are all equal to 1. Clearly λ_p is the characteristic function of L_p if $p\neq 2$; also, for $\xi\in V$ we have $\lambda(\xi)=1$ if and only if $\xi-2^{-1}u\in L$.

Given $0\le k\in\mathbf{Z}$, put $h=2k+(n/4)$. Then

$$N(h,L+2^{-1}u)=T_n(k).\tag{4.4}$$

Indeed, if $\varphi[\alpha]=h$ and $\alpha=\xi+2^{-1}u$ with $\xi\in\mathbf{Z}^n$, then $2k+(n/4)=\varphi[\xi+2^{-1}u]=(n/4)+\sum_{i=1}^n(\xi_i^2+\xi_i)$, so that $k=\sum_{i=1}^n(\xi_i^2+\xi_i)/2$. Conversely, if we have the last equality, then $\varphi[\xi+2^{-1}u]=h$. Thus we obtain (4.4).

Let us now show that with this λ we have

$$b_2(2k+n/4,\ \lambda,\ s)=1+2^{-s}\quad \textit{for every}\quad k\in\mathbf{Z}.\tag{4.5}$$

Indeed, putting $h=2k+n/4$ and $\psi[x]=\sum_{i=1}^n(x_i^2+x_i)$, we have

$$\begin{aligned}b_2(h,\ \lambda,\ s)&=\int_{\mathbf{Q}_2}\nu(\sigma)^{-s}\int_{V_2}\mathbf{e}_2\big(\{\varphi[x]-h\}\sigma\big)\lambda_2(x)dxd\sigma\\&=\int_{\mathbf{Q}_2}\nu(\sigma)^{-s}\int_{L_2}\mathbf{e}_2\big(\{\varphi[x+u/2]-h\}\sigma\big)dxd\sigma\end{aligned}$$

$$= \int_{\mathbf{Q}_2} \nu(\sigma)^{-s} \int_{L_2} \mathbf{e}_2(\{\psi[x] - 2k\}\sigma)dxd\sigma$$

$$= \int_{\mathbf{Q}_2} \nu(\sigma)^{-s}\mathbf{e}_2(-2k\sigma) \int_{L_2} \mathbf{e}_2(\psi[x]\sigma)dxd\sigma$$

$$= \int_{\mathbf{Q}_2} \nu(\sigma)^{-s}\mathbf{e}_2(-2k\sigma)\tau_1(\sigma)^n d\sigma.$$

By (4.2), the last integral over $\mathbf{Q}_2$ can be reduced to the integral over $2^{-1}\mathbf{Z}_2$. Therefore we obtain (4.5).

THEOREM 4.4. *Given $0 \le k \in \mathbf{Z}$, put $h = 2k + n/4$ and $h = 2^\nu g$ with an integer ν and an odd positive integer g; let $g = r^2 s$ with a positive integer r and a squarefree integer s; let $r = \prod_p p^{\lambda_p}$ be the prime decomposition of r. Then $R(h, \lambda)/\mathfrak{m}(\lambda)$ for λ as in §4.3 can be given as follows:*

(I) $n \in 4\mathbf{Z}$:
$$\frac{2^{\nu(\mu-1)+2}\mu}{(2^\mu - 1)|B_\mu|} \cdot \sum_{d|g} d^{\mu-1}.$$

(II) $n - 2 \in 4\mathbf{Z}$:
$$\frac{c_n 2^3}{E_{(\mu-1)/2}} \cdot \sum_{d|g} \eta(g/d)d^{\mu-1}.$$

(III) n *odd*:
$$\frac{2^3 \cdot \mu!(2\pi)^{-\mu}(8k+n)^{\mu-1/2}L(\mu, \psi')}{(2^\mu + \varepsilon)|B_{n-1}|}$$
$$\cdot \prod_{p|r} \left\{ \sum_{j=0}^{\lambda_p} p^{j(2-n)} - \psi'(p) \sum_{i=1}^{\lambda_p} p^{i(2-n)+(n-3)/2} \right\}.$$

Here $\mu = [n/2]$, $c_n = 1$ if $n > 2$ and $c_2 = 1/2$; B_* is the Bernoulli number, E_* is the Euler number, η is the primitive Dirichlet character modulo 4, ψ' is the primitive Dirichlet character corresponding to $\mathbf{Q}((-1)^{\mu/2}(8k+n)^{1/2})$, and $\varepsilon = (-1)^{(n^2-1)/8}$.

Proof. Define $\mathfrak{e}'$ and $\mathfrak{e}_1$ as in Theorem 1.5. We have $L = 2\widetilde{L}$, so that $[\widetilde{L} : L] = 2^n$ and $\mathfrak{e}' = \mathfrak{e}_1 = 2\mathbf{Z}$. Thus, specializing the formula of Theorem 1.5, we obtain

$$\frac{R(h, \lambda)}{\mathfrak{m}(\lambda)} = c_n \pi^{n/2} \Gamma(n/2)^{-1} h^{(n-2)/2} b_2(h, \lambda, 0) \prod_{p|g} \gamma_p(n/2)$$

$$\cdot \begin{cases} \left(1 - \psi(2)2^{-\mu}\right)^{-1} L(\mu, \psi)^{-1} & \text{if } n = 2\mu \in 2\mathbf{Z}, \\ \dfrac{(1 - \psi_h(2)2^{-\mu})L(\mu, \psi_h)}{(1 - 2^{-2\mu})\zeta(2\mu)} & \text{if } n = 2\mu + 1 \notin 2\mathbf{Z}. \end{cases}$$

From (4.5) we obtain $b_2(h, \lambda, 0) = 2$. Notice that $\nu = -1$ if $n - 2 \in 4\mathbf{Z}$ and $\nu = -2$ if $n \notin 2\mathbf{Z}$. Therefore, using (1.18), (1.19), and (1.20), we obtain

$$\frac{R(h, \lambda)}{\mathfrak{m}(\lambda)} = g^{\mu-1} \prod_{p|g} \gamma_p(\mu) \cdot \begin{cases} \dfrac{2^{\nu(\mu-1)+2}\mu}{(2^\mu - 1)|B_\mu|} & \text{if } n \equiv 0 \pmod 4, \\[2ex] \dfrac{c_n 2^3}{E_{(\mu-1)/2}} & \text{if } n \equiv 2 \pmod 4. \end{cases}$$

In view of (4.3), from the first and fifth formulas for γ_p in §1.6 we obtain

$$\text{(4.6)} \qquad g^{\mu-1} \textstyle\prod_{p|g} \gamma_p(\mu) = \begin{cases} \displaystyle\sum_{d|g} d^{\mu-1} & \text{if } n \equiv 0 \pmod 4, \\ \displaystyle\sum_{d|g} \eta(g/d) d^{\mu-1} & \text{if } n \equiv 2 \pmod 4. \end{cases}$$

Combining these, we obtain the formulas of our theorem for even n.

Next, suppose n is odd. By (4.3), $t_p = 1$ for every $p \neq 2$, and so $\gamma_p(n/2)$ can be given by the second formula of §1.6. Since $4h = 8k + n$, we easily see that $\psi_h(2) = (-1)^{(n^2-1)/8}$. Writing ψ' for ψ_h, we obtain our formula for odd n.

4.5. In Section 6 we shall prove that

$$\text{(4.7)} \quad \theta(\lambda, g, 1) = \mathfrak{m}(\lambda)^{-1} \textstyle\sum_{h \in \mathbf{Q}} R(h, \lambda)\mathbf{e}(hz) \text{ for } \lambda \text{ of §4.3 if } 2 \le n \le 8.$$

This combined with (4.4) implies that $R(h, \lambda)/\mathfrak{m}(\lambda)$ of Theorem 4.4 gives $T_n(k)$ if $0 \le n \le 8$. For $n > 8$ the quantity $R(h, \lambda)/\mathfrak{m}(\lambda)$ gives $T_n(k)$ asymptotically, as explained at the end of Section 2.

Let us now state the formulas explicitly. For $n = 2$ and $n = 4$, our results coincide with (3.1) and (3.3). For other values of $n \le 8$, from the above theorem we obtain

$$\text{(4.8)} \qquad T_6(k) = 8 \sum_{d|4k+3} (-1)^{(d+1)/2} d^2,$$

$$\text{(4.9)} \qquad T_8(k) = 2^{3\nu+5} \sum_{d|g} d^3,$$

where $2k + 2 = 2^\nu g$ with $0 < \nu \in \mathbf{Z}$ and odd g;

$$\text{(4.10)} \qquad T_3(k) = 2^4 \cdot 3(2\pi)^{-1}(8k+3)^{1/2} L(1, \psi') \cdot \prod_{p|r} \Bigl\{ \sum_{j=0}^{\lambda_p} p^{-j} - \psi'(p) \sum_{i=1}^{\lambda_p} p^{-i} \Bigr\},$$

$$T_5(k) = 2^5 \cdot 5(2\pi)^{-2}(8k+5)^{3/2}L(2, \psi') \cdot \prod_{p|r}\left\{\sum_{j=0}^{\lambda_p} p^{-3j} - \psi'(p)\sum_{i=1}^{\lambda_p} p^{1-3i}\right\}, \tag{4.11}$$

$$T_7(k) = 2^5 \cdot 7(2\pi)^{-3}(8k+7)^{5/2}L(3, \psi') \cdot \prod_{p|r}\left\{\sum_{j=0}^{\lambda_p} p^{-5j} - \psi'(p)\sum_{i=1}^{\lambda_p} p^{2-5i}\right\}, \tag{4.12}$$

where r, λ_p, and ψ' are as in Theorem 4.4. Notice that in the product $\prod_{p|r}$ of (4.10) we can restrict p to those for which $\psi'(p) \neq 1$, since the factor becomes 1 if $\psi'(p) = 1$. These formulas are true for $0 \leq k \in \mathbf{Z}$. As for the value of $L(\mu, \psi')$, see (1.17).

In our recent paper [S8, §§1.7 and 3.9] we gave explicit formulas for $S_n(k)$ when $2 \leq n \leq 8$; those for even n are classical. Comparing them with the above formulas for $T_n(k)$, we find the following identities of which the first three are true for $0 \leq k \in \mathbf{Z}$:

$$7 \cdot T_5(k) = 2 \cdot S_5(8k+5), \tag{4.13}$$

$$5 \cdot T_6(k) = 2 \cdot S_6(4k+3), \tag{4.14}$$

$$37 \cdot T_7(k) = 2 \cdot S_7(8k+7), \tag{4.15}$$

$$T_8(k) = 2^4 \cdot S_8(k+1) \quad \text{if} \quad 0 \leq k \in 2\mathbf{Z}. \tag{4.16}$$

Since $T_n(k)$ is a multiple of 2^n, from (4.15) we see that $S_7(8k+7)$ is always divisible by $2^6 \cdot 37$. There may be an elementary proof of this fact. Also, there is a possibility that any of the above four equalities can be derived from some identities between theta series.

5. Pentagonal numbers.

5.1. This time we take $\lambda(x) = \prod_p \lambda_p(x_p)$ for $x \in V_{\mathbf{h}}$ with the characteristic function of $\mathbf{Z}_p^n - 6^{-1}u$ as λ_p, where u is the element of $\mathbf{Z}^n$ whose components are all equal to 1. Clearly λ_p is the characteristic function of $\mathbf{Z}_p^n$ if $p \nmid 6$.

Given $0 \leq k \in \mathbf{Z}$, put $h = (2k/3) + (n/36)$. Then

$$N(h, \mathbf{Z}^n - 6^{-1}u) = P_n(k). \tag{5.1}$$

This can be verified in the same manner as for (4.4).

5.2. With λ of §5.1 and $h = (2k/3) + (n/36)$, $k \in \mathbf{Z}$, we have

$$b_2(h, \lambda, s) = 1 + 2^{-s}, \tag{5.2a}$$

$$b_3(h, \lambda, s) = \frac{1}{3}. \tag{5.2b}$$

Indeed, putting $\omega[x] = \sum_{i=1}^{n}(3x_i^2 - x_i)$, we have

$$\begin{aligned} b_2(h, \lambda, s) &= \int_{\mathbf{Q}_2} \nu(\sigma)^{-s} \int_{V_2} \mathbf{e}_2\big(\{\varphi[x] - h\}\sigma\big)\lambda_2(x)dxd\sigma \\ &= \int_{\mathbf{Q}_2} \nu(\sigma)^{-s} \int_{L_2} \mathbf{e}_2\big(\{\varphi[x - u/6] - h\}\sigma\big)dxd\sigma \\ &= \int_{\mathbf{Q}_2} \nu(\sigma)^{-s} \int_{L_2} \mathbf{e}_2\big(\{\omega[x] - 2k\}\sigma/3\big)dxd\sigma \\ &= \int_{\mathbf{Q}_2} \nu(\sigma)^{-s}\mathbf{e}_2(-2k\sigma/3) \int_{L_2} \mathbf{e}_2\big(\omega[x]\sigma/3\big)dxd\sigma \\ &= \int_{\mathbf{Q}_2} \nu(\sigma)^{-s}\mathbf{e}_2(-2k\sigma/3)\tau_3(\sigma/3)^n d\sigma, \end{aligned}$$

where τ_3 is defined by (4.1) with $b = 3$. By (4.2), the last integral over $\mathbf{Q}_2$ can be reduced to the integral over $2^{-1}\mathbf{Z}_2$. Therefore we obtain (5.2a).

As for b_3, the same type of calculation shows that

$$b_3(h, \lambda, s) = \int_{\mathbf{Q}_3} \nu(\sigma)^{-s}\mathbf{e}_3(-2k\sigma/3)\rho(\sigma/3)^n d\sigma$$

$$\text{with} \quad \rho(\sigma) = \int_{\mathbf{Z}_3} \mathbf{e}_3\big(\sigma(3x^2 - x)\big)dx.$$

To determine $\rho(\sigma/3)$, put $\sigma/3 = 3^{-m}a$ with $a \in \mathbf{Z}_3$ and $0 < m \in \mathbf{Z}$. Then

$$\rho(\sigma/3) = 3^{-m} \sum_{x \in \mathbf{Z}/3^m\mathbf{Z}} \mathbf{e}_3\big(3^{-m}a(3x^2 - x)\big) = 3^{-m} \sum_{y \in \mathbf{Z}/3^m\mathbf{Z}} \mathbf{e}_3(3^{-m}ay)$$

by Lemma 4.1 (ii). This is 0 or 1 according as $a \notin 3^m\mathbf{Z}_3$ or $a \in 3^m\mathbf{Z}_3$. Therefore

$$b_3(h, \lambda, s) = \int_{3\mathbf{Z}_3} \nu(\sigma)^{-s}\mathbf{e}_3(-2k\sigma/3)d\sigma = \frac{1}{3},$$

which is (5.2b).

THEOREM 5.3. *Given $0 \le k \in \mathbf{Z}$, put $h = (2k/3) + (n/36)$ and $h = 2^a 3^b g$ with integers a, b and an integer g prime to 6; let $g = r^2 s$ with a positive integer r and a squarefree integer s; let $r = \prod_p p^{\lambda_p}$ be the prime decomposition of r. Then $R(h, \lambda)/\mathfrak{m}(\lambda)$ for λ as in §5.1 can be given as follows:*

$$\text{(I)}\quad n \in 4\mathbf{Z}: \qquad \frac{2^{a(\mu-1)+2}3^{(b+1)(\mu-1)}\mu}{(2^\mu - 1)(3^\mu - 1)|B_\mu|} \cdot \sum_{d|g} d^{\mu-1}.$$

$$\text{(II)}\quad n - 2 \in 4\mathbf{Z}: \qquad \frac{c_n 2^3 3^{(b+1)(\mu-1)}}{(3^\mu + 1)E_{(\mu-1)/2}} \cdot \sum_{d|g} \eta(g/d) d^{\mu-1}.$$

$$\text{(III)}\quad n \ \textit{odd}: \qquad \frac{2^3 \cdot \mu!(2\pi)^{-\mu}(24k+n)^{\mu-1/2}L(\mu, \psi')}{(2^\mu + \varepsilon)|B_{n-1}|} \cdot \frac{1 - \psi'(3)3^{-\mu}}{3^{2\mu} - 1}$$

$$\cdot \prod_{p|r} \left\{ \sum_{j=0}^{\lambda_p} p^{j(2-n)} - \psi'(p) \sum_{i=1}^{\lambda_p} p^{i(2-n)+(n-3)/2} \right\}.$$

Here $\mu = [n/2]$, $c_n = 1$ *if* $n > 2$ *and* $c_2 = 1/2$; B_* *is the Bernoulli number,* E_* *is the Euler number,* η *is the primitive Dirichlet character modulo 4,* ψ' *is the primitive Dirichlet character corresponding to* $\mathbf{Q}((-1)^{\mu/2}(24k+n)^{1/2})$, *and* $\varepsilon = (-1)^{(n^2-1)/8}$.

Proof. Define $\mathfrak{e}'$ and $\mathfrak{e}_1$ as in Theorem 1.5. This time we have $\mathfrak{e}' = 2\mathbf{Z}$ and $\mathfrak{e}_1 = 6\mathbf{Z}$. Specializing the formulas of that theorem, we obtain the desired results in the same manner as in the proof of Theorem 4.4.

In the present case the analogue of (4.7) is

(5.3) $\theta(\lambda, g, 1) = \mathfrak{m}(\lambda)^{-1} \sum_{h \in \mathbf{Q}} R(h, \lambda)\mathbf{e}(hz)$ for λ of §4.3 if n is 2 or 3.

Thus $R(h, \lambda)/\mathfrak{m}(\lambda)$ gives the exact formulas for $P_n(k)$ only for $n = 2$ and 3. That for $P_2(k)$ is nothing but (3.4). As for P_3, the above theorem gives

$$(5.4)\quad P_3(k) = \frac{3 - \psi'(3)}{\pi} \cdot (24k+3)^{1/2} L(1, \psi') \prod_{p|r} \left\{ \sum_{j=0}^{\lambda_p} p^{-j} - \psi'(p) \sum_{i=1}^{\lambda_p} p^{-i} \right\}.$$

Since the right-hand side of (5.4) is nonzero, we thus obtain:

THEOREM 5.4. *Every nonnegative integer is the sum of three integers each of which is of the form* $x(3x-1)/2$ *with* $x \in \mathbf{Z}$.

Also, (5.4) combined with the formula for $S_3(m)$ given in [S8, §1.7] yields (3.5).

6. Nonexistence of certain cusp forms.

6.1. We first consider the following three theta functions:

$$\theta(z) = \sum_{m \in \mathbf{Z}} \mathbf{e}(m^2 z),$$

$$\theta_1(z) = \sum_{m\in\mathbf{Z}} \mathbf{e}\big((2m+1)^2z/4\big),$$
$$\theta_2(z) = \sum_{m\in\mathbf{Z}} \mathbf{e}\big((6m+1)^2z/12\big).$$

As shown in [S2] there is a factor of automorphy h such that

(6.1) $$\theta\big(\gamma(z)\big) = h(\gamma, z)\theta(z) \quad \text{for every} \quad \gamma \in \Gamma_0(4).$$

Observe that θ_1 and θ_2 are the functions of [S2, (2.0)] obtained by taking (A, N, h, P) there to be $(2, 2, 1, 1)$ and $(6, 6, 1, 1)$, respectively. Also observe that we obtain the same functions by taking $h = -1$ instead of $h = 1$. Therefore by [S2, Proposition 2.1] and the remark after it, we obtain

$$\theta_1\big(\gamma(z)\big) = \mathbf{e}(a_\gamma b_\gamma/4)h(\gamma, z)\theta_1(z) \quad \text{for every} \quad \gamma \in \Gamma_0(4),$$
$$\theta_2\big(\gamma(z)\big) = \mathbf{e}(a_\gamma b_\gamma/12)\left(\frac{3}{d}\right)h(\gamma, z)\theta_2(z) \quad \text{for every} \quad \gamma \in \Gamma_0(12),$$

where $\left(\frac{3}{d}\right)$ should be understood in the sense of [S2, p. 442], and $\gamma = \begin{bmatrix} a_\gamma & b_\gamma \\ & * \end{bmatrix}$.

6.2. We prove (4.7) by showing that there is no nontrivial cusp form satisfying the same transformation formula as θ_1^n if $n \le 8$. Let us now discuss the problem in a more general setting. We consider the set $\mathfrak{G}$ consisting of all the pairs $\big(\alpha, p(z)\big)$ formed by $\alpha = \begin{bmatrix} * & * \\ c & d \end{bmatrix} \in SL_2(\mathbf{Q})$ and a holomorphic function $p(z)$ on H such that $p(z)^2 = t\cdot(cz+d)$ with a root of unity t. We can make $\mathfrak{G}$ a group with a law of multiplication

$$\big(\alpha, p(z)\big)\big(\beta, q(z)\big) = \big(\alpha\beta, p\big(\beta(z)\big)q(z)\big).$$

Given $\kappa \in \mathbf{Z}$, $\xi = \big(\alpha, p(z)\big)$, and a function f on H, we define $f\|_\kappa\xi$ to be a function on H defined by

$$(f\|_\kappa\xi)(z) = p(z)^{-\kappa}f\big(\alpha(z)\big).$$

For simplicity we let $\mathfrak{G}$ act on H by putting $\xi(z) = \alpha(z)$.

We consider a subgroup Δ of $\mathfrak{G}$ with the following two properties: (i) the projection map of $\mathfrak{G}$ to $SL_2(\mathbf{Q})$ gives an isomorphism of Δ onto a congruence subgroup Γ of $SL_2(\mathbf{Q})$; (ii) if $\big(\gamma, p(z)\big) \in \Delta$ with $\gamma \in \Gamma$, then $p(z)$ coincides with $h(\gamma, z)$ of (6.1) for every γ in a congruence subgroup of Γ. We then denote by $\mathscr{M}_\kappa(\Delta)$ the set of all holomorphic functions f on H such that $f\|_\kappa\gamma = f$ for every $\gamma \in \Delta$ and that f^m is a modular form of weight $\kappa m/2$ for a suitable even integer m. This is consistent with our definition in Section 2. Further we denote by $\mathscr{S}_\kappa(\Delta)$ the space of all cusp forms in $\mathscr{M}_\kappa(\Delta)$.

Let $H^* = H \cup \mathbf{Q} \cup \{\infty\}$. Then $\Delta\backslash H^*$ for any Δ as above has a structure of a compact Riemann surface. For $0 \neq f \in \mathscr{M}_\kappa(\Delta)$ we can define the divisor of f, written $\mathrm{div}(f)$, as follows. For each point $P \in \Delta\backslash H^*$ we define the order of zero $O_f(P)$ of f at P and put $\mathrm{div}(f) = \sum_P O_f(P)P$. To explain how to define $O_f(P)$, let us assume, for simplicity, that Δ has no elliptic elements. If P is represented by a point b of H, then $O_f(P)$ is exactly the order of zero of f at b.

To treat the case of a cusp, let $s \in \mathbf{Q} \cup \{\infty\}$; let ρ be an element of $\mathfrak{G}$ such that $\rho(s) = \infty$. Then for every $\pi \in \Delta$ such that $\pi(s) = s$, we have

$$\rho\pi\rho^{-1} = \left(\pm\begin{bmatrix}1 & h\\ 0 & 1\end{bmatrix}, \zeta\right) \tag{6.2}$$

with $h \in \mathbf{Q}$ and a root of unity ζ. Fixing s, take π so that h is the smallest positive number, and put $\zeta = \mathbf{e}(r_s)$ with $r_s \in \mathbf{Q}$. If $f \in \mathscr{M}_\kappa(\Delta)$, then $f\|_\kappa\rho^{-1}$ has an expansion

$$(f\|_\kappa\rho^{-1})(z) = \textstyle\sum_{\xi\in\mathbf{Q}} c_\xi \mathbf{e}(\xi z/h), \tag{6.3}$$

such that $c_\xi \neq 0$ only if $\xi - \kappa r_s \in \mathbf{Z}$. Let P_s denote the point of $\Delta\backslash H^*$ represented by s. Then $O_f(P_s)$ is the smallest $\xi \geq 0$ such that $c_\xi \neq 0$. Thus

$$O_f(P_s) \equiv \kappa r_s \pmod{\mathbf{Z}}. \tag{6.4}$$

In general, by a divisor we understand a formal finite $\mathbf{Q}$-linear combination $A = \sum_P a_P P$ with $a_P \in \mathbf{Q}$ and $P \in \Delta\backslash H^*$; we put $\deg(A) = \sum_P a_P$, and write $A \succ 0$ if $a_P \geq 0$ for every P; for two divisors A and B we write $A \succ B$ if $A - B \succ 0$. If $\kappa = 0$, then $\mathrm{div}(f)$ is the divisor of a modular function viewed as a meromorphic function on $\Delta\backslash H^*$.

If Δ has no elliptic elements and $0 \neq f \in \mathscr{M}_\kappa(\Delta)$, then we have

$$\deg\bigl(\mathrm{div}(f)\bigr) = (\kappa/4)(c + 2g - 2), \tag{6.5}$$

where c is the number of inequivalent cusps of Δ and g the genus of $\Delta\backslash H^*$. Since $\mathrm{div}(f^m) = m \cdot \mathrm{div}(f)$ for $m \in \mathbf{Z}$, it is sufficient to show (6.5) for a form of an even weight in the classical sense. The formula in such a case is given in [S1, (2.6.3)].

Now, given κ and s, take $t_{\kappa,s}$ so that $0 < t_{\kappa,s} \leq 1$ and $t_{\kappa,s} - \kappa r_s \in \mathbf{Z}$. In view of (6.4) we easily see that an element f of $\mathscr{M}_\kappa(\Delta)$ is a cusp form if and only if

$$\mathrm{div}(f) \succ \textstyle\sum_s t_{\kappa,s} P_s, \tag{6.6}$$

where the sum is taken over all inequivalent cusps of Δ.

6.3. To prove (4.7), let us take Δ to be the set of the pairs $\bigl(\gamma, \theta_1(\gamma(z))/\theta_1(z)\bigr)$ for all $\gamma \in \Gamma_0(4)$. In this case the genus is 0 and there are exactly three inequivalent cusps represented by 0, $1/2$, and ∞. Therefore, from (6.5) we obtain $\deg\bigl(\mathrm{div}(\theta_1)\bigr) = 4^{-1}$. Clearly $O_{\theta_1}(P_\infty) = 4^{-1}$, and hence $\mathrm{div}(\theta_1) = 4^{-1}P_\infty$; consequently $r_s \in \mathbf{Z}$ for $s = 0, 1/2$. Put $f = \theta_1^\kappa g$ with a meromorphic function g on $\Delta\backslash H^*$. Then from (6.6) we see that $f \in \mathscr{S}_\kappa(\Delta)$ if and only if

$$\operatorname{div}(g) \succ P_0 + P_{1/2} + 4^{-1}(\kappa_0 - \kappa)P_\infty,$$

where κ_0 is the smallest positive integer such that $\kappa - \kappa_0 \in 4\mathbf{Z}$. The dimension of the vector space consisting of all such g can easily be obtained from the Riemann-Roch theorem. Indeed, we have

$$\text{(6.7)} \qquad \dim\big(\mathscr{S}_\kappa(\Delta)\big) = \begin{cases} 0 & (1 \le \kappa \le 8), \\ (\kappa - \kappa_0)/4 - 1 & (\kappa > 8). \end{cases}$$

Now $\theta_1(z)^n = \theta(z;\, \lambda,\, 1)$ with λ of §4.3. By Lemma 2.2 (4), the form (2.2) in the present case is a cusp form belonging to $\mathscr{S}_n(\Delta)$. Therefore, by (6.7) it must be 0 if $n \le 8$. This proves (4.7).

6.4. Next, to treat the case of pentagonal numbers, we take Δ to be the set of the pairs $\big(\gamma,\, \theta_2(\gamma(z))/\theta_2(z)\big)$ for all $\gamma \in \Gamma_0(12)$. The genus is 0 and there are exactly six inequivalent cusps represented by $0, 1/2, 1/3, 1/4, 1/6$ and ∞. In this case the determination of r_s at each cusp requires tedious, if not very complicated, calculations. Since we do not get results as good as in the case of triangular numbers, we merely state two facts without detailed proof:

$$\text{(6.8)} \qquad \operatorname{div}(\theta_2) = 3^{-1}P_{1/3} + 4^{-1}P_{1/4} + 3^{-1}P_{1/6} + 12^{-1}P_\infty,$$

$$\text{(6.9)} \qquad \dim\big(\mathscr{S}_\kappa(\Delta)\big) = 0 \ \text{ if } \kappa \le 3, \qquad \dim\big(\mathscr{S}_\kappa(\Delta)\big) > 0 \ \text{ if } \kappa > 3.$$

We have $\theta_2(z)^n = \theta(3z,\, \lambda,\, 1)$ for λ of §5.1, and so $\theta(z,\, \lambda,\, 1)$ belongs to the conjugate of Δ under $z \mapsto 3z$. Therefore by Lemma 2.2 (4) we obtain (5.3) from (6.9).

DEPARTMENT OF MATHEMATICS, PRINCETON UNIVERSITY, PRINCETON, NJ 08544-1000

REFERENCES

[D] L. E. Dickson, *History of the Theory of Numbers*, vol. II, Carnegie Institution, 1920.
[H] G. H. Hardy, *Ramanujan*, Cambridge Univ. Press, 1940.
[S1] G. Shimura, *Introduction to the Arithmetic Theory of Automorphic Functions, Publ. Math. Soc. of Japan*, no. 11, Iwanami Shoten and Princeton Univ. Press, 1971.
[S2] ———, On modular forms of half integral weight, *Ann. of Math.* **97** (1973), 440–481.
[S3] ———, On Hilbert modular forms of half-integral weight, *Duke Math. J.* **55** (1987), 765–838.
[S4] ———, *Euler Products and Eisenstein Series, CBMS Reg. Conf. Ser. Math.*, no. 93, Amer. Math. Soc., Providence, RI, 1997.
[S5] ———, An exact mass formula for orthogonal groups, *Duke Math. J.* **97** (1999), 1–66.

[S6] ———, The number of representations of an integer by a quadratic form, *Duke Math. J.* **100** (1999), 59–92.

[S7] ———, *Arithmeticity in the Theory of Automorphic Forms, Math. Surveys Monog.*, vol. 82, Amer. Math. Soc., Providence, RI, 2000.

[S8] ———, The representation of integers as sums of squares, *Amer. J. Math.* **124** (2002), 1059–1081.

[Si1] C. L. Siegel, Über die analytische Theorie der quadratishen Formen I, *Ann. of Math.* **36** (1935), 527–606; II, **37** (1936), 230–263.

[Si2] ———, Indefinite quadratische Formen und Funktionentheorie I, *Math. Ann.* **124** (1951), 17–54; II (1952), 364–387.

[W] A. Weil, Sur la formule de Siegel dans la théorie des groupes classiques, *Acta Math.* **113** (1965), 1–87.

QUADRATIC DIOPHANTINE EQUATIONS AND ORDERS IN QUATERNION ALGEBRAS

By GORO SHIMURA

Abstract. We first reformulate the main theorem on quadratic Diophantine equations given in our recent book in terms of the even Clifford group. In the quaternary case the group is the multiplicative group of a quaternion algebra, and the reformulated theorem connects a class of primitive solutions of a quadratic equation with a class of an order in the algebra. In the indefinite case each class is essentially a class of binary hermitian forms. In the definite case the result is an analogue of the result of Gauss on the sums of three squares.

Introduction. There are two classical facts which form the origin of our work: one by Dedekind on the correspondence between the classes of binary quadratic forms over $\mathbf{Z}$ and the ideal classes in an order in a quadratic field; the other by Gauss on the connection of such classes with primitive representations of an integer as sums of three squares. In our recent book [S4] we presented a general principle, applicable to a quadratic form φ on a vector space V of any dimension over an arbitrary algebraic number field F, which includes those two facts as special cases. Namely, fixing a lattice L in V, $q \in F^\times$, and a fractional ideal $\mathfrak{b}$ in F, we put

$$L[q, \mathfrak{b}] = \{h \in V \mid \varphi[h] = q,\ \varphi(h, L) = \mathfrak{b}\}, \tag{0.1}$$

$$\Gamma(L) = \{\gamma \in SO^\varphi(V) \mid L\gamma = L\}, \tag{0.2}$$

where $SO^\varphi(V)$ is the special orthogonal group of φ; we view φ as a symmetric F-bilinear form and put $\varphi[h] = \varphi(h, h)$. Given $h \in L[q, \mathfrak{b}]$, we take the orthogonal complement W of Fh in V. To each element of $L[q, \mathfrak{b}]/\Gamma(L)$ we can assign a "class" of the orthogonal group $SO^\varphi(W)$ of the restriction of φ to W. In those two classical cases we take our form φ to be $4ac - b^2$ and $a^2 + b^2 + c^2$, which are typical indefinite and definite ternary forms. In [S4] we proved generalizations of the results of Dedekind and Gauss for an arbitray ternary form over F of unit discriminant by simpler and more conceptual methods.

These and many other results in [S4] were given in odd-dimensional cases for some technical reasons. In the present paper we treat quaternary forms as

Manuscript received May 21, 2004; revised September 9, 2004.
Shimura, Goro. "Quadratic Diophantine equations and orders in quaternion algebras." American Journal of Mathematics 128:2 (2006), 481–518.

exemplary even-dimensional cases, and show that there are clear-cut results in both definite and indefinite cases which are analogues of the results for ternary forms. We start our investigations by reformulating the main theorem on Diophantine quadratic equations in [S4] in terms of the even Clifford group $G^+(V)$ in a somewhat general form. There is a natural subgroup $\Delta(L)$ of $G^+(V)$ similar to $\Gamma(L)$ such that $\sum_i \#\{L_i[q, \mathfrak{b}]/\Delta(L_i)\}$ equals the class number of the even Clifford group $G^+(W)$ of the subspace W of V defined above, where $\{L_i\}$ is a certain set of representatives for the classes in the genus of L (Theorem 1.6).

We apply this principle to the case where V is the set of all matrices of the form

$$(0.3) \qquad h = \begin{bmatrix} a & e \\ e^\rho & c \end{bmatrix}.$$

Here $a, c \in F$, e belongs to a quadratic extension K of F, and ρ is conjugation of K over F. We define a quadratic form φ on V by $\varphi[h] = \det(h)$, and put

$$(0.4) \qquad \mathcal{G} = \{\alpha \in GL_2(K) \mid \det(\alpha) \in F^\times\}, \qquad \Delta = GL_2(\mathfrak{r}) \cap \mathcal{G},$$

where $\mathfrak{r}$ is the maximal order of K. We let an element α of $\mathcal{G}$ act on V by the rule

$$(0.5) \qquad h \mapsto \det(\alpha)^{-1} \cdot {}^t\alpha^\rho h \alpha.$$

This map belongs to $SO^\varphi(V)$. We define the class of h to be the orbit of h under Δ, and call h *primitive* if $\mathfrak{g}a + \mathfrak{g}c + \mathrm{Tr}_{K/F}(e\mathfrak{r}) = \mathfrak{g}$, where $\mathfrak{g}$ is the maximal order of F. Then we can prove:

Given $q \in \mathfrak{g} \cap F^\times$, define a quaternion algebra R over F containing K by $R = K + \varepsilon K$ with an element ε such that $\varepsilon^2 = -q$ and $\varepsilon a = a^\rho \varepsilon$ for every $a \in K$; put $\mathfrak{o} = \mathfrak{r} + \varepsilon\mathfrak{r}$; denote by c_F the class number of F and by $c(\mathfrak{o})$ the class number of $\mathfrak{o}$. Then the number of classes of primitive h of determinant q equals $c(\mathfrak{o})/c_F$, if we exclude the case in which F is totally real, K is totally imaginary, and q is totally positive (Theorem 2.11).

This follows from a more general result applicable even to that exceptional case (Theorem 2.10). It should be noted that $\mathcal{G} = G^+(V)$ and $SO^\varphi(V)$ consists of the maps of (0.5) for all $\alpha \in \mathcal{G}$. There is an obvious parallelism of this case of hermitian forms with the case of binary quadratic forms treated in [S4, Theorem 12.9]; namely an order in a quadratic field is replaced here by an order in a quaternion algebra.

As a special case we obtain the following:

Let $F = \mathbf{Q}$ and $K = \mathbf{Q}(\sqrt{m})$ with a squarefree positive or negative integer m, and let $0 \neq q \in \mathbf{Z}$. Suppose $q < 0$ if $m < 0$, and suppose also that q is prime to *every odd factor of m. Then* $c(\mathfrak{o}) = 1$ *if* $8 \nmid q$ *or* $m - 2 \notin 8\mathbf{Z}$, *and* $c(\mathfrak{o}) = 2$ *otherwise. Consequently if* $8 \nmid q$ *or* $m - 2 \notin 8\mathbf{Z}$, *then every primitive h of determinant q is equivalent to* diag $[1, q]$ *under* Δ (Proposition 2.12).

This does not apply to the case where $q > 0$ and $m < 0$. In such a case R is a definite quaternion algebra over $\mathbf{Q}$. To deal with this case, we take a three-dimensional space S on which $SL_2(\mathbf{C})$ acts. Then to each h of determinant $q > 0$, we can associate a point $\zeta(h)$ of S so that $\zeta(\gamma h \cdot {}^t\gamma^\rho) = \gamma(\zeta(h))$ for every $\gamma \in \Delta_1 = SL_2(\mathfrak{r})$. The question of obtaining a complete set of representatives for the classes of primitive h of a fixed determinant can then be reduced to the question of finding certain points of $\Delta_1 \backslash S$. This is similar to the reduction process of Lagrange for binary quadratic forms and also to reduction theory of Minkowski. We shall find a fundamental domain for $\Delta_1 \backslash S$ in an exact form (that is, with no overlapping even at the boundary points) when $K = \mathbf{Q}(\sqrt{-3})$ and $K = \mathbf{Q}(\sqrt{-1})$, which enables us to state an algorithm for obtaining the desired set of representatives. It seems that a fundamental domain in an exact form has never been given even for such K's.

To discuss the definite case, we take a quaternion algebra B over F, denote by ι the main involution of B, consider a four-dimensional V of the form $V = F \oplus T$, where $T = \{x \in B \mid x^\iota = -x\}$, and define φ on V by $\varphi[(a, x)] = a^2 + \gamma x x^\iota$ for $a \in F$ and $x \in T$ with a fixed $\gamma \in F^\times$. It can be shown that any quaternary form is equivalent to a constant multiple of such a φ. If B is a definite quaternion algebra over $\mathbf{Q}$ and γ is a positive integer, then φ is definite. We again connect the class number of an order $\mathfrak{o}$ in a quaternion algebra R with $\#\{L[q, \mathfrak{b}]/\Delta(L)\}$, but the nature of the problem depends on whether or not γ is a square. If γ is a square, then R is always isomorphic to B (Theorem 1.11); otherwise many nonisomorphic R can appear in the same manner as in the case of ternary forms. In §4.4 we shall give four explicit examples of definite φ of class number 1 with nonsquare determinant, but state the result on the connection of $L[q, 2^{-1}\mathbf{Z}]$ with the class number of $\mathfrak{o}$ only in one special case (Theorem 4.7). The same type of difference between square and nonsquare discriminants occurs in any even dimension (Theorem 1.12).

In the last part of the paper we shall make some additional comments related to the mass formula and $\#L[q, 2^{-1}\mathbf{Z}]$.

1. The main theorem on quadratic Diophantine equations formulated in terms of the Clifford group.

1.1. Throughout the paper F denotes a field of characteristic different from 2. Let V be a finite-dimensional vector space over F, and let $\varphi\colon V \times V \to F$ be a nondegenerate F-valued F-bilinear symmetric form. We put $\varphi[x] = \varphi(x, x)$ for $x \in V$. We denote by $A(V, \varphi)$ or simply by $A(V)$ the Clifford algebra of (V, φ). We define the canonical automorphism $x \mapsto x'$ of $A(V)$ and the canonical involution

$x \mapsto x^*$ of $A(V)$ by the property that $v = -v' = v^*$ for every $v \in V$. We then put

$$A^+(V) = A^+(V,\varphi) = \{\alpha \in A(V) \mid \alpha' = \alpha\},$$
$$A^-(V) = A^-(V,\varphi) = \{\alpha \in A(V) \mid \alpha' = -\alpha\}.$$

In particular, $A^+(V)$ is the even Clifford algebra of (V,φ). We also put

$$O^\varphi = O^\varphi(V) = \{\alpha \in GL(V,F) \mid \varphi[x\alpha] = \varphi[x], \text{ for every } x \in V\},$$
$$SO^\varphi = SO^\varphi(V) = SL(V,F) \cap O^\varphi(V),$$
$$G(V) = G(V,\varphi) = \{\alpha \in A(V)^\times \mid \alpha^{-1}V\alpha = V\},$$
$$G^+(V) = G^+(V,\varphi) = G(V) \cap A^+(V).$$

Notice that we let $GL(V,F)$ act on the right of V, so that $x\alpha$ is the image of x under α.The group $G(V)$ and $G^+(V)$ are called the Clifford group and the even Clifford group of (V,φ). We can define a homomorphism ν: $G^+(V) \to F^\times$ and a homomorphism τ: $G(V) \to O^\varphi$ by $\nu(\alpha) = \alpha\alpha^*$ and $x\tau(\alpha) = \alpha^{-1}x\alpha$ for $x \in V$. The map τ gives an isomorphism of $G^+(V)/F^\times$ onto SO^φ. For this and other basic properties of $A(V)$, $G^+(V)$, ν, and τ, etc., the reader is referred to any standard book on this topic, and in particular to [S4], as there are several facts which can be found only in that volume.

Let X be a vector subspace of V on which φ is nondegenerate. We use the letter φ for its restriction to X; we can then define the Clifford algebra and the even Clifford group of (X,φ), which we denote by $A(X)$ and $G^+(X)$. We identify $A(X)$ with the F-subalgebra of $A(V)$ generated by F and the elements of X; then $G^+(X) = G^+(V) \cap A(X)$. If we put

$$X^\perp = \{y \in V \mid \varphi(y,X) = 0\},$$

then $xy = -yx$ for every $x \in X$ and $y \in X^\perp$.

LEMMA 1.2. *Given an element k of V such that $k^2 \neq 0$, put $W = (Fk)^\perp$. Then*

$$A^+(W) = \{\alpha \in A^+(V) \mid \alpha k = k\alpha\}, \textit{ and } G^+(W) = \{\alpha \in G^+(V) \mid \alpha k = k\alpha\}.$$

Moreover, if $\dim(V) = 4$, *then $A^+(W)$ is a quaternion algebra over F, $G^+(W) = A^+(W)^\times$, and the canonical involution of $A(W)$ restricted to $A^+(W)$ coincides with the main involution of $A^+(W)$.*

Proof. We have $ku = -uk$ for $u \in W$, and hence $A^\pm(W) = \{x \in A(W) \mid kx = \pm xk\}$. We also see that $A^+(V) = A^+(W) \oplus kA^-(W)$. For $x \in A^+(W)$ and $y \in A^-(W)$ we have $(x + ky)k = k(x - ky)$, from which we obtain the first equality of our lemma. Then the second equality follows immediately. The assertions in the case $\dim(V) = 4$ follow from the general principles explained in [S4, §7.3]. □

1.3. We consider the field of quotients F of a Dedekind domain $\mathfrak{g}$ in the following two cases: (i) F is an algebraic number field of finite degree and $\mathfrak{g}$ is the ring of all algebraic integers in F; (ii) F and $\mathfrak{g}$ are the localizations of those in Case (i) at a nonarchimedean prime. We call a field of type (i) a *global field*, and that of type (ii) a *local field*. In this paper, we employ the terms global and local fields only in these senses. Thus, a local field always means a nonarchimedean one. In both local and global cases, by a $\mathfrak{g}$-*lattice* in a finite-dimensional vector space V over F, we mean a finitely generated $\mathfrak{g}$-module in V that spans V over F. When F is a global field, we denote by $\mathbf{a}$ and $\mathbf{h}$ the sets of archimedean primes and nonarchimedean primes of F respectively, and put $\mathbf{v} = \mathbf{a} \cup \mathbf{h}$. Given an algebraic group G defined over F, we define G_v for each $v \in \mathbf{v}$ and the adelization $G_{\mathbf{A}}$ as usual, and view G and G_v as subgroups of $G_{\mathbf{A}}$. (We do not use the symbol $G_{\mathbf{Q}}$ or G_F.) We then denote by $G_{\mathbf{a}}$ and $G_{\mathbf{h}}$ the archimedean and nonarchimedean factors of $G_{\mathbf{A}}$, respectively. In particular, the adelization of the field F and the multiplicative group $F^\times$ are denoted by $F_{\mathbf{A}}$ and $F_{\mathbf{A}}^\times$, which are the adele ring and the idele group of F. For $x \in G_{\mathbf{A}}$ we denote by x_v, $x_{\mathbf{a}}$, and $x_{\mathbf{h}}$ its projections to G_v, $G_{\mathbf{a}}$, and $G_{\mathbf{h}}$.

1.4. Let V be a finite-dimensional vector space over F and L a $\mathfrak{g}$-lattice in V; let G be an algebraic subgroup of $GL(V, F)$ defined over F. We can naturally let G act on V on the right (see §1.1). Given $x \in G_{\mathbf{A}}$, we denote by Lx the $\mathfrak{g}$-lattice determined by the property that $(Lx)_v = L_v x_v$ for every $v \in \mathbf{h}$. In particular, given a fractional ideal $\mathfrak{a}$ and $t \in F_{\mathbf{A}}^\times$, we denote by $t\mathfrak{a}$ or $\mathfrak{a}t$ the fractional ideal such that $(t\mathfrak{a})_v = t_v\mathfrak{a}_v$ for every $v \in \mathbf{h}$. By the *G-class* (resp. *G-genus*) of L we mean the set of all $\mathfrak{g}$-lattices in V of the form $L\alpha$ with $\alpha \in G$ (resp. with $\alpha \in G_{\mathbf{A}}$). Clearly the G-genus of L can be decomposed into a disjoint union of G-classes. Now put $D = \{\alpha \in G_{\mathbf{A}} \mid L\alpha = L\}$. Then we see that $Dx \mapsto Lx$ gives a bijection of $D\backslash G_{\mathbf{A}}$ onto the genus of L, and therefore it gives a bijection of $D\backslash G_{\mathbf{A}}/G$ onto the set of G-classes contained in that genus. In general, given an open subgroup U of $G_{\mathbf{A}}$, we call $\#(U\backslash G_{\mathbf{A}}/G)$ the *class number of G with respect to U*. Clearly this is equal to $\#(G\backslash G_{\mathbf{A}}/U)$. If U is D defined above, then the class number is exactly the number of G-classes in the G-genus of L.

1.5. Let (V, φ) be as in §1.1 defined over a local or global field F. For a $\mathfrak{g}$-lattice Λ in V, an element q in $F^\times$, and a fractional ideal $\mathfrak{b}$ in F, we put

$$\Lambda[q, \mathfrak{b}] = \{x \in V \mid \varphi[x] = q, \quad \varphi(x, \Lambda) = \mathfrak{b}\}. \tag{1.1}$$

The point of this definition is that we investigate the equation $\varphi[x] = q$ under the condition $\varphi(x, \Lambda) = \mathfrak{b}$. This idea leads us to several clear-cut results, as we have shown in [S4, Chapter II]. We are going to reformulate the main theorem on $\Lambda[q, \mathfrak{b}]$ obtained there in terms of $G^+(V)$ in a somewhat generalized form. We now take F to be global, fix a $\mathfrak{g}$-maximal lattice L in V in the sense of [S4, §6.1], and put

$$C = \{\gamma \in SO^{\varphi}(V)_{\mathbf{A}} \mid L\gamma = L\}. \tag{1.2}$$

Let h be an element of V such that $\varphi[h] \neq 0$. By a basic principle stated in [S4, Lemma 1.5(ii)] we have

$$\{k \in V \mid \varphi[k] = \varphi[h]\} = h \cdot SO^{\varphi}(V) = h \cdot \tau(G^+(V)), \tag{1.3}$$

provided $\dim(V) > 1$. Now, for every $\sigma \in G^+(V)_{\mathbf{A}}$, the symbol $h\tau(\sigma)$ is meaningful as an element of $V_{\mathbf{A}}$, so that for any subset S of $G^+(V)_{\mathbf{A}}$ the symbol $h\tau(S)$ is meaningful as a subset of $V_{\mathbf{A}}$. Therefore $V \cap h\tau(S)$ is a well-defined subset of V.

Theorem 1.6. *Assuming* $\dim(V) > 1$, *for a fixed* $h \in V$ *such that* $\varphi[h] \neq 0$ *put* $W = (Fh)^{\perp}$, $G = G^+(V)$, *and* $H = G^+(W)$. *Let* D *be an open subgroup of* $G_{\mathbf{A}}$ *containing* $G_{\mathbf{a}}$ *and such that* $\tau(D) \subset C$; *put* $\Delta_y = G \cap yDy^{-1}$ *for* $y \in G_{\mathbf{A}}$. *Then the following assertions hold.*

(i) *For* $y \in G_{\mathbf{A}}$ *we have* $H_{\mathbf{A}} \cap GyD \neq \emptyset$ *if and only if* $V \cap h\tau(Dy^{-1}) \neq \emptyset$.

(ii) *Fixing* $y \in G_{\mathbf{A}}$, *for every* $\varepsilon \in H_{\mathbf{A}} \cap GyD$ *take* $\alpha \in G$ *so that* $\varepsilon \in \alpha yD$. *Then the map* $\varepsilon \mapsto h\tau(\alpha)$ *gives a bijection of* $H\backslash(H_{\mathbf{A}} \cap GyD)/(H_{\mathbf{A}} \cap D)$ *onto* $[V \cap h\tau(Dy^{-1})]/\tau(\Delta_y)$.

(iii) *Take* $\{y_i\}_{i\in I} \subset G_{\mathbf{A}}$ *so that* $G_{\mathbf{A}} = \bigsqcup_{i\in I} Gy_iD$, *and put* $\Delta_i = G \cap y_iDy_i^{-1}$. *Then*

$$\#\{H\backslash H_{\mathbf{A}}/(H_{\mathbf{A}} \cap D)\} = \sum_{i\in I} \#\{[V \cap h\tau(Dy_i^{-1})]/\tau(\Delta_i)\}. \tag{1.4}$$

(iv) *Suppose* $\dim(V) > 2, \tau(D) = C$, *and also that if n is odd, then* $\det(\varphi)\mathfrak{g}$ *is a square in the ideal group of* F; *put* $q = \varphi[h]$ *and* $\mathfrak{b} = \varphi(h, L)$. *Then for every* $y \in G_{\mathbf{A}}$ *we have*

$$V \cap h\tau(Dy^{-1}) = (L\tau(y)^{-1})[q, \mathfrak{b}]. \tag{1.5}$$

Proof. By Lemma 1.2, $H = \{\xi \in G \mid h\tau(\xi) = h\}$. Let y, ε, and α be as in (ii); then $\varepsilon = \alpha y\gamma$ with $\gamma \in D$, and $h = h\tau(\varepsilon)$, as $\varepsilon \in H_{\mathbf{A}}$. Thus $h\tau(\alpha) = h\tau(\gamma^{-1}y^{-1}) \in h\tau(Dy^{-1})$, and so $h\tau(\alpha) \in V \cap h\tau(Dy^{-1})$. If $\eta\varepsilon\zeta \in \beta yD$ with $\eta \in H$, $\zeta \in H_{\mathbf{A}} \cap D$, and $\beta \in G$, then $\beta^{-1}\eta\alpha \in G \cap yDy^{-1} = \Delta_y$, and hence $h\tau(\alpha) = h\tau(\eta\alpha) \in h\tau(\beta\Delta_y)$. Thus our map is well defined. Next let $k \in V \cap h\tau(Dy^{-1})$. Then $k = h\tau(\delta y^{-1})$ with $\delta \in D$, and moreover, by (1.3), $k = h\tau(\xi)$ with $\xi \in G$. Then $h = h\tau(\xi y\delta^{-1})$, so that $\xi y\delta^{-1} \in H_{\mathbf{A}}$. Thus $\xi y\delta^{-1} \in H_{\mathbf{A}} \cap GyD$. This shows that k is the image of an element of $H_{\mathbf{A}} \cap GyD$. To prove that the map is injective, suppose $\varepsilon \in \alpha yD \cap H_{\mathbf{A}}$ and $\delta \in \beta yD \cap H_{\mathbf{A}}$ with $\alpha, \beta \in G$ and $h\tau(\alpha) = h\tau(\beta\zeta)$ with $\zeta \in \Delta_y$. Put $\omega = \beta\zeta\alpha^{-1}$. Then $h\tau(\omega) = h$, so that $\omega \in H$. Since $\zeta \in yDy^{-1}$, we have $\beta yD = \beta\zeta yD = \omega\alpha yD$, and hence $\delta \in \beta yD \cap H_{\mathbf{A}} = \omega\alpha yD \cap H_{\mathbf{A}} = \omega(\alpha yD \cap H_{\mathbf{A}}) = \omega(\varepsilon D \cap H_{\mathbf{A}}) = \omega\varepsilon(D \cap H_{\mathbf{A}}) \subset H\varepsilon(D \cap H_{\mathbf{A}})$. This proves the injectivity, and completes the proof of (ii). At the same time we obtain (i).

Since $H_{\mathbf{A}} = \bigsqcup_{i \in I} (H_{\mathbf{A}} \cap Gy_iD)$, we can derive (iii) immediately from (ii).

As for (iv), clearly $V \cap hC \subset L[q, \mathfrak{b}]$. Conversely, if $x \in L[q, \mathfrak{b}]$, then $x \in hC$ by [S4, Theorem 10.5]. Thus

$$V \cap hC = L[q, \mathfrak{b}]. \tag{1.6}$$

Suppose $\tau(D) = C$. If $k \in V \cap h\tau(Dy^{-1})$, then $h\tau(Dy^{-1}) = k\tau(yDy^{-1})$. Now $\tau(yDy^{-1}) = \tau(y)C\tau(y)^{-1}$ and $\varphi(k, L\tau(y)^{-1}) = \varphi(h, L) = \mathfrak{b}$. Taking $k, \tau(y)C\tau(y)^{-1}$, and $L\tau(y)^{-1}$ in place of h, C, and L in (1.6), we obtain $V \cap k\tau(yDy^{-1}) = (L\tau(y)^{-1})[q, \mathfrak{b}]$. This proves (1.5) when $V \cap h\tau(Dy^{-1}) \neq \emptyset$. To prove the remaining case, suppose $\ell \in (L\tau(y)^{-1})[q, \mathfrak{b}]$; then $\varphi(\ell_v\tau(y)_v, L_v) = \mathfrak{b}_v = \varphi(h, L)_v$ for every $v \in \mathbf{h}$. Therefore, by [S4, Theorem 10.5], $\ell\tau(y) \in hC$. Now taking ℓ, $\tau(y)C\tau(y)^{-1}$, and $L\tau(y)^{-1}$ in place of h, C, and L in (1.6), we obtain

$$\emptyset \neq (L\tau(y)^{-1})[q, \mathfrak{b}] = V \cap \ell\tau(y)C\tau(y)^{-1} = V \cap h\tau(Dy^{-1}).$$

This shows that $(L\tau(y)^{-1})[q, \mathfrak{b}] = \emptyset$ if $V \cap h\tau(Dy^{-1}) = \emptyset$, and hence (1.5) holds for every $y \in G_{\mathbf{A}}$. This completes the proof. □

We see that (ii) and (iii) above contain reformulations of the results given in [S4, Theorem 11.6(iv) and (11.9)]. It should be noted that Δ_i is defined in [S4, p. 106, line 7] erroneously; it must be defined as $\Delta_i = G \cap y_iQy_i^{-1}$; also $J \cup Jy$ on line 10 of the same page should be $\bigcup_{\nu \in \mathbf{Z}} Jy^{\nu}$.

In view of (i) we can restrict the indices i on the right-hand side of (1.4) to those for which $H_{\mathbf{A}} \cap Gy_iD \neq \emptyset$. If I' denotes the set of all such i's, then $H_{\mathbf{A}} = \bigsqcup_{j \in I'} (H_{\mathbf{A}} \cap Gy_jD)$, and [S4, Theorem 11.6(iv) and (11.9)] were stated with such I'.

Notice also that if $\tau(D) = C$, then $F_{\mathbf{A}}^{\times}D = \{\xi \in G_{\mathbf{A}} \mid L\tau(\xi) = L\}$. Therefore in such a case $\Delta_y = \Delta_z$ for $y, z \in G_{\mathbf{A}}$ if $L\tau(y)^{-1} = L\tau(z)^{-1}$.

1.7. Let $G = G^{+}(V)$ as above. A natural choice for D is

$$J = G_{\mathbf{a}} \prod_{v \in \mathbf{h}} J_v, \qquad J_v = \{\gamma \in G_v \mid \tau(\gamma) \in C,\ \nu(\gamma) \in \mathfrak{g}_v^{\times}\}. \tag{1.7}$$

A subgroup Q of $G_{\mathbf{A}}$ such that $\tau(Q) = C$ can be obtained in the form $Q = G_{\mathbf{a}} \prod_{v \in \mathbf{h}} Q_v$ with $Q_v = \bigcup_{\nu \in \mathbf{Z}} \omega_v^{\nu} J_v$, where ω_v is a certain element of G_v; J_v is a normal subgroup of Q_v and $[C_v\colon \tau(J_v)] \leq 2$; see [S4, Theorem 8.9]. For almost all $v \in \mathbf{h}$ we can take $\omega_v = 1$, but we need nontrivial ω_v for certain v. Now it can often happen that $h\tau(Q) = h\tau(J)$. In such a case we have, by (1.5),

$$V \cap h\tau(Jy^{-1}) = V \cap h\tau(Qy^{-1}) = (L\tau(y)^{-1})[q, \mathfrak{b}] \tag{1.8}$$

for every $y \in G_{\mathbf{A}}$, and so from (1.4) we obtain

$$\#\{H\backslash H_{\mathbf{A}}/(H_{\mathbf{A}} \cap J)\} = \sum_{i\in I} \#\{(L\tau(y_i)^{-1})[q, \mathfrak{b}]/\tau(\varDelta_i)\} \tag{1.9}$$

with $\varDelta_i = G \cap y_i J y_i^{-1}$. This has an advantage that $H_{\mathbf{A}} \cap J$ is easier to handle than $H_{\mathbf{A}} \cap Q$. Fix any $v \in \mathbf{h}$ such that $Q_v \neq J_v$. Then we have

(1.10) *If $k_v C_v = k_v \tau(J_v)$ for some $k_v \in hC_v$, then $hC_v = h\tau(J_v)$ and we can take $Q_v = \bigcup_{\nu\in\mathbf{Z}} \alpha_v^\nu J_v$ with an element α_v of $G^+(W)_v$.*

Indeed, for such v and k_v take $\beta, \omega \in Q_v$ so that $k_v = h\tau(\beta)$ and $Q_v = \bigcup_{\nu\in\mathbf{Z}} \omega^\nu J_v$. We have $k_v\tau(\omega) = k_v\tau(\gamma)$ with $\gamma \in J_v$. Put $\alpha_v = \beta\omega\gamma^{-1}\beta^{-1}$. Then $h\tau(\alpha_v) = h$, so that $\alpha_v \in G^+(W)_v$. Since J_v is normal in Q_v, Q_v is generated by J_v and α_v, and so $hC_v = h\tau(Q_v) = h\tau(J_v)$. This proves (1.10).

1.8. Before proceeding further, let us recall several basic facts on quaternion algebras. Let B be a quaternion algebra over a local or global field F, and ι the main involution of B. We put $\mathrm{Tr}_{B/F}(x) = x + x^\iota$ and $N_{B/F}(x) = xx^\iota$ for $x \in B$. Taking F to be global, let $\mathfrak{o}$ be an order in B containing $\mathfrak{g}$. For $\xi \in B_{\mathbf{A}}^\times$ the principle of §1.4 enables us to define $\mathfrak{o}\xi$ as a $\mathfrak{g}$-lattice in B. We call a $\mathfrak{g}$-lattice in B of this form a *proper left* $\mathfrak{o}$*-ideal*. Thus the $B^\times$-genus of $\mathfrak{o}$ consists of all proper left $\mathfrak{o}$-ideals, and the genus is stable under right multiplication by the elements of $B^\times$, and so a $B^\times$-class of proper left $\mathfrak{o}$-ideals is meaningful. Define a subgroup X of $B_{\mathbf{A}}^\times$ by $X = B_{\mathbf{a}}^\times \prod_{v\in\mathbf{h}} \mathfrak{o}_v^\times$. Then $\#(B^\times\backslash B_{\mathbf{A}}^\times/X)$ gives the number of $B^\times$-classes in this genus, which we call *the class number of* $\mathfrak{o}$. For $y \in B_{\mathbf{A}}^\times$ we can define a $\mathfrak{g}$-lattice $y\mathfrak{o}y^{-1}$ in B by the property that $(y\mathfrak{o}y^{-1})_v = y_v\mathfrak{o}_v y_v^{-1}$ for every $v \in \mathbf{h}$. We easily see that $y\mathfrak{o}y^{-1}$ is an order in B, and the class number of $y\mathfrak{o}y^{-1}$ is the same as that of $\mathfrak{o}$. We call the set of the orders $y\mathfrak{o}y^{-1}$ for all $y \in B_{\mathbf{A}}^\times$ the *genus* of the order $\mathfrak{o}$. If $\mathfrak{o}$ is maximal, then the class number of $\mathfrak{o}$ is independent of the choice of $\mathfrak{o}$, and so we call it *the class number of* B. For $v \in \mathbf{v}$ we say that B is ramified at v and also that v is ramified in B if B_v is a division algebra.

Given an order $\mathfrak{o}$ in B over a local or global F, put $\widetilde{\mathfrak{o}} = \{x \in B \mid \mathrm{Tr}_{B/F}(x\mathfrak{o}) \subset \mathfrak{g}\}$. Then the $\mathfrak{g}$-module $\widetilde{\mathfrak{o}}/\mathfrak{o}$ is isomorphic to a finite direct sum $\bigoplus_\kappa (\mathfrak{g}/\mathfrak{a}_\kappa)$ with integral ideals $\mathfrak{a}_\kappa$. It can be shown that the product $\prod_\kappa \mathfrak{a}_\kappa$ depends only on $\mathfrak{o}$, and independent of the choice of the $\mathfrak{a}_\kappa$, and that it is a square of an integral ideal $\mathfrak{b}$. We call $\mathfrak{b}$ *the discriminant of* $\mathfrak{o}$. In the local case, if B is a division algebra, then $\mathfrak{o}$ is maximal if and only if the disriminant is the prime ideal of $\mathfrak{g}$. In the global case, if $\mathfrak{o}$ is maximal, then the discriminant of $\mathfrak{o}$ is a squarefree ideal independent of the choice of $\mathfrak{o}$. It is actually the product of all the prime ideals where B is ramified. Thus we call it the *discriminant of* B. It is well known that the isomorphism class of B is completely determined by the discriminant and the set of the archimedean primes v ramified in B. If $F = \mathbf{Q}$, then writing the discriminant of $\mathfrak{o}$ or of B in the form $b\mathbf{Z}$ with a positive integer b, we call b *the discriminant of* $\mathfrak{o}$ *or of* B. We call B *definite* if B is ramified at the archimedean prime. Let us quote here two results due to Eichler, the first of which is a local result ([E3, Satz 3]):

(1.11) *If F is local, B is isomorphic to $M_2(F)$, and the discriminant of an order $\mathfrak{o}$ in B is the prime ideal $\mathfrak{p}$ of $\mathfrak{g}$, the there is an isomorphism of B onto $M_2(F)$ that maps $\mathfrak{o}$ onto*

$$\left\{ \begin{bmatrix} a & b \\ c & d \end{bmatrix} \in M_2(\mathfrak{g}) \mid c \in \mathfrak{p} \right\}. \tag{1.11a}$$

To state the next result, let B be a definite quaternion algebra over $\mathbf{Q}$, and $\mathfrak{o}$ an order in B whose discriminant d is squarefree. Then $d = tr$, where t is the discriminant of B and r is a squarefree positive integer prime to t. Then the class number of $\mathfrak{o}$ can be given by the following formula ([E3, (64)]):

$$\begin{aligned} \frac{1}{12}\prod_{p|t}(p-1)\prod_{p|r}(p+1) &+ \frac{1}{4}\prod_{p|t}\left\{1-\left(\frac{-4}{p}\right)\right\}\prod_{p|r}\left\{1+\left(\frac{-4}{p}\right)\right\} \\ &+ \frac{1}{3}\prod_{p|t}\left\{1-\left(\frac{-3}{p}\right)\right\}\prod_{p|r}\left\{1+\left(\frac{-3}{p}\right)\right\}. \end{aligned} \tag{1.12}$$

The case $r = 1$ was given earlier in [E1]. Once an order $\mathfrak{o}$ of discriminant tr is fixed, then by (1.11) every other order in B of discriminant tr is of the form $y\mathfrak{o}y^{-1}$ for some $y \in B_{\mathbf{A}}^{\times}$, and so the class number is independent of the choice of $\mathfrak{o}$. Notice that the class number of B is 1 exactly when $t = 2, 3, 5, 7$, or 13.

The indefinite case is easier. We state the fact in a somewhat generalized form that suits our later applications.

LEMMA 1.9. *Let S be a quaternion algebra over a global field F such that S_v is isomorphic to $M_2(F_v)$ for at least one $v \in \mathbf{a}$; define a map N: $S_{\mathbf{A}}^{\times} \to F_{\mathbf{A}}^{\times}$ by $N(x) = xx^{\iota}$ for $x \in S_{\mathbf{A}}^{\times}$ and a subgroup $\mathfrak{G}$ of $S^{\times}$ by $\mathfrak{G} = \{x \in S^{\times} \mid N(x) \in M^{\times}\}$ with a fixed subfield M of F; let Y be an open subgroup of $\mathfrak{G}_{\mathbf{A}}$. Then*

$$\mathfrak{G}\alpha Y = \{\beta \in \mathfrak{G}_{\mathbf{A}} \mid N(\beta) \in M^{\times}N(\alpha Y)\} \quad \textit{for every} \quad \alpha \in \mathfrak{G}_{\mathbf{A}}, \tag{1.13}$$

and the map $\alpha \mapsto M^{\times}N(\alpha Y)$ for $\alpha \in \mathfrak{G}_{\mathbf{A}}$ gives a bijection of $\mathfrak{G}\backslash\mathfrak{G}_{\mathbf{A}}/Y$ onto $M_{\mathbf{A}}^{\times}/[M^{\times}N(Y)]$, so that $\#(\mathfrak{G}\backslash\mathfrak{G}_{\mathbf{A}}/Y) = [M_{\mathbf{A}}^{\times} \colon M^{\times}N(Y)]$.

Proof. Clearly our map is well defined. Given $b \in M_{\mathbf{A}}^{\times}$, we can find an element c of $M^{\times}$ such that $(c^{-1}b)_{\mathbf{a}}$ belongs to the identity component of $F_{\mathbf{a}}^{\times}$. Since

$N(S_v^\times) = F_v^\times$ for every $v \in \mathbf{h}$ and $N(S_\mathbf{a}^\times)$ contains the identity component of $F_\mathbf{a}^\times$, we have $c^{-1}b = N(d)$ with $d \in S_\mathbf{A}^\times$. Then $d \in \mathfrak{G}_\mathbf{A}$, and hence $M_\mathbf{A}^\times = M^\times N(\mathfrak{G}_\mathbf{A})$. This shows that our map is surjective. Next, put $\mathfrak{G}^1 = \{x \in S^\times \mid N(x) = 1\}$. Our condition on S_v for $v \in \mathbf{a}$ implies that strong approximation holds on $\mathfrak{G}^1$, by [E2]. Suppose $\alpha, \beta \in \mathfrak{G}_\mathbf{A}$ and $N(\beta\alpha^{-1}) = xN(y)$ with $x \in M^\times$ and $y \in Y$. Then $x_v \in N(S_v^\times)$ for every $v \in \mathbf{v}$, so that $x = N(\xi)$ with $\xi \in S^\times$. Then $\xi \in \mathfrak{G}$ and $N(\xi^{-1}\beta y^{-1}\alpha^{-1}) = 1$. By strong approximation on $\mathfrak{G}^1$, we have $\xi^{-1}\beta y^{-1}\alpha^{-1} \in \mathfrak{G}^1\alpha Y\alpha^{-1}$, so that $\xi^{-1}\beta y^{-1}\alpha^{-1} = \gamma\alpha z\alpha^{-1}$ with $\gamma \in \mathfrak{G}^1$ and $z \in Y$. Then $\beta = \xi\gamma\alpha z y \in \mathfrak{G}\alpha Y$. This proves (1.13) as well as the injectivity, and completes the proof. □

1.10. To illustrate the principles of Theorem 1.6 and §1.7, we consider an easy case in which φ is a quaternion norm form. Let B and ι be as in §1.8 with a global F. We take $V = B$ and define φ on V by $\varphi[x] = xx^\iota$; then $2\varphi(x, y) = \mathrm{Tr}_{B/F}(xy^\iota)$. We again put $G = G^+(V)$. As shown in [S4, §7.4], we have $A^+(V) = B \times B$,

$$G = \{(a, b) \in B^\times \times B^\times \mid aa^\iota = bb^\iota\}, \tag{1.14}$$

and $x\tau(\alpha) = a^{-1}xb$ and $\nu(\alpha) = aa^\iota$ for $x \in V$ and $\alpha = (a, b) \in G$. For $h \in V$ such that $\varphi[h] \neq 0$ put $W = (Fh)^\perp$ and $H = G^+(W)$. From Lemma 1.2 we obtain

$$A^+(W) = \{(a, h^{-1}ah) \mid a \in B\} \quad \text{and} \quad H = A^+(W)^\times. \tag{1.15}$$

Thus $A^+(W)$ is always isomorphic to B. Let $\mathfrak{O}$ be a maximal order in B. Then $\mathfrak{O}$ is a $\mathfrak{g}$-maximal lattice in V and $J_v = G_v \cap (\mathfrak{O}_v^\times \times \mathfrak{O}_v^\times)$. For $v \in \mathbf{h}$ put $Q_v = J_v$ if B is unramified at v; if B is ramified at v, then put $Q_v = \bigcup_{\nu\in\mathbf{Z}} \omega_v^\nu J_v$ with an element $\omega_v \in G_v$ such that $\nu(\omega_v)$ is a prime element of F_v. By [S4, Theorem 8.9], $\tau(Q_v) = C_v$ for every $v \in \mathbf{h}$ and hence $\tau(Q) = C$.

Take $q \in F^\times$ such that $q \in \mathfrak{g}_v$ for every $v \in \mathbf{h}$ where B is unramified. Suppose $h \in L[q, 2^{-1}\mathfrak{b}]$ with $L = \mathfrak{O}$ and an integral ideal $\mathfrak{b}$ divisible only by the prime ideals ramified in B. Then $hC = h\tau(J)$. To show this, it is sufficient to prove that $hC_v = h\tau(J_v)$ for every v where B is ramified. For such a v, take $\zeta \in \mathfrak{O}_v$ such that $\zeta\zeta^\iota$ is a prime element of F_v. Then we can take $(\zeta, h^{-1}\zeta h)$ as ω_v. Since $\omega_v \in G^+(W)_v$, we have $h\tau(\omega_v) = h$, which proves the desired fact. Therefore (1.8) and (1.9) are applicable.

Identify $A^+(W)$ with B through the map $(a, h^{-1}ah) \mapsto a$; put $\mathfrak{o} = \mathfrak{O} \cap h\mathfrak{O}h^{-1}$. Then $H = B^\times$ and $H_\mathbf{A} \cap J = B_\mathbf{a}^\times \prod_{v\in\mathbf{h}} \mathfrak{o}_v^\times$. Thus the left-hand side of (1.9) is the class number of $\mathfrak{o}$. If B is ramified at v, then $\mathfrak{o}_v = \mathfrak{O}_v$. Suppose B is unramified at v. Then we can find an isomorphism of B_v onto $M_2(F_v)$ that maps $\mathfrak{O}_v$ onto $M_2(\mathfrak{g}_v)$. Let $k_v = \mathrm{diag}[1, q]$. Then $k_v \in L_v[q, 2^{-1}\mathfrak{g}_v]$, and hence $k_v \in h\tau(J_v)$ by

[S4, Theorem 10.5]. From this we easily see that $\mathfrak{o}_v$ is conjugate to

$$(1.16) \qquad \left\{ \begin{bmatrix} a & b \\ c & d \end{bmatrix} \in M_2(\mathfrak{g}_v) \mid c \in q\mathfrak{g}_v \right\}.$$

Thus the genus of $\mathfrak{o}$ is determined by q. Taking $F = \mathbf{Q}$, we can now state a result which explains the meaning of (1.4) and (1.9) in the simplest case:

THEOREM 1.11. *Let B be a definite quaternion algebra over $\mathbf{Q}$ of class number 1, and $\mathfrak{O}$ a maximal order in B. Let s be a positive integer divisible only by the primes ramified in B, and q an arbitrary positive integer. Further let $\mathfrak{o}$ be an order in B such that $\mathfrak{o}_p = \mathfrak{O}_p$ if p is ramified in B and $\mathfrak{o}_p$ can be transformed to an order of type (1.16) for every p unramified in B. Then the class number of $\mathfrak{o}$ equals the number of elements $x \in B$ such that $xx^\iota = q$ and $\mathrm{Tr}_{B/F}(x\mathfrak{O}) = s\mathbf{Z}$, counted modulo the maps $x \mapsto a^{-1}xb$ for all $a, b \in \mathfrak{O}^\times$, provided such an x exists.*

Proof. We consider $L[q, 2^{-1}s\mathbf{Z}]$ with $L = \mathfrak{O}$. Let $(x, y) \in G_\mathbf{A}$. As B has class number 1, $x\mathfrak{O} = \alpha\mathfrak{O}$ and $y\mathfrak{O} = \beta\mathfrak{O}$ with $\alpha, \beta \in B^\times$. Then $\alpha\alpha^\iota = \pm\beta\beta^\iota$. Since B is definite, we have $\alpha\alpha^\iota = \beta\beta^\iota$, so that $(\alpha, \beta) \in G$. Thus $G_\mathbf{A} = GJ$, and hence the right-hand side of (1.9) is $\#\{L[q, 2^{-1}s\mathbf{Z}]/\tau(\Delta)\}$ with $\Delta = G \cap J$. The left-hand side is the class number of $\mathfrak{o}$, as shown in §1.10. Therefore we obtain our theorem. □

Decompose q in the form $q = q_0q_1$ with q_0 divisible only by the primes ramified in B and q_1 divisible only by the primes unramified in B. If q_1 is squarefree, then the class number of $\mathfrak{o}$ can be given by (1.12). Thus (1.12) with $r = q_1$ gives $\#\{L[q, 2^{-1}s\mathbf{Z}]/\tau(\Delta)\}$. If q_1 is not squarefree, then (1.12) does not apply, and therefore we can view our result as a new method of finding the class number of $\mathfrak{o}$. For example, if $t = 2$ and $q = 9$, we easily find that $L[q, 2^{-1}\mathbf{Z}]/\tau(\Delta)$ can be represented by a single element, and so the class number of $\mathfrak{o}$ is 1.

The above theorem is an anlogue of the result of Gauss that connects the primtive representations of an integer as sums of three squares with the class number of a binary form. There is a notable difference, however. In the case of three squares, many different quadratic fields appear, but $\mathfrak{o}$ in the above theorem is contained in the same quaternion algebra B, once the quadratic form φ is fixed. This is caused by our choice of φ, which has square determinant. In fact, as will be shown later, we have an analogue even for a quaternary form φ with nonsquare determinant, and we obtain the class number of an order $\mathfrak{o}$ in many different quaternion algebras. The peculiarity of a quaternary form of square determinant is a special case of the phenomena which occur in any even dimension as follows:

THEOREM 1.12. *Suppose F is a local or global field,* $\dim(V) = n \in 2\mathbf{Z}$, *and* $(-1)^{n/2}\det(\varphi)$ *is a square in F; let $q \in F^\times$ and $W = (Fh)^\perp$ with an element $h \in V$ such that $\varphi[h] = q$. Then the isomorphism class of $(W, q\varphi)$ is completely determined by φ and the signatures of q at the real archimedean primes of F.*

Proof. By (1.3) the isomorphism class of (W, φ) depends only on q. Thus we can take any h such that $\varphi[h] = q$. In view of the Hasse principle, it is sufficient to discuss the problem when F is local or $F = \mathbf{R}$. The matter is trivial if $F = \mathbf{R}$; so suppose F is local. Suppose also φ is anisotropic. Since $(-1)^{n/2}\det(\varphi)$ is a square, we have $\dim(V) = 4$ and φ is a quaternion norm form. In this case the fact is proved in [S4, §11.3]. Next suppose φ is isotropic. Then V has a weak Witt decomposition $V = Fe + Ff + U$. Taking $h = qe + f$, we find that $W = Fk + U$ with $k = qe - f$. The discriminant field of (U, φ) is F, and so $(U, q\varphi)$ is isomorphic to (U, φ) by [S4, Theorem 7.13]. Since $q\varphi[k] = -q^2$, we see that $(W, q\varphi)$ does not depend on q. This completes the proof. □

2. Binary hermitian forms over a local or global field.

2.1. We take a couple (K, ρ) consisting of an F-algebra K of rank 2 and a nontrivial F-linear automorphism ρ of K belonging to the following two types:

(I) K is a quadratic extension of F and ρ is the generator of $\mathrm{Gal}(K/F)$;

(II) $K = F \times F$ and $(x, y)^\rho = (y, x)$ for $(x, y) \in F \times F$.

We need K of type (II) principally because it appears as a localization of a global quadratic extension.

Let us now consider the algebra $M_2(K)$ with K belonging to these two types, and define an involution σ of $M_2(K)$, a vector space V over F, a quadratic form φ on V, and a group $\mathcal{G}$ by

$$x^\sigma = {}^t x^\rho \quad \text{for} \quad x \in M_2(K), \tag{2.1a}$$

$$V = \{h \in M_2(K) \mid h^\sigma = h\}, \quad \varphi[h] = \det(h) \qquad (h \in V), \tag{2.1b}$$

$$\mathcal{G} = \{\alpha \in GL_2(K) \mid \det(\xi) \in F^\times\}. \tag{2.1c}$$

Thus V consists of all ρ-hermitian matrices of size 2 with coefficients in K. Denote by ι the main involution of $M_2(K)$. Then

$$2\varphi(h, k) = \mathrm{tr}(hk^\iota) \qquad (h, k \in V). \tag{2.1d}$$

Now $\mathcal{G}$ is essentially $G^+(V, \varphi)$. More precisely we have:

Proposition 2.2. *There is an isomorphism Ξ of $G^+(V)$ onto $\mathcal{G}$ such that if $\alpha \in G^+(V)$ and $\xi = \Xi(\alpha)$, then $\nu(\alpha) = \det(\xi)$ and $h\tau(\alpha^{-1}) = \det(\xi)^{-1}\xi h \xi^\sigma$ for every $h \in V$, where τ is the map of $G^+(V)$ to $SO^\varphi(V)$ defined in §1.1. Consequently $SO^\varphi(V)$ consists of the maps $h \mapsto \det(\xi)^{-1}\xi h \xi^\sigma$ for all $\xi \in \mathcal{G}$, and $SO^\varphi(V)$ is isomorphic to $\mathcal{G}/F^\times$.*

2.3. To construct such a map Ξ, we first put

$$e = \begin{bmatrix} 1 & 0 \\ 0 & 0 \end{bmatrix}, \qquad f = \begin{bmatrix} 0 & 0 \\ 0 & 1 \end{bmatrix}, \qquad g = \begin{bmatrix} 0 & 1 \\ 1 & 0 \end{bmatrix}, \tag{2.2a}$$

$$U = K_0 g, \qquad K_0 = \{\mathrm{diag}\,[b^\rho, b] \mid b \in K\}. \tag{2.2b}$$

We view U as a left K-module by putting $bg = \mathrm{diag}\,[b^\rho, b]g$ for $b \in K$. Then e, f, and bg belong to V, $U = Kg$, $\varphi[g] = -1$, and

$$\varphi[ae + cf + bg] = \det \begin{bmatrix} a & b^\rho \\ b & c \end{bmatrix} = ac - bb^\rho \qquad (a, c \in F, b \in K). \tag{2.3}$$

By the general principle on $A(U)$ in the case $\dim(U) = 2$ discussed in [S4, §7.2], we can identify K with $A^+(U)$ and put $A(U) = K + Kg$; moreover, $gx = x^\rho g$ for every $x \in K$, and bg is the product of b and g in the ring $A(U)$. Now $V = Fe + Ff + Kg$, which is a weak Witt decomposition. Therefore, as shown in [S4, §2.4], we can define an F-linear isomorphism Ψ: $A(V) \to M_2(A(U))$ such that

$$\Psi(ae + cf + bg) = \begin{bmatrix} bg & a \\ c & -bg \end{bmatrix} \tag{2.4}$$

for a, b, c as in (2.3). We then put

$$\Xi(\alpha) = \mathrm{diag}\,[g^{-1}, 1]\Psi(\alpha)\,\mathrm{diag}\,[g, 1] \qquad (\alpha \in A(V)). \tag{2.5}$$

This is a special case of the map defined in [S4, §2.5]. In particular, by [S4, (2.15)] we have $\Xi(A^+(V)) = M_2(A^+(U)) = M_2(K)$. In [S4, (7.3c)] we showed that if $\alpha \mapsto \alpha^*$ is the canonical involution of $A(V)$, then $\Xi(\alpha^*) = \Xi(\alpha)^\iota$ for every $\alpha \in A^+(V)$. Moreover, by [S4, (7.3d)] we have

$$\Xi(G^+(V)) = \mathcal{G} \quad \text{and} \quad \nu(\alpha) = \det(\xi) \quad \text{if} \quad \xi = \Xi(\alpha),\ \alpha \in G^+(V). \tag{2.6}$$

Consequently $\Xi(G^1(V)) = SL_2(K)$. Since $b = -bg \cdot g$, we can easily verify that

$$\Xi(b) = \mathrm{diag}\,[b^\rho, b] \qquad (b \in K). \tag{2.7}$$

Now we have $\Xi(ae + cf + bg) = \begin{bmatrix} gb & -ag \\ cg & -bg \end{bmatrix}$, and hence

$$h = \Xi(h)\delta \quad \text{with} \quad \delta = \begin{bmatrix} 0 & -g \\ g & 0 \end{bmatrix} \qquad (h \in V). \tag{2.8}$$

Notice that $\delta = \Xi(f - e), \delta^2 = 1$, and

$$x^\sigma = \delta x^\iota \delta^{-1} \quad \text{for every} \quad x \in M_2(K). \tag{2.9}$$

Given $\alpha \in G^+(V)$, put $\xi = \Xi(\alpha)$; then by (2.9), for $h, k \in V$ we have

$$h = k\tau(\alpha) \Longleftrightarrow \xi h = k\xi^{\iota\sigma}. \tag{2.10}$$

In particular, we obtain

$$h\tau(\alpha^{-1}) = \det(\xi)^{-1} \cdot \xi h \xi^\sigma \quad \text{for every} \quad h \in V \text{ if } \alpha \in G^+(V) \text{ and } \xi = \Xi(\alpha). \tag{2.11}$$

Combining this with (2.6), we obtain Proposition 2.2.

2.4. Fix an element h of V such that $\varphi[h] \neq 0$; put $W = (Fh)^\perp$ and

$$R = \{\xi \in M_2(K) \mid h\xi^{\iota\sigma} = \xi h\}. \tag{2.12}$$

By Lemma 1.2 and (2.10) we have

$$\Xi(A^+(W)) = R, \qquad \Xi(G^+(W)) = R^\times, \tag{2.13}$$

R is a quaternion algebra, and the main involution of R corresponds to the canonical involution of $A(W)$ restricted to $A^+(W)$. Let us consider the special case where $h = e + qf = \mathrm{diag}[1, q]$ with $q \in F$. Put $k = e - qf$. Then we easily see that

$$R = \left\{ \begin{bmatrix} a^\rho & -b \\ qb^\rho & a \end{bmatrix} \mid a, b \in K \right\}, \qquad W = Fk + Kg. \tag{2.14}$$

From (2.7) and (2.14) we see that

$$R = \Xi(K) + \varepsilon\Xi(K) \quad \text{with} \quad \varepsilon = \begin{bmatrix} 0 & -1 \\ q & 0 \end{bmatrix}. \tag{2.15}$$

Clearly $\varepsilon^2 = -q$ and $\Xi(c)\varepsilon = \varepsilon\Xi(c^\rho)$ for every $c \in K$.

2.5. We now consider (K, ρ) of §2.1 with local or global F. We then denote by $\mathfrak{r}$ the ring of all elements of K integral over $\mathfrak{g}$. We have $\mathfrak{r} = \mathfrak{g} \times \mathfrak{g}$ if $K = F \times F$. Next define (V, φ) as in §2.1 with such F and K, and put

$$L = \mathfrak{g}e + \mathfrak{g}f + \mathfrak{r}g, \tag{2.16a}$$

$$\Delta = \{\alpha \in GL_2(\mathfrak{r}) \mid \det(\alpha) \in \mathfrak{g}^\times\}. \tag{2.16b}$$

By [S4, Lemma 6.3], L is a $\mathfrak{g}$-maximal lattice in the sense of [S4, §6.1]. From (2.1d) we obtain $2\varphi(h, L) = a\mathfrak{g} + c\mathfrak{g} + \mathrm{Tr}_{K/F}(b\mathfrak{r})$ for $h = ae + cf + bg$, and hence

$$\{x \in V \mid 2\varphi(x, L) \subset \mathfrak{g}\} = \mathfrak{g}e + \mathfrak{g}f + \mathfrak{d}^{-1}g, \tag{2.17}$$

where $\mathfrak{d}$ is the different of K relative to F. Thus we have

$$\begin{bmatrix} a & b^\rho \\ b & c \end{bmatrix} \in L[q, 2^{-1}\mathfrak{g}] \iff \begin{cases} q = ac - bb^\rho, \\ a\mathfrak{g} + c\mathfrak{g} + \mathrm{Tr}_{K/F}(b\mathfrak{r}) = \mathfrak{g}, \end{cases} \tag{2.18}$$
$$\implies a \in \mathfrak{g},\ c \in \mathfrak{g}, b \in \mathfrak{d}^{-1}.$$

We call an element of V of determinant q *primitive* if it belongs to $L[q, 2^{-1}\mathfrak{g}]$; similarly we call a binary hermitian form *primitive* if it is represented by a primitive hermitian matrix. The group Δ, being a subgroup of $\mathcal{G}$, acts on V, and leaves $L[q, 2^{-1}\mathfrak{g}]$ stable. Then $L[q, 2^{-1}\mathfrak{g}]/\Delta$ represents the classes of primitive hermitian forms of discriminant q, if we define *the discriminant* to be the determinant of the matrix representing the form.

Let $A(L)$ denote the subring of $A(V)$ generated by L and $\mathfrak{g}$, and put $A^+(L) = A(L) \cap A^+(V)$. Now $\Xi(ef)$, $\Xi(fe)$, $\Xi(ge)$, and $\Xi(gf)$ are the standard matrix units of $M_2(K)$. Therefore, in view of (2.7) we have $\Xi(A^+(L)) = M_2(\mathfrak{r})$. Thus $A^+(L)$ is a maximal order in $A^+(V)$, and

$$\Xi(A^+(L)^\times \cap G^+(V)) = \Delta, \tag{2.19a}$$

$$\Xi(A^+(L) \cap A^+(W)) = R \cap M_2(\mathfrak{r}). \tag{2.19b}$$

LEMMA 2.6. *In the global case with a quadratic extension K of F, define a quaternion algebra R over F by $R = K + \varepsilon K$ with an element ε such that $\varepsilon^2 = -s \in F^\times \cap \mathfrak{g}$ and $\varepsilon a = a^\rho \varepsilon$ for every $a \in K$; put $\mathfrak{o} = \mathfrak{r} + \varepsilon\mathfrak{r}$. Let $\mathfrak{d}$ be the different of K relative to F and let $d(K/F) = N_{K/F}(\mathfrak{d})$. Then the following assertions hold:*

(i) *The discriminant of $\mathfrak{o}$ equals $s \cdot d(K/F)$, and consequently if $v \nmid d(K/F)s$, then R is unramified at v, and $\mathfrak{o}_v$ is a maximal order.*

(ii) *If a prime $v \in \mathbf{h}$ remains prime in K and s is a prime element of F_v, then R_v is a division algebra and $\mathfrak{o}_v$ is its maximal order.*

(iii) *R is ramified at a prime $v \in \mathbf{v}$ if and only if $-s \notin N_{K/F}(K_v^\times)$. If such a v belongs to $\mathbf{h}$, $v \nmid s$, and $d(K/F)_v$ is the prime ideal of $\mathfrak{g}_v$, then $\mathfrak{o}_v$ is maximal.*

(iv) *Suppose that both $s\mathfrak{g}$ and $d(K/F)$ are squarefree, $-s \notin N_{K/F}(K_v^\times)$ for every $v|d(K/F)$, and every prime factor of $s\mathfrak{g}$ remains prime in K. Then $\mathfrak{o}$ is a maximal order in R, and R is ramified at a prime $v \in \mathbf{h}$ if and only if $v|d(K/F)s$.*

Proof. Let $\tilde{\mathfrak{o}} = \{x \in R \mid \mathrm{Tr}_{R/F}(x\mathfrak{o}) \subset \mathfrak{g}\}$. Then $\tilde{\mathfrak{o}} = \mathfrak{d}^{-1} + s^{-1}\varepsilon\mathfrak{d}^{-1}$. Therefore if $v \nmid d(K/F)s$, then $\tilde{\mathfrak{o}}_v = \mathfrak{o}_v$, so that we obtain (i). Assertion (ii) and the first part of (iii) follow from well known facts on quaternion algebras over a local field.

If v is as in the last part of (iii), then the discriminant of $\mathfrak{o}_v$ is the prime ideal of $\mathfrak{g}_v$, and hence $\mathfrak{o}_v$ is maximal. Combining (i), (ii), and (iii), we obtain (iv). □

2.7. In the setting of §2.1 let us now assume that F is a global field and K is a quadratic extension of F; we put then $G = G^+(V)$ and define a subgroup C of $SO^\varphi(V)_\mathbf{A}$ by (1.2) and a subgroup J of $G_\mathbf{A}$ by (1.7). Then $J_v = A^+(L)_v^\times \cap G_v$ for every $v \in \mathbf{h}$ by [S4, Proposition 8.8(ii)]. For each $v \in \mathbf{h}$ ramified in K, take a prime element ω_v of $K_v = A^+(U)_v$. Then $\omega_v \in G^+(U)_v \subset G_v$, $\nu(\omega_v) = \omega_v\omega_v^\rho$, and

$$(2.20) \qquad (ae + cf + bg)\tau(\omega_v) = ae + cf + (b\omega_v^\rho/\omega_v)g.$$

This shows that $\tau(\omega_v) \in C_v$. With $d(K/F)$ as in Lemma 2.6, we put

$$(2.21) \qquad Q = G_\mathbf{a}\prod_{v\in\mathbf{h}} Q_v, \qquad Q_v = \begin{cases} J_v & \text{if } v \nmid d(K/F), \\ \bigcup_{m\in\mathbf{Z}} J_v\omega_v^m & \text{if } v | d(K/F). \end{cases}$$

By (2.19a) we have $\Xi(J_v) = \{\alpha \in GL_2(\mathfrak{r}_v) \mid \det(\alpha) \in \mathfrak{g}_v^\times\}$. Also, by [S4, Proposition 8.8(iii) and Theorem 8.9] we have $\nu(J_v) = \mathfrak{g}_v^\times$ and $\tau(Q_v) = C_v$ for every $v \in \mathbf{h}$. Therefore $\tau(Q) = C$, and consequently $F_\mathbf{A}^\times Q = \{x \in G_\mathbf{A} \mid L\tau(x) = L\}$.

The $SO^\varphi(V)$-genus of L consists of all $\mathfrak{g}$-maximal lattices in V; see [S4, §9.7]. Since $\tau(G_\mathbf{A}) = SO^\varphi(V)_\mathbf{A}$ as proven in [S4, Lemma 9.11], the genus consists of $L\tau(\alpha^{-1})$ for all $\alpha \in G_\mathbf{A}$.

LEMMA 2.8. (i) *For each $\mathfrak{g}$-lattice M in the genus of L take $\alpha \in G_\mathbf{A}$ so that $M = L\tau(\alpha)^{-1}$. Then the map $M \mapsto \nu(\alpha)$ gives a bijection of the set of all classes in the genus of L onto $F_\mathbf{A}^\times/[F^\times\nu(F_\mathbf{A}^\times Q)]$.*

(ii) *$\#(G\backslash G_\mathbf{A}/J)$ equals the class number of F.*

(iii) *Let $\mathcal{I}_F$ denote the ideal group of F, and $\mathcal{I}^\varphi$ the subgroup of $\mathcal{I}_F$ generated by the principal ideals of F, the squares of the ideals of F, and the prime ideals in F ramified in K. Then $[\mathcal{I}_F\colon \mathcal{I}^\varphi] = [F_\mathbf{A}^\times\colon F^\times\nu(F_\mathbf{A}^\times Q)] = \#(G\backslash G_\mathbf{A}/F_\mathbf{A}^\times Q)$, which is the number of G-classes in the genus of L.*

Proof. Since $F_\mathbf{A}^\times Q = \{x \in G_\mathbf{A} \mid L\tau(x) = L\}$, the map $\alpha \mapsto L\tau(\alpha)^{-1}$ gives a bijection of $G\backslash G_\mathbf{A}/F_\mathbf{A}^\times Q$ onto the classes in the genus of L. Take $M_2(K)$, F, K, and $F_\mathbf{A}^\times Q$ as S, M, F, and Y in Lemma 1.9. Then $\mathfrak{G}$ there corrresponds to G here, and we obtain (i) as an immediate consequence of that lemma. Put $E = F_\mathbf{a}^\times\prod_{v\in\mathbf{h}}\mathfrak{g}_v^\times$. Then $\nu(J) = E$, and hence $\nu(Q)$ is the subgroup of $F_\mathbf{A}^\times$ generated by E and the prime elements of F_v for all $v|d(K/F)$. Now we easily see that $a \mapsto a\mathfrak{g}$ for $a \in F_\mathbf{A}^\times$ gives the bijection of $F_\mathbf{A}^\times/[F^\times\nu(F_\mathbf{A}^\times Q)]$ onto $\mathcal{I}_F/\mathcal{I}^\varphi$. This proves (iii), which is essentially a special case of [S4, Theorem 9.26]. Take J as Y in Lemma 1.9. Then we obtain (ii), as $[F_\mathbf{A}^\times\colon F^\times E]$ is the class number of F. □

LEMMA 2.9. *Let the notation be as in Theorem 1.6 with the present (V, φ) of (2.1b) as (V, φ) there. Then $H_{\mathbf{A}} \cap GyJ \neq \emptyset$ for every $y \in G_{\mathbf{A}}$, and consequently the G-genus of L coincides with the H-genus of L.*

Proof. In view of (1.3), $A^+(W)$ is isomorphic to the algebra defined with a special choice $h = e + qf$ as in §2.4. Thus it is isomorphic to R of (2.16), and $G^+(W)$ to $R^\times$. By Lemma 2.6(iii), R_v is a division algebra if and only if $-q \notin N_{K/F}(K_v^\times)$. Since H_v is essentially $R_v^\times$, we have, by a well-known fact (see [S4, Proposition 5.15]), $\nu(H_v) = F_v^\times$ for every $v \in \mathbf{h}$. Let Z be the set of all $z \in F_{\mathbf{A}}^\times$ such that $z_{\mathbf{a}}$ belongs to the identity component of $F_{\mathbf{a}}^\times$. Then $Z \subset \nu(H_{\mathbf{A}})$. Given $y \in G_{\mathbf{A}}$, we can find $a \in F^\times$ such that $a\nu(y) \in Z$. Take $\xi \in H_{\mathbf{A}}$ so that $a\nu(y) = \nu(\xi)$. Then $\xi \in GyJ$ by (1.13), and hence $H_{\mathbf{A}} \cap GyJ \neq \emptyset$ for every $y \in G_{\mathbf{A}}$. This proves our lemma. □

THEOREM 2.10. *Given $q \in F^\times \cap \mathfrak{g}$, define R and $\mathfrak{o}$ as in Lemma 2.6 with q as s there. Take $\{y_i\}_{i \in I} \subset G_{\mathbf{A}}$ so that $G_{\mathbf{A}} = \bigsqcup_{i \in I} Gy_iJ$ and put $\Delta_i = G \cap y_iJy_i^{-1}$ and $L_i = L\tau(y_i)^{-1}$. Then $\#(I)$ is the class number of F, and the class number of $\mathfrak{o}$ (see §1.4 for the definition) equals $\sum_{i \in I} \#\{L_i[q, 2^{-1}\mathfrak{g}]/\tau(\Delta_i)\}$.*

Proof. The assertion concerning $\#(I)$ is Lemma 2.8(ii). Put $h = \operatorname{diag}[1, q]$; then $\varphi[h] = q$ and $2\varphi(h, L) = \mathfrak{g}$, so that $h \in L[q, 2^{-1}\mathfrak{g}]$. From (2.20) we see that $h\tau(\omega_v) = h$, and hence $h\tau(Q) = h\tau(J)$. Thus (1.9) is valid in the present situation. Define R and $\mathfrak{o}$ as in Lemma 2.6. We can identify R with R of (2.15), so that Ξ maps $A^+(W)$ and $G^+(W)$ onto R and $R^\times$. Then we have $\Xi(H_v) = R_v^\times$ for every $v \in \mathbf{v}$ and $\Xi(H_v \cap J_v) = \mathfrak{o}_v^\times$ for every $v \in \mathbf{h}$ by (2.19b). Therefore the left-hand side of (1.9) is the class number of $\mathfrak{o}$, and consequently we obtain our theorem. □

The above theorem gives an equality between the class number of an order in a quaternion algebra and the sum of the numbers of classes of "primitive binary hermitian forms." If the hermitian forms are indefinite, we obtain clear-cut results:

THEOREM 2.11. *The notation and assumptions being as in Theorem 2.10, exclude the case in which F is totally real, K is totally imaginary, and q is totally positive. Let c_F be the class number of F, and $c(\mathfrak{o})$ the class number of $\mathfrak{o}$. Then the following assertions hold:*

(i) $\#\{L_i[q, 2^{-1}\mathfrak{g}]/\tau(\Delta_i)\} = c(\mathfrak{o})/c_F$ *for every $i \in I$.*

(ii) *Let λ be the number of prime ideals $\mathfrak{p}$ in F ramified in K such that $\mathfrak{p}|q$ or $\mathfrak{p}^2$ divides the discriminant of K relative to F. Then $c(\mathfrak{o})/c_F$ divides 2^λ.*

Proof. By Lemma 2.9 we can take $\{y_i\}_{i \in I} \subset H_{\mathbf{A}}$. Then by (1.8) we have $V \cap h\tau(Jy_i^{-1}) = L_i[q, 2^{-1}\mathfrak{g}]$. Thus, by Theorem 1.6(ii), we have

$$(*) \qquad \#\{L_i[q, 2^{-1}\mathfrak{g}]/\tau(\Delta_i)\} = \#\{H\backslash(H_{\mathbf{A}} \cap Gy_iJ)/(H_{\mathbf{A}} \cap J)\}.$$

By (1.13), $Gy_iJ = \{x \in G_{\mathbf{A}} \mid \nu(x) \in F^\times \nu(y_iJ)\}$, and hence $H_{\mathbf{A}} \cap Gy_iJ = \{x \in H_{\mathbf{A}} \mid \nu(x) \in F^\times \nu(y_iJ)\}$. Our condition on F and q implies that R_v is not a division algebra for some $v \in \mathbf{a}$. Put $X = H_{\mathbf{A}} \cap J$. Taking (H, X) as $(\mathfrak{G}, Y)$ in Lemma 1.9, we see that the map $z \mapsto F^\times \nu(zX)$ gives a bijection of $H\backslash H_{\mathbf{A}}/X$ onto $F_{\mathbf{A}}^\times/[F^\times \nu(X)]$. Thus $c(\mathfrak{o}) = [F_{\mathbf{A}}^\times \colon F^\times \nu(X]$ and

$$(**) \qquad \#\{H\backslash(H_{\mathbf{A}} \cap Gy_iJ)/X\} = [F^\times \nu(J) \colon F^\times \nu(X)] = [F_{\mathbf{A}}^\times \colon F^\times \nu(X]/[F_{\mathbf{A}}^\times \colon F^\times \nu(J)].$$

We have $[F_{\mathbf{A}}^\times \colon F^\times \nu(J)] = [F_{\mathbf{A}}^\times \colon F^\times E] = c_F$, where $E = F_{\mathbf{a}}^\times \prod_{v \in \mathbf{h}} \mathfrak{g}_v^\times$. Combining this with $(*)$ and $(**)$, we obtain (i). To prove (ii), put $T = F^\times F_{\mathbf{a}}^\times$, $E_{\mathbf{h}} = \prod_{v \in \mathbf{h}} \mathfrak{g}_v^\times$, and $E_0 = \prod_{v \in \mathbf{h}} \nu(\mathfrak{o}_v^\times)$. Since $F_{\mathbf{a}}^\times \subset F^\times \nu(R_{\mathbf{a}}^\times)$, we have $F^\times \nu(X) = F^\times F_{\mathbf{a}}^\times E_0 = TE_0$. Thus $[F^\times E \colon F^\times \nu(X)] = [TE_{\mathbf{h}} \colon TE_0] = [E_{\mathbf{h}} \colon TE_0 \cap E_{\mathbf{h}}]$. Let $e \in TE_0 \cap E_{\mathbf{h}}$. Then $e = abc$ with $a \in F^\times, b \in F_{\mathbf{a}}^\times$, and $c \in E_0$. Clearly $a \in \mathfrak{g}^\times$ and $e = a_{\mathbf{h}}c$. Let $E_1 = \{a_{\mathbf{h}} \mid a \in \mathfrak{g}^\times\}$. Then we have $TE_0 \cap E_{\mathbf{h}} = E_1E_0$, and hence

$$(***) \qquad [F^\times E \colon F^\times \nu(X)] = [E_{\mathbf{h}} \colon E_1E_0].$$

Since $\mathfrak{r}_v^\times \subset \mathfrak{o}_v^\times$, we have $N_{K/F}(\mathfrak{r}_v^\times) \subset \nu(\mathfrak{o}_v^\times) \subset \mathfrak{g}_v^\times$. If K is unramified at v, then $N_{K/F}(\mathfrak{r}_v^\times) = \mathfrak{g}_v^\times$, and hence $\nu(\mathfrak{o}_v^\times) = \mathfrak{g}_v^\times$. Thus the question is about the primes v ramified in K. Let $\mathfrak{d}$ denote the different of K relative to F. Let us now prove:

(2.22) If $q \in \mathfrak{g}_v^\times$ and $\mathfrak{d}_v$ is the maximal ideal of $\mathfrak{r}_v$, then $\nu(\mathfrak{o}_v^\times) = \mathfrak{g}_v^\times$.

Assuming such conditions on q and v, define $\widetilde{\mathfrak{o}}$ as in the proof of Lemma 2.6. Then $\widetilde{\mathfrak{o}}_v/\mathfrak{o}_v$ is isomorphic to $\mathfrak{g}/\mathfrak{p} \oplus \mathfrak{g}/\mathfrak{p}$, where $\mathfrak{p}$ is the prime ideal of $\mathfrak{g}$ at v. If R_v is a division algebra, then $\mathfrak{o}_v$ must be maximal, in which case $\nu(\mathfrak{o}_v^\times) = \mathfrak{g}_v^\times$. Suppose R_v is not a division algebra. Then the property of $\widetilde{\mathfrak{o}}_v/\mathfrak{o}_v$ implies, by (1.11), that $\mathfrak{o}_v$ is isomorphic to the order given there. Therefore $\nu(\mathfrak{o}_v^\times) = \mathfrak{g}_v^\times$. This proves (2.22). Now, denote by $\mathbf{x}$ the set of all $v \in \mathbf{h}$ as in (ii) of our theorem, and put $E' = \prod_{v \in \mathbf{x}} \mathfrak{g}_v^\times$ and $E_0' = \prod_{v \in \mathbf{x}} \nu(\mathfrak{o}_v^\times)$. Then $E_{\mathbf{h}}/E_1E_0$ is isomorphic to $E'/E_1'E_0'$, where E_1' is the projection of $\mathfrak{g}^\times$ to E'. Since $[\mathfrak{g}_v^\times \colon \nu(\mathfrak{o}_v^\times)] \leq [\mathfrak{g}_v^\times \colon N_{K/F}(\mathfrak{r}_v^\times)] = 2$ for $v \in \mathbf{x}$ and $\#(\mathbf{x}) = \lambda$, we obtain (ii). This completes the proof. □

If we take the effect of E_1' and $\nu(\varepsilon)$ into account, we can state a result more clear-cut than (ii) above. For simplicity we treat only the case $F = \mathbf{Q}$.

PROPOSITION 2.12. *Let $F = \mathbf{Q}$ and $K = \mathbf{Q}(\sqrt{m})$ with a squarefree positive or negative integer $m \neq 1$, and let $0 \neq q \in \mathbf{Z}$. Suppose $q < 0$ if $m < 0$, and suppose also that q is prime to every odd factor of m. Then $c(\mathfrak{o}) = 1$ if $8 \nmid q$ or $m - 2 \notin 8\mathbf{Z}$, and $c(\mathfrak{o}) = 2$ otherwise. Consequently if $8 \nmid q$ or $m - 2 \notin 8\mathbf{Z}$, then every element of $L[q, 2^{-1}\mathbf{Z}]$ is equivalent to* diag $[1, q]$ *under Δ.*

Proof. We apply Theorem 2.11(ii) to the present setting. Define $\mathbf{x}$ as in the above proof. If $m-1 \in 4\mathbf{Z}$, then $\mathbf{x} = \emptyset$, and so $2^\lambda = 1$. Suppose $m-1 \notin 4\mathbf{Z}$; then $\mathbf{x}$ consists of the prime 2, $2^\lambda = 2$, and $c(\mathfrak{o}) = [\mathbf{Z}_2^\times : \{\pm 1\}\nu(\mathfrak{o}_2^\times)]$. Now $-1 \notin N_{K/\mathbf{Q}}(\mathfrak{r}_2^\times)$ if $m-1 \notin 4\mathbf{Z}$ and $m-2 \notin 8\mathbf{Z}$. In such a case $\mathbf{Z}_2^\times = \{\pm 1\}N_{K/\mathbf{Q}}(\mathfrak{r}_2^\times) \subset \{\pm 1\}\nu(\mathfrak{o}_2^\times)$, so that $c(\mathfrak{o}) = 1$. Finally suppose $m-2 \in 8\mathbf{Z}$; then $N_{K/\mathbf{Q}}(\mathfrak{r}_2^\times) = \{a \in \mathbf{Z}_2^\times \mid a \pm 1 \in 8\mathbf{Z}_2\}$. Now $\nu(a+b\varepsilon) = aa^\rho + qbb^\rho$ for $a, b \in \mathfrak{r}_2$. If $8|q$, then this shows that $\nu(\mathfrak{o}_2^\times) = N_{K/\mathbf{Q}}(\mathfrak{r}_2^\times)$, and hence $c(\mathfrak{o}) = 2$. If $8 \nmid q$, we can find an element x of $\mathfrak{o}_2^\times$ such that $\nu(x) \notin N_{K/\mathbf{Q}}(\mathfrak{r}_2^\times)$ as follows. If $q-4 \in 8\mathbf{Z}$, take $b \in \mathfrak{r}_2$ such that $bb^\rho = -1$; then $\nu(1+\varepsilon b)+3 \in 8\mathbf{Z}_2$. If $q-2 \in 4\mathbf{Z}$, then $\nu(1+\varepsilon\sqrt{m})+3 \in 8\mathbf{Z}$. If $q \pm 1 \in 4\mathbf{Z}$, then $\nu(a+\varepsilon\sqrt{m}) \mp 3 \in 8\mathbf{Z}_2$ with an element $a \in \mathfrak{r}_2$ such that $aa^\rho = \pm 1$. In all these cases $\mathbf{Z}_2^\times = \nu(\mathfrak{o}_2^\times)$, and hence $c(\mathfrak{o}) = 1$. This completes the proof. □

3. A fundamental domain for $SL_2(\mathfrak{r})$.

3.1. Given a positive definite real symmetric matrix $p = \begin{bmatrix} a & b/2 \\ b/2 & c \end{bmatrix}$ of determinant q, we define a point $\omega(p)$ of the upper half complex plane $\mathfrak{H} = \{z \in \mathbf{C} \mid \mathrm{Im}(z) > 0\}$ by $\omega(p) = (2c)^{-1}(b + 2i\sqrt{q})$. It can easily be seen that $\omega(\gamma p \cdot {}^t\gamma) = \gamma(\omega(p))$ for every $\gamma = \begin{bmatrix} r & s \\ t & u \end{bmatrix} \in SL_2(\mathbf{R})$, where $\gamma(z) = (rz+s)/(tz+u)$ for $z \in \mathfrak{H}$. When $a, b, c \in \mathbf{Z}$, we can associate with p a binary quadratic form $ax^2 + bxy + cy^2$. By means of the map ω, we can send each class of binary forms of a fixed discriminant $4ac - b^2$ to a point on $SL_2(\mathbf{Z})\backslash\mathfrak{H}$. Then, using the classical fundamental domain for $SL_2(\mathbf{Z})\backslash\mathfrak{H}$, we can easily determine a complete set of representatives for the classes of positive definite primitive binary forms of a given discriminant, and consequently the number of classes. We are going to do the same for positive definite binary hermitian forms of Section 2 when $F = \mathbf{Q}$. Instead of $\mathfrak{H}$, we employ a three-dimensional space on which $SL_2(\mathbf{C})$ acts. Let us now recall some elementary properties of the space and action.

Let $\mathbf{H}$ denote the division ring of Hamilton quaternions, whose standard quaternion units are denoted by 1, $\mathbf{i}$, $\mathbf{j}$, $\mathbf{k}$. For $w \in \mathbf{H}$ we denote by $\overline{w}$ the quaternion conjugate of w and put $|w| = \sqrt{w\overline{w}}$. We identify $\mathbf{R}+\mathbf{R}\mathbf{i}$ with $\mathbf{R}+\mathbf{R}i = \mathbf{C}$ as usual, and define a subset S of $\mathbf{H}$ by

$$S = \{z + v\mathbf{j} \mid z \in \mathbf{C}, \quad 0 < v \in \mathbf{R}\}. \tag{3.1}$$

We can let $SL_2(\mathbf{C})$ act on S. More precisely, for $\alpha \in SL_2(\mathbf{C})$ and $\mathfrak{z} \in S$ we can define $\alpha(\mathfrak{z}) \in S$ and a factor of automorphy $\lambda_\alpha(\mathfrak{z}) \in \mathbf{H}^\times$ by the formula

$$\alpha \begin{bmatrix} \mathfrak{z} \\ 1 \end{bmatrix} = \begin{bmatrix} \alpha(\mathfrak{z}) \\ 1 \end{bmatrix} \lambda_\alpha(\mathfrak{z}). \tag{3.2}$$

Clearly $\lambda_\alpha(\mathfrak{z}) = r\mathfrak{z} + s$ if $[r \quad s]$ is the lower half of α. If we define η: $S \to \mathbf{R}$ by $\eta(z + v\mathbf{j}) = v$ for $z \in \mathbf{C}$ and $0 < v \in \mathbf{R}$, then

$$\eta(\alpha(\mathfrak{z})) = \eta(\mathfrak{z})|\lambda_\alpha(\mathfrak{z})|^{-2} \qquad (\alpha \in SL_2(\mathbf{C}),\ \mathfrak{z} \in S). \tag{3.3}$$

Hereafter whenever we use the expression of type $z + v\mathbf{j}$ for a point of S, we always take $z \in \mathbf{C}$ and $0 < v \in \mathbf{R}$. We shall often write $\alpha\mathfrak{z}$ for $\alpha(\mathfrak{z})$.

Let us now find a connection of S with the set

$$P_q = \{h \in GL_2(\mathbf{C}) \mid h^* = h > 0,\ \det(h) = q\}. \tag{3.4}$$

Here q is a fixed positive real number; we put $x^* = {}^t\bar{x}$ for a complex matrix x, and write $h > 0$ if h is positive definite. Given $h \in P_q$ we take an element $\xi \in SL_2(\mathbf{C})$ so that $q^{-1/2}h = \xi\xi^*$, and put $\zeta(h) = \xi(\mathbf{j})$. Then ζ is well defined, and gives a bijection of P_q onto S such that $\zeta(\gamma h\gamma^*) = \gamma\zeta(h)$ for every $\gamma \in SL_2(\mathbf{C})$.

3.2. Let $K = \mathbf{Q}(\sqrt{-d})$ with $0 < d \in \mathbf{Z}$. We normalize d so that $-d$ is the discriminant of K, denote by $\mathfrak{r}$ the maximal order of K, and put $\mathfrak{d} = \mathfrak{r}\sqrt{-d}$. We consider V and L of Section 2 with this K. Then $L = V \cap M_2(\mathfrak{r})$. We view K as a subfield of $\mathbf{C}$ by identifying $\sqrt{-d}$ with $i\sqrt{d}$. We write an element $h \in V$ in the form

$$h = \begin{bmatrix} a & -b/\sqrt{-d} \\ \bar{b}/\sqrt{-d} & c \end{bmatrix}. \tag{3.5}$$

Then we easily see that

$$h \in L[q, 2^{-1}\mathbf{Z}] \iff \begin{cases} q = ac - d^{-1}|b|^2, \\ a\mathbf{Z} + c\mathbf{Z} + \mathrm{Tr}_{K/\mathbf{Q}}(b\mathfrak{d}^{-1}) = \mathbf{Z}, \end{cases} \tag{3.6}$$
$$\implies a \in \mathbf{Z},\ c \in \mathbf{Z}, b \in \mathfrak{r}.$$

For the moment we do not assume that $q \in \mathbf{Z}$.

We hereafter consider the case $q > 0$. Let us now put

$$\Delta = \{\gamma \in GL_2(\mathfrak{r}) \mid \det(\gamma) = \pm 1\}, \qquad \Delta_1 = SL_2(\mathfrak{r}). \tag{3.7}$$

Then $\Delta = G \cap J$ in the present case, and so our task is to investigate $L[q, 2^{-1}\mathbf{Z}]/\Delta$. If $\det(h) = q$, then h is positive or negative definite. Now $\Delta = \Delta_1 \cap \Delta_1\sigma$ with $\sigma = \mathrm{diag}[1, -1]$, and $h \mapsto \det(\sigma)\sigma h\sigma^*$ changes a negative definite h into a positive definite matrix. Therefore $L[q, 2^{-1}\mathbf{Z}]/\Delta$ is equivalent to $L_+[q, 2^{-1}\mathbf{Z}]/\Delta_1$, where

$$L_+[q, 2^{-1}\mathbf{Z}] = \{h \in L[q, 2^{-1}\mathbf{Z}] \mid h > 0\}. \tag{3.8}$$

To find $\zeta(h)$ for $h \in P_q$ of the form (3.5), take $\xi = \begin{bmatrix} v^{1/2} & zv^{-1/2} \\ 0 & v^{-1/2} \end{bmatrix} \in SL_2(\mathbf{C})$ with $z \in \mathbf{C}$ and $0 < v \in \mathbf{R}$. Then $\xi(\mathbf{j}) = z + v\mathbf{j}$. We have $h = q^{1/2}\xi\xi^*$ if we take $v = q^{1/2}/c$ and $z = -b/(c\sqrt{-d})$. Thus

$$\zeta(h) = c^{-1}(ibd^{-1/2} + q^{1/2}\mathbf{j}) \text{ and } |\zeta(h)|^2 = a/c. \tag{3.9}$$

Our next problem is to find, for a given $h \in L_+[q, 2^{-1}\mathbf{Z}]$, an element $\gamma \in \Delta_1$ such that $\gamma h \gamma^*$ is of a certain reduced form. We do this by finding a fundamental domain for $\Delta_1 \backslash S$. The problem is similar to, but more complicated than, that for the upper half plane modulo $SL_2(\mathbf{Z})$. Let us now prove

(3.10) *For a fixed $\beta \in P_q$, the minimum value of $a^*\beta a$ for all $a \in \mathfrak{r}^2$, $\neq 0$, is $\leq (8dq)^{1/2}/\pi$.*

To show this, put $B = \{x \in \mathbf{C}^2 \mid x^*\beta x \leq t^2\}$ with $0 < t \in \mathbf{R}$. We can put $\beta = q^{1/2}\eta^*\eta$ with $\eta \in SL_2(\mathbf{C})$. Then the map $x \mapsto \eta x$ sends B onto a 4-dimensional Euclidean ball of radius $q^{-1/4}t$. Thus the volume of B is $\pi^2 t^4 q^{-1}/2$. Now $\mathrm{vol}(\mathbf{C}^2/\mathfrak{r}^2) = d/4$. Take t so that $\pi^2 t^4 q^{-1}/2 > 4d$, that is, $t^2 > (8dq)^{1/2}/\pi$. By Minkowski's principle, B contains a nonzero element x of $\mathfrak{r}^2$. This proves (3.10).

LEMMA 3.3. *Let $\mathcal{P}$ be a subset of $\mathbf{C}$ that represents $\mathbf{C}/\mathfrak{r}$, and let*

$$\mathcal{F} = \{\mathfrak{z} = z + v\mathbf{j} \in S \mid z \in \mathcal{P},\ |\mathfrak{z}| \geq 1\}. \tag{3.11a}$$

Then, given $\mathfrak{z} \in S$, there exists an element $\gamma \in \Delta_1$ such that $\gamma(\mathfrak{z}) \in \mathcal{F}$. Moreover, if the class number of K is 1, then $\mathcal{F}$ can be replaced by

$$\mathcal{F}' = \{\mathfrak{z} = z + v\mathbf{j} \in \mathcal{F} \mid v \geq (8d)^{-1/2}\pi\}. \tag{3.11b}$$

Proof. We first observe that

$$|z + w\mathbf{j}|^2 = |z|^2 + |w|^2 \text{ for } z,\ w \in \mathbf{C}. \tag{*}$$

We are going to prove, for a fixed $\mathfrak{z} \in S$,

(3.12a) $\mathrm{Max}\{\eta(\gamma\mathfrak{z}) \mid \gamma \in \Delta_1\}$ *exists;*

(3.12b) *If $\eta(\gamma\mathfrak{z})$ for a $\gamma \in \Delta_1$ takes the maximum value, then $|\gamma(\mathfrak{z})| \geq 1$.*

To prove these, let $\mathfrak{z} = z + v\mathbf{j} \in S$ and $\gamma = \begin{bmatrix} * & * \\ r & s \end{bmatrix} \in \Delta_1$. Now $(t, u) \mapsto t\mathfrak{z} + u$ is an $\mathbf{R}$-linear isomorphism of $\mathbf{C} \times \mathbf{C}$ onto $\mathbf{H}$, which maps $\mathfrak{r} \times \mathfrak{r}$ onto a $\mathbf{Z}$-lattice in $\mathbf{H}$. Therefore the minimum value of $|r\mathfrak{z} + s|$ for $\gamma \in \Delta_1$ as above exists. This proves (3.12a), since $\eta(\gamma\mathfrak{z}) = v|r\mathfrak{z} + s|^{-2}$. Put $\mathfrak{w} = \gamma\mathfrak{z}$ with γ for which $\eta(\gamma\mathfrak{z})$ takes the maximum value. Let $\varepsilon = \begin{bmatrix} 0 & -1 \\ 1 & 0 \end{bmatrix}$. Then $\varepsilon\gamma \in \Delta_1$, so that $\eta(\mathfrak{w}) \geq \eta(\varepsilon\gamma(\mathfrak{z})) = \eta(\varepsilon\mathfrak{w}) = \eta(\mathfrak{w})|\mathfrak{w}|^{-2}$, and hence $|\mathfrak{w}| \geq 1$, which is (3.12b). Now take $\alpha = \begin{bmatrix} 1 & u \\ 0 & 1 \end{bmatrix} \in \Delta_1$. Then for $\mathfrak{w}$ as above, we have $\alpha\mathfrak{w} = \mathfrak{w} + u$ and $\eta(\alpha\mathfrak{w}) = \eta(\mathfrak{w})$, so that $|\alpha\mathfrak{w}| \geq 1$ by (3.12b). Thus $|\mathfrak{w} + u| \geq 1$ for every $u \in \mathfrak{r}$. We can choose u so that $\mathfrak{w} + u \in \mathcal{F}$. This shows that $\mathcal{F}$ represents $\Delta_1\backslash S$. Next, suppose that K has class number 1. Then (3.10) is true even if we restrict a to the type (r, s) such that $r\mathfrak{r}+s\mathfrak{r} = \mathfrak{r}$. Put $\beta = \begin{bmatrix} |z|^2 + v^2 & \bar{z} \\ z & 1 \end{bmatrix}$. Then $\beta \in P_{v^2}$, and for $a = (r, s) \in \mathfrak{r}^2$ we have $a^*\beta a = |r\mathfrak{z}+s|^2$. Therefore if $\eta(\gamma z)$ is the maximum value in the above sense, then by (3.10), $|r\mathfrak{z}+s|^2 \leq (8d)^{1/2}v/\pi$, so that $\eta(\gamma\mathfrak{z}) \geq (8d)^{-1/2}\pi$. Therefore we can replace $\mathcal{F}$ by $\mathcal{F}'$. This completes the proof. □

3.4. The above lemma gives merely the first approximation to the fundamental domain. To obtain a better one, we first consider the question of how to choose a set $\mathcal{P}$ that represents $\mathbf{C}/\mathfrak{r}$. Take w so that $\mathfrak{r} = \mathbf{Z} + \mathbf{Z}w$. Since $|\operatorname{Im}(w)| = \sqrt{d}/2$, we can take $\mathcal{P}$ to be the rectangle

$$\mathcal{P} = \{z \in \mathbf{C} \mid |\operatorname{Re}(z)| \leq 1/2,\ |\operatorname{Im}(z)| \leq \sqrt{d}/4\}. \tag{3.13a}$$

A conceptually more natural choice is

$$\mathcal{H} = \{z \in \mathbf{C} \mid |z + u| \geq |z| \quad \text{for every} \quad u \in \mathfrak{r}\}. \tag{3.13b}$$

Indeed, for every fixed $z \in \mathbf{C}$, the minimum value of $|z + s|$ for $s \in \mathfrak{r}$ exists, and hence the set $\mathcal{H}$ of (3.13b) represents $\mathbf{C}/\mathfrak{r}$. This has the advantage of being stable under the map $z \mapsto \kappa z$ for every root of unity κ in K.

Now, for a fixed $u \in \mathfrak{r}$, the set of $z \in \mathbf{C}$ such that $|z + u| \geq |z|$ is the half plane determined by the inequality

$$\bar{u}z + u\bar{z} \geq -|u|^2. \tag{3.14}$$

Thus $\mathcal{H}$ of (3.13b) is the intersection of all such half planes.

If $4|d$, then $\mathfrak{r} = \mathbf{Z} + \mathbf{Z}\sqrt{-d/4}$, and (3.13b) coincides with (3.13a). If $d \equiv 3 \pmod 4$, then $\mathfrak{r} = \mathbf{Z}+\mathbf{Z}w$ with $w = (1+\sqrt{-d})/2$, and checking (3.14) for $u = \pm 1$, $u = \pm w$, and $u = \pm\bar{w}$, we find that $\mathcal{H}$ is the hexagon

$$\mathcal{H} = \{x + iy \in \mathbf{C} \mid |x| \leq 1/2, \quad x - (d+1)/4 \leq \sqrt{d}y \leq x + (d+1)/4, \\ x - (d+1)/4 \leq -\sqrt{d}y \leq x + (d+1)/4\}, \tag{3.15}$$

whose vertices are all on the circle $|z| = (d+1)/(4\sqrt{d})$. Thus $|z| \leq (d+1)/(4\sqrt{d})$ for every $z \in \mathcal{H}$.

LEMMA 3.5. *Define $\mathcal{F}$ by (3.11a) with $\mathcal{H}$ of (3.13b) as $\mathcal{P}$. Suppose $\mathfrak{z} = z + v\mathbf{j} \in \mathcal{F}$, $\mathfrak{z}' = z' + v'\mathbf{j} \in \mathcal{F}$, $v' \geq v$, and $\mathfrak{z}' = \gamma(\mathfrak{z})$ with $\gamma = \begin{bmatrix} p & q \\ r & s \end{bmatrix} \in \Delta_1$, $r \neq 0$. Then the following assertions hold:*

(i) *If $|r| = 1$, then $v' = v$, $|z| = |z + r^{-1}s| = |z'| = |z' - r^{-1}p|$, and $|\mathfrak{z}| = |\mathfrak{z}'| = 1$.*

(ii) *If $v > 1/\mu$, then $|r| = 1$, where $\mu = \mathrm{Min}\{|x| \mid x \in \mathfrak{r},\ |x| > 1\}$.*

Proof. Since $v \leq v' = v|r\mathfrak{z}+s|^{-2}$, we have $|r\mathfrak{z}+s| \leq 1$. Then $|rz+s|^2+|rv|^2 = |r\mathfrak{z} + s|^2 \leq 1$, so that $|rv| \leq 1$. If $v > 1/\mu$, then $|r| < \mu$. From our definition of μ we see that $|r| = 1$. This proves (ii). To prove (i), suppose $|r| = 1$ (without assuming $v > 1/\mu$); put $u = r^{-1}s$. Then $u \in \mathfrak{r}$ and

$$(*) \qquad |z + u|^2 + v^2 = |\mathfrak{z} + u|^2 = |r\mathfrak{z} + s|^2 \leq 1 \leq |\mathfrak{z}|^2 = |z|^2 + v^2,$$

and hence $|z+u| \leq |z|$. Since $z \in \mathcal{H}$, our definition of $\mathcal{H}$ implies that $|z+u| = |z|$, which together with $(*)$ shows that $|\mathfrak{z}| = |r\mathfrak{z} + s| = 1$, so that $v' = v$. We have $\mathfrak{z} = \gamma^{-1}(\mathfrak{z}')$ with $\gamma^{-1} = \begin{bmatrix} s & -q \\ -r & p \end{bmatrix}$. Thus, exchanging $\mathfrak{z}$ and $\mathfrak{z}'$, we obtain $|\mathfrak{z}'| = 1$ and $|z| = |z'| = |z' - r^{-1}p|$. This completes the proof. □

3.6. Let us now consider the case $d = 3$. In this case (3.15) takes the form

$$(3.15\text{a}) \quad \mathcal{H} = \{x + iy \in \mathbf{C} \mid |x| \leq 1/2, x - 1 \leq \sqrt{3}y \leq x + 1,\ -x - 1 \leq \sqrt{3}y \leq 1 - x\}.$$

This is a regular hexagon with 0 as its center and $i/\sqrt{3}$ as one of its vertices. Therefore $|z|^2 \leq 1/3$ for every $z \in \mathcal{H}$. Define $\mathcal{F}$ by (3.11a) with $\mathcal{H}$ as $\mathcal{P}$. If $\mathfrak{z} = z + v\mathbf{j} \in \mathcal{F}$, then $|z|^2 + v^2 \geq 1$, and hence $v^2 \geq 1 - |z|^2 \geq 2/3$. Thus

$$(3.16) \qquad z + v\mathbf{j} \in \mathcal{F} \Longrightarrow |z|^2 \leq 1/3 \quad \text{and} \quad v^2 \geq 2/3.$$

Let us hereafter put $\omega = (1 + \sqrt{-3})/2$. This is a primitive sixth root of unity. Then for $\alpha = \mathrm{diag}[\omega, \omega^{-1}]$, $z \in \mathbf{C}$ and $v > 0$ we have $\alpha(z + v\mathbf{j}) = \omega^2 z + v\mathbf{j}$, and hence α maps $\mathcal{F}$ onto itself. Thus, if we define a lozenge $\mathcal{L}$ by

$$(3.17) \qquad \mathcal{L} = \{x + iy \in \mathcal{H} \mid x \geq 0, \quad \sqrt{3}y \geq -x\},$$

then $\Delta_1 \backslash S$ can be represented by

$$(3.18) \qquad \mathcal{F}_1 = \{\mathfrak{z} = z + v\mathbf{j} \in S \mid z \in \mathcal{L},\ |\mathfrak{z}| \geq 1\}.$$

Now the vertices of $\mathcal{L}$ are

$$w_0 = 0, \quad w_1 = 2^{-1} - i/(2\sqrt{3}), \quad w_2 = 2^{-1} + i/(2\sqrt{3}), \quad w_3 = i/\sqrt{3}. \tag{3.19}$$

Denoting by $[w_iw_j]$ the closed line segment whose end points are w_i and w_j, put

$$\mathcal{L}_0 = \{w_0, w_2\} \cup \{z \in \mathcal{L} \mid z \notin [w_0w_3] \cup [w_2w_3]\}. \tag{3.20}$$

This is the union of $[w_0w_1]$, $[w_1w_2]$, and the interior of $\mathcal{L}$. We also need the regular triangle $\mathcal{T}$ whose vertices are w_0, w_1, and w_2. This can be given in the form

$$\mathcal{T} = \{x + iy \in \mathcal{L}_0 \mid \sqrt{3}y \le x\}. \tag{3.21}$$

LEMMA 3.7. (i) *Let A be the group of transformations $z \mapsto \omega^{2m}z+s$ with $m \in \mathbf{Z}$ and $s \in \mathfrak{r}$. Then $\mathcal{L}_0$ represents $A\backslash\mathbf{C}$, and no two distinct points of $\mathcal{L}_0$ are equivalent under A.*

(ii) *For $0 \ne z \in \mathcal{T}$ and $u \in \mathfrak{r}$ we have $|z + u| = |z|$ only in the following three cases:* (a) $u = -1$ *and* $z \in [w_1w_2]$; (b) $u = \omega^2$ *and* $z = w_1$; (c) $u = \omega^4$ *and* $z = w_2$.

Proof. The map $z \mapsto \omega^2 z+1$ sends $[w_2w_3]$ onto $[w_2w_1]$; also $z \mapsto \omega^2 z$ sends $[w_0w_1]$ onto $[w_0w_3]$. Thus $\mathcal{L}_0$ represents $A\backslash\mathbf{C}$. Suppose $z, z' \in \mathcal{L}_0$ and $z' = \omega^{2m}z+s$ with $m \in \mathbf{Z}$ and $s \in \mathfrak{r}$. Our task is to show that $z' = z$. The matter is simple if $s = 0$. For $0 \ne t \in \mathfrak{r}$ the map $z \mapsto z + t$ sends the interior of $\mathcal{H}$ to the outside of $\mathcal{H}$, and so $\omega^{2m}z + s \in \mathcal{L}_0$ with $s \ne 0$ only when z lies on the boundary of $\mathcal{H}$, that is, $z \in [w_1w_2]$. We easily see that this is so only when z is w_1 or w_2, and $z' = z$ in both cases. This proves (i). As for (ii), looking at the translations of $\mathcal{T}$ under $z \mapsto z + u$ for all $u \in \mathfrak{r}$, we can easily verify the desired fact. □

Let us now put

$$\mathcal{F}_0 = \{\mathfrak{z} = z + v\mathbf{j} \in S \mid z \in \mathcal{L}_0,\ |\mathfrak{z}| > 1\} \cup \{\mathfrak{z} = z + v\mathbf{j} \in S \mid z \in \mathcal{T},\ |\mathfrak{z}| = 1\}. \tag{3.22}$$

THEOREM 3.8. *For $d = 3$ the set $\mathcal{F}_0$ of (3.22) represents $\Delta_1\backslash S$, and no two distinct points of $\mathcal{F}_0$ are equivalent under Δ_1.*

Proof. We have seen that $\mathcal{F}_1$ represents $\Delta_1\backslash S$. Let $\alpha = \begin{bmatrix} \omega^m & \omega^{-m}s \\ 0 & \omega^{-m} \end{bmatrix}$ with $m \in \mathbf{Z}$ and $s \in \mathfrak{r}$; let $\mathfrak{z} = z + v\mathbf{j} \in \mathcal{F}_1$. Then $\alpha(\mathfrak{z}) = \omega^{2m}z + s + v\mathbf{j}$. Therefore, by Lemma 3.7(i) we can replace $\mathcal{L}$ by $\mathcal{L}_0$. Suppose $z \in \mathcal{L}_0$ and $|\mathfrak{z}| = 1$; let $\varepsilon = \begin{bmatrix} 0 & -\omega^{-1} \\ \omega & 0 \end{bmatrix}$. Then $\varepsilon(\mathfrak{z}) = -\omega^{-1}\bar{\mathfrak{z}}\omega^{-1} = \omega\bar{z} + v\mathbf{j}$ and $|\varepsilon(\mathfrak{z})| = 1$. If $z \notin \mathcal{T}$, then we see that $\omega\bar{z} \in \mathcal{T}$. This shows that $\mathcal{F}_0$ represents $\Delta_1\backslash S$.

Suppose $\mathfrak{z} = z + v\mathbf{j} \in \mathcal{F}_0$, $\mathfrak{z}' = z' + v'\mathbf{j} \in \mathcal{F}_0$, and $\mathfrak{z}' = \gamma(\mathfrak{z})$ with $\gamma = \begin{bmatrix} p & q \\ r & s \end{bmatrix} \in \Delta_1$. Our task is to show that $\mathfrak{z}' = \mathfrak{z}$. If $r = 0$, then $p \in \mathfrak{r}^\times$, and so $z' = p^2 z + pq$ and $v' = v$. By Lemma 3.7(i) we have $z' = z$, so that $\mathfrak{z}' = \mathfrak{z}$. Thus we assume that $r \neq 0$. Since $\gamma^{-1} = \begin{bmatrix} s & -q \\ -r & p \end{bmatrix}$ we may assume, changing $\mathfrak{z}$ for $\mathfrak{z}'$ if necessary, that $v' \geq v$. Now μ of Lemma 3.5(ii) is $\sqrt{3}$ in the present case. By (3.16), $v^2 \geq 2/3 > 1/3$, and so Lemma 3.5 is applicable. Therefore $|r| = |\mathfrak{z}| = |\mathfrak{z}'| = 1$, $v = v'$, and $|z + u| = |z|$ with $u = r^{-1}s$. In view of our definition of $\mathcal{F}_0$, we have $z, z' \in \mathcal{T}$.

Our verification of $\mathfrak{z} = \mathfrak{z}'$ depends on the vanishing or nonvanishing of p and s. First suppose $p = s = 0$; then $\gamma = \begin{bmatrix} 0 & -r^{-1} \\ r & 0 \end{bmatrix}$, and $z' = -r^{-2}\bar{z}$. Looking at the images of $\mathcal{T}$ under $z \mapsto -\bar{z}$ and $z \mapsto r^2 z$, we find that $z' = -r^{-2}\bar{z}$ can happen only in the following three cases: (1) $z \in [w_0 w_1]$ and $r^2 = \omega^4$; (2) $z \in [w_0 w_2]$ and $r^2 = \omega^2$; (3) $z = 0$ and $r^2 = 1$. In all cases $z' = z$ as expected.

Next suppose $s \neq 0$ and $p = 0$; put $u = r^{-1}s$ as above. Then $\gamma = \begin{bmatrix} 0 & -r^{-1} \\ r & ru \end{bmatrix}$, $|z + u| = |z|$, $|\mathfrak{z} + u| = 1$, and $z' = -r^{-2}(\bar{z} + \bar{u})$. By Lemma 3.7(ii) there are three possibilities for such (z, u). Checking those cases one by one, we see that $z' = z$.

The case $p \neq 0$ and $s = 0$ can be handled in the same way by exchanging $\mathfrak{z}$ and $\mathfrak{z}'$. Therefore it remains to examine the case $ps \neq 0$. By Lemma 3.5 we have $|r| = 1$, $|z + u| = |z|$ with $u = r^{-1}s$, and $|z' + y| = |z'|$ with $y = -r^{-1}p$. Thus $\gamma = \begin{bmatrix} -ry & q \\ r & ru \end{bmatrix}$, and so $q = -r^{-1} - ryu$. Since $|r\mathfrak{z} + ru| = 1$, we have $(r\mathfrak{z} + ru)^{-1} = \bar{\mathfrak{z}}\bar{r} + \overline{ru}$, and hence $\mathfrak{z}' = (-ry\mathfrak{z} - r^{-1} - ryu)(\bar{\mathfrak{z}}\bar{r} + \overline{ru})$. Now from the equality $|z + u| = |z|$ we obtain $\bar{u}z + u\bar{z} + u\bar{u} = 0$. Therefore a direct calculation shows that $z' + y = -r^{-2}(\bar{z} + \bar{u})$. The pairs (z, u) and (z', y) belong to the three types of Lemma 3.7(ii), and so there are 6 combinations. Checking these one by one, we find that $z' + y = -r^{-2}(\bar{z} + \bar{u})$ only when $z' = z$. This completes the proof. □

3.9. The case $d = 4$ can be treated in a similar but somewhat different manner. In this case $\mathcal{H}$ coincides with the rectangle of (3.13a), which is a square. Put $w_0 = 0$, $w_1 = 1/2$, $w_2 = (1 + i)/2$, $w_3 = i/2$, $w_4 = (i - 1)/2$, $w_5 = -1/2$, and

(3.23a) $$\mathcal{H}_0 = \{w_0, w_3\} \cup \{z \in \mathcal{H} \mid \mathrm{Im}(z) \geq 0,\ z \notin [w_3 w_4] \cup [w_4 w_5] \cup [w_5 w_0]\},$$

(3.23b) $$\mathcal{H}_1 = \{z \in \mathcal{H}_0 \mid \mathrm{Re}(z) \geq 0\},$$

(3.23c) $$\mathcal{F}_0 = \{\mathfrak{z} = z + v\mathbf{j} \in S \mid z \in \mathcal{H}_0,\ |\mathfrak{z}| > 1\} \cup \{\mathfrak{z} = z + v\mathbf{j} \in S \mid z \in \mathcal{H}_1,\ |\mathfrak{z}| = 1\}.$$

Notice that $|z|^2 \le 1/2$ for every $z \in \mathcal{H}$, and so $v^2 \ge 1/2$ if $\mathfrak{z} = z + v\mathbf{j} \in \mathcal{F}_0$. Let $\gamma = \begin{bmatrix} t & t^{-1}s \\ 0 & t^{-1} \end{bmatrix}$ with $t \in \mathfrak{r}^\times$ and $s \in \mathfrak{r}$. Then for $\mathfrak{z} = z + v\mathbf{j} \in S$ we have $\gamma(\mathfrak{z}) = t^2 z + s + v\mathbf{j}$. Observe that every point of $\mathcal{H}$ can be transformed to a point of $\mathcal{H}_0$ by $z \mapsto -z$, or $z \mapsto z + 1$, or $z \mapsto i - z$, or by a combination of these maps. Thus, for the purpose of representing $\Delta_1 \backslash S$, we can replace $\mathcal{P}$ by $\mathcal{H}_0$ in (3.11a). Suppose $\mathfrak{z} = z + v\mathbf{j} \in S$, $z \in \mathcal{H}_0$, and $|\mathfrak{z}| = 1$; let $\delta = \begin{bmatrix} 0 & -1 \\ 1 & 0 \end{bmatrix}$. Then $\delta(\mathfrak{z}) = -\mathfrak{z}^{-1} = -\bar{\mathfrak{z}} = -\bar{z} + v\mathbf{j}$. Therefore, changing z for $-\bar{z}$ if necessary, we may assume that $\mathrm{Re}\,(z) \ge 0$ if $|\mathfrak{z}| = 1$. This proves the first part of:

THEOREM 3.10. *For $d = 4$ the set $\mathcal{F}_0$ of (3.23c) represents $\Delta_1 \backslash S$, and no two distinct points of $\mathcal{F}_0$ are equivalent under Δ_1.*

Proof. This can be proved by the same technique as for Theorem 3.8, with some modifications. Suppose $\mathfrak{z} = z + v\mathbf{j} \in \mathcal{F}_0$, $\mathfrak{z}' = z' + v'\mathbf{j} \in \mathcal{F}_0$, and $\mathfrak{z}' = \gamma(\mathfrak{z})$ with $\gamma = \begin{bmatrix} p & q \\ r & s \end{bmatrix} \in \Delta_1$. If $r = 0$, then $p \in \mathfrak{r}^\times$, and so $p^2 = \pm 1$, $z' = p^2 z + pq$ and $v' = v$. Since $z, z' \in \mathcal{H}_0$, we easily see that $z' = z$, and so $\mathfrak{z}' = \mathfrak{z}$. Thus we assume that $r \ne 0$. Since $\gamma^{-1} = \begin{bmatrix} s & -q \\ -r & p \end{bmatrix}$ we may assume, changing $\mathfrak{z}$ for $\mathfrak{z}'$ if necessary, that $v' \ge v$. We have $v^2 \ge 1/2$ and $|rv|^2 \le |rz + s|^2 + |rv|^2 = |r\mathfrak{z} + s|^2 = v/v' \le 1$, and so $|r|^2 \le |v|^{-2} \le 2$. Since $r \in \mathfrak{r}$, we have $|r| = 1$ or $|r|^2 = 2$.

Let us first treat the case $|r|^2 = 2$. Then $v^2 = 1/2$, and so $rz + s = 0$ and $|r\mathfrak{z} + s| = 1$; consequently $v' = v$. Since $|z|^2 \le 1/2$ and $|z|^2 + v^2 = |\mathfrak{z}|^2 \ge 1$, we have $|z|^2 = 1/2$ and $|\mathfrak{z}| = 1$. Now $z = -r^{-1}s$, and so $|s| = |rz| = 1$. Thus there are exactly four possibilities for z: $z = (1 \pm i)/2$ and $(-1 \pm i)/2$. Since $|\mathfrak{z}| = 1$ and $\mathfrak{z} \in \mathcal{F}_0$, we have $\mathrm{Re}\,(z) \ge 0$ and $\mathrm{Im}\,(z) \ge 0$, and so $z = (1 + i)/2$. We have $\mathfrak{z} = \gamma^{-1}(\mathfrak{z}')$ and $v' = v$, and hence we can exchange $\mathfrak{z}$ and $\mathfrak{z}'$ and find that $z' = (1 + i)/2$. Thus $\mathfrak{z} = \mathfrak{z}'$, and the case $|r|^2 = 2$ is settled.

Next suppose $|r| = 1$; then Lemma 3.5(i) shows that $|r| = |\mathfrak{z}| = |\mathfrak{z}'| = 1$, $v = v'$, $|z + u| = |z|$ with $u = r^{-1}s$, and $|z' + y| = |z'|$ with $y = -r^{-1}p$. In view of our definition of $\mathcal{F}_0$, we have $z, z' \in \mathcal{H}_1$. Our verification of $\mathfrak{z} = \mathfrak{z}'$ can be done in the same way as in the case $d = 3$. If $p = s = 0$, then $z' = \bar{z}$ or $z' = -\bar{z}$; in either case we have $z' = z$. If $s \ne 0$, we have $|z + u| = |z|$ only in the following three cases: $z \in [w_1 w_2]$ and $u = -1$; $z \in [w_2 w_3]$ and $u = -i$; $z = w_2$ and $u = -1 - i$. Suppose $p = 0$; then $z' = \pm(\bar{z} + \bar{u})$. Then we can easily verify that $z' = z$ in the first two cases, and the third case cannot happen. The case $p \ne 0$ and $s = 0$ can be handled in the same way by exchanging $\mathfrak{z}$ and $\mathfrak{z}'$. It remains to examine the case $ps \ne 0$. Then the same calculation as in the case $d = 3$ shows that $z' + y = \pm(\bar{z} + \bar{u})$. The pairs (z, u) and (z', y) belong to the three types just mentioned. Checking all combinations, we always obtain $\mathfrak{z} = \mathfrak{z}'$. This completes the proof. □

THEOREM 3.11. *Given an imaginary quadratic field* $K = \mathbf{Q}(\sqrt{-d})$ *as in §3.2 and a positive integer* q, *define a quaternion algebra* R *over* $\mathbf{Q}$ *by* $R = K + \varepsilon K$ *with an element* ε *such that* $\varepsilon^2 = -q$ *and* $\varepsilon a = a^\rho \varepsilon$ *for every* $a \in K$; *put* $\mathfrak{o} = \mathfrak{r} + \varepsilon\mathfrak{r}$. *Then* $\#\{L_+[q,\, 2^{-1}\mathbf{Z}]/\Delta_1\}$ *equals the class number of* $\mathfrak{o}$. *Moreover, if* $d = 3$ *or* 4, *then it equals* $\#\{h \in L_+[q,\, 2^{-1}\mathbf{Z}] \mid \zeta(h) \in \mathcal{F}_0\}$, *where* $\mathcal{F}_0$ *is defined by (3.22) or (3.23c) according as* $d = 3$ *or* $d = 4$.

Proof. The first assertion is merely a special case of Theorem 2.10, since we can replace $L[q, 2^{-1}\mathbf{Z}]$ by $L_+[q, 2^{-1}\mathbf{Z}]$ as observed in §3.2. The map ζ of P_q onto S commutes with the action of Δ_1, and therefore we obtain the second assertion. □

If dq (with an arbitrary $\mathbf{Q}(\sqrt{-d})$) is squarefree, then $\#\{L_+[q, 2^{-1}\mathbf{Z}]/\Delta_1\}$ can be given by (1.12). We need the decomposition $dq = tr$, which can be done easily by the ideas of Lemma 2.6.

3.12. If $d > 4$, it is not easy to find a simple subset of S equivalent to $\Delta_1\backslash S$. Even so, we can find in many cases a complete set of representatives for $L_+[q, 2^{-1}\mathbf{Z}]/\Delta_1$, and hence the class number of $\mathfrak{o}$, at least when d and q are small. Let us first state a few useful inequalities. Take $\mathcal{H}$ of (3.13b); define $\mathcal{F}$ by (3.11a) with $\mathcal{H}$ as $\mathcal{P}$, and $\mathcal{F}'$ by (3.11b). Let $M = \mathrm{Max}_{z\in\mathcal{H}}|z|^2$. Take $h \in L_+[q, 2^{-1}\mathbf{Z}]$ as in (3.5) with $q \in \mathbf{Z}$, and put $e = -b/\sqrt{-d}$. Then $e \in \mathfrak{r}$ and $q = ac - |e|^2$. Suppose $\zeta(h) = z + v\mathbf{j} \in \mathcal{F}'$; then by (3.9) we have $a/c = |\zeta(h)|^2 \geq 1$ and $e/c \in \mathcal{H}$, and hence $|e/c|^2 \leq M$; also, $q/c^2 = v^2 \geq \pi^2/8d$ if K has class number 1. Thus

$$|\,\mathrm{Re}\,(e/c)| \leq 1/2, \quad |e|^2 \leq Mc^2, \quad \text{and} \quad a \geq c; \tag{3.24a}$$

$$c^2 \leq 8dq/\pi^2 \quad \text{if } K \text{ has class number 1.} \tag{3.24b}$$

If $d \leq 11$, then $1 - M > 0$, and $q/c^2 = v^2 = |\mathfrak{z}|^2 - |z|^2 \geq 1 - M$. Thus

$$c^2 \leq q/(1 - M) \quad \text{if} \quad d \leq 11. \tag{3.24c}$$

For $d \leq 11$ we see that (3.24c) is stronger than (3.24b), which means that $\mathcal{F} = \mathcal{F}'$ if $d \leq 11$. We have $|e|^2 \leq 8dMq/\pi^2$ if K has class number 1, and $|e|^2 \leq qM/(1 - M)$ if $d \leq 11$. Therefore, once q is fixed, there is only a finite number of values for $|e|$ and c. This fact combined with the conditions $q + |e|^2 = ac$ and $e/c \in \mathcal{H}$ restricts (a, c, e) to a relatively small set.

Let us now consider the case $d = 3$. To determine $h \in L_+[q, 2^{-1}\mathbf{Z}]$ such that $\zeta(h) \in \mathcal{F}_0$, write h in the form (3.5) and put $b = \beta_1 - \beta_2\omega$ with $\beta_1, \beta_2 \in \mathbf{Q}$ and $\omega = (1 + \sqrt{-3})/2$. Then $\mathrm{Tr}_{K/\mathbf{Q}}(b\mathfrak{d}^{-1}) = \beta_1\mathbf{Z} + \beta_2\mathbf{Z}$. From (3.9) we see that $ib/(c\sqrt{3}) \in \mathcal{L}_0$ if $a > c$, and $ib/c\sqrt{3} \in \mathcal{T}$ if $a = c$. The map $z \mapsto -\sqrt{-3}z$ sends

$\mathcal{L}$ onto the lozenge $\{x - y\omega \mid 0 \le x \le 1, 0 \le y \le 1\}$. Therefore $h \in L_+[q, 2^{-1}\mathbf{Z}]$ and $\zeta(h) \in \mathcal{F}_0$ if and only if the following conditions are satisfied:

(3.25a) $$a\mathbf{Z} + c\mathbf{Z} + \beta_1\mathbf{Z} + \beta_2\mathbf{Z} = \mathbf{Z};$$

(3.25b) $$3q = 3ac - |b|^2, \quad |b| \le c \le a, \quad c^2 \le 3q/2;$$

(3.25c) $$\beta_1^2 - \beta_1\beta_2 + \beta_2^2 = |b|^2, \quad 0 \le \beta_1 \le c, \quad 0 \le \beta_2 \le c;$$

(3.25d) $$\beta_1 = 0 \text{ if } a > c \text{ and } \beta_2 = 0; \ \beta_2 = c \text{ if } a > c \text{ and } \beta_1 = c;$$

(3.25e) $$\beta_1 \le \beta_2 \text{ if } a = c.$$

In particular, if $q \in \mathbf{Z}$, then we can put $b = -e\sqrt{-3}$ with $e \in \mathfrak{r}$. Since $M = 1/3$, from (3.24a, c) we obtain $3|e|^2 \le c^2 \le 3q/2$.

In the case $d = 4$ we can similarly state the conditions for (a, b, c) such that $\zeta(h)$ belongs to $\mathcal{F}_0$ of (3.23c), but we leave the details to the reader. In either case it should be easy to make a relatively simple computer program by which we can determine the class number.

Example 1. $d = 3$, $q = 9$. From (3.24a, c) and the equality $ac = q+|e|^2$ we see that the only possible $(a, c, |e|)$ are $(9, 1, 0)$, $(5, 2, 1)$, and $(4, 3, \sqrt{3})$. Checking the above conditions, we find that (a, c, b) must be $(9, 1, 0)$, $(5, 2, \sqrt{-3})$, $(4, 3, -3\omega)$, or $(4, 3, 3 - 3\omega)$, which form a complete set of representatives for primitive hermitian forms in this case. Thus the class number of $\mathfrak{o}$ is 4. This case cannot be covered by Eichler's formula (1.12), since the discriminant of $\mathfrak{o}$ is 27.

Example 2. $d = 3$, $q = 19$. We have $3|e|^2 \le c^2 \le 57/2$, and so $c \le 5$. Then we easily see that $(19, 1, 0)$, $(10, 2, 1)$, and $(5, 4, 1)$ are the only possible $(a, c, |e|)$. Checking the above conditions, we find that $(19, 1, 0)$, $(10, 2, 1)$, $(5, 4, 1)$, and $(5, 4, \omega)$ as (a, c, e) give a complete set of representatives for primitive hermitian forms in this case. Thus the class number of $\mathfrak{o}$ is 4, which agrees with what is determined by (1.12) with $t = 3$ and $r = 19$.

Example 3. $d = 4$, $q = 5$. In this case we can put $b = -e\sqrt{-4}$ with $e \in \mathfrak{r}$.Then the only possible $(a, c, |e|)$ are $(5, 1, 0)$, $(3, 2, 1)$, and $(3, 3, 2)$. Checking the condition $ie/2 \in \mathcal{H}_0$ in the second case, and $ib/6 \in \mathcal{H}_1$ in the third case, we eventually find that $(5, 1, 0)$, $(3, 2, 1)$, $(3, 2, i)$, $(3, 3, 2)$, and $(3, 3, 2i)$ as (a, c, e) give a complete set of representatives for primitive hermitian forms in this case. Thus the class number of $\mathfrak{o}$ is 5.

3.13. Let us next consider the case $d > 4$. If $a = c$, then $h \mapsto \delta h \delta^\sigma$ with $\delta = \begin{bmatrix} 0 & -1 \\ 1 & 0 \end{bmatrix}$ changes (a, b, a) into $(a, -\bar{b}, a)$. This corresponds to $\mathfrak{z} \mapsto -\mathfrak{z}^{-1}$ on

S. Thus, in the case $a = c$, we can reduce $\mathcal{H}$ to

$$\mathcal{H}_1 = \{z \in \mathcal{H} \mid \mathrm{Re}\,(z) \geq 0\}. \tag{3.26}$$

In this way we can restrict (a, c, e) to those satisfying (3.24a, b, c) and

$$a\mathbf{Z} + c\mathbf{Z} + \mathrm{Tr}_{K/\mathbf{Q}}(e\mathfrak{r}) = \mathbf{Z}; \tag{3.27a}$$

$$a \geq c, \quad ac = q + |e|^2, \quad e/c \in \mathcal{H}; \tag{3.27b}$$

$$e/c \in \mathcal{H}_1 \ \text{ if } \ a = c. \tag{3.27c}$$

Example 4. $d = 11$, $q = 4$. We have $M = 9/11$. By (3.24c), $c^2 \leq 22$, and so $c \leq 4$; hence $|e|^2 \leq 144/11$ by (3.24a). Thus $(4, 1, 0)$, $(3, 3, \sqrt{5})$, and $(4, 4, \sqrt{12})$ are the only possible values for $(a, c, |e|)$. The last case must be eliminated, as it does not satisfy (3.27a). In the case $a = c = 3$ we have $e/3 \in \mathcal{H}_1$. Since $|(1 + \sqrt{-11})/2|^2 = 3$, μ of Lemma 3.5(ii) is $\sqrt{3}$ in the present case. Now $\zeta(h) = c^{-1}(e + 2\mathbf{j})$, to which Lemma 3.5(ii) is applicable if $c = 3$. Therefore we see that $(4, 1, 0)$ is not equivalent to $(3, 3, e)$. Since $|e|^2 = 5$ and $e/3 \in \mathcal{H}_1$, e must be $(3 \pm \sqrt{-11})/2$. Put $e = (3 + \sqrt{-11})/2$, $e' = (3 - \sqrt{-11})/2$, $z = e/3$, $z' = e'/3$, $\mathfrak{z} = z + (2/3)\mathbf{j}$, and $\mathfrak{z}' = z' + (2/3)\mathbf{j}$; suppose $\mathfrak{z}' = \gamma(\mathfrak{z})$ with $\gamma = \begin{bmatrix} p & * \\ r & s \end{bmatrix} \in \Delta_1$. If $r = 0$, then $\gamma = \pm \begin{bmatrix} 1 & t \\ 0 & 1 \end{bmatrix}$ with $t \in \mathfrak{r}$ and $z' = z + t$, which cannot happen, as $z' - z = -\sqrt{-11}/3 \notin \mathfrak{r}$. Therefore $r \neq 0$. By Lemma 3.5 we have $r = \pm 1$. Changing γ for $-\gamma$ if necessary, we may assume that $r = 1$. Suppose $s = 0$. Then $\gamma = \begin{bmatrix} p & -1 \\ 1 & 0 \end{bmatrix}$, and $\gamma(\mathfrak{z}) = (p\mathfrak{z} - 1)\bar{\mathfrak{z}} = p - \bar{\mathfrak{z}}$, as $|\mathfrak{z}| = 1$. Thus $z' = p - \bar{z}$, which is impossible, since $z' + \bar{z} \notin \mathfrak{r}$. Thus $s \neq 0$, and $|z + s| = |z|$ by Lemma 3.5(i). This can happen only if $s = -1$. Similarly exchanging $(\mathfrak{z}, \gamma)$ and $(\mathfrak{z}', -\gamma^{-1})$, we see that $p = 1$, and so $\gamma = \begin{bmatrix} 1 & -2 \\ 1 & -1 \end{bmatrix}$. Since $|\mathfrak{z}| = |\mathfrak{z} - 1| = 1$, we have $\gamma(\mathfrak{z}) = (\mathfrak{z} - 2)(\bar{\mathfrak{z}} - 1) = 3 - z - 2\bar{z} + (2/3)\mathbf{j}$, which is not equal to $\mathfrak{z}'$. Therefore $\mathfrak{z}$ and $\mathfrak{z}'$ are not equivalent under Δ_1.

Thus $(4, 1, 0)$, $(3, 3, (3 + \sqrt{-11})/2)$, and $(3, 3, (3 - \sqrt{-11})/2)$, as (a, c, e), give inequivalent representatives for $L_+[4, 2^{-1}\mathbf{Z}]/\Delta_1$, and consequently the class number of $\mathfrak{o}$ in this case is 3. Since the discriminat of $\mathfrak{o}$ is 44, this case cannot be covered by Eichler's formula (1.12).

Example 5. $d = 19$, $q = 3$. We can take at least $(3, 1, 0)$, $(2, 2, 1)$, and $(4, 2, f)$ with $f = (1 \pm \sqrt{-19})/2$ as possible (a, c, e). For $K = \mathbf{Q}(\sqrt{-19})$ the number μ of Lemma 3.5(ii) is 2, and so we have $\sqrt{q}/c > 1/\mu$ for all these four choices.

Suppose one of these with $c = 2$ is equivalent to another under $\gamma = \begin{bmatrix} * & * \\ r & * \end{bmatrix} \in \Delta_1$. It can easily be seen that $r \neq 0$, and hence by Lemma 3.5 we have $|r| = 1$ and $a/c = 1$, which is so only for $(2, 2, 1)$. Therefore the above four are inequivalent under Δ_1. In this case our quaternion algebra is ramified only at 3, and so $\mathfrak{o}$ is of the same type as in Example 2. We have seen there that the classs number is 4. Thus the above four form a complete set of representatives for primitive hermitian forms in this case.

4. Definite quaternary forms.

4.1. We shall eventually consider a definite quaternary form over $\mathbf{Q}$ whose discriminant is not a square, as we said after Theorem 1.11, but we start with a somewhat general setting. We take a quaternion algebra B over F and (K, ρ) as in §2.1; we then put

$$\mathfrak{B} = B \otimes_F K. \tag{4.1}$$

We denote by ι the K-linear extension of the main involution of B to $\mathfrak{B}$. We identify K with the subset $F + F\varepsilon$ of $M_2(F)$ with ε given by

$$\varepsilon = \begin{bmatrix} 0 & 1 \\ \gamma & 0 \end{bmatrix}. \tag{4.2}$$

Here γ is an element of $F^\times$, chosen so that $F + F\varepsilon$ is isomorphic to K. If K is a field, then K is isomorphic to $F(\gamma^{1/2})$. We identify B with the set of all $b1_2 \in M_2(B)$ for $b \in B$. Then $\mathfrak{B} = B \otimes_F K = B + B\varepsilon$, which can be identified with the set of elements $b + c\varepsilon = \begin{bmatrix} b & c \\ \gamma c & b \end{bmatrix}$ of $M_2(B)$ with $b, c \in B$.

Define an involution $x \mapsto x^*$ of $M_2(B)$ by

$$\begin{bmatrix} p & q \\ r & s \end{bmatrix}^* = \begin{bmatrix} 1 & 0 \\ 0 & \gamma \end{bmatrix} \begin{bmatrix} p^\iota & r^\iota \\ q^\iota & s^\iota \end{bmatrix} \begin{bmatrix} 1 & 0 \\ 0 & \gamma \end{bmatrix}^{-1} = \begin{bmatrix} p^\iota & \gamma^{-1} r^\iota \\ \gamma q^\iota & s^\iota \end{bmatrix} \tag{4.3}$$

for $p, q, r, s \in B$. Then $(b + c\varepsilon)^* = b^\iota + c^\iota \varepsilon$, so that $x^* = x^\iota$ for $x \in \mathfrak{B}$. Put now

$$T = \{x \in B \mid x^\iota = -x\}, \tag{4.4}$$

$$V = \left\{ \begin{bmatrix} a & x \\ \gamma x^\iota & -a \end{bmatrix} \mid a \in F,\ x \in T \right\} = F\delta + \delta\varepsilon T, \qquad \delta = \mathrm{diag}[1, -1]. \tag{4.5}$$

We can define a quadratic form φ on V by $\varphi[w]1_2 = w^2$ for $w \in V$, or rather,

$$\varphi[a\delta + \delta\varepsilon x] = a^2 + \gamma x x^\iota \quad \text{for } a \in F \text{ and } x \in T. \tag{4.6}$$

Now we can easily verify that V and F generate the whole $M_2(B)$. Thus we can identify $A(V, \varphi)$ with $M_2(B)$. Moreover, the canonical involution of $A(V)$ coincides with what we defined in (4.3), since the fact is true for every element of V. We also see that K is the discriminant algebra of (V, φ) in the sense of [S4, §1.6].

Let $x \mapsto x'$ be the canonical automorphism of $A(V)$. Then $x' = \varepsilon x \varepsilon^{-1}$ for every $x \in A(V)$, since the equality holds for every element of V. Therefore $A^+(V) = \{x \in A(V) \mid x\varepsilon = \varepsilon x\}$. Then we see that

$$A^+(V) = B + B\varepsilon = \mathfrak{B}. \tag{4.7}$$

In [S4, (7.3b)] we showed that $G^+(V) = \{\alpha \in A^+(V) \mid \alpha\alpha^* \in F^\times\}$ whenever $\dim(V) = 4$. We have also seen that $\alpha^* = \alpha^\iota$ for $\alpha \in \mathfrak{B}$. Therefore we obtain

$$G^+(V) = \{\alpha \in \mathfrak{B}^\times \mid \alpha\alpha^\iota \in F^\times\}, \tag{4.8}$$

and $\nu(\alpha) = \alpha\alpha^* = \alpha\alpha^\iota$ for $\alpha \in G^+(V)$. Hereafter we put $G = G^+(V)$.

Our (V, φ) may look special, but it is not so. In fact, an arbitrary (V_1, φ_1) with $\dim(V_1) = 4$ is isomorphic to $(V, c\varphi)$ with some $c \in F^\times$ and (V, φ) defined as above. The proof can be given by employing [S4, Theorems 7.4c and 7.5(i)]; see also [S4, Theorem 7.13]. This gives a conceptual raison d'être for our (V, φ). Also, taking $B = M_2(F)$, we obtain (V, φ) of (2.1b) as a special case.

4.2. By the *core dimension* of (V, φ) we mean the dimension of a subspace Z of V such that V is the direct sum of Z and a hyperbolic subspace; see [S4, p. 11]. If K is a local quadratic extension of F, then the core dimension is 2, since K is the discriminant field of φ. If K is isomorphic to $F \times F$, then (4.6) shows that φ is equivalent to the norm form of B. Therefore φ is anisotropic if and only if K is not a field and B is a division algebra.

We now take K to be a quadratic extension of a global field F. Denote by $\mathfrak{r}$ the maximal order of K. For $v \in \mathbf{h}$ let t_v be the core dimension of $(V, \varphi)_v$. From the above observations we see that $t_v = 2$ if and only if K_v is a field; $t_v = 4$ if and only if v splits in K and B_v is a division algebra; $t_v = 0$ in the remaining cases.

Fixing a $\mathfrak{g}$-maximal lattice L in V, define C by (1.2) and J by (1.7). Take $q \in \mathfrak{g} \cap F^\times$ and $h \in L[q, 2^{-1}\mathfrak{g}]$; put $W = (Fh)^\perp$. Our immediate problem is the structure of $A(W)$.

Suppose $t_v = 4$; then φ_v is equivalent to the norm form of B_v, and so $A^+(W)_v$ is given by (1.15), which is isomorphic to B_v; see also [S4, §7.3].

Suppose $t_v \neq 4$. Then we have a weak Witt decomposition $V_v = F_v e + F_v f + K_v g$ such that $\varphi[ae + cf + bg] = ac + d_v bb^\rho$ for $a, c \in F_v$ and $b \in K_v$. Here $d_v \in \mathfrak{g}_v^\times$ except when K_v is an unramified quadratic extension of F_v, in which case $d_v = 1$ or d_v is a prime element of F_v. For the purpose of determining the structure of $A^+(W)_v$ we can take $L_v = \mathfrak{g}_v e + \mathfrak{g}_v f + \mathfrak{r}_v g$ and $h = qe + f$. (The set $K_v g$ is defined only locally; there is nothing global whose localization becomes $K_v g$.) Then $W_v = F_v(qe - f) + K_v g$. We have $A^+(K_v g) = K_v$, $A^-(K_v g) = K_v g$, and $A(K_v g) = K_v + K_v g$, which is a quaternion algebra. (For these, see [S4, §7.2].) As explained in [S4, §2.4], there is an isomorphism Ψ of $A(V)_v$ onto $M_2(K_v + K_v g)$ such that $\Psi(h) = \begin{bmatrix} 0 & q \\ 1 & 0 \end{bmatrix}$ and $\Psi(b) = \mathrm{diag}\,[b, b]$ for every $b \in K_v$. By Lemma 1.2 and [S4, (2.13)] we can easily verify that $\Psi(A^+(W)_v) = K_v + K_v \eta_v$, where $\eta_v = \Psi(h)\,\mathrm{diag}\,[g, g]$ and we identify $\Psi(K_v)$ with K_v. Notice that $\eta_v^2 = d_v q$ and $\eta_v b = b^\rho \eta_v$ for every $b \in K_v$. This much is true even for $v \in \mathbf{a}$ if $t_v \neq 4$. Thus $A^+(W)_v$ for every $v \in \mathbf{v}$ can be determined.

For $v \in \mathbf{h}$ let $\mathfrak{r}_v = \mathfrak{g}_v + \mathfrak{g}_v \xi$ with $\xi \in \mathfrak{r}_v$, and let φ_v be the matrix representing φ with respect to $\{e, f, g, \xi g\}$. Then we easily find that

$$d_v^2(\xi - \xi^\rho)^2 = 2^4 \det(\varphi_v). \tag{4.9}$$

4.3. We define a subgroup Q of $G_{\mathbf{A}}$ by (2.21) for the present G with $\omega_v \in G_v$ chosen suitably for each v as described in [S4, Theorem 8.9]; then $\tau(Q) = C$ and $[C_v : \tau(J_v)] \leq 2$. For the reason explained in §1.7, we have to check whether $hC = h\tau(J)$. Now $Q_v \neq J_v$ only in the following three cases: (i) $t_v = 4$; (ii) v is ramified in K; (iii) v remains prime in K and d_v is a prime element of F_v. In Case (i) we have seen that $A^+(W)_v$ is isomorphic to B_v. Therefore we can find an element ω_v of $A^+(W)_v^\times = G^+(W)_v$ such that $\nu(\omega_v)$ is a prime element of F_v. Then $h\tau(\omega_v) = h$ and Q_v is generated by J_v and ω_v. Thus $hC_v = h\tau(J_v)$. In Case (ii) we can take a prime element of K_v as ω_v, as we have seen in §2.7. Let $k_v = qe + f$; then from (2.20) we obtain $k_v \tau(\omega_v) = k_v$, so that $k_v C_v = k_v \tau(J_v)$. Since $k_v \in L_v[q, 2^{-1}\mathfrak{g}_v]$, we have $k_v \in hC_v$ by [S4, Theorem 10.5] and so $hC_v = h\tau(J_v)$ by (1.10).

In case (iii) put $\mathfrak{t}_v = \mathfrak{r}_v + \mathfrak{r}_v g$. Since $g^2 = d_v$ and d_v is a prime element of F_v, we see that $A(K_v g)$ is a division quaternion algebra over F_v, and $\mathfrak{t}_v$ is its maximal order. Let S_v be the order in $A(V)_v$ such that $\Psi(S_v) = M_2(\mathfrak{t}_v)$. This is the order defined in [S4, Theorem 8.6(i)], and so $J_v = S_v^\times \cap G^+(V)_v$ by [S4, (8.4b) and Proposition 8.8(ii)]. Put $k_v = qe + f$ and $\omega_v = g(qe - f)$. Then $\omega_v \in G_v$, $\nu(\omega_v) = -qd_v$, and $(ae + cf + bg)\tau(\omega_v) = qce + q^{-1}af - b^\rho g$ for $a, c \in F_v$ and $b \in K_v$. Suppose $q \in \mathfrak{g}_v^\times$; then $\tau(\omega_v) \in C_v$, but $\omega_v \notin F_v^\times J_v$, since $\nu(\omega_v)$ is a prime element. Therefore if we put $Q_v = \bigcup_{\nu \in \mathbf{Z}} \omega_v^\nu J_v$, then by [S4, Theorem 8.9], we have $\tau(Q_v) = C_v$. Now $k_v \in L_v[q, 2^{-1}\mathfrak{g}_v]$ and $k_v \tau(\omega_v) = k_v$, and hence we obtain $hC_v = h\tau(J_v)$ for the same reason as in case (ii). Thus we have proved:

(4.10) $$hC_v = h\tau(J_v) \text{ if } q \in \mathfrak{g}_v^\times \text{ in case (iii).}$$

4.4. We are going to apply Theorem 1.6 to the present setting. To make a clear-cut statement, we consider a special case in which $F = \mathbf{Q}$, $0 < \gamma \in \mathbf{Z}$, and B is a definite quaternion algebra over $\mathbf{Q}$ whose norm form is the sum of four squares. Thus $V = \mathbf{Q}^4$ and φ is a quaternary form $x_0^2 + \gamma(x_1^2 + x_2^2 + x_3^2)$. Let us first prove

(4.11) $$SO^\varphi(V)_\mathbf{A} = SO^\varphi(V)C \text{ and } G_\mathbf{A} = GJ \text{ if } \gamma = 2, 3, 5, \text{ or } 7.$$

We start with a five-dimensional space $X = \sum_{i=1}^5 \mathbf{Q}e_i$ and define a quadratic form ψ on X by $\psi[\sum_{i=1}^5 x_i e_i] = \sum_{i=1}^5 x_i^2$. Put $\Lambda_0 = \mathbf{Z}\ell + \sum_{i=2}^4 \mathbf{Z}e_i$ with $\ell = 2^{-1}\sum_{i=1}^4 e_i$ and $\Lambda = \Lambda_0 + \mathbf{Z}e_5$. We see that Λ_0 is a $\mathbf{Z}$-maximal lattice in $\sum_{i=1}^4 \mathbf{Q}e_i$, and so Λ is a $\mathbf{Z}$-maximal lattice in X by [S4, Lemma 11.14]; see [S4, §12.7].

Take a squarefree positive integer γ. Then $\Lambda[\gamma, \mathbf{Z}] \neq \emptyset$ by [S4, Lemma 12.13]. For any fixed $m \in \Lambda[\gamma, \mathbf{Z}]$ let V be the orthogonal complement of $\mathbf{Q}m$ in X. Then the restriction of ψ to V is represented by $\mathrm{diag}[1, \gamma 1_3]$ and $\Lambda \cap V$ is $\mathbf{Z}$-maximal in V; see [S4, Theorem 12.14(i), (ii)]. Thus we can take this V as our space and put $L = \Lambda \cap V$. Now 2 is the only nonarchimedean prime where B is ramified. For $\gamma = 2, 3, 5$, or 7, the prime 2 does not split in $K = \mathbf{Q}(\sqrt{\gamma})$. Thus $t_p = 4$ cannot happen; $t_p = 0$ if p splits in K and $t_p = 2$ otherwise.

Moreover, by [S4, Theorem 12.14(v), (vi)], the class number of $SO^\varphi(V)$ with respect to C equals $\lambda = \#\{\Lambda[\gamma, \mathbf{Z}]/\Gamma(\Lambda)\}$, and the idea of determining λ is explained in [S4, §12.15]. We use that idea for the proof of (4.11). Once we have $SO^\varphi(V)_\mathbf{A} = SO^\varphi(V)C$, then $G_\mathbf{A} = GQ$, since $\tau(Q) = C$ and $\mathbf{Q}_\mathbf{A}^\times \subset \mathbf{Q}^\times J$. We naturally write J_p and Q_p with a prime p instead of J_v and Q_v. To prove that $G_\mathbf{A} = GJ$, it is sufficient to show that $Q_p \subset GJ$ whenever $Q_p \neq J_p$.

Case $\gamma = 2$. We can take $m = e_1 + e_2$, and easily see that $\lambda = 1$; we can also take $\{\ell - e_2, e_3, e_4, e_5\}$ as a $\mathbf{Z}$-basis of $L = \Lambda \cap V$. Then φ is represented by

(4.12a) $$\zeta_2 = \begin{bmatrix} 1 & 1/2 & 1/2 & 0 \\ 1/2 & 1 & 0 & 0 \\ 1/2 & 0 & 1 & 0 \\ 0 & 0 & 0 & 1 \end{bmatrix}$$

with respect to that basis; $\det(\zeta_2) = 1/2$. From (4.9) we see that case (iii) of §4.3 does not occur. Thus $\mathbf{Q}_p \neq J_p$ only if $p = 2$. The element ε of (4.2) belongs to G and $\nu(\varepsilon) = \gamma = 2$. We can easily verify that $\tau(\varepsilon) = -1$. Thus $\tau(\varepsilon) \in C$, and $\varepsilon \in J_p$ for $p \neq 2$; also $\varepsilon \notin J_2$. Therefore $Q_2 \subset \bigcup_{\nu \in \mathbf{Z}} \varepsilon^\nu J \subset GJ$ as desired. This proves (4.11) for $\gamma = 2$.

Case $\gamma = 3$. We can take $m = e_1 + e_2 + e_5$ and find that $\lambda = 1$. We can also take $\{\ell - e_2, e_1 - e_5, e_3, e_4\}$ as a $\mathbf{Z}$-basis of $L = \Lambda \cap V$. Then φ is represented by

$$\text{(4.12b)} \qquad \zeta_3 = \begin{bmatrix} 1 & 1/2 & 1/2 & 1/2 \\ 1/2 & 2 & 0 & 0 \\ 1/2 & 0 & 1 & 0 \\ 1/2 & 0 & 0 & 1 \end{bmatrix}$$

with respect to that basis; $\det(\zeta_3) = 3/4$. In view of (4.9), case (iii) of §4.3 does not occur. Thus $\mathbf{Q}_p \neq J_p$ only if $p = 2$ or $p = 3$. We can prove that $Q_3 \subset \bigcup_{\nu\in\mathbf{Z}} \varepsilon^\nu J \subset GJ$ in the same manner as in the case $\gamma = 2$. To deal with Q_2, put $\omega = e_3(e_1 - e_2)$. Then we can easily verify that $\omega \in G$, $\nu(\omega) = 2$, and $L\tau(\omega) = L$. Therefore $Q_2 \subset \bigcup_{\nu\in\mathbf{Z}} \omega^\nu J \subset GJ$. This proves (4.11) for $\gamma = 3$.

Case $\gamma = 5$. We can take $m = 2e_1 + e_5$, and easily find that $\lambda = 1$. We easily see that $\{e_2, e_3, e_4, \ell - e_5\}$ gives a $\mathbf{Z}$-basis of $L = \Lambda \cap V$. Then φ is represented by

$$\text{(4.12c)} \qquad \zeta_5 = \begin{bmatrix} 1 & 0 & 0 & 1/2 \\ 0 & 1 & 0 & 1/2 \\ 0 & 0 & 1 & 1/2 \\ 1/2 & 1/2 & 1/2 & 2 \end{bmatrix}$$

with respect to that basis; $\det(\zeta_5) = 5/4$. Let p be a prime number such that $t_p = 2$. Define d_p as in §4.2 with p as v there. From (4.9) we see that $d_p \notin \mathbf{Z}_p^\times$ only if $p = 2$ and we can take $d_2 = 2$. Thus $Q_p \neq J_p$ only for $p = 2$ or $p = 5$. The case $p = 5$ can be handled by the method in the case $\gamma = 2$; we find that $Q_5 \subset \bigcup_{\nu\in\mathbf{Z}} \varepsilon^\nu J \subset GJ$ as desired. Next take $p = 2$. Let $\beta = (e_2 + e_3)e_4$. Then $\beta \in G$ and $\nu(\beta) = 2$. We can easily verify that $\tau(\beta)$ sends e_2, e_3, e_4, and $e_1 - 2e_5$ to $-e_3$, $-e_2$, $-e_4$, and $e_1 - 2e_5$. Since $2(\ell - e_5) = e_1 - 2e_5 + \sum_{i=2}^4 e_i$, we obtain$(\ell - e_5)\tau(\beta) = \ell - e_5 - \sum_{i=2}^4 e_i$. Thus $L\tau(\beta) = L$, and so $\tau(\beta) \in C$. Since $\nu(\beta) = 2$, we have $\beta \in J_p$ for $p \neq 2$, and $\beta \notin J_2$. Therefore $Q_2 \subset \bigcup_{\nu\in\mathbf{Z}} \beta^\nu J \subset GJ$ as expected. Thus we obtain (4.11) for $\gamma = 5$.

Case $\gamma = 7$. We can take $m = 2e_1 + e_2 + e_3 + e_5$, and find that $\lambda = 1$; also $\{\ell - e_2 - e_3, e_5 - e_2, e_5 - e_3, e_4\}$ gives a $\mathbf{Z}$-basis of $L = \Lambda \cap V$. Then

$$\text{(4.12d)} \qquad \zeta_7 = \begin{bmatrix} 1 & 1/2 & 1/2 & 1/2 \\ 1/2 & 2 & 1 & 0 \\ 1/2 & 1 & 2 & 0 \\ 1/2 & 0 & 0 & 1 \end{bmatrix}$$

represents φ with respect to that basis; $\det(\zeta_7) = 7/4$. We have $Q_p \neq J_p$ only for $p = 2$ or $p = 7$. The case $p = 7$ can be handled by the method in the case $\gamma = 2$.

To deal with the case $p = 2$, put $\beta = e_4(e_2 - e_3)$; then $\beta \in G$ and $\nu(\beta) = 2$. We can easily verify that $L\tau(\beta) = L$, and obtain the desired fact in the same manner as in the case $\gamma = 5$ and $p = 2$. This completes the proof of (4.11).

LEMMA 4.5. *Given (V, φ) over $\mathbf{Q}$, let L be a $\mathbf{Z}$-maximal lattice in V and q a squarefree odd integer prime to* $\det(2\varphi_0)$, *where φ_0 is the matrix representing φ with respect to a $\mathbf{Z}$-basis of L. Then*

$$\{h \in L \mid \varphi[h] = q\} = L[q, \mathbf{Z}] \cup L[q, 2^{-1}\mathbf{Z}]. \tag{4.13}$$

Moreover, if $\dim(V) \geq 4$, *φ is positive definite, and the genus of L consists of a single class, then the set of (4.13) is nonempty for every $q \in \mathbf{Z}$, > 0.*

Proof. The right-hand side is contained in the left-hand side by virtue of [S4, Lemma 6.2(3)]. Conversely, suppose $h \in L$ and $\varphi[h] = q$. We can put $2\varphi(h, L) = e\mathbf{Z}$ with $e \in \mathbf{Z}$. Then $e|2q$. Let p be an odd prime factor of e. Then $p|q$, and so p is prime to $\det(2\varphi_0)$, and hence $h \in eL_p \subset pL_p$. Thus $q \in p^2\mathbf{Z}_p$, which is a contradiction, as q is squarefree. Therefore $e|2$, which proves (4.13). Suppose $\dim(V) \geq 4$ and φ is positive definite. For every $q \in \mathbf{Z}$, > 0, we can find an element k of V such that $\varphi[k] = q$; see [S4, Corollary 7.12]. We can also find a $\mathbf{Z}$-maximal lattice L' in V containing k. If the class number is 1, then we can find an element γ of $SO^{\varphi}(V)$ such that $L'\gamma = L$. Then $k\gamma$ is contained in the left-hand side of (4.13). This completes the proof. □

4.6. Put $W = (\mathbf{Q}h)^{\perp}$ with a fixed $h \in L[q, 2^{-1}\mathbf{Z}]$ in the setting of §4.4. Let us now determine the structure of $A(W)$ and $G^+(W) \cap J$. To make our discussion simpler, we consider here only the case in which $\gamma = 5$ and q is a positive squarefree integer prime to 10, though our technique is applicable to other cases. Thus $K = \mathbf{Q}(\sqrt{5})$. We first note that $A^+(W)$ is a definite quaternion algebra over $\mathbf{Q}$. For each prime p we have a weak Witt decomposition of V_p and we can put $L_p = \mathbf{Z}_p e + \mathbf{Z}_p f + \mathfrak{r}_p g$ as in §4.2. In order to determine $A(W)_p$ and $G^+(W)_p \cap J_p$, we can put $h = qe + f$. Then, as shown in §4.2, $A^+(W)_p$ is isomorphic to $K_p + K_p\eta_p$ with $\eta_p^2 = d_p q$. Thus $A(W)_p$ is a division algebra if and only if K_p is a field and $d_p q \notin N_{K/\mathbf{Q}}(K_p^{\times})$. If $p = 5$, then we can take $d_5 = 1$, and so $A(W)_5$ is a division algebra if and only if q is a quadratic nonresidue modulo 5. We have seen that $d_2 = 2$, and so $A^+(W)_2$ is a division algebra. For $p \nmid 10$ we can take $d_p = 1$, and so $A(W)_p$ is a division algebra if and only if p remains prime in K and $p|q$.

We have $d_p = 1$ for $p \neq 2$, and so if we define Ξ: $A(V)_p \to M_2(A(K_p g))$ by $\Xi(x) = \mathrm{diag}[g^{-1}, 1]\Psi(x)\,\mathrm{diag}[g, 1]$, then $\Xi(A^+(V)_p) = M_2(K_p)$ as shown in [S4, (2.15)]. Also, we have $\Xi(A^+(L)_p) = M_2(\mathfrak{r}_p)$; see [S4, Theorem 8.6(ii), (vi)]. Then $\Xi(A^+(L)_p \cap A^+(W)_p) = \mathfrak{r}_p + \mathfrak{r}_p\eta_p$, which is an order of discriminant $5q\mathbf{Z}_p$ in $A^+(W)_p$. This is maximal if and only if p is ramified in $A^+(W)$ or $p \nmid 5q$. If $p = 2$, then using S_v and $\mathfrak{t}_v$ of §4.3, we find that

$$\Psi(A^+(W)_2 \cap S_2) = \left\{ \begin{bmatrix} a & qbg \\ bg & a \end{bmatrix} \mid a, b \in \mathfrak{r}_2 \right\}, \tag{4.14}$$

which can be written $\mathfrak{r}_2 + \mathfrak{r}_2\eta_2$. This is a maximal order whose discriminant is $2\mathbf{Z}_p$. Define an order $\mathfrak{o}'$ in $A^+(W)$ by $\mathfrak{o}'_p = A^+(L)_p \cap A^+(W)_p$ for $p \neq 2$ and $\mathfrak{o}'_2 = S_2 \cap A^+(W)_2$. Then $G^+(W)_p \cap J_p = (\mathfrak{o}')^\times_p$ for every p.

THEOREM 4.7. *Let q be an odd integer of the form $q = p_1 \cdots p_m p'_1 \cdots p'_n$ with different prime numbers p_i and p'_i such that $\left(\dfrac{p_i}{5}\right) = -1$ and $\left(\dfrac{p'_i}{5}\right) = 1$. Let R be a definite quaternion algebra over $\mathbf{Q}$ whose discriminant is $2p_1 \cdots p_m$ or $10p_1 \cdots p_m$ according as m is even or odd. In the setting of §4.4 with $\gamma = 5$, (that is, with $\varphi = \zeta_5$ and $L = \mathbf{Z}^4$) let $W = (\mathbf{Q}h)^\perp$ with $h \in L[q, 2^{-1}\mathbf{Z}]$. Then there exists an order $\mathfrak{o}$ in R of discriminant $10q$ and an isomorphism of $A^+(W)$ onto R that maps $G^+(W)_p \cap J_p$ onto $\mathfrak{o}^\times_p$ for every prime p. Moreover, put $\Delta = \{\alpha \in G \mid L\tau(\alpha) = L,\ \nu(\alpha) = \pm 1\}$. Then $\#\{L[q, 2^{-1}\mathbf{Z}]/\tau(\Delta)\}$ equals the class number of $\mathfrak{o}$ which can be given by (1.12) with the discriminant of R as t and $r = 10q/t$.*

Proof. The assertion concerning an isomorphism of $A^+(W)$ onto R is already proved in §4.6. Now $hC_p = h\tau(J_p)$ for every p as shown in §4.3. Case (iii) there occurs only for $p = 2$, and (4.10) is applicable, as q is odd. Thus $hC = h\tau(J)$, and so (1.8) and (1.9) are valid. By (4.11) we have $G_{\mathbf{A}} = GJ$, and hence the right-hand side of (1.9) is $\#\{L[q, 2^{-1}\mathbf{Z}]/\tau(\Delta)\}$ and $\Delta = G \cap J$. By our isomorphism of $A^+(W)$ onto R, we see that the left-hand side is $\#\{R^\times \backslash R^\times_{\mathbf{A}}/R^\times_{\mathbf{a}} \prod_p \mathfrak{o}^\times_p\}$, which is the class number of $\mathfrak{o}$. This completes the proof. □

Employing the formula $\varphi[ae + cf + bg] = ac + 2bb^\rho$ on V_2, we can easily show that $L[q, \mathbf{Z}] = \emptyset$ if q is odd. Therefore, from (4.13) we obtain, when $\gamma = 5$,

$$\emptyset \neq \{h \in L \mid \varphi[h] = q\} = L[q, 2^{-1}\mathbf{Z}] \text{ if } q \text{ is squarefree and prime to } 10. \tag{4.15}$$

4.8. Put $\Gamma = \{\alpha \in SO^\varphi(V) \mid L\alpha = L\}$ and $\Delta = G \cap J$ in the setting of §4.4. By means of the formula of [S1, Theorem 5.8] we find the numerical value of the mass of $SO^\varphi(V)$ with respect to C, which must be $\#(\Gamma)^{-1}$, since the class number of $SO^\varphi(V)$ is 1 as noted in (4.11). We obtain $\#(\Gamma) = 48$ for $\gamma = 2, 3, 5$, and $\#(\Gamma) = 12$ for $\gamma = 7$. Now the argument of §4.3 in the case $\gamma = 5$ shows that if we denote by $\widetilde{\Delta}$ the group generated by Δ and the elements β and ε there, then $\Gamma = \tau(\widetilde{D})$. (Since $\mathbf{Q}^\times_{\mathbf{A}} \subset \mathbf{Q}J$, we have $\mathbf{Q}^\times_{\mathbf{A}} J \cap G = \mathbf{Q}^\times \Delta$.) From this we can derive that $\#(\Delta) = 24$. The same technique is applicable to the other cases. Thus $\#(\Delta) = 48$ if $\gamma = 2$; $\#(\Delta) = 24$ if $\gamma = 3$ or 5; $\#(\Delta) = 6$ if $\gamma = 7$.

4.9. Still in the setting of §4.8 we note

$$\#L[q, s\mathbf{Z}] = \#\Gamma \cdot \mathfrak{m}(SO^{\varphi}(W),\ SO^{\varphi}(W)_{\mathbf{A}} \cap C) \tag{4.16}$$

for every $s \in \mathbf{Q}^{\times}$, provided the left-hand side is nonzero. Here the second factor of the right-hand side is the mass of $SO^{\varphi}(W)$ with respect to the open subgroup $SO^{\varphi}(W)_{\mathbf{A}} \cap C$, as defined in [S4, §13.11] and [S1], and W is defined with a fixed $h \in L[q, s\mathbf{Z}]$. This is a special case of [S4, (13.17b)]; the left-hand side in this case consists of a single term because of (4.11). To compute the right-hand side, take a quaternion algebra R isomorphic to $A^{+}(W)$ and an order $\mathfrak{o}$ in R such that $\mathfrak{o}_p^{\times}$ can be identified with $A^{+}(W)_p \cap J_p$ for every p. Put $Y = \{x \in R \mid x^{\iota} = -x\}$. As shown in [S3, §7.3], there exists a $\mathbf{Q}$-linear bijection f of Y onto W and an element d of $\mathbf{Q}^{\times}$ such that $\varphi[yf] = dyy^{\iota}$ for every $y \in Y$. Take a maximal order $\mathfrak{O}$ in R containing $\mathfrak{o}$; put $M = (\mathfrak{O} \cap Y)f$ and $D = \{x \in SO^{\varphi}(W)_{\mathbf{A}} \mid Mx = M\}$. Then M is a $\mathbf{Z}$-maximal lattice in W with respect to $d^{-1}\varphi$. In [S3, Theorem 6.4] we gave an explicit formula for $\mathfrak{m}(SO^{\varphi}(W), D)$, which is a special case of [S2, Theorem 5.8]. Now,

$$\mathfrak{m}(U) = [D : U]\mathfrak{m}(D) \tag{4.17}$$

for every open subgroup U of D, where we put $\mathfrak{m}(U) = \mathfrak{m}(SO^{\varphi}(W), U)$ for simplicity; see [S1, Lemma 24.2(1)]. Therefore

$$\mathfrak{m}(SO^{\varphi}(W)_{\mathbf{A}} \cap C) = \mathfrak{m}(D)[D : U]/[SO^{\varphi}(W)_{\mathbf{A}} \cap C : U] \tag{4.18}$$

with $U = \tau(R_{\mathbf{a}}^{\times} \prod_p \mathfrak{o}_p^{\times})$. Thus we obtain the quantity of (4.16) by calculating the group indices in (4.18), which is feasible. On the other hand, we obtained in [S3] explicit formulas for the weighted average of the numbers of representations of an integer by φ on the lattices belonging to the genus of maximal lattices. (The formulas involve certain functions γ_p; if $F = \mathbf{Q}$, the formulas for γ_p given in [S5] are easier to use than those in [S3].) If the class number is 1, this gives a formula for $\#\{h \in L \mid \varphi[h] = q\}$ whose nature is completely different from (4.16), as explained in [S4, §13.13]. Suppose we have obtained $\#L[q, \mathbf{Z}]$ or $\#L[q, 2^{-1}\mathbf{Z}]$ by (4.16). Then that value, together with (4.13) and the formula for $\#\{h \in L \mid \varphi[h] = q\}$, determines the numbers of all three sets of (4.13). For example, take the case $\gamma = 2$; namely, take φ to be ζ_2 of (4.12a). Then for $L = \mathbf{Z}^4$ and every odd squarefree positive integer q, we eventually obtain

$$\#\{h \in L \mid \zeta_2[h] = q\} = 2\left\{8 - \left(\frac{2}{q}\right)\right\} \prod_{p|q} \left\{p + \left(\frac{2}{p}\right)\right\}, \tag{4.19a}$$

$$\#L[q, 2^{-1}\mathbf{Z}] = 2\left\{2 + \left(\frac{2}{q}\right)\right\} \cdot \#L[q, \mathbf{Z}] = 12 \prod_{p|q} \left\{p + \left(\frac{2}{p}\right)\right\}. \tag{4.19b}$$

If $\gamma = 5$ and φ is ζ_5 of (4.12c), then for every odd squarefree positive integer q prime to 5 we have

$$(4.20)\quad \#\{h \in L \mid \zeta_5[h] = q\} = \#L[q, 2^{-1}\mathbf{Z}] = \left\{5 + \left(\frac{5}{q}\right)\right\} \prod_{p|q} \left\{p + \left(\frac{5}{p}\right)\right\}.$$

In [S4, Theorem 13.14] we gave formulas for $\#L[q, \mathbf{Z}]$ and $\#L[q, 2^{-1}\mathbf{Z}]$ when φ is the sum of 5, 7, or 9 squares. Formulas (4.19b) and (4.20) are their quaternary analogues.

4.10. We conclude this paper by noting a natural question. In Theorem 1.6 and its consequences we always have to start with an element h of V, but we cannot start from $L[q, \mathfrak{b}]$, as we do not know whether the set is nonempty. Indeed, $L[q, \mathbf{Z}]$ can be empty as shown in (4.15) and [S4, Theorem 13.14]; also, $L[q, 2^{-1}\mathbf{Z}]$ seems less likely to be empty. Thus we can ask: *Are there practical criteria by which we can determine when* $L[q, \mathfrak{b}] \neq \emptyset$?

DEPARTMENT OF MATHEMATICS, PRINCETON UNIVERSITY, PRINCETON, NJ

REFERENCES

[E1] M. Eichler, Über die Idealklassenzahl total definiter Quaternionen-Algebren, *Math. Z.* **43** (1937), 102–109.

[E2] ———, Allgemeine Kongruenzklasseneinteilungen der Ideale einfacher Algebren über algebraischen Zahlkörpern und ihre L-Reihen, *J. Reine Angew. Math.* **179** (1938), 227–251.

[E3] ———, Zur Zahlentheorie der Quaternionen-Algebren, *J. Reine Angew. Math.* **195** (1956), 127–151.

[S1] G. Shimura, *Euler Products and Eisenstein Series*, *CBMS Regional Conference Series in Mathematics*, No. 93, American Mathematical Society, Providence, RI, 1997.

[S2] ———, An exact mass formula for orthogonal groups, *Duke Math. J.* **97** (1999), 1–66 (= *Collected Papers*, IV, 509–574).

[S3] ———, The number of representations of an integer by a quadratic form, *Duke Math. J.* **100** (1999), 59–92 (= *Collected Papers*, IV, 575–608).

[S4] ———, Arithmetic and analytic theories of quadratic forms and Clifford groups, *Math. Surveys and Monogr.*, vol. 109, American Mathematical Society, Providence, RI, 2004.

[S5] ———, Inhomogeneous quadratic forms and triangular numbers, *Amer. J. Math.* **126** (2004), 191–214.

QUADRATIC DIOPHANTINE EQUATIONS, THE CLASS NUMBER, AND THE MASS FORMULA

GORO SHIMURA

1. The basic setting and two ternary cases

We take a finite-dimensional vector space V over a field F and take also an F-bilinear symmetric form $\varphi : V \times V \to F$. We then put $\varphi[x] = \varphi(x, x)$ for $x \in V$, thus using the same letter φ for the quadratic form and the corresponding symmetric form. By a *quadratic Diophantine equation* we mean an equation of the type

$$\varphi[x] = q \tag{1}$$

with a given $q \in F^\times$. In particular, in the classical case with $F = \mathbf{Q}$ and $V = \mathbf{Q}^n$, we usually assume that φ is $\mathbf{Z}$-valued on $\mathbf{Z}^n$ and $q \in \mathbf{Z}$. The purpose of the present article is to present some new ideas on various arithmetical questions on such an equation. We start with some of our basic symbols and terminology. For a set X we denote by $\#X$ or $\#\{X\}$ the number ($\leq \infty$) of elements of X. For an associative ring R with identity element, we denote by $R^\times$ the group of invertible elements of R and by $M_n(R)$ the ring of all square matrices of size n with entries in R. We then put $GL_n(R) = M_n(R)^\times$ and denote by 1_n the identity element of $M_n(R)$. For two square matrices A and B of size m and n we denote by $\mathrm{diag}[A, B]$ the square matrix of size $m + n$ with A and B in diagonal blocks and zeros in the remaining blocks.

Now, given (V, φ) as above, we always assume that φ is nondegenerate. We also put $n = \dim(V)$ and define, as usual, the orthogonal group $O^\varphi(V)$ and the special orthogonal group $SO^\varphi(V)$ by

$$O^\varphi(V) = \{\alpha \in GL(V) \mid \varphi[x\alpha] = \varphi[x] \text{ for every } x \in V\},$$
$$SO^\varphi(V) = O^\varphi(V) \cap SL(V), \quad \text{written also } SO(\varphi) \text{ and } SO(V, \varphi).$$

We let $GL(V)$ act on V on the right, so that $x\alpha$ is the image of x under α.

As the base field F we take, for the moment, an algebraic number field or its completion at a nonarchimedean prime. We denote by $\mathfrak{g}$ the ring of algebraic integers in the former case and the ring of local integers in the latter case. Those who are not much interested in the general case may assume that F is $\mathbf{Q}$ or the p-adic number field $\mathbf{Q}_p$ for any prime number p, and $\mathfrak{g}$ is $\mathbf{Z}$ or the ring $\mathbf{Z}_p$ of p-adic integers. By a $\mathfrak{g}$-*lattice* (simply a *lattice*) in V we mean a finitely generated

---

Received by the editors October 26, 2005.
2000 Mathematics Subject Classification. Primary 11E12; Secondary 11D09, 11E41
Shimura, Goro. "Quadratic Diophantine equations, the class number, and the mass formula." Bulletin of the American Mathematical Society (N.S.) 43 (2006), 285–304.

$\mathfrak{g}$-submodule of V that spans V over F. We call a lattice L in V *integral* (with respect to φ) if $\varphi[x] \in \mathfrak{g}$ for every $x \in L$. Given a lattice L in V, we put

$$\Gamma(L) = \{\alpha \in SO^{\varphi}(V) \mid L\alpha = L\}. \tag{2}$$

In the simplest case we can take $V = \mathbf{Q}^n$, $L = \mathbf{Z}^n$, and $\varphi[x] = x\Phi \cdot {}^t x$ with a matrix $\Phi = (c_{ij})$ with $c_{ij} \in \mathbf{Q}$ and a row vector $x = (x_i)_{i=1}^n$. Clearly L is integral if and only if $2c_{ij} \in \mathbf{Z}$ and $c_{ii} \in \mathbf{Z}$ for every i and j; $\Gamma(L) = SO^{\varphi}(V) \cap SL_n(\mathbf{Z})$.

Let us now consider some questions about equation (1). Taking a lattice L in V, we can ask, for example, the number of elements in the set

$$L[q] = \{x \in L \mid \varphi[x] = q\}, \tag{3a}$$

which is an old problem investigated by many number theorists, especially when $\varphi[x] = \sum_{i=1}^n x_i^2$ and $L = \mathbf{Z}^n$. This question is preceded by a famous statement of Fermat that every natural number is the sum of at most three triangular numbers, and also of four squares, of five pentagonal numbers, etc. Though this problem is of a nature quite different from the type of problems we are going to discuss, it must be remembered that it was also a question about quadratic forms and that many mathematicians in the 18th and early 19th centuries were conscious of this problem and expended considerable effort toward its solution.

In addition to $L[q]$, there is another set which has a long history as an object of study. Namely, assuming $L = \mathbf{Z}^n$, we put

$$L^0[q] = \Big\{x = \{x_i\}_{i=1}^n \in L[q] \,\Big|\, \sum_{i=1}^n x_i\mathbf{Z} = \mathbf{Z}\Big\} \tag{3b}$$

and call an element of $L^0[q]$ a *primitive* solution of equation (1). In terms of matrices, the equation has the form $x\Phi \cdot {}^t x = q$, and $x = \{x_i\}_{i=1}^n$ is primitive if the x_i have no nontrivial common divisor. Notice that both $L[q]$ and $L^0[q]$ are stable under $\Gamma(L)$.

This condition of primitivity is fundamental in our theory, as will be explained later. For the moment, let us just say that the nature of the question about $L^0[q]$ is quite different from that about $L[q]$: roughly speaking, the former is more conceptual, and the latter more computational. We shall eventually explain this difference and present a new interpretation of $L^0[q]$, which leads to certain class number formulas for orthogonal groups and new mass formulas for such groups. We will also consider an equation of the type $\xi\Phi \cdot {}^t\xi = \Psi$ with another symmetric matrix Ψ of size m (instead of a scalar q) and an $(m \times n)$-matrix ξ and will develop a theory parallel to that for $\varphi[x] = q$.

We now recall an ancient problem: to find a systematic method of obtaining $x \in L[q]$ when $F = \mathbf{Q}$, $n = 2$, and $L = \mathbf{Z}^2$. Such was discussed by many mathematicians, Fermat and Euler, in particular. But all those works were restricted to some special binary forms. The case of an *arbitrary* binary form was settled by Lagrange [12] and later reformulated by Gauss [7]. In this article, however, we are *not* interested in their methods of how to find x. Indeed, what concerns us is the "conceptual meaning" of each solution of (1) when $n > 2$. Still, there are some aspects of Lagrange's work relevant to our topic. We first note that he introduced the notion of a *class* of binary forms. To be precise, take $F = \mathbf{Q}$ and $\mathfrak{g} = \mathbf{Z}$ and consider a quadratic form $\xi u^2 + \zeta uv + \eta v^2$ of two variables u, v with coefficients ξ, η, $\zeta \in \mathbf{Z}$. We then put $q = 4\xi\eta - \zeta^2$ and call $-q$ the *discriminant* of the binary form; we naturally assume that $q \neq 0$. If we represent the form by the matrix

$$h = \begin{bmatrix} \xi & \zeta/2 \\ \zeta/2 & \eta \end{bmatrix},$$

then $\det(2h) = q$. We say that another form represented by a similar matrix h' belongs to the same *class* as h if

$$(4) \qquad h' = \det(\alpha)^{-1} \cdot \alpha h \cdot {}^t\alpha \quad \text{with } \alpha \in GL_2(\mathbf{Z}).$$

Here we deviate from the traditional definition $h' = \alpha h \cdot {}^t\alpha$ with $\alpha \in SL_2(\mathbf{Z})$, because (4) with $GL_2(\mathfrak{g})$ in place of $GL_2(\mathbf{Z})$ is the best definition in the case of an arbitrary number field.

Now Lagrange showed that the number of classes of the forms with a given discriminant is finite. For this and other results of Lagrange and his predecessors, the reader is referred to Weil's book [31]. For an obvious technical reason, it is natural to assume, as Lagrange and later researchers did, that the binary form is *primitive*, which means that ξ, η, ζ have no nontrivial common divisor.

In order to interprete the notion of class in a different way, we consider a *ternary* form φ defined on $\mathbf{Q}^3$ by

$$(5a) \qquad \varphi[(x,\, y,\, z)] = 4xy - z^2.$$

If we fix q, then a primitive binary form of discriminat $-q$ corresponds to a primitive solution $h = (\xi,\, \eta,\, \zeta) \in \mathbf{Z}^3$ of the equation $\varphi[h] = q$ with φ of (5a), that is, an element of $L^0[q]$ with $L = \mathbf{Z}^3$. This much is trivial, but here is a nontrivial fact: *$\Gamma(\mathbf{Z}^3)$ defined by (2) in the present case consists of all the maps $h \mapsto h'$ of (4).* (It seems that the previous researchers did not connect $SO(\varphi)$ for φ of (5a) with binary forms, and the last fact on $\Gamma(\mathbf{Z}^3)$ is not a well known old result. Indeed, the corresponding fact in the case of an arbitrary number field F depends on the nature of the ideal class group of F; see [22, Lemma 12.10].) Therefore the classes of binary forms of discriminant $-q$ correspond bijectively to $L^0[q]/\Gamma(L)$ defined with respect to φ of (5a). Denoting by $c(q)$ the number of classes of primitive binary forms of discriminant $-q$, we thus obtain

$$(5b) \qquad \#\{L^0[q]/\Gamma(L)\} = c(q).$$

Before discussing another ternary form, let us state here a basic result of Dedekind. Assuming that $-q$ is not a square in $\mathbf{Q}$, define a quadratic extension K of $\mathbf{Q}$ by $K = \mathbf{Q}(\sqrt{-q}\,)$. Then we can put $-q = f^2 d$ with $0 < f \in \mathbf{Z}$ and the discriminant d of K. Denote by $\mathfrak{r}$ the ring of algebraic integers in K and define a subring $\mathfrak{o}$ of $\mathfrak{r}$ by $\mathfrak{o} = \mathbf{Z} + f\mathfrak{r}$. Then Dedekind [4, §187] showed that there is a bijection from the set of all ideal classes of proper $\mathfrak{o}$-ideals onto the set of all classes of primitive binary forms of discriminat $-q$. If $-q = d$, then $f = 1$, $\mathfrak{o} = \mathfrak{r}$, and a proper $\mathfrak{o}$-ideal is a fractional ideal in K. As to later expositions of this result of Dedekind, the reader is referred to [10, §53] and [3, Chapter II, §7.5], for example. By virtue of this result, $c(q)$ is the class number of such an $\mathfrak{o}$ and, in particular, the class number of K if $-q = d$. Though the theory of proper $\mathfrak{o}$-ideals is not so complicated, we will not go into details here, as such is not essential for the understanding of our main ideas.

Next we turn to a more difficult problem concerning

$$(6a) \qquad \varphi[(x,\, y,\, z)] = x^2 + y^2 + z^2.$$

We naturally look for primitive solutions h of the equation $\varphi[h] = q$ for a given $q \in \mathbf{Z}, > 0$. In this case, the permutations of $(x,\, y,\, z)$ and the diagonal elements of $GL_3(\mathbf{Z})$ generate a group of order 48 which has $\Gamma(\mathbf{Z}^3)$ as a subgroup of index 2. Now Gauss showed in [7] that such a primitive solution exists if and only if $q = b^2m$ with an odd positive integer b and a squarefree positive integer m such that $m \not\equiv 7 \pmod 8$. From this he derived by a short elementary argument that every natural number is the sum of at most three triangular numbers. (It may be added that Lagrange solved the case of four squares some 30 years earlier.) These were already wonderful achievements, but he proved a far deeper result on $L^0[q]$ for such a q, which can be reformulated as follows:

$$\text{(6b)} \qquad \#\{L^0[q]/\Gamma(L)\} = c(q) \cdot \begin{cases} 1 & \text{if } m \equiv 3 \pmod 4 \text{ or } q \le 2, \\ 2^{-1} & \text{otherwise.} \end{cases}$$

From this we can easily derive a formula valid for $q = b^2m > 3$:

$$\text{(6c)} \qquad \#L^0[q] = c(q) \cdot \begin{cases} 24 & \text{if } m \equiv 3 \pmod 4, \\ 12 & \text{otherwise.} \end{cases}$$

In fact, Gauss stated something equivalent to (6c), but did not give a clear-cut statement in the form (6b). His proof is long and roundabout. In order to study a ternary equation $x\Phi \cdot {}^t x = q$, he considered the representation of a binary form by another ternary form, namely an equation $\xi\Psi \cdot {}^t\xi = \eta$, where Ψ is such that $\Phi = -\det(\Psi)\Psi^{-1}$ and η is a (2×2)-matrix that represents a given binary form. This technique was initiated earlier by Legendre in [13] for $\Psi = 1_3$, and he almost proved the results of Gauss, including the one on three triangular numbers. Though Gauss rigorously proved them, it must be remembered that he was not so original in this respect. In any case, there is a simpler method, as we will show later, and so we are tempted to say that in a sense Gauss was misled by Legendre's work. Without [13] he might have found a different, possibly shorter, proof, but of course we cannot change history, and, after all, the idea worked.

Gauss was not the only person "misled" by Legendre; indeed, Eisenstein and Minkowski employed the same method in dealing with $\#L^0[q]$ when φ is the sum of five squares. But our aim is to present new ideas which are more straightforward than their approach, and so we will not explain how they obtained their results, which are splendid in spite of the seeming awkwardness of the method. The reader is referred to Bachmann's book [1] for an exposition of this subject; see also [22, p. 137, lines 18–23] for an interpretation of their methods.

History aside, our point is that there is an unmistakable parallelism between (5b) and (6b). Therefore we can expect the existence of a general principle of which (5b) and (6b) are special cases. In the above two cases, $n = 3$ and binary forms have two variables, and so the expected principle may be of the following type:

(7) $\#\{L^0[q]/\Gamma(L)\}$ equals the class number of an object in dimension $n - 1$.

But things are not that simple, though the idea of (7) is basically right. Also, we can seek an n-dimensional generalization of (6c), but that requires the concept of "mass". To make precise statements, we first put

$$\text{(8)} \qquad L[q, \mathbf{Z}] = \{x \in V \mid \varphi[x] = q,\ \varphi(x, L) = \mathbf{Z}\}.$$

This will replace $L^0[q]$. Notice that $L[q, \mathbf{Z}]$ is not defined as a subset of $L[q]$. If $\varphi[x] = x\Phi \cdot {}^t x$ with $\Phi = (c_{ij})$ and $L = \mathbf{Z}^n$, then $\varphi(x, L) = \sum_{i=1}^n \bigl(\sum_{j=1}^n c_{ij}x_j\bigr)\mathbf{Z}$. In particular, if $\varphi[x] = \sum_{i=1}^n x_i^2$ and $L = \mathbf{Z}^n$, then $L^0[q] = L[q, \mathbf{Z}]$. Next, we call

L *φ-maximal,* or simply *maximal,* if L is the only integral lattice containing itself. The lattice $\mathbf{Z}^3$ is maximal with respect to φ of (6a). As for φ of (5a), if we denote by Λ the set of all $(\xi, \eta, \zeta) \in \mathbf{Q}^3$ such that $2\xi, 2\eta, \zeta \in \mathbf{Z}$, then we easily see that Λ is maximal with respect to (5a) and $(\mathbf{Z}^3)^0[q] = \Lambda[q, \mathbf{Z}]$. Thus one of our key ideas is to consider $L[q, \mathbf{Z}]$ with a maximal L in place of $L^0[q]$.

Now the precise version of (7) is the first of the following two formulas, which form the principal results whose explanation is the main objective of this article:

$$\sum_{i \in I} \#\{L_i[q, \mathbf{Z}]/\Gamma(L_i)\} = \text{the class number of a group } H, \tag{9a}$$

$$\sum_{i \in I} \#\{L_i[q, \mathbf{Z}]\}/\#\Gamma(L_i) = \text{the mass of } H \text{ if } \varphi \text{ is definite,} \tag{9b}$$

provided L is maximal and $L[q, \mathbf{Z}] \neq \emptyset$. Formula (9b) is a generalization of (6c).

We have to explain what $\{L_i\}_{i \in I}$, H, the class number, and the mass are. To define H, we first pick $h \in L[q, \mathbf{Z}]$ and take the orthogonal complement W of $\mathbf{Q}h$, which we write $(\mathbf{Q}h)^\perp$ and which is defined by

$$W = (\mathbf{Q}h)^\perp = \{y \in V \mid \varphi(h, y) = 0\}. \tag{10}$$

Clearly $\dim(W) = n - 1$. Now the group H in (9a, b) is given by $H = SO^\varphi(W)$, where we use φ also for its restriction to W. We view H as a subgroup of $SO^\varphi(V)$ by putting

$$H = SO^\varphi(W) = \{\alpha \in SO^\varphi(V) \mid h\alpha = h\}. \tag{11}$$

In the next section we shall explain what $\{L_i\}_{i \in I}$, the class number, and the mass are. Before doing so, we should mention a natural question: Is $\#\{L^0[q]/\Gamma(L)\}$ or $\#\{L[q, \mathbf{Z}]/\Gamma(L)\}$ really finite? If φ is positive or negative definite, then $\Gamma(L)$, $L[q]$, and $L[q, \mathbf{Z}]$ are finite sets, as can easily be seen, and so the matter is trivial. But the finiteness of $\#\{L[q]/\Gamma(L)\}$ for indefinite φ is nontrivial. There is a similar question about the equation $\xi\Phi \cdot {}^t\xi = \Psi$, mentioned above, over a number field. Namely, fixing a lattice Λ in the space of $(m \times n)$-matrices, we ask whether the orbit of all such $\xi \in \Lambda$ under the stabilizer of Λ in $SO(\Phi)$ is finite. We can even ask the same type of question over a nonarchimedean completion of F. The answer is indeed affirmative in both global and local cases; see [22, Theorems 13.2 and 13.3].

Such finiteness results, when $F = \mathbf{Q}$, were known to Siegel; also they are implicit in the convergence of the Siegel-Weil formula, though that requires the convergence condition. He always said that the fact followed from reduction theory ([27, p. 399], for example), which is true, but he never bothered himself with its proof until very late. Indeed, in [28], he treated the question of finding ξ satisfying $\xi\Phi \cdot {}^t\xi = \Psi$ over $\mathbf{Z}$ in the classical style of Lagrange; his answer includes the finiteness. It is as if he did not wish to cause later researchers to complain about this point. In any case, the facts are nontrivial, and the local case does not seem to have been previously stated.

There is another natural question: why do we consider $L^0[q]$ or $L[q, \mathbf{Z}]$ instead of $L[q]$? Before answering this question, we note here a result about $\#L[q]$ for φ of (6a) given in [21, p. 1067]. Let $q = 2^\nu r^2 t$ with $0 \leq \nu \in \mathbf{Z}$ and odd positive integers

r and t, t squarefree. Let $r = \prod_p p^{\lambda_p}$ be the prime decomposition of r. Then

$$(*)\quad \#L[q] = \begin{cases} 0 & \text{if } \nu \in 2\mathbf{Z} \text{ and } t - 7 \in 8\mathbf{Z}, \\ A\pi^{-1}L(1,\,\psi)rt^{1/2}\prod_p^*\left\{\sum_{k=0}^{\lambda_p} p^{-k} - \psi(p)\sum_{i=1}^{\lambda_p} p^{-i}\right\} & \text{otherwise,} \end{cases}$$

where A is 24 or $24\sqrt{2}$ according to whether $\nu \in 2\mathbf{Z}$ or $\nu \notin 2\mathbf{Z}$, ψ is the primitive Dirichlet character corresponding to $\mathbf{Q}(\sqrt{-q})$, and $\prod_p^*$ is the product over all the prime factors p of r such that $\psi(p) \neq 1$. Notice that $L(1,\,\psi)$ is the class number of $\mathbf{Q}(\sqrt{-q})$ times an elementary factor (as shown by (13a) below), and so $(*)$ is similar to (6c) to some extent but far more complex. In general the formula for $\#L[q]$ for a definite φ, when n is odd, is as complicated as, and often more complicated than, $(*)$, whereas we have simpler formulas for $L^0[q]$, as can be seen from (9a). Besides, the proof of (9a) gives a conceptual explanation why the class number is involved, but we cannot do that for $L[q]$ except in some special cases. This is the main reason for our discussing $L[q,\,\mathbf{Z}]$ and $L^0[q]$.

2. The class number and mass of an orthogonal group

We are going to define the *class* and *genus* of a lattice in V. Fixing $(V,\,\varphi)$ over $\mathbf{Q}$, put $G = SO^{\varphi}(V)$. For each prime number p we put $V_p = V \otimes_{\mathbf{Q}} \mathbf{Q}_p$, extend φ to a symmetric form $\varphi_p : V_p \times V_p \to \mathbf{Q}_p$, and put $G_p = SO(V_p,\,\varphi_p)$. For a lattice L in V, we denote by L_p the $\mathbf{Z}_p$-linear span of L in V_p, which is a $\mathbf{Z}_p$-lattice in V_p.

Two lattices L and M in V are said to belong to the same G-*class* if $M = L\alpha$ with $\alpha \in G$. They belong to the same G-*genus* if $M_p = L_p\alpha_p$ with some $\alpha_p \in G_p$ for every prime number p. Now here is a basic fact:

(12) The G-genus of L consists of a finite number of G-classes of lattices.

This follows from reduction theory of Minkowski. We can define the genus and class of a $\mathfrak{g}$-lattice in the case of an arbitrary number field F in a similar way by taking nonarchimedean completions of F in place of $\mathbf{Q}_p$. Then (12) holds. We can define the G-class and G-genus for an algebraic group G over F of a more general type acting on V. Then (12) holds; see [2] for the result in the most general case.

Originally the genus and class were defined in terms of matrices; see [25], for example. They can easily be translated into those in terms of lattices defined as above, provided F has class number 1. The reader is referred to [22, §9.27] for the explanation.

Now the number of classes in the genus is called the *class number of the genus*. But this depends on the choice of L, or rather, of a genus of lattices. However, all maximal lattices form a single G-genus, and so the number of G-classes in that genus is well defined. Therefore we may call it *the class number* of $(V,\,\varphi)$, or even *the class number* of G. But the latter definition is somewhat misleading, as $SO^{\varphi}(V) = SO^{c\varphi}(V)$ for every $c \in F^{\times}$, and a φ-maximal lattice may not be $c\varphi$-maximal. A better alternative is to define the class number in terms of a certain subgroup of the adelization of G, which will be explained in Section 4.

Returning to (9a), $\{L_i\}_{i \in I}$ is a complete set of representatives for the G-classes in the G-genus of a fixed maximal lattice L, where $G = SO^{\varphi}(V)$. The right-hand side is actually the number of H-classes in the H-genus of L. Since the group H acts on the vector space W of (10), taking a maximal lattice M in W, we can consider the number of H-classes in the H-genus of M, but this may not be equal to the number of H-classes in the H-genus of L, and so (9a) is not a precise statement, though the two class numbers often coincide.

If the class number of G is 1, then the left-hand side of (9a) or (9b) consists of a single term, which is the case for the forms of (5a) and (6a), and so (5b) and (6b) are special cases of (9a). In those two cases the group H is given by $H = \{x \in K^\times \mid N_{K/\mathbf{Q}}(x) = 1\}$ with $K = \mathbf{Q}(\sqrt{-q})$, and there is a simple relation between the class number of H and that of K.

The proof of (9a) is not so short, but it is conceptually straightforward; we can at least say that it is not as painful as the proof of (6c) by Gauss. But postponing the proof, let us now explain what the mass means.

The mass formula for an orthogonal group when $n = 2$ is actually Dirichlet's formulas for the class number h_K of a quadratic field K. Let us state them here in the forms given in modern textbooks, not in the original style of Dirichlet. Let $\chi(a) = \left(\frac{d}{a}\right)$ for $a \in \mathbf{Z}$, where d is the discriminant of K (and so $K = \mathbf{Q}(d^{1/2})$), and let $L(s, \chi) = \sum_{m=1}^{\infty} \chi(m)m^{-s}$ with a complex variable s. This series is convergent for $\mathrm{Re}(s) > 1$ and can be continued to an entire function. We have then

$$h_K \cdot w^{-1} = (2\pi)^{-1}|d|^{1/2} \cdot L(1, \chi) \quad \text{if} \quad d < 0, \tag{13a}$$

$$h_K \cdot \log \varepsilon = 2^{-1}|d|^{1/2} \cdot L(1, \chi) \quad \text{if} \quad d > 0. \tag{13b}$$

Here w is the number of roots of unity in K. For $d > 0$, we fix an embedding of K into $\mathbf{R}$ and take a fundamental unit $\varepsilon > 1$ in K. We can view these as special cases of the formula for the residue of the Dedekind zeta function at 1, due to Dedekind himself, which is also a kind of mass formula, but let us restrict ourselves to orthogonal groups.

The idea of the mass of a genus for $n > 2$ goes back to Eisenstein and Minkowski. We state here the mass formula essentially in Siegel's style, though we employ the definition of the mass introduced in [17] and [18]. Given a lattice L in V, we define *the mass of* $G = SO^{\varphi}(V)$ relative to L (or the mass of the genus of L with respect to G) to be the quantity given by

$$\mathfrak{m}(G, L) = \sum_{i \in I} \nu(\Gamma_i), \quad \Gamma_i = \Gamma(L_i), \tag{14a}$$

$$\nu(\Gamma) = \begin{cases} [\Gamma : 1]^{-1} & \text{if } \varphi \text{ is definite,} \\ \mathrm{vol}(\Gamma\backslash\mathcal{Z})/\#(\Gamma \cap \{\pm 1\}) & \text{otherwise.} \end{cases} \tag{14b}$$

Here L is not necessarily maximal, $\{L_i\}_{i \in I}$ is a complete set of representatives for the G-classes in the G-genus of L, and $\mathcal{Z}$ is the symmetric space on which G has a natural action; we fix a G-invariant measure on $\mathcal{Z}$. Siegel showed (for an integral $L = \mathbf{Z}^n$) in [25] and [26] that

$$\mathfrak{m}(G, L) = d_\infty \prod_p d_p, \tag{15}$$

$$d_p = \lim_{m \to \infty} p^{-mn(n-1)/2} \#\{\alpha \in GL_n(\mathbf{Z}_p/(p^m)) \mid \alpha\varphi \cdot {}^t\alpha \equiv \varphi \pmod{p^m}\}. \tag{16}$$

In fact, the right-hand side of (16) becomes constant for sufficiently large m. The number d_p is called *the representation density* at p; d_∞ is defined suitably; see [26, pp. 410–412]. But Siegel did not give a precise form of the right-hand side of (15), though he computed d_∞ and d_p for p not dividing $2\det(\varphi)$. An exact formula for

$\mathfrak{m}(G, L)$ when L is maximal was given in [18]. To state it, we need a few basic facts on the classification of (V, φ) over $\mathbf{Q}$. We first put $\delta(\varphi) = (-1)^{n(n-1)/2}\det(\varphi)$ and call it the *discriminant* of φ. This is determined modulo the squares of the elements of $\mathbf{Q}^\times$. Next, the signature of φ as a real quadratic form is an obvious invariant. In addition, for each prime number p, once n and $\delta(\varphi)$ are given, there exist only a finite number of quadratic spaces over $\mathbf{Q}_p$, which can be explicitly described. Now (V, φ) over $\mathbf{Q}$ is completely determined by the signature of φ and (V_p, φ_p) for all p, a fact called *the Hasse principle.* These results can be found in almost any book on quadratic forms, [6] or [22, Section 7], for example.

To state the formula for $\mathfrak{m}(G, L)$, fix a maximal lattice L and denote by φ_0 the matrix that represents φ with respect to a $\mathbf{Z}$-basis of L. Then [18, Theorem 5.8] gives

$$\mathfrak{m}(G, L) = A \cdot \det(2\varphi_0)^{\mu} \prod_p \lambda_p \cdot \begin{cases} 1 & \text{if } n \notin 2\mathbf{Z}, \\ L(n/2, \psi) & \text{if } n \in 2\mathbf{Z}. \end{cases} \tag{17}$$

Here A is an elementary constant depending only on n, $\delta(\varphi)$, the signature of φ, and the choice of the measure on $\mathcal{Z}$; $\mu = (n-1)/2$, p runs over the prime factors of $\det(2\varphi_0)$, λ_p is a rational number determined by (V_p, φ_p), and ψ is the primitive Dirichlet character corresponding to $\mathbf{Q}(\delta^{1/2})$, $\delta = (-1)^{n/2}\det(\varphi_0)$. The proof of (17) relies on the fact that the mass is the residue of a certain zeta function on G, and that residue can be obtained by computing the residue of an Eisenstein series on a split orthogonal group. This may be viewed as a byproduct of the general theory of such zeta functions and is different from Siegel's proof of (15).

Suppose that the $\nu(\Gamma_i)$ are the same for all i. (This is the case if $n = 2$, as G becomes commutative.) Then we obtain

$$\#I \cdot \nu(\Gamma_1) = \text{the right-hand side of (17).}$$

If $n = 2$, this is essentially the same as (13a) or (13b), as $\#I = 2^{2-t}h_K$ if $d > 0$ and $N(\varepsilon) = 1$, and $\#I = 2^{1-t}h_K$ otherwise, where t is the number of prime factors of d and ε is a fundamental unit of K; see [22, 9.16].

In general $\nu(\Gamma_i)$ can take many different values, and so we cannot prove a clear-cut formula for $\#(I)$. We have formulas for their sum which generalize (13a, b).

If φ is indefinite, then $\mathrm{vol}(\Gamma\backslash\mathcal{Z})$ plays the role of $\log\varepsilon$. In that sense (14a) combined with (15) is of the same nature as (13a, b), though (13a, b) are simpler. In any case, for $n > 2$ there is a formula for $\mathfrak{m}(G, L)$ like (17), but no such formula exists for the class number. In general, it is difficult to find the class number of an orthogonal group. However, by means of formula (9a), we can determine the class number in certain cases by counting the number $\#\{L[q, \mathbf{Z}]/\Gamma(L)\}$. We will give examples in Theorem 3.

But before discussing this, let us note another result of Siegel concerning $\#L[q]$. To make our exposition easier, we introduce the notion of the mass of a subset X of V, written $\mathfrak{m}(X)$. We take a set X of the form $X = \bigsqcup_{j=1}^{k} h\beta_j\Gamma$ with $\Gamma = \Gamma(L)$ and a finite subset $\{\beta_j\}$ of G. Then we put

$$\mathfrak{m}(X) = \sum_{j=1}^{k} \nu(\Delta_j)/\nu(\Gamma), \quad \Delta_j = \beta_j\Gamma\beta_j^{-1} \cap H. \tag{18}$$

It can easily be seen that $\mathfrak{m}(X) = \#X$ if φ is definite. Now, for $L = \mathbf{Z}^n$ Siegel showed in [25] and [26] that

$$\text{(19)}\qquad \begin{aligned} &\mathfrak{m}(G,\, L)^{-1} \sum_{i \in I} \nu(\Gamma_i)\mathfrak{m}\big(L_i[q]\big) = e_\infty \prod_p e_p, \quad \Gamma_i = \Gamma(L_i), \\ &e_p = \lim_{m \to \infty} p^{m(1-n)} \#\big\{x \in \big(\mathbf{Z}/p^m\mathbf{Z}\big)^n \;\big|\; \varphi[x] - q \in p^m\mathbf{Z}\big\}. \end{aligned}$$

The right-hand side becomes constant for sufficiently large m. The number e_p is called *the representation density of q by φ at p*; e_∞ is a similar density at ∞ defined by means of certain volumes. Thus, if φ is definite, then

$$\text{(20)}\qquad \left\{\sum_{i \in I} [\Gamma_i : 1]^{-1}\right\}^{-1} \sum_{i \in I} [\Gamma_i : 1]^{-1} \#L_i[q] = e_\infty \prod_p e_p.$$

In particular, if $\#I = 1$, then $\#L[q] = e_\infty \prod_p e_p$. If $\#I > 1$, however, only the weighted average of $\big\{\#L_i[q]\big\}_{i \in I}$ can be given.

How can one compute $e_\infty \prod_p e_p$? Siegel did a few easy cases. An explicit formula was given in [19] when F is totally real, φ is totally definite, and L is maximal:

$$\text{(21)}\qquad e_\infty \prod_p e_p = Bq^{(n-2)/2} \prod_p{}'' \varepsilon_p \cdot \begin{cases} L(n/2,\, \psi)^{-1} & (n \in 2\mathbf{Z}), \\ L((n-1)/2, \psi_q) & (n \notin 2\mathbf{Z}). \end{cases}$$

Here B is an elementary factor depending only on φ and independent of q; ε_p is a rational expression in $p^{n/2}$; $\prod_p''$ is the product over all prime factors p of $q \cdot \det(2\varphi_0)$, where φ_0 is as in (17); ψ is the same as in (17); ψ_q is the primitive Dirichlet character corresponding to $\mathbf{Q}(\kappa^{1/2})$, where $\kappa = q\delta(\varphi)$. The lattice L must be maximal. We prove (21) by using the fact that $e_\infty \prod_p e_p$ is a Fourier coefficient of a certain Eisenstein series. Thus the weighted average is computable by means of (21), but there is no formula for each individual $\#L_i[q]$ in general (unless $\#I = 1$). In fact, however, the quantity of (21) is a good approximation to $\#L_i[q]$. These points will be explained in Section 6. Also, it should be mentioned that (21) does not apply to the case of the sum of n squares with $L = \mathbf{Z}^n$ if $n > 3$, as $\mathbf{Z}^n$ is not maximal in such cases. Even so, the problem can be handled, as will be explained in Section 6, (R1).

3. Precise forms of the main formulas

Let us now present precise forms of (9a, b). We take $(V,\, \varphi)$ over an arbitrary number field F. (Of course the reader may assume that $F = \mathbf{Q}$.) For a $\mathfrak{g}$-lattice Λ in V, an element q in $F^\times$, and a fractional ideal $\mathfrak{b}$ in F, we put

$$\text{(22)}\qquad \Lambda[q,\, \mathfrak{b}] = \big\{\, x \in V \,|\, \varphi[x] = q, \quad \varphi(x,\, \Lambda) = \mathfrak{b}\,\big\}.$$

If $F = \mathbf{Q}$, we take $\mathfrak{b} = r\mathbf{Z}$ with $r \in \mathbf{Q}^\times$; thus (22) includes (8) as a special case. It should be noted that $\Lambda[q,\, \mathfrak{b}]$ is not necessarily contained in Λ, even when $\mathfrak{b} \subset \mathfrak{g}$.

We fix a maximal lattice L, take $h \in V$ such that $q = \varphi[h] \neq 0$, and put $\mathfrak{b} = \varphi(h,\, L)$. (This means: take q and $\mathfrak{b}$ so that $L[q,\, \mathfrak{b}] \neq \emptyset$, and take $h \in L[q,\, \mathfrak{b}]$.) Put $W = (Fh)^\perp$, that is,

$$\text{(23)}\qquad W = \big\{x \in V \,\big|\, \varphi(h,\, x) = 0\big\}.$$

Then put $G = SO^{\varphi}(V)$ and $H = SO^{\varphi}(W)$, where we use the letter φ also for its restriction to W. We view H as a subgroup of G as in (11). From Witt's theorem on quadratic forms we easily obtain

$$\#\{\{x \in V \mid \varphi[x] = q\}/G\} \leq 1 \text{ if } q \neq 0 \text{ and } n > 1. \tag{24}$$

Since H acts on V, we can define the H-class and the H-genus of a lattice in V. Now (9a, b) can be given in stronger and more precise forms as follows.

Theorem 1. *Let L, h, $\mathfrak{b}$, W, and H be as above. Let $\{L_i\}_{i\in I}$ be a complete set of representatives for the G-classes in the G-genus of L. Then there exists a bijection from $\bigsqcup_{i\in I}\{L_i[q, \mathfrak{b}]/\Gamma(L_i)\}$ onto the set of H-classes in the H-genus of L, and therefore*

$$\sum_{i\in I}\#\{L_i[q, \mathfrak{b}]/\Gamma(L_i)\} = \textit{the number of } H\textit{-classes in the } H\textit{-genus of } L. \tag{25}$$

$$\sum_{i\in I}\#L_i[q, \mathfrak{b}]/\#\Gamma(L_i) = \textit{the mass of } H \textit{ relative to } L \textit{ if } \varphi \textit{ is totally definite.} \tag{26}$$

Here the total definiteness means that F is totally real and φ is positive or negative definite at all archimedean primes of F. The mass of H relative to L is $\sum_{j\in J}[\Delta(M_j) : 1]^{-1}$, where $\{M_j\}_{j\in J}$ is a complete set of representatives for the H-classes in the H-genus of L and $\Delta(M_j) = \{\alpha \in H \mid M_j\alpha = M_j\}$. This is similar to (14a).

If $\{L_i\}_{i\in I}$ consists only of L (which is the case in the setting of (5a, b) or (6a, b)), the left-hand side of (25) is $\#\{L[q, \mathfrak{b}]/\Gamma(L)\}$, and equality (25) in this case is a precise version of (7). In general, $L_i[q, \mathfrak{b}]$ may be empty for some $i \in I$.

The bijection in the theorem can be given as follows. Given $k \in L_i[q, \mathfrak{b}]$, we can find, by (24), an element α of G such that $k = h\alpha$. Then we can show that $L_i\alpha^{-1}$ belongs to the H-genus of L, and so we assign the H-class of $L_i\alpha^{-1}$ to k. This gives the desired bijection.

The crucial point in the proof of Theorem 1 is the following local result.

Theorem 2. *Let L be a maximal lattice in V and p a prime number. If $h, k \in V_p$, $\varphi[h] = \varphi[k]$, and $\varphi(h, L_p) = \varphi(k, L_p)$, then there exists an element γ of $SO(V_p, \varphi_p)$ such that $h\gamma = k$ and $L_p\gamma = L_p$.*

In other words, the local analogue of $L[q, \mathfrak{b}]$, if nonempty, consists of a single orbit under the local analogue of $\Gamma(L)$. The fact can be generalized to the case of a number field. This theorem is stated in [22, Theorem 10.5] under the condition that $\det(\varphi_p)$ is a square times a unit if n is odd, but actually that condition is unnecessary, as will be shown in [24]. In any case, the proof of Theorem 2 is long. We first classify the structures (V, φ, L) over a local field F as follows. Denoting by $\mathfrak{g}$ the valuation ring of F, we can put

$$V = \sum_{i=1}^{r}(Fe_i + Ff_i) \oplus Z, \qquad L = \sum_{i=1}^{r}(\mathfrak{g}e_i + \mathfrak{g}f_i) \oplus M,$$

$$2\varphi(e_i, f_j) = \delta_{ij}, \quad \varphi(e_i, e_j) = \varphi(f_i, f_j) = 0, \quad 0 \leq n - 2r = t = \dim(Z) \leq 4.$$

$$\varphi[z] \neq 0 \text{ if } 0 \neq z \in Z, \quad M \text{ is a maximal lattice in } Z.$$

The structure (V, φ, L) is determined by (r, Z). If we denote by ζ the restriction of φ to Z, then (Z, ζ) is determined by $\delta(\zeta)$ if $t \neq 2$; for $t = 2$, there are exactly two types of (Z, ζ) with the same $\delta(\zeta)$. (For these, see [6] and [22, §§6 and 7].) Then we prove Theorem 2 (in a generalized form) for each type of (V, φ, L); we have to treat the cases separately according to the nature of the ideal $q\mathfrak{g}$ and also to whether or not the prime ideal $\mathfrak{p}$ divides 2. For instance, take the simplest case in which $\varphi[h] = q \in \mathbf{Z}_p^\times$, $p \neq 2$, and $\varphi(h, L_p) = \mathbf{Z}_p$. Then we have $L_p = \mathbf{Z}_p h \oplus (L_p \cap (\mathbf{Q}_p h)^\perp)$, from which we easily obtain the desired fact.

Once Theorem 2 is established, Theorem 1 can be proved easily; see the proof of [22, Theorems 11.6 and 13.12]. Also, Theorem 2 is analogous to the fact that every local ideal is a principal ideal; that is, the local class number is always 1. It may be emphasized that the proof of Theorem 1 is purely arithmetic and involves no analysis. Formulas (5b) and (6b) are special cases of (25), and (6c) is a special case of (26). In such cases, $H = \{x \in K \mid N_{K/\mathbf{Q}}(x) = 1\}$ with a quadratic field K.

If $F = \mathbf{Q}$ and $\mathfrak{g} = \mathbf{Z}$, it is natural to consider $L[q, \mathbf{Z}]$, but the matter is not so simple. First of all, it is a highly nontrivial question to determine for which value of q the set $L[q, \mathbf{Z}]$ is not empty, as can be seen from the case of the sum of three squares. Indeed, it is nontrivial, if elementary, to settle this point and to show that the right-hand side of (25) in this case can be given as in (6b); the reader is referred to [22, pp. 118–119]. Next, it is not always best to formulate the result with respect to a $\mathbf{Z}$-basis of L. We will illustrate these points by numerical examples in Theorem 6.

Let us now state several consequences of Theorem 1 in the following four theorems.

Theorem 3. (i) *Suppose $n \geq 7$ and the class number of F in the narrow sense is odd. Suppose also that φ represents 0 nontrivially at an archimedean prime of F, and the same holds for the restriction of φ to W. Then $\#\{L[q, \mathfrak{b}]/\Gamma(L)\} = 1$.*

(ii) *Suppose $F = \mathbf{Q}$ and $\varphi[x] = \sum_{i=1}^n x_i^2$ with $n = 5, 7,$ or 9. Let L be a maximal $\mathbf{Z}$-lattice in $V = \mathbf{Q}^n$ and q an odd prime number. If $n > 5$, suppose $q = |d_K|$, where d_K is the discriminant of $K = \mathbf{Q}(\sqrt{\kappa})$, $\kappa = (-1)^{(n-1)/2}q$. Then $L[q, \mathbf{Z}] \neq \emptyset$, and $\#\{L[q, \mathbf{Z}]/\Gamma(L)\}$ equals the class number of $SO(\psi)$ with respect to the stabilizer of a maximal lattice in $\mathbf{Q}^{n-1}$, where $\psi = \mathrm{diag}[1_{n-4}, q1_3]$.*

For the proof, see [22, Lemma 12.13, Theorems 12.1 and 12.14]. We note here only that (i) follows from (25) combined with strong approximation.

It seems that the result of Dedekind mentioned in the paragraph after (5b) has not been generalized in a clear-cut form over an arbitrary algebraic number field F. Therefore let us now consider binary forms $ax^2 + bxy + cy^2$ over F and present a generalization of (5b). We consider (V, φ) by taking $V = \{h \in M_2(F) \mid {}^t h = h\}$ and by putting $\varphi[h] = \det(h)$ for $h \in V$. We associate the element

$$h = \begin{bmatrix} a & b/2 \\ b/2 & c \end{bmatrix}$$

of V to the form $ax^2+bxy+cy^2$; then $-\varphi[2h] = b^2-4ac$, which is the discriminant of the binary form. For $\Lambda = M_2(\mathfrak{g}) \cap V$, an element $q \in F^\times$, and a fractional ideal $\mathfrak{b}$ we consider the elements $h \in V$ such that $\varphi[h] = q$ and $\varphi(h, \Lambda) = \mathfrak{b}$, which constitute the set $\Lambda[q, \mathfrak{b}]$. Let Δ be the group of all transformations $x \mapsto \det(\alpha)^{-1} \cdot {}^t\alpha x \alpha$ with $\alpha \in GL_2(\mathfrak{g})$. Then we can prove:

Theorem 4. *Suppose $\Lambda[q, \mathfrak{b}] \neq \emptyset$ and $-q$ is not a square in F. Then there exists an order $\mathfrak{f}$ of discriminant $q\mathfrak{b}^{-2}$ in the field $F(\sqrt{-q})$. Moreover, $\#\{\Lambda[q, \mathfrak{b}]/\Delta\} =$*

$\varepsilon c(\mathfrak{f})/c_F$, where $c(\mathfrak{f})$ is the class number of $\mathfrak{f}$, c_F is the class number of F, $\varepsilon = 2$ if all archimedean and nonarchimedean primes of F are unramified in $F(\sqrt{-q})$, and $\varepsilon = 1$ otherwise.

Though this is essentially a special case of (25), we need to reformulate it in terms of the even Clifford group of (V, φ); see [22, Theorem 12.9]. If $F = \mathbf{Q}$ and $\mathfrak{b} = \mathbf{Z}$, Theorem 4 gives the result of Dedekind and also (5b). It can be shown that $[\Gamma(\Lambda) : \Delta]$ equals the number of the ideal classes in F whose squares are the principal class; see [22, Lemma 12.10]. Also, if $q\mathfrak{b}^{-2}$ is the discriminant of an order in $F(\sqrt{-q})$ and c_F is odd, then $\Lambda[q, \mathfrak{b}] \neq \emptyset$, but in general $\Lambda[q, \mathfrak{b}]$ can be empty even when such an order exists.

Theorem 5. *Let n, φ, and L be as in Theorem 3 (ii); let $K = \mathbf{Q}(\sqrt{\kappa})$, $\kappa = (-1)^{(n-1)/2}q$, with a squarefree positive integer q; let $L(s, \chi)$ be the L-function of the primitive Dirichlet character χ corresponding to K. Then*

$$\#L[q, 2^{-1}\mathbf{Z}] = A_n(q)q^{(n-2)/2}(2\pi)^{-m}L(m, \chi),$$
$$\#L[q, \mathbf{Z}] = c_n(q) \cdot \#L[q, 2^{-1}\mathbf{Z}].$$

Here $m = (n-1)/2$, $0 < A_n(q) \in \mathbf{Z}$, and $0 \leq c_n(q) \in \mathbf{Q}$; the numbers $A_n(q)$ and $c_n(q)$ are determined explicitly by q (mod 8) and n. For example,

$$A_7(q) = 2^9 \cdot 3^2 \cdot 7 \quad \text{and} \quad c_7(q) = 0 \quad \text{if} \quad q - 3 \notin 4\mathbf{Z},$$
$$A_9(q) = 2^9 \cdot 3^2 \cdot 5 \cdot 17 \quad \text{and} \quad c_9(q) = 1/136 \quad \text{if} \quad q - 5 \in 8\mathbf{Z}.$$

This theorem will be explained in Section 6, (R2).

As we said earlier, it is not always best to formulate the result with respect to a $\mathbf{Z}$-basis of L. To see this, first take the standard basis $\{e_i\}_{i=1}^n$ of $\mathbf{Z}^n = L$ and define a matrix Φ by $\Phi = [\varphi(e_i, e_j)]_{i,j=1}^n$; put $f_i = e_i\Phi^{-1}$ and

$$\widetilde{L} = \{x \in V \mid 2\varphi(x, L) \subset \mathbf{Z}\}.$$

Then $2\widetilde{L} = \sum_{i=1}^n \mathbf{Z}f_i$ and $\Gamma(\widetilde{L}) = \Gamma(L)$. Let $h \in L[q, \mathbf{Z}]$ with $q \in \mathbf{Q}^\times$. Then $h \in 2\widetilde{L}$, and so $h = \sum_{i=1}^n a_if_i$ with $a_i \in \mathbf{Z}$. We easily see that $h \in L[q, \mathbf{Z}]$ if and only if $\sum_{i=1}^n a_i\mathbf{Z} = \mathbf{Z}$, and also that $\Phi^{-1} = [\varphi(f_i, f_j)]_{i,j=1}^n$. This means that

$$L[q, \mathbf{Z}] = \left\{h = \sum_{i=1}^n a_if_i \mid a\Phi^{-1} \cdot {}^ta = q, \ \sum_{i=1}^n a_i\mathbf{Z} = \mathbf{Z}\right\}, \quad a = (a_i)_{i=1}^n.$$

Therefore if we follow the traditional definition of primitivity, we have to use the matrix Φ^{-1} instead of Φ. Take, for example, $n = 3$ and

$$2\Phi = \operatorname{diag}\left[2, \begin{bmatrix} 4 & -1 \\ -1 & 2 \end{bmatrix}\right], \qquad 7\Phi^{-1} = \operatorname{diag}\left[7, \begin{bmatrix} 4 & 2 \\ 2 & 8 \end{bmatrix}\right].$$

Then the result about $L[q, \mathbf{Z}]$ can be stated in terms of the equation $a\Phi^{-1} \cdot {}^ta = q = s/7$ with $s \in \mathbf{Z}$ and $a \in \mathbf{Z}^3$ as follows.

Theorem 6. *For $0 < s \in \mathbf{Z}$ put $K = \mathbf{Q}(\sqrt{-s})$. Then a vector (a, b, c) such that*

$$7a^2 + 4(b^2 + bc + 2c^2) = s \quad \text{and} \quad a\mathbf{Z} + b\mathbf{Z} + c\mathbf{Z} = \mathbf{Z}$$

exists if and only if $s = r^2m$ with a positive integer r and a squarefree positive integer m such that 7 does not split in K, $7 \nmid r$, and $2 | r$ if $m + 1 \notin 4\mathbf{Z}$. Moreover, *the number of such* (a, b, c) *equals*

$$\frac{2^{3-\mu}h_K}{w}\cdot f\prod_{p|f}\left\{1-\left(\frac{-m}{p}\right)p^{-1}\right\}.$$

Here w *is the number of roots of unity in* K, $\mu = 1$ *if* $7|m$ *and* $\mu = 0$ *otherwise,* h_K *is the class number of* K, $f = r$ *if* $m+1 \in 4\mathbf{Z}$ *and* $f = r/2$ *otherwise.*

In fact, this concerns one of the 30 ternary forms for which we can determine exactly when $L[q, \mathbf{Z}] \neq \emptyset$ and also $\#L[q, \mathbf{Z}]$. The details will be given in [24].

4. Formulation in terms of adeles and generalizations

Let us now formulate various things in terms of the adelization $G_{\mathbf{A}}$ of G, assuming that the reader is familiar with that notion. An easy introduction to this topic can be found in [17, Section 8]. We denote by $\mathbf{a}$ and $\mathbf{h}$ the sets of archimedean primes and nonarchimedean primes of F respectively, and put $\mathbf{v} = \mathbf{a} \cup \mathbf{h}$. Given an algebraic group G defined over F, we define G_v for each $v \in \mathbf{v}$ and the adelization $G_{\mathbf{A}}$ as usual and view G and G_v as subgroups of $G_{\mathbf{A}}$. (We do not use the symbol $G_{\mathbf{Q}}$ or G_F.) We then denote by $G_{\mathbf{a}}$ and $G_{\mathbf{h}}$ the archimedean and nonarchimedean factors of $G_{\mathbf{A}}$, respectively. For $x \in G_{\mathbf{A}}$ we denote by x_v its projection to G_v.

Take $G = SO(\varphi)$, though we can take G to be an arbitrary reductive algebraic subgroup of $GL(V)$. For $x \in G_{\mathbf{A}}$ and a lattice L in V we denote by Lx the lattice in V such that $(Lx)_v = L_v x_v$ for every $v \in \mathbf{h}$. Define a subgroup U of $G_{\mathbf{A}}$ by $U = \{x \in G_{\mathbf{A}} \mid Lx = L\}$. This means $U = G_{\mathbf{a}}\prod_{v\in\mathbf{h}} U_v$, $U_v = \{x \in G_v \mid L_v x = L_v\}$. Clearly $xU \mapsto Lx^{-1}$ gives a surjection of $G_{\mathbf{A}}/U$ onto the genus of L. Therefore the map gives a bijection of $G\backslash G_{\mathbf{A}}/U$ onto the set of classes in that genus. Thus $\#(G\backslash G_{\mathbf{A}}/U)$ is the class number of the genus of L.

More generally, given an open subgroup D of $G_{\mathbf{A}}$ containing $G_{\mathbf{a}}$ and such that $D \cap G_{\mathbf{h}}$ is compact, we put $\Gamma^a = G \cap aDa^{-1}$ for every $a \in G_{\mathbf{A}}$. Let $C_{\mathbf{a}}$ be a maximal compact subgroup of $G_{\mathbf{a}}$ and let $\mathcal{Z} = G_{\mathbf{a}}/C_{\mathbf{a}}$. Then $\mathcal{Z}$ is a symmetric space on which G acts through its projection into $G_{\mathbf{a}}$. Taking a complete set of representatives $\mathcal{B}$ for $G\backslash G_{\mathbf{A}}/D$, we put

$$\mathfrak{m}(G, D) = \mathfrak{m}(D) = \sum_{a\in\mathcal{B}} \nu(\Gamma^a). \tag{27}$$

Here $\nu(\Gamma)$ is a quantity defined by

$$\nu(\Gamma) = \begin{cases} [\Gamma : 1]^{-1} & \text{if } G_{\mathbf{a}} \text{ is compact,} \\ \mathrm{vol}(\Gamma\backslash\mathcal{Z})/\#(\Gamma \cap T) & \text{otherwise,} \end{cases} \tag{27a}$$

where $T = \{\alpha \in G \mid \alpha = \text{id. on } \mathcal{Z}\}$, and we fix a Haar measure on $G_{\mathbf{a}}$, which determines a unique $G_{\mathbf{a}}$-invariant measure on $\mathcal{Z}$ by a well known principle; see (37) below. (We can give a uniform definition of $\nu(\Gamma)$ by understanding that $\mathrm{vol}(\Gamma\backslash\mathcal{Z}) = [\Gamma : \Gamma \cap T]^{-1}$ if $G_{\mathbf{a}}$ is compact.) We easily see that $\mathfrak{m}(D)$ does not depend on the choice of $\mathcal{B}$. We call $\mathfrak{m}(G, D)$ *the mass of* G *relative to* D. This coincides with the mass defined by (14a) if $D = \{x \in G_{\mathbf{A}} \mid Lx = L\}$. Also, $\#\mathcal{B}$ may be called the class number of G with respect to D. In Theorem 9 we shall give more conceptual definitions of $\mathfrak{m}(D)$ and $\nu(\Gamma)$. If $G_{\mathbf{a}}$ is compact, we have

$$\mathfrak{m}(G, D) = \mathfrak{m}(D) = \sum_{a\in\mathcal{B}} [\Gamma^a : 1]^{-1}. \tag{28}$$

Let us now present generalizations of (25) and (26) to the case of a Diophantine equation of the form $\xi\Phi\cdot{}^t\xi = \Psi$ with a nonscalar Ψ mentioned in Section 1. Instead of a matrix Ψ, take a quadratic space (X, ψ) over F of dimension m $(< n)$, and put $\mathcal{V} = \mathrm{Hom}(X, V)$. For $u \in \mathcal{V}$ we define $\varphi[u]$ to be a quadratic form on X defined by $\varphi[u][x] = \varphi[xu]$ for every $x \in X$. (In other words, $\varphi[u] = u\varphi\cdot{}^tu$.) We look for u such that $\varphi[u] = \psi$. Fixing $h \in \mathcal{V}$ such that $\varphi[h] = \psi$, put $W = (Xh)^{\perp}$, $G = SO^{\varphi}(V)$, and $H = SO^{\varphi}(W)$. Clearly $\dim(W) = n - m$. We have

$$\{k \in \mathcal{V} \mid \varphi[k] = \varphi[h]\} = h\cdot G. \tag{29}$$

This is a generalization of (24) and follows easily from the Witt theorem. We identify H with $\{\alpha \in G \mid h\alpha = h\}$. For $x \in G_{\mathbf{A}}$ and $h \in \mathcal{V}$ the symbol hx is meaningful as an element of $\mathcal{V}_{\mathbf{A}}$, and so $h\Xi$ is meaningful as a subset of $\mathcal{V}_{\mathbf{A}}$ for any subset Ξ of $G_{\mathbf{A}}$. Then $\mathcal{V}\cap h\Xi$ is meaningful.

Theorem 7. *The symbols h, W, G and H being as above, let $D = D_0G_{\mathbf{a}}$ with an open compact subgroup D_0 of $G_{\mathbf{h}}$. Then the following assertions hold.*

(i) *For $y \in G_{\mathbf{A}}$ we have $H_{\mathbf{A}} \cap GyD \neq \emptyset$ if and only if $\mathcal{V}\cap hDy^{-1} \neq \emptyset$.*

(ii) *Fixing $y \in G_{\mathbf{A}}$, for every $\varepsilon \in H_{\mathbf{A}} \cap GyD$ take $\alpha \in G$ so that $\varepsilon \in \alpha yD$. Then the map $\varepsilon \mapsto h\alpha$ gives a bijection of $H\backslash(H_{\mathbf{A}}\cap GyD)/(H_{\mathbf{A}}\cap D)$ onto $(\mathcal{V}\cap hDy^{-1})/\Gamma^y$, where $\Gamma^y = G\cap yDy^{-1}$.*

(iii) *Take $Y \subset G_{\mathbf{A}}$ so that $G_{\mathbf{A}} = \bigsqcup_{y\in Y} GyD$. Then*

$$\#\{H\backslash H_{\mathbf{A}}/(H_{\mathbf{A}}\cap D)\} = \sum_{y\in Y}\#\{(\mathcal{V}\cap hDy^{-1})/\Gamma^y\}. \tag{30}$$

Proof. It is not very difficult to prove (i) and (ii); the only essential point is (29). The details of the proof will be given in [24]. Once (ii) is established, (iii) follows from the fact that $H_{\mathbf{A}} = \bigsqcup_{y\in Y}(H_{\mathbf{A}}\cap GyD)$. □

We are tempted to call the set $\mathcal{V}\cap hD$ *the genus* of h. Then $\#\{(\mathcal{V}\cap hD)/\Gamma\}$ may be called the *number of classes* in that genus.

Now (30) is a generalization of (25). Before explaining that point, let us present a generalization of (26). This requires a generalization of $\mathfrak{m}(X)$ of (18). We consider a subset S of $\mathcal{V}$ of the form $S = \bigsqcup_{\zeta\in Z} h\zeta\Gamma$, where h is as above, Z is a finite subset of G, and $\Gamma = G\cap D$ with D as in the above theorem. Then we put

$$\mathfrak{m}(S) = \sum_{\zeta\in Z}\nu(\Delta_\zeta)/\nu(\Gamma), \quad \Delta_\zeta = H\cap\zeta\Gamma\zeta^{-1} \tag{31}$$

and call $\mathfrak{m}(S)$ the *mass* of S. Here, to define $\nu(\Delta_\zeta)$, we need to fix a measure on $H_{\mathbf{a}}$. Thus $\mathfrak{m}(S)$ depends on the choice of measures on $G_{\mathbf{a}}$ and $H_{\mathbf{a}}$, but $\mathfrak{m}(S)$ is independent of the choice of Z and Γ, as will be shown in Section 5. (Let $\mathfrak{Y} = \{x \in \mathcal{V}_{\mathbf{a}} \mid \varphi[x] = q\}$. Since $\mathfrak{Y}$ can be identified with $H_{\mathbf{a}}\backslash G_{\mathbf{a}}$, once a measure on $G_{\mathbf{a}}$ is fixed, a measure on $\mathfrak{Y}$ determines a measure on $H_{\mathbf{a}}$, and vice versa. The replacement of h by an element of hG changes the group H, but that does not change $\mathfrak{m}(S)$ if we start with a fixed measure on $\mathfrak{Y}$. Notice also that an identification of $\mathcal{V}_{\mathbf{a}}$ with a Euclidean space determines a $G_{\mathbf{a}}$-invariant measure on $\mathfrak{Y}$.) Since the left-hand side of (30) is finite, we see that $(\mathcal{V}\cap hDy^{-1})/\Gamma^y$ is a finite set for every $y \in G_{\mathbf{A}}$. Thus we can define $\mathfrak{m}(\mathcal{V}\cap hDy^{-1})$ for every $y \in G_{\mathbf{A}}$.

If $G_{\mathbf{a}}$ is compact, we naturally take the measures of $G_{\mathbf{a}}$ and $H_{\mathbf{a}}$ to be 1. Then $\mathfrak{m}(S)$ can be defined in a unique way. We easily see that $S = \bigsqcup_{\zeta\in Z}\bigsqcup_{\gamma\in\Delta'_\zeta\backslash\Gamma} h\zeta\gamma$, where $\Delta'_\zeta = \zeta^{-1}\Delta_\zeta\zeta$, and so $\#S = \sum_{\zeta\in Z}[\Gamma : \Delta'_\zeta] = \mathfrak{m}(S)$ if $G_\mathbf{a}$ is compact. Thus we obtain

(32) $$\mathfrak{m}(S) = \#(S) \text{ if } G_\mathbf{a} \text{ is compact.}$$

Theorem 8. *The notation being the same as in Theorem 7, we have*

(33) $$\mathfrak{m}(H,\, H_\mathbf{A}\cap D) = \sum_{y\in Y}\nu(\Gamma^y)\mathfrak{m}(\mathcal{V}\cap hDy^{-1}).$$

Proof. Let $E_y = H\backslash(H_\mathbf{A}\cap GyD)/(H_\mathbf{A}\cap D)$. For each $\varepsilon\in E_y$ take $\zeta_\varepsilon\in G$ so that $\varepsilon\in\zeta_\varepsilon yD$. By (ii) of Theorem 7 we obtain $\mathcal{V}\cap hDy^{-1} = \bigsqcup_{\varepsilon\in E_y} h\zeta_\varepsilon\Gamma^y$. Then we obtain (33) by several lines of easy calculations; see [24] for details. □

To explain that (30) and (33) are generalizations of (25) and (26), take $m=1$, $X=F$, and $\psi[u] = qu^2$ for $u\in F$ with a fixed $q\in F^\times$. Then every element k of V defines an element of $\mathcal{V} = \mathrm{Hom}(F,\, V)$ that maps u to uk. Thus we can put $\mathcal{V} = V$. Fix $h\in V$ such that $\varphi[h] = q\in F^\times$, put $G = SO^\varphi(V)$, $W = (Fh)^\perp$, and $H = \{\alpha\in G \mid h\alpha = h\} = SO^\varphi(W)$. Take $\{y_i\}_{i\in I}\subset G_\mathbf{A}$ so that $G_\mathbf{A} = \bigsqcup_{i\in I} Gy_iD$, and put $\Gamma_i = G\cap y_iDy_i^{-1}$. Then (30) and (33) can be written

(34a) $$\#\{H\backslash H_\mathbf{A}/(H_\mathbf{A}\cap D)\} = \sum_{i\in I}\#\{[V\cap hDy_i^{-1}]/\Gamma_i\},$$

(34b) $$\mathfrak{m}(H,\, H_\mathbf{A}\cap D) = \sum_{i\in I}\#\{V\cap hDy_i^{-1}\}/\#\Gamma_i \quad \text{if } G_\mathbf{a} \text{ is compact.}$$

Here D is arbitrary, but now we fix a maximal lattice L in V and take $D = \{x\in G_\mathbf{A} \mid Lx = L\}$. Then Theorem 2 implies that if $h\in L[q,\, \mathfrak{b}]$, then $L[q,\, \mathfrak{b}] = V\cap hD$. More generally we can prove (see [24] for the proof)

(35) $$V\cap hDy^{-1} = (Ly^{-1})[q,\, \mathfrak{b}] \text{ for every } y\in G_\mathbf{A} \text{ if } n>2 \text{ and } h\in L[q,\, \mathfrak{b}].$$

Combining this result with (34a, b) and putting $L_i = Ly_i^{-1}$, we obtain

(36a) $$\#\{H\backslash H_\mathbf{A}/(H_\mathbf{A}\cap D)\} = \sum_{i\in I}\#\{L_i[q,\, \mathfrak{b}]/\Gamma_i\},$$

(36b) $$\mathfrak{m}(H,\, H_\mathbf{A}\cap D) = \sum_{i\in I}\#\{L_i[q,\, \mathfrak{b}]\}/\#\Gamma_i \quad \text{if } G_\mathbf{a} \text{ is compact.}$$

These are exactly (25) and (26). Now the sum of the left-hand side of (20) is similar to the sum of (36b): we have $L_i[q,\, \mathfrak{b}]$ in (36b) instead of $L_i[q]$ in (20). But there is another essential difference between (20) and (36b): namely the right-hand side of (20) is the infinite product of type (19), but the right-hand side of (36b) is the mass of an $(n-1)$-dimensional group H. It can happen that $H_\mathbf{A}\cap D = \{y\in H_\mathbf{A} \mid \Lambda y = \Lambda\}$ with a maximal lattice Λ in W. In such a case, the left-hand side of (36b) can be given by (17) with $(H,\, \Lambda)$ in place of $(G,\, L)$. Actually $\mathfrak{m}(H,\, H_\mathbf{A}\cap D)$ for a general type of D is $\mathfrak{m}(H,\, \Lambda)$ times a certain group index, as will be shown in Theorem 10.

5. Some more comments on the mass

Take G to be $SO^\varphi(V)$ or more generally a reductive algebraic subgroup of $GL(V)$; define $\mathcal{Z}$ as in Section 4. We fix a Haar measure dg on $G_\mathbf{a}$ and a measure dw on $\mathcal{Z}$ so that

(37) $$\int_{G_\mathbf{a}} f(g(\mathbf{1}))dg = \int_{\mathcal{Z}} f(w)dw$$

for every integrable function f on $\mathcal{Z}$, where $\mathbf{1}$ is the origin of $\mathcal{Z}$ represented by 1. Now we fix a Haar measure on $G_{\mathbf{A}}$ by the condition that the total measure of $G\backslash G_{\mathbf{A}}$ is one. (Recall that G is a discrete subgroup of $G_{\mathbf{A}}$.) Then we have a well defined measure on $G_{\mathbf{h}}$, as $G_{\mathbf{A}} = G_{\mathbf{a}}G_{\mathbf{h}}$. For a measurable subset S of $G_{\mathbf{a}}$, $G_{\mathbf{h}}$, or $\mathcal{Z}$ we denote by $\mathrm{vol}(S)$ the measure of S.

Theorem 9. *Let D be a subgroup of $G_{\mathbf{A}}$ of the form $D = D_0G_{\mathbf{a}}$ with an open compact subgroup D_0 of $G_{\mathbf{h}}$, and let $\Gamma = G \cap D$. Then*

$$\mathfrak{m}(G,\, D) = \mathrm{vol}(D_0)^{-1}, \tag{38}$$

$$\nu(\Gamma) = \mathrm{vol}(\Gamma\backslash G_{\mathbf{a}}). \tag{39}$$

Proof. The last formula is obvious if $G_{\mathbf{a}}$ is compact, so we assume that $G_{\mathbf{a}}$ is not compact. Let $\mathfrak{F}$ be a fundamental domain for $\Gamma\backslash\mathcal{Z}$ and let $Y = \{g \in G_{\mathbf{a}} \,|\, g(\mathbf{1}) \in \mathfrak{F}\}$. By (37) we have $\mathrm{vol}(\mathfrak{F}) = \mathrm{vol}(Y)$. Now we easily see that $(\Gamma \cap T)\backslash Y$ gives $\Gamma\backslash G_{\mathbf{a}}$, where T is as in (27a), and so $\mathrm{vol}(\Gamma\backslash G_{\mathbf{a}}) = [\Gamma \cap T : 1]^{-1}\mathrm{vol}(Y)$, which proves (39). (To be precise, we have to eliminate nontrivial fixed points of Γ from $\mathfrak{F}$ and ignore subsets of measure 0. Alternatively, we can replace Γ by a sugroup with no nontrivial fixed points. Then we prove (iii) of the following theorem, which can be done in an elementary way; see [17, Lemma 24.2 (2)].) Next, take $\mathcal{B} \subset G_{\mathbf{h}}$ so that $G_{\mathbf{A}} = \bigsqcup_{a\in\mathcal{B}} GaD$. Then $G\backslash G_{\mathbf{A}} = \bigsqcup_{a\in\mathcal{B}}(G\backslash GaD)$, and $G\backslash GaD$ is represented by $\Gamma^a\backslash aD$, which can be given by $(\Gamma^a\backslash G_{\mathbf{a}}) \times aD_0$. Therefore $\mathrm{vol}(G\backslash GaD) = \mathrm{vol}(\Gamma^a\backslash G_{\mathbf{a}})\mathrm{vol}(D_0)$. We have thus $1 = \mathrm{vol}(G\backslash G_{\mathbf{A}}) = \sum_{a\in\mathcal{B}} \mathrm{vol}(G\backslash GaD) = \sum_{a\in\mathcal{B}} \mathrm{vol}(\Gamma^a\backslash G_{\mathbf{a}})\mathrm{vol}(D_0)$, which combined with (27) and (39) proves (38). □

Theorem 10. *Let D be as above, and let $D' = D_0'G_{\mathbf{a}}$ with an open compact subgroup D_0' of $G_{\mathbf{h}}$; put $\Gamma = G \cap D$ and $\Gamma' = G \cap D'$. Then*

(i) $[\Gamma : \Gamma'] \leq [D : D'] < \infty$ *if* $D' \subset D$.

(ii) $[D : D \cap D']\mathfrak{m}(G, D) = [D' : D \cap D']\mathfrak{m}(G, D')$.

(iii) $[\Gamma : \Gamma \cap \Gamma']\nu(\Gamma) = [\Gamma' : \Gamma \cap \Gamma']\nu(\Gamma')$.

(iv) $\nu(\Gamma) = \nu(\alpha\Gamma\alpha^{-1})$ *for every* $\alpha \in G$ *and* $\mathfrak{m}(D) = \mathfrak{m}(xDx^{-1})$ *for every* $x \in G_{\mathbf{A}}$.

Proof. If $D' \subset D$, we have $[\Gamma : \Gamma'] = [\Gamma D' : D'] \leq [D : D'] = [D_0 : D_0'] < \infty$, as D_0' is open and D_0 is compact. This proves (i). The remaining three assertions follow immediately from (38) and (39). □

Let us now show that $\mathfrak{m}(S)$ of (31) can be defined independently of the choice of ζ and Γ. First of all, we easily see that ζ can be replaced by any element of $H\zeta\Gamma$ with the same value of $\nu(\Delta_\zeta)$. Now let $\Gamma' = G \cap D'$ with D' as in Theorem 10 contained in D. Then $h\zeta\Gamma = \bigsqcup_{\alpha\in A} h\zeta\alpha\Gamma$ with $A = \zeta^{-1}\Delta_\zeta\zeta\backslash\Gamma/\Gamma'$. Put $E_\alpha = H \cap \zeta\alpha\Gamma'\alpha^{-1}\zeta^{-1}$. By (iii) above we have

$$\sum_{\alpha\in A} \nu(E_\alpha)/\nu(\Gamma') = \sum_{\alpha\in A} [\Delta_\zeta : E_\alpha]\nu(\Delta_\zeta)/\nu(\Gamma') = \frac{\nu(\Delta_\zeta)}{\nu(\Gamma)[\Gamma : \Gamma']} \sum_{\alpha\in A} [\Delta_\zeta : E_\alpha]. \tag{$*$}$$

We easily see that $\zeta^{-1}\Delta_\zeta\zeta\alpha\Gamma' = \bigsqcup_{\beta\in B_\alpha} \beta\alpha\Gamma'$ with $B_\alpha = \zeta^{-1}\Delta_\zeta\zeta\backslash\zeta^{-1}E_\alpha\zeta$. Thus $\sum_\alpha [\Delta_\zeta : E_\alpha] = \sum_\alpha \#B_\alpha = [\Gamma : \Gamma']$, and so the quantity of $(*)$ equals $\nu(\Delta_\zeta)/\nu(\Gamma)$. This proves that $\mathfrak{m}(S)$ defined with respect to Γ' gives the same value as (31) as expected.

6. Additional comments on formula (21) and some remarks

Let us now discuss how to obtain (21) and also add various related facts. Assuming $F = \mathbf{Q}$ and φ to be definite, take a $\mathbf{Z}$-lattice L in V and take $\{L_i\}_{i\in I}$ as in (14a). We then put

$$\theta_i(z) = \sum_{x\in L_i} \exp\left(2\pi i\varphi[x]\right) = \sum_{q\in\mathbf{Q}} \#L_i[q]e^{2\pi iqz}, \quad z\in\mathbf{C},\ \mathrm{Im}(z) > 0, \tag{40}$$

$$f(z) = \Big\{\sum_{i\in I}[\Gamma_i:1]^{-1}\Big\}^{-1}\sum_{i\in I}[\Gamma_i:1]^{-1}\theta_i(z) = \sum_{q\in\mathbf{Q}} b(q)e^{2\pi iqz}, \tag{41}$$

where $\Gamma_i = \Gamma(L_i)$. Then θ_i is a modular form of weight $n/2$; see [15]. Formula (20) shows that $b(q)$, if nonzero, equals $e_\infty \prod_p e_p$.

Theorem 11. (i) *f is an Eisenstein series.*
(ii) *$\theta_i - f$ is a cusp form.*

Proof. For $n > 4$ Siegel proved (i) in [25, Satz 3]; (ii) was also shown in [25, p. 376]. An easier proof of (ii) valid even in the Hilbert modular case is given in [20, Theorem A4.3 (2)]. To prove (i) in general, we first note that

$$f(z) = \int_{G\backslash G_{\mathbf{A}}} \theta(z, g)dg, \qquad \left(\int_{G\backslash G_{\mathbf{A}}} dg = 1\right), \tag{42}$$
$$\text{where}\quad \theta(z, g) = \sum_{x\in Lg^{-1}} \exp\left(2\pi i\varphi[x]\right) \qquad (g\in G_{\mathbf{A}}).$$

Indeed, let $D = \{x \in G_{\mathbf{A}} \mid Lx = L\}$ and $G_{\mathbf{A}} = \bigsqcup_{i\in I} Ga_iD$. Then we can take La_i^{-1} as L_i. Putting $\Gamma^a = G\cap aDa^{-1}$ as in Section 4, we have $\Gamma^{a_i} = \Gamma_i$. Since $G\backslash G_{\mathbf{A}} = \bigsqcup_i (G\backslash Ga_iD)$ and $G\backslash GaD$ can be given by $\Gamma^a\backslash aD$ as observed in the proof of Theorem 9, we have

$$\int_{G\backslash G_{\mathbf{A}}} \theta(z, g)dg = \sum_{i\in I}\int_{\Gamma_i\backslash a_iD} \theta(z, g)dg.$$

Clearly $\theta(z, g) = \theta_i(z)$ if $g\in a_iD$, and so the last sum equals $\sum_i \mathrm{vol}(\Gamma_i\backslash a_iD)\theta_i(z)$ $= \sum_i \mathrm{vol}(D_0)\mathrm{vol}(\Gamma_i\backslash G_{\mathbf{a}})\theta_i(z)$ for the reason explained in the proof of Theorem 9. This combined with (38) and (39) gives (42). Now by virtue of the Siegel-Weil formula [30] we have, for $n > 4$,

$$\int_{G\backslash G_{\mathbf{A}}} \theta(z, g)dg = \text{an Eisenstein series explicitly depending on}(\varphi, L). \tag{43}$$

This proves (i) for $n > 4$. This is true even for $2 \le n \le 4$ by virtue of [11] and [14]; see also the explanation of this point in [19, pp. 86–87]. □

Now there is a technique of expressing each Fourier coefficient of an Eisenstein series as an infinite product of local integrals; see [17], [20], and the author's papers cited there. This gives a proof of the equality $b(q) = e_\infty \prod_p e_p$, with each e_p given as a local integral; for details, see [19]. Thus, as we said, the number of (21) is computable in that sense, but not $\#L_i[q]$ (unless $\#I = 1$). Next, here are two basic facts on an arbitrary elliptic modular form $h(z) = \sum_{q\in\mathbf{Q}} a_q e^{2\pi iqz}$ of weight k, where $0 < k \in 2^{-1}\mathbf{Z}$:

(A) *There is a constant* M *depending only on* h *such that* $|a_q| \le Mq^k$ *for every* $q \in \mathbf{Q}, > 0$. *Moreover, if* h *is a cusp form, then* M *can be taken so that* $|a_q| \le Mq^{k/2}$ *for every* $q \in \mathbf{Q}, > 0$.

(B) *There exist an Eisenstein series* h_0 *and a cusp form* h_1 *such that* $h = h_0 + h_1$. *Moreover, such* h_0 *and* h_1 *are unique for* h.

These are due to Hecke when $k \in \mathbf{Z}$. The proof for an arbitrary $k \in 2^{-1}Z$, even in the Hilbert modular case, can be found in [20, Proposition A6.4] and [16].

Thus the decomposition $\theta_i = f + (\theta_i - f)$ is an expression of type (B); also the estimate given in (A) means that f is "the dominant part" of θ_i. Consequently $b(q)$ gives $\#L_i[q]$ asymptotically.

Let us now add some remarks and a few problems.

(R1) The exact formulas (17) and (21) have been obtained when L is maximal. Now take $\varphi[x] = \sum_{i=1}^n x_i^2$. Then the lattice $\mathbf{Z}^n$ is maximal if and only if $n \le 3$, and therefore (21) is not applicable if $n > 3$. However, the problem can be reduced to the case of maximal lattices for such a φ, and so we can give exact formulas for $b(q)$ not only when L is maximal but also when $L = \mathbf{Z}^n$. It is classically known that the genus of $\mathbf{Z}^n$ has class number 1 for $1 \le n \le 8$. It can also be shown that the genus of maximal lattices has class number 1 for $1 \le n \le 9$. Thus, in all such cases we have explicit (and computable) formulas for $\#L[q]$. For details the reader is referred to [21]. Also, we can take, instead of a lattice L, a coset $M = L + y$ with $y \in V$ and define the genus, class, and mass with respect to M. Moreover, we can define the analogues of θ_i and f by taking M in place of L. Then Theorem 11 holds in that case. Formula (20) and what we said about $b(q)$ can be generalized too. The results can be used for the question about the representation of integers as sums of polygonal numbers. For all these, see [23].

(R2) Still with $\varphi[x] = \sum_{i=1}^n x_i^2$, let us now explain how to obtain $\#L[q, \mathbf{Z}]$ and $\#L[q, 2^{-1}\mathbf{Z}]$ as stated in Theorem 5. Here L is maximal. For $n \le 9$ formula (36b) shows that $\#L[q, \mathfrak{b}]/\#\Gamma(L) = \mathfrak{m}(H, H_\mathbf{A} \cap D)$, provided $L[q, \mathfrak{b}] \ne \emptyset$. Take q to be an odd prime number and $n = 5$, 7, or 9. Then $H_\mathbf{A} \cap D$ is the stabilizer of a maximal lattice in W, and so $\mathfrak{m}(H, H_\mathbf{A} \cap D)$ is computable by means of (17) with H as G there. Also $L[q] = L[q, \mathbf{Z}] \cup L[q, 2^{-1}\mathbf{Z}]$ if q is a squarefree positive integer. Now the formulas of Theorem 5 can be obtained by combining these facts and using Theorem 10 (ii); for details the reader is referred to [22, Theorem 13.14], where all the values of $A_n(q)$ and $c_n(q)$ are given.

(R3) The determination of $\#\mathbf{Z}^n[q]$ for $\varphi[x] = \sum_{i=1}^n x_i^2$ is a subject investigated by many mathematicians. We mention here only the paper [8] by Hardy, in which the problem was treated by the circle method. Also, the introduction of the paper includes a short history of the problem. The case of even n can be handled without difficulties, but for odd n he obtained only $\#(\mathbf{Z}^n)^0[q]$ for $n = 5$ and 7, but not $\#\mathbf{Z}^n[q]$. His idea was all right, but there were nontrivial mistakes, as Stanley pointed out in her paper [29] on $\#(\mathbf{Z}^n)^0[q]$ for $n = 7$. Hardy's own corrections were published in [9]. In his later articles on this topic he cited only [8] and never mentioned [9] or [29].

(R4) There is an old and natural method of treating the sum of three squares. Namely we consider a quaternion algebra $B = \mathbf{Q} + \mathbf{Q}i + \mathbf{Q}j + \mathbf{Q}k$ with the standard Hamilton quaternion units i, j, k. If $0 \ne q = a^2 + b^2 + c^2$ with $a, b, c \in \mathbf{Z}$, then $ai + bj + ck$ generates a subfield of B isomorphic to $\mathbf{Q}(\sqrt{-q})$. For a prime p, B_p is a division algebra if and only if $p = 2$, and so a well known principle says that the prime 2 cannot split in $\mathbf{Q}(\sqrt{-q})$. This is the easiest way to find the condition for the representability of q as a sum of three squares. Moreover, the problem about the number of representations can be reduced to that of counting the embeddings of $\sqrt{-q}$ into a maximal order of B. We can similarly treat some other ternary forms even over an algebraic number field by the same technique. This gives satisfactory results in some special cases, but in general we cannot obtain explicit formulas as good as what is given by (20). We mention here only [5] by Donkar, in which the basic ideas are explained and several references before 1975 are given.

(R5) As we already said, it is a highly nontrivial problem to determine when $L[q, \mathfrak{b}] \neq \emptyset$. For instance, in the setting of Theorem 5 we have $L[q, 2^{-1}\mathbf{Z}] \neq \emptyset$, but $L[q, \mathbf{Z}] = \emptyset$ can happen. Also, as Theorem 6 shows, the expected necessary and sufficient condition is not simple. Still we can ask if there exists a reasonably simple necessary or sufficient condition for $L[q, 2^{-1}\mathbf{Z}] \neq \emptyset$.

(R6) At present, formulas (30) and (33) are the best we can say about the problem for an arbitrary $m < n$, though we can add something more when $m = n - 1$. If $m = 1$, we have (35), which explains the meaning of "the genus" of h. We naturally ask: can we characterize the set $\mathcal{V} \cap hDy^{-1}$ in a similar way when $m > 1$?

(R7) As shown in Theorem 11 (ii) and explained after the proof, the weighted average of (20) gives an asymptotic value of $\#L[q]$. Is there any analogue of this phenomenon for $\#L[q, \mathfrak{b}]$? Namely, does the quantity

$$\left\{\sum_j (\#\Gamma_j)^{-1}\right\}^{-1} \sum_j \#L_j[q, \mathfrak{b}]/\#\Gamma_j$$

give the value $\#L[q, \mathfrak{b}]$ asymptotically? Here the notation is as in (36b), and j runs over the indices for which $L_j[q, \mathfrak{b}] \neq \emptyset$.

Acknowledgments

This article is an expanded version of my Gill Distinguished Lectures at the Department of Mathematics, University of California, Los Angeles, May 17, 19, and 20, 2005. The author thanks the school for giving him the chance to organize the ideas and also for their hospitality during his stay there.

References

[1] P. Bachmann, Die Arithmetik der Quadratischen Formen, Leipzig, Teubner, 1898.
[2] A. Borel, Some finiteness properties of adele groups over number fields, Publ. Math. IHES, No. 16, 1963. MR0202718 (34:2578)
[3] Z. I. Borevich and I. R. Shafarevich, Number Theory, Academic Press, New York and London, 1966. MR0195803 (33:4001)
[4] L. Dirichlet, Vorlesungen über Zahlentheorie, supplement by R. Dedekind, Braunschweig, 1894.
[5] E. N. Donkar, On sums of three integral squares in algebraic number fields, Amer. J. Math. 99 (1977), 1297–1328. MR0460286 (57:280)
[6] M. Eichler, Quadratische Formen und orthogonale Gruppen, Springer, Berlin, 1952; 2nd ed., 1974. MR0351996 (50:4484)
[7] C. F. Gauss, Disquisitiones Arithmeticae; 1801; English translation by A. A. Clarke, Yale Univ. Press, 1966. MR0197380 (33:5545)
[8] G. H. Hardy, On the representation of a number as the sum of any number of squares, and in particular of five, Trans. Amer. Math. Soc. 21 (1920), 255–284. MR1501144
[9] Hardy, Errata, Trans. Amer. Math. Soc. 29 (1927), 845–847. MR1500501

[10] E. Hecke, Vorlesungen über die Theorie der Algebraischen Zahlen, Leipzig, 1923; Chelsea, New York, 1948. MR0352036 (50:4524)
[11] S. Kudla and S. Rallis, On the Weil-Siegel formula, J. Reine Angew. Math. 387 (1988), 1–68. MR0946349 (90e:11059)
[12] Lagrange, Recherches d'Arithmétique, 1773–75, Oeuvres III, 695–758.
[13] Legendre, Recherches d'Analyses Indéterminée, 1785.
[14] S. Rallis, *L*-functions and the oscillator representation, Lecture Notes in Math., No. 1245, Springer, 1987. MR0887329 (89b:11046)
[15] G. Shimura, On modular forms of half integral weight, Ann. of Math. 97 (1973), 440–481 (= Collected Papers II, 532–573). MR0332663 (48:10989)
[16] G. Shimura, On the Eisenstein series of Hilbert modular groups, Revista Matemática Iberoamericana 1 (1985), 1–42 (= Collected Papers III, 644–685). MR0836282 (87h:11038)
[17] G. Shimura, Euler Products and Eisenstein Series, CBMS Regional Conference Series in Mathematics, No. 93, Amer. Math. Soc., 1997. MR1450866 (98h:11057)
[18] G. Shimura, An exact mass formula for orthogonal groups, Duke Math. J. 97 (1999), 1–66 (= Collected Papers IV, 509–574). MR1681092 (2000a:11073)
[19] G. Shimura, The number of representations of an integer by a quadratic form, Duke Math. J. 100 (1999), 59–92 (= Collected Papers IV, 575–608). MR1714755 (2001f:11057)
[20] G. Shimura, Arithmeticity in the theory of automorphic forms, Mathematical Surveys and Monographs, vol. 82, Amer. Math. Soc., 2000. MR1780262 (2001k:11086)
[21] G. Shimura, The representation of integers as sums of squares, Amer. J. Math. 124 (2002), 1059–1081. MR1925343 (2003i:11049)
[22] G. Shimura, Arithmetic and analytic theories of quadratic forms and Clifford groups, Mathematical Surveys and Monographs, vol. 109, Amer. Math. Soc., 2004. MR2027702 (2004m:11068)
[23] G. Shimura, Inhomogeneous quadratic forms and triangular numbers, Amer. J. Math. 126 (2004), 191–214. MR2033567 (2005a:11047)
[24] G. Shimura, Integer-valued quadratic forms and quadratic Diophantine equations, preprint 2005.
[25] C. L. Siegel, Über die analytische Theorie der quadratischen Formen, Ann. of Math. 36 (1935), 527–606 (= Gesammelte Abhandlungen I, 326–405). MR1503238
[26] C. L. Siegel, Über die analytische Theorie der quadratischen Formen II, Ann. of Math. 37 (1936), 230–263 (= Gesammelte Abhandlungen I, 410–443). MR1503276
[27] C. L. Siegel, Über die Zetafunktionen indefiniter quadratischer Formen II, Math. Zeitschr. 44 (1939), 398–426 (= Gesammelte Abhandlungen II, 68–96). MR1545778
[28] C. L. Siegel, Zur Theorie der quadratischen Formen, Nachr. Akad. Wiss. Göttingen, Math.-phys. Klasse, 1972, Nr. 3, 21–46 (= Gesammelte Abhandlungen IV, 224–249). MR0311578 (47:140)
[29] G. K. Stanley, On the representation of a number as the sum of seven squares, J. London Math. Soc. 2 (1927), 91–96.
[30] A. Weil, Sur la formule de Siegel dans la théorie des groupes classiques, Acta Math. 113 (1965), 1–87 (= Collected Papers III, 71–157). MR0223373 (36:6421)
[31] A. Weil, Number Theory, Birkhäuser, Boston, Basel, Stuttgart, 1984. MR0734177 (85c:01004)

Department of Mathematics, Princeton University, Princeton, NJ 08544-0001

Integer-Valued Quadratic Forms and Quadratic Diophantine Equations

Goro Shimura

Received: September 12, 2006

Communicated by Don Blasius

Abstract. We investigate several topics on a quadratic form Φ over an algebraic number field including the following three: (A) an equation $\xi\Phi \cdot {}^t\xi = \Psi$ for another form Ψ of a smaller size; (B) classification of Φ over the ring of algebraic integers; (C) ternary forms. In (A) we show that the "class" of such a ξ determines a "class" in the orthogonal group of a form Θ such that $\Phi \approx \Psi \oplus \Theta$. Such was done in [S3] when Ψ is a scalar. We will treat the case of nonscalar Ψ, and prove a class number formula and a mass formula, both of new types. In [S5] we classified all genera of $\mathbf{Z}$-valued Φ. We generalize this to the case of an arbitrary number field, which is topic (B). Topic (C) concerns some explicit forms of the formulas in (A) when Φ is of size 3 and Ψ is a scalar.

2000 Mathematics Subject Classification: 11E12 (primary), 11D09, 11E41 (secondary)

Introduction

A quadratic Diophantine equation in the title means an equation of the form $\xi\Phi \cdot {}^t\xi = \Psi$, where Φ and Ψ are symmetric matrices of size n and m, and ξ is an $(m \times n)$-matrix. We assume that $n > m$, $\det(\Phi)\det(\Psi) \neq 0$, and all these matrices have entries in an algebraic number field F. The purpose of this paper is to present various new ideas and new results on such an equation. In the simplest case $m = 1$, we take a vector space V of dimension n over F, take also a nondegenerate symmetric F-bilinear form $\varphi : V \times V \to F$, and put $\varphi[x] = \varphi(x, x)$ for $x \in V$. Then the equation can be written $\varphi[x] = q$ with $q \in F$, $\neq 0$. In our recent book [S3] we formulated a new arithmetic framework of such an equation. In the present paper we attempt to extend the theory so that equations of the type $\xi\Phi \cdot {}^t\xi = \Psi$ can be treated along the same line of ideas, and give essential improvements on the results in [S3], as well as some new results.

Shimura, Goro. "Integer-valued quadratic forms and quadratic Diophantine equations." Documenta Mathematica 11 (2006), 333–367.

Let us first recall the basic ideas of [S3] on this topic. Let $\mathfrak{g}$ denote the ring of algebraic integers in F. For a $\mathfrak{g}$-lattice L in V and a fractional ideal $\mathfrak{b}$ in F we put

$$L[q,\,\mathfrak{b}] = \{\, x \in V \mid \varphi[x] = q,\ \varphi(x,\,L) = \mathfrak{b}\,\},$$
$$\Gamma(L) = \{\, \gamma \in SO^{\varphi}(V) \mid L\gamma = L\},$$
$$SO^{\varphi}(V) = \{\, \alpha \in SL(V,\,F) \mid \varphi[x\alpha] = \varphi[x] \quad \text{for every} \quad x \in V\}.$$

We call L *integral* if $\varphi[x] \in \mathfrak{g}$ for every $x \in L$, and call an integral lattice *maximal* if it is the only integral lattice containing itself.

Given $h \in L[q,\,\mathfrak{b}]$, put $G = SO^{\varphi}(V)$, $W = (Fh)^{\perp}$, that is,

$$W = \{x \in V \mid \varphi(x,\,h) = 0\},$$

and $H = SO^{\varphi}(W)$, where we use φ also for its restriction to W. Then in [S3] we proved, for a maximal L, that

(1a) *There is a bijection of* $\bigsqcup_{i\in I}\{L_i[q,\,\mathfrak{b}]/\Gamma(L_i)\}$ *onto* $H\backslash H_{\mathbf{A}}/(H_{\mathbf{A}}\cap C)$, *and consequently*

(1b) $$\sum_{i\in I}\#\{L_i[q,\,\mathfrak{b}]/\Gamma(L_i)\} = \#\{H\backslash H_{\mathbf{A}}/(H_{\mathbf{A}}\cap C)\}.$$

Here the subscript $\mathbf{A}$ means adelization, $C = \{x \in G_{\mathbf{A}} \mid Lx = L\}$, and $\{L_i\}_{i\in I}$ is a complete set of representatives for the classes in the genus of L with respect to G. If $F = \mathbf{Q}$, $\mathfrak{g} = \mathfrak{b} = \mathbf{Z}$, $L = \mathbf{Z}^3$, and φ is the sum of three squares, then the result is a reformulation of the result of Gauss that connects the number of primitive representations of an integer q as sums of three squares with the class number of primitive binary forms of discriminant $-q$. This result can be formulated as a statement about $\#L[q,\,\mathbf{Z}]$ for such φ and L as Gauss did, and also as another statement concerning $\#\{L[q,\,\mathfrak{b}]/\Gamma(L)\}$ as in (1b), which Gauss did not present in a clear-cut form. Though Eisenstein and Minkowski investigated $\#L[q,\,\mathbf{Z}]$ when φ is the sum of five squares, neither (1a) nor (1b) appears in their work. It seems that the idea of $L[q,\,\mathfrak{b}]/\Gamma(L)$ was taken up for the first time in [S3].

Now the major portion of this paper consists of several types of new results, which are the fruits of the ideas developed from (1a) and (1b). More explicitly, they can be described as follows.

I. The main theorems about $\#\{L[q,\,\mathfrak{b}]/\Gamma(L)\}$ in [S3], formula (1b) in particular, were proved under the following condition: if n is odd, then $\det(\varphi)\mathfrak{g}$ is a square ideal. We will show that this condition is unnecessary.

II. The same types of problems about the equation $\xi\Phi\cdot{}^t\xi = \Psi$ for $m > 1$ were discussed in [S3] to some extent, but mainly restricted to the case $m = n-1$. In Theorem 2.2 we will prove more general results for an arbitrary m formulated from a new viewpoint. Namely, we present the principle that the class of a solution ξ determines a class in the orthogonal group in dimension $n - m$, which is complementary to that of Ψ. We then obtain generalizations of (1a) and (1b).

III. In [S3] we proved a certain mass formula which connects the "mass" of the set $L[q, \mathfrak{b}]$ with the mass of H with respect to a subgroup of $H_{\mathbf{A}}$. If F is totally real and φ is totally definite, this can be given in the form

$$\sum_{i \in I} \#L_i[q, \mathfrak{b}]/\#\Gamma(L_i) = \sum_{\varepsilon \in E} [H \cap \varepsilon C \varepsilon^{-1} : 1]^{-1}, \tag{2}$$

where E is a subset of $H_{\mathbf{A}}$ such that $H_{\mathbf{A}} = \bigsqcup_{\varepsilon \in E} H\varepsilon(H_{\mathbf{A}} \cap C)$. The right-hand side may be called the mass of H with respect to $H_{\mathbf{A}} \cap C$. In Theorem 3.2 we will generalize this to the case of $\xi\Phi \cdot {}^t\xi = \Psi$ with $m \geq 1$ and definite or indefinite φ. Our formulas are different from any of the mass formulas of Siegel. See the remark at the end of Section 3 for more on this point.

IV. For $n = 3$, the right-hand side of (1b) can be written $[H_{\mathbf{A}} : H(H_{\mathbf{A}} \cap C)]$. In [S3] we gave an explicit formula for this index under the condition mentioned in I. In Theorem 5.7 of the present paper we prove the result without that condition. In general it is difficult to determine when $L[q, \mathfrak{b}] \neq \emptyset$. We will investigate this problem for a ternary form.

Every ternary space is isomorphic to a space of type (B°, β) obtained from a quaternion algebra B over F as follows. Denote by ι the main involution of B and put $B^\circ = \{x \in B \mid x^\iota = -x\}$ and $\beta[x] = dxx^\iota$ for $x \in B^\circ$ with $d \in F^\times$. Then we will determine for a maximal L exactly when $L[q, \mathbf{Z}] \neq \emptyset$ and give an explicit formula for $\#L[q, \mathbf{Z}]$ for (B°, β) over $\mathbf{Q}$ under the following conditions: (i) B is definite and the genus of maximal lattices in B° consists of a single class; (ii) the discriminant e of B is a prime number; (iii) d is one of the following four types: $d = 1$, $d = e$, d is a prime $\neq e$, d/e is a prime $\neq e$ (Theorems 6.6 and 6.7). In fact there are exactly 30 positive definite ternary forms over $\mathbf{Q}$ satisfying these three conditions, including of course the case of the sum of three squares. If we drop conditions (ii) and (iii), then there are exactly 64 ternary quadratic spaces over $\mathbf{Q}$ of type (i). Though our methods are applicable to those 64 spaces, we impose the last two conditions in order to avoid complicated analysis.

Indeed, though the results about $L[q, \mathbf{Z}]$ are not so complicated, they are not of the type one can easily guess, even under all three conditions. Also, to make transparent statements, it is better to consider an equation $\xi \cdot \lambda\Phi^{-1} \cdot {}^t\xi = s = \lambda q$ with a suitable λ, where Φ is the matrix representing φ with respect to a $\mathbf{Z}$-basis of L. Take, for example, a ternary form $2x^2 + 3y^2 - yz + z^2$. Then we consider the equation $\xi \cdot \lambda\Phi^{-1} \cdot {}^t\xi = s$ for $s \in \mathbf{Z}$ with $\lambda = 22$, which can be written

$$11x^2 + 8(y^2 + yz + 3z^2) = s. \tag{3}$$

We can show that there is a bijection of $L[s/22, \mathbf{Z}]$ for this φ onto the set of solutions (x, y, z) of (3) such that $x\mathbf{Z} + y\mathbf{Z} + z\mathbf{Z} = \mathbf{Z}$. Moreover, such a solution exists if and only if $s = r^2 m$ with a squarefree positive integer m such that $m - 3 \in 8\mathbf{Z}$ and an odd positive integer r such that $11|r$ if 11 remains prime in K, where $K = \mathbf{Q}(\sqrt{-m})$. Thus $\#L[s/22, \mathbf{Z}]$ equals the number of such solutions of (3), and can be given as

$$\frac{2^{3-\mu_0-\mu_1}c}{w}\cdot r\prod_{p|r}\left\{1-\left(\frac{-m}{p}\right)\right\}. \tag{4}$$

Here $\mu_0 = 0$ if 2 is unramified in K and $\mu_0 = 1$ otherwise; $\mu_1 = 0$ if $11|r$ or $11\nmid m$, and $\mu_1 = 1$ otherwise; c is the class number of K; w is the number of roots of unity in K; $\prod_{p|r}$ is the product over all prime factors p of r. We can state similar results in the 30 cases mentioned above, and the same can be said, in principle, even in the 64 cases too, though we will not do so in the present paper. The condition that the class number is 1 is necessary, as the left-hand side of (2) has more than one term otherwise.

A much smaller portion of this paper, Section 4, is devoted to the classification of Φ over $\mathfrak{g}$. In [S5] we classified the genera of matrices that represent reduced $\mathbf{Z}$-valued quadratic forms. Here a quadratic form is called *reduced* if it cannot be represented nontrivially over $\mathbf{Z}$ by another $\mathbf{Z}$-valued quadratic form. This is different from Eisenstein's terminology for ternary forms. We will treat the same type of problem over the ring of algebraic integers of an arbitrary algebraic number field. For this we first have to define the genus of a symmetric matrix in a proper way, so that every genus of maximal lattices can be included. The formulation requires new concepts, and the classification has some interesting features. It should be noted that the obvious definition of a genus employed by Siegel applies only to a special case.

As a final remark we mention the article [S6], in which the reader will find a historical perspective of this topic that we do not include here. For example, [S6] contains a more detailed account of the work of Gauss and his predecessors, Lagrange and Legendre, and also comparisons of our formulas with Siegel's mass formulas. Therefore, [S6] is complementary to the present paper in that sense.

1. Basic symbols and a crucial local result

1.1. Throughout the paper we denote by V a finite-dimensional vector space over a field F and by φ a nondegenerate F-bilinear symmetric form $V\times V \to F$; we put then $\varphi[x] = \varphi(x, x)$ for $x \in V$, $n = \dim(V)$, and

$$O^\varphi(V) = \{\alpha \in GL(V)\,|\,\varphi[x\alpha] = \varphi[x] \text{ for every } x \in V\},$$
$$SO^\varphi(V) = O^\varphi(V) \cap SL(V).$$

For every subspace U of V on which φ is nondegenerate, we denote the restriction of φ to U also by φ, and use the symbols $O^\varphi(U)$ and $SO^\varphi(U)$. We denote by $A(V)$ the Clifford algebra of (V, φ), and define the *canonical automorphism* $\alpha \mapsto \alpha'$ and the *canonical involution* $\alpha \mapsto \alpha^*$ of $A(V)$ by the condition $-x' = x^* = x$ for every $x \in V$. We then put

$$A^+(V) = A^+(V, \varphi) = \{\alpha \in A(V)\,|\,\alpha' = \alpha\},$$
$$G^+(V) = \{\alpha \in A^+(V)^\times\,|\,\alpha^{-1}V\alpha = V\}.$$

We can define a homomorphism $\nu : G^+(V) \to F^\times$ by $\nu(\alpha) = \alpha\alpha^*$ and also a surjective homomorphism $\tau : G^+(V) \to SO^\varphi(V)$ by $x\tau(\alpha) = \alpha^{-1}x\alpha$ for $\alpha \in G^+(V)$ and $x \in V$. Then $\mathrm{Ker}(\tau) = F^\times$. We denote by $\delta(\varphi)$ the coset of $F^\times/F^{\times 2}$ represented by $(-1)^{n(n-1)/2}\det(\varphi)$, where $F^{\times 2} = \{a^2 \mid a \in F^\times\}$.

We note here an easy fact [S3, Lemma 1.5 (ii)]:

(1.1) $\{k \in V \mid \varphi[k] = q\} = h\cdot SO^\varphi(V)$ if $n > 1$, $h \in V$, and $\varphi[h] = q \in F^\times$.

1.2. We now consider symbols F and $\mathfrak{g}$ in the following two cases: (i) F is an algebraic number field of finite degree and $\mathfrak{g}$ is its maximal order; (ii) F and $\mathfrak{g}$ are the completions of those in Case (i) at a nonarchimedean prime. We call a field of type (i) a *global field*, and that of type (ii) a *local field*. In this paper, we employ the terms global and local fields only in these senses. If F is a local field, we denote by $\mathfrak{p}$ the maximal ideal of $\mathfrak{g}$. In both local and global cases, by a $\mathfrak{g}$-*lattice* (or simply a *lattice*) in a finite-dimensional vector space V over F, we mean a finitely generated $\mathfrak{g}$-module in V that spans V over F. When F is a global field, we denote by $\mathbf{a}$ and $\mathbf{h}$ the sets of archimedean primes and nonarchimedean primes of F respectively, and put $\mathbf{v} = \mathbf{a} \cup \mathbf{h}$. For each $v \in \mathbf{v}$ we denote by F_v the v-completion of F. Given an algebraic group G defined over F, we define an algebraic group G_v over F_v for each $v \in \mathbf{v}$ and the adelization $G_\mathbf{A}$ as usual, and view G and G_v as subgroups of $G_\mathbf{A}$. We then denote by $G_\mathbf{a}$ and $G_\mathbf{h}$ the archimedean and nonarchimedean factors of $G_\mathbf{A}$, respectively. In particular, $F_\mathbf{A}^\times$ is the idele group of F. For $v \in \mathbf{v}$ and $x \in G_\mathbf{A}$ we denote by x_v the v-component of x.

Given a $\mathfrak{g}$-lattice L in V and another $\mathfrak{g}$-lattice M contained in L in both local and global cases, we can find a finite set $\{\mathfrak{a}\}$ of integral ideals $\mathfrak{a}$ such that L/M as a $\mathfrak{g}$-module is isomorphic to $\bigoplus_{\mathfrak{a}\in\{\mathfrak{a}\}} \mathfrak{g}/\mathfrak{a}$. We then put $[L/M] = \prod_{\mathfrak{a}\in\{\mathfrak{a}\}} \mathfrak{a}$. For $x \in G_\mathbf{A}$ with G acting on V in the global case, we denote by Lx the $\mathfrak{g}$-lattice in V such that $(Lx)_v = L_v x_v$ for every $v \in \mathbf{h}$. We call the set of all such Lx the G-*genus* of L, and call the set of $L\alpha$ for all $\alpha \in G$ the G-*class* of L.

In our later treatment we will often use a quaternion algebra over a local or global field F. Whenever we deal with such an algebra B, we always denote by ι the main involution of B; we then put

(1.2) $$B^\circ = \{\, x \in B \mid x^\iota = -x \,\},$$

$\mathrm{Tr}_{B/F}(x) = x + x^\iota$, and $N_{B/F}(x) = xx^\iota$ for $x \in B$.

Given (V, φ) as in §1.1 over a local or global F, we call a $\mathfrak{g}$-lattice L in V *integral* if $\varphi[x] \in \mathfrak{g}$ for every $x \in L$, and call such an L *maximal* if L is the only integral lattice containing L. For an integral lattice L in V we denote by $A(L)$ the subring of $A(V)$ generated by $\mathfrak{g}$ and L, and put $A^+(L) = A(L) \cap A^+(V)$. For a $\mathfrak{g}$-lattice Λ in V, an element q in $F^\times$, and a fractional ideal $\mathfrak{b}$ in F, we put

(1.3a) $$\widetilde{\Lambda} = \{\, y \in V \mid 2\varphi(y, \Lambda) \subset \mathfrak{g} \,\},$$

(1.3b) $$\Lambda[q, \mathfrak{b}] = \{\, x \in V \mid \varphi[x] = q,\ \varphi(x, \Lambda) = \mathfrak{b} \,\},$$

(1.3c) $D(\Lambda) = \{\gamma \in O^{\varphi}(V) \mid \Lambda\gamma = \Lambda\}, \quad C(\Lambda) = D(\Lambda) \cap SO^{\varphi}(V) \quad (F \text{ local}).$

THEOREM 1.3. *If F is local, L is a maximal lattice in V, and $n > 2$, then* $\#\{L[q, \mathfrak{b}]/C(L)\} \le 1$.

This was given in [S3, Theorem 10.5] under the condition that *if n is odd, then $\delta(\varphi) \cap \mathfrak{g}^{\times} \neq \emptyset$*; see [S3, (8.1)]. Our theorem says that this condition is unnecesary. It is sufficient to prove the case $\mathfrak{b} = 2^{-1}\mathfrak{g}$, since $cL[q, \mathfrak{b}] = L(c^2q, c\mathfrak{b}]$ for every $c \in F^{\times}$. We devote §§1.4 and 1.5 to the proof. To avoid possible misunderstandings, we note that a result which looks similar to the above theorem was stated in [E1, Satz 10.4]. This result of Eichler has no relevance to our theory, as it does not consider the set $\varphi(x, L)$, an essential ingredient of our theorem. Besides, we have $\#\{L[q, \mathfrak{b}]/C(L)\} = 2$ for certain nonmaximal L; see [S7, Theorem 4.2 and (4.3)].

1.4. Given (V, φ) over a local field F and a maximal lattice L in V, by Lemma 6.5 of [S3] we can find decompositions

$$V = Z \oplus U, \quad U = \sum_{i=1}^{r}(Fe_i + Ff_i), \quad \varphi(Z, U) = 0, \tag{1.4a}$$

$$L = M \oplus R, \quad R = \sum_{i=1}^{r}(\mathfrak{g}e_i + \mathfrak{g}f_i), \quad M = \{z \in Z \mid \varphi[z] \in \mathfrak{g}\}, \tag{1.4b}$$

$$2\varphi(e_i, f_j) = \delta_{ij}, \quad \varphi(e_i, e_j) = \varphi(f_i, f_j) = 0, \quad \varphi[z] \neq 0 \text{ for } 0 \neq z \in Z. \tag{1.4c}$$

We put $t = \dim(Z)$, and call it *the core dimension* of (V, φ). To prove Theorem 1.3, we assume hereafter until the end of §1.6 that n is odd and $\delta(\varphi)$ contains a prime element of F. Then t is 1 or 3. We denote by π any fixed prime element of F. We first note a few auxiliary facts:

(1.5) $\{x \in \mathfrak{g} \mid x - 1 \in 4\mathfrak{p}^m\} = \{a^2 \mid a - 1 \in 2\mathfrak{p}^m\}$ *if* $0 < m \in \mathbf{Z}$.

(1.6) *If $k \in U$ and $2\varphi(k, R) = \mathfrak{g}$, then $k\alpha = e_1 + sf_1$ with some $\alpha \in D(R)$ and $s \in \mathfrak{g}$; α can be taken from $C(R)$ if $r > 1$.*

(1.7) *If $h \in L[q, 2^{-1}\mathfrak{g}]$, $\notin Z$, then there exists an element α of $C(L)$ such that $h\alpha = \pi^m(e_1 + sf_1) + z$ with $0 \le m \in \mathbf{Z}$, $s \in \mathfrak{g}$, and $z \in M$.*

The first of these is [S3, Lemma 5.5 (i)]; (1.6) is proven in [S3, §10.7, (A1), (A2)]. Notice that the proof there is valid even when $\varphi[k] = 0$. Now let $h = k + w \in L[q, 2^{-1}\mathfrak{g}]$ with $k \in U$ and $w \in Z$. If $h \notin Z$, then $k \neq 0$, and we can put $2\varphi(k, R) = \mathfrak{p}^m$ with $0 \le m \in \mathbf{Z}$. By (1.6), there exists an element $\alpha_0 \in D(R)$ such that $\pi^{-m}k\alpha_0 = e_1 + sf_1$ with $s \in \mathfrak{g}$. Since $D(M)$ contains an element of determinant -1, we can extend α_0 to an element α of $C(L)$ such that $M\alpha = M$. This proves (1.7).

1.5. By virtue of (1.7), it is sufficient to prove Theorem 1.3 when $r \le 1$. If $r = 0$, then $L = M$ and $SO^{\varphi}(V) = C(L)$ by [S3, Lemma 6.4]. Therefore, if $n = t = 3$ and $\varphi[z] = \varphi[w]$ with $z, w \in V$, then by (1.1), $z = w\alpha$ with $\alpha \in C(L)$, which gives the desired fact. For this reason we hereafter assume that

$r = 1$, and write e and f for e_1 and f_1. We represent the elements of $O^\varphi(V)$ by $(n \times n)$-matrices with respect to $\{e, g_1, , \ldots, g_t, f\}$, where $\{g_1, , \ldots, g_t\}$ is a $\mathfrak{g}$-basis of M.

(A) Case $t = 1$. In this case $M = \mathfrak{g}g$ with an element g such that $\varphi[g]$ is a prime element of F. Thus we can take $\varphi[g]$ as π.

(a1) Suppose $L[q, 2^{-1}\mathfrak{g}] \cap Z \neq \emptyset$; let $w \in L[q, 2^{-1}\mathfrak{g}] \cap Z$ and $h \in L[q, 2^{-1}\mathfrak{g}]$. Then we can put $w = [0 \;\; c \;\; 0]$ with $c \in F$ such that $2c\mathfrak{p} = \mathfrak{g}$. By (1.1), $h = w\alpha$ with $\alpha \in SO^\varphi(V)$. Now $SO^\varphi(V) = PC(L)$ with the subgroup P of $SO^\varphi(V)$ consisting of the upper triangular elements; see [S3, Theorem 6.13 (ii)]. Put $\alpha = \beta\gamma$ with $\beta \in P$ and $\gamma \in C(L)$. The second row of β is of the form $[0 \;\; 1 \;\; j]$ with $j \in F$. Thus $h\gamma^{-1} = w\beta = [0 \;\; c \;\; cj]$, and so $cj \in \mathfrak{g}$. Since $2c\mathfrak{p} = \mathfrak{g}$, we can find an element p of $\mathfrak{g}$ such that $2c\pi p = cj$. Take the matrix

$$\eta = \begin{bmatrix} 1 & -p & -\pi p^2 \\ 0 & 1 & 2\pi p \\ 0 & 0 & 1 \end{bmatrix}. \tag{1.8}$$

Then $\eta \in C(L)$ and $w\eta = w\beta = h\gamma^{-1}$, which gives the desired fact.

(a2) Thus we assume that $L[q, 2^{-1}\mathfrak{g}] \cap Z = \emptyset$. Let $h \in L[q, 2^{-1}\mathfrak{g}]$. By (1.7) we can put $h = \pi^m(e + sf) + cg$ with $0 \leq m \in \mathbf{Z}$ and $s, c \in \mathfrak{g}$. Then $q = \pi^{2m}s + \pi c^2$ and $\mathfrak{p}^m + 2c\mathfrak{p} = \mathfrak{g}$. Put $d = 2\pi c$. Suppose $d \in \mathfrak{g}^\times$. Then by (1.5), we can put $d^2 + 4\pi^{2m+1}s = d_1^2$ with $d_1 \in \mathfrak{g}^\times$. We easily see that $(2\pi)^{-1}d_1 g \in L[q, 2^{-1}\mathfrak{g}]$, a contradiction, as $g \in Z$. Thus $d \in \mathfrak{p}$, so that $m = 0$ and $h = e + sf + (2\pi)^{-1}dg$, and $q = s + (4\pi)^{-1}d^2$. Suppose we have another element $h_1 = e + s_1 f + (2\pi)^{-1}d_1 g$ with $s_1 \in \mathfrak{g}$ and $d_1 \in \mathfrak{p}$ such that $\varphi[h_1] = q$. Then $d_1^2 - d^2 = 4\pi(s - s_1) \in 4\mathfrak{p}$, and so $d_1 - d \in 2\mathfrak{p}$ or $d_1 + d \in 2\mathfrak{p}$. Since $-2d \in 2\mathfrak{p}$, we have $d_1 - d \in 2\mathfrak{p}$ in both cases. Let α be the element of (1.8) with $p = (d_1 - d)/(2\pi)$. Then $\alpha \in C(L)$ and $h = h_1\alpha \in h_1 C(L)$. This completes the proof when $t = 1$.

(B) Case $t = 3$. As shown in [S3, §§7.3 and 7.7 (III)], there is a division quaternion algebra B over F with which we can put $Z = B^\circ$ and $2\varphi(x, y) = d\mathrm{Tr}_{B/F}(xy^\iota)$ for $x, y \in Z$, where d is an element of $F^\times$ that represents the determinant of the restriction of φ to Z. Thus we can take a prime element π of F as d; then $M = \{x \in Z \,|\, \pi x x^\iota \in \mathfrak{g}\}$. Let K be an unramified quadratic extension of F, $\mathfrak{r}$ the maximal order of K, and ρ the generator of $\mathrm{Gal}(K/F)$. We can put $B = K + K\eta$ with an element η such that $\eta^2 = \pi$ and $x\eta = \eta x^\rho$ for every $x \in K$. Take $u \in \mathfrak{r}$ so that $\mathfrak{r} = \mathfrak{g}[u]$ and put $\sigma = u - u^\rho$. Then $M = \mathfrak{g}\sigma + \mathfrak{r}\eta^{-1}$ and $\widetilde{M} = (2\mathfrak{p})^{-1}\sigma + \mathfrak{r}\eta^{-1}$. Since $r = 1$, φ is represented by a matrix

$$\begin{bmatrix} 0 & 0 & 2^{-1} \\ 0 & \zeta & 0 \\ 2^{-1} & 0 & 0 \end{bmatrix},$$

where ζ is an element of $\mathfrak{g}_3^3$ that represents the restriction of φ to Z. Define $\lambda = (\lambda_{ij}) \in \mathfrak{g}_3^3$ by $\lambda_{ii} = \zeta_{ii}$, $\lambda_{ij} = 2\zeta_{ij}$ if $i < j$ and $\lambda_{ij} = 0$ if $i > j$. Then $\lambda + {}^t\lambda = 2\zeta$. Let P be the subgroup of $SO^\varphi(V)$ consisting of the elements that send Ff onto itself, Then $SO^\varphi(V) = PC(L)$; see [S3, Theorem 6.13 (ii)].

(b1) Suppose $L[q, 2^{-1}\mathfrak{g}] \cap Z \neq \emptyset$. Let $h \in L[q, 2^{-1}\mathfrak{g}]$ and $w \in L[q, 2^{-1}\mathfrak{g}] \cap Z$. Identify w with a row vector $w = [0 \quad y \quad 0]$, where $y = (y_i)_{i=1}^3$ with $y_i \in \mathfrak{g}$. By (1.1), $h = w\alpha$ with $\alpha \in SO^\varphi(V)$. Put $\alpha = \beta\gamma$ with $\beta \in P$ and $\gamma \in C(L)$. The middle three rows of β can be written in the form $[0 \quad \varepsilon \quad j]$ with $\varepsilon \in SO^\varphi(Z)$ and a column vector j. Replacing γ by $\mathrm{diag}[1, \varepsilon, 1]\gamma$, we may assume that $\varepsilon = 1$. Then $h\gamma^{-1} = w\beta = [0 \quad y \quad yj]$ and $yj \in \mathfrak{g}$. Since $w \in L[q, 2^{-1}\mathfrak{g}]$, we see that $2y\zeta$ is primitive, and so we can find $p \in \mathfrak{g}_3^1$ such that $-2y\zeta \cdot {}^tp = yj$. Let

$$\psi = \begin{bmatrix} 1 & p & -p\lambda \cdot {}^tp \\ 0 & 1 & -2\zeta \cdot {}^tp \\ 0 & 0 & 1 \end{bmatrix}. \tag{1.9}$$

Then $\psi \in C(L)$ and $w\psi = w\beta = h\gamma^{-1}$, and so $h \in wC(L)$ as expected.

(b2) Let $h \in L[q, 2^{-1}\mathfrak{g}], \notin Z$. By (1.7) we may assume that $h = \pi^m(e + sf) + z$ with $0 \leq m \in \mathbf{Z}$, $s \in \mathfrak{g}$, and $z \in Z$. Put $z = (2\pi)^{-1}c\sigma + x\eta^{-1}$ with $c \in F$ and $x \in K$. Then $\mathfrak{g} = \mathfrak{p}^m + c\mathfrak{g} + \mathrm{Tr}_{K/F}(x\mathfrak{r})$, $q = \pi^{2m}s - \pi z^2$, and $\pi z^2 = (4\pi)^{-1}c^2\sigma^2 + xx^\rho$. Suppose $m > 0$; then $c \in \mathfrak{g}^\times$ or $x \in \mathfrak{r}^\times$. If $c \in \mathfrak{g}^\times$, then by (1.5) we can put $c^2 - 4\pi^{2m+1}\sigma^{-2}s = b^2$ with $b \in \mathfrak{g}^\times$. Put $w = (2\pi)^{-1}b\sigma + x\eta^{-1}$. Then we see that $w \in L[q, 2^{-1}\mathfrak{g}] \cap Z$. Next suppose $x \in \mathfrak{r}^\times$; then $xx^\rho - \pi^{2m}s \in \mathfrak{g}^\times$. Since $N_{K/F}(\mathfrak{r}^\times) = \mathfrak{g}^\times$, we can find $x_1 \in \mathfrak{r}^\times$ such that $x_1x_1^\rho = xx^\rho - \pi^{2m}s$. Put $x = (2\pi)^{-1}c\sigma + x_1\eta^{-1}$. Then $x \in L[q, 2^{-1}\mathfrak{g}] \cap Z$. Therefore, if $m > 0$, then the problem can be reduced to case (b1), which has been settled.

(b3) Thus if $L[q, 2^{-1}\mathfrak{g}] \cap Z = \emptyset$ and $h \in L[q, 2^{-1}\mathfrak{g}]$, then we may assume that $h = e + sf + z$ with $s \in \mathfrak{g}$ and $z \in Z$. Take another element $h_1 = e + s_1f + z_1 \in L[q, 2^{-1}\mathfrak{g}]$ with $s_1 \in \mathfrak{g}$ and $z_1 \in Z$. We have $z = (2\pi)^{-1}c\sigma + x\eta^{-1}$ and $z_1 = (2\pi)^{-1}c_1\sigma + x_1\eta^{-1}$ with $c, c_1 \in \mathfrak{g}$ and $x, x_1 \in \mathfrak{r}$. Since $s - \pi z^2 = s_1 - \pi z_1^2$, we have $(4\pi)^{-1}(c^2 - c_1^2) \in \mathfrak{g}$, so that $c^2 - c_1^2 \in 4\mathfrak{p}$. Then $c - c_1 \in 2\mathfrak{p}$ or $c + c_1 \in 2\mathfrak{p}$. Now the map $x\sigma + y\eta^{-1} \mapsto -x\sigma + y^\rho\eta^{-1}$ is an element of $C(M)$. Therefore, replacing z_1 by its image under this map if necessary, we may assume that $c - c_1 \in 2\mathfrak{p}$, which implies that $z - z_1 \in M$. Define ψ by (1.9) by taking p to be the vector that represents $z - z_1$. Then $h_1\psi = h$, which gives the desired fact.

1.6. We can now remove condition (8.1) from [S3, Lemma 10.8], which concerns the case where $t = 1$ and $r > 0$. To be precise, that lemma can be restated, without condition (8.1), as follows:

Let $L = \mathfrak{g}e_1 + \mathfrak{g}f_1 + \mathfrak{g}g$ with g such that $\mathfrak{p} \subset \varphi[g]\mathfrak{g} \subset \mathfrak{g}$, and let $h \in L[q, 2^{-1}\mathfrak{g}]$ with $q \in F^\times$. Then there exists an element γ of $C(L)$ such that $h\gamma = cg$ with c satisfying $2c\varphi[g]\mathfrak{g} = \mathfrak{g}$ or $h\gamma = ae_1 + f_1 + cg$ with $a \in \mathfrak{g}$ and $c \in 2^{-1}\mathfrak{g}$.

This follows from the discussions of (a1) and (a2) combined with (1.7).

1.7. Still asssuming F to be local, take the symbols as in (1.4a, b, c) with even or odd n, put $C = C(L)$ for simplicity, and define subgroups T and J of $G^+(V)$ by

$$T = \{ \alpha \in G^+(V) \mid \tau(\alpha) \in C \}, \tag{1.10}$$

$$J = \{ \gamma \in T \mid \nu(\gamma) \in \mathfrak{g}^\times \}. \tag{1.11}$$

In [S3], Proposition 8.8, Theorems 8.6, and 8.9 we discussed the structure of $A(L)$, $A^+(L)$, $\nu(J)$, and the connection of $\tau(J)$ with C under the condition that $\delta(\varphi) \cap \mathfrak{g}^\times \neq \emptyset$ if n is odd. Let us now prove the results in the cases in which the condition is not satisfied.

THEOREM 1.8. *Suppose F is local, $1 < n - 1 \in 2\mathbf{Z}$, and $\delta(\varphi)$ contains a prime element. Then the following assertions hold.*

(i) *If $t = 1$, then there exist an order $\mathfrak{O}$ in $M_2(F)$ of discriminant $\mathfrak{p}$ (see §5.1 for the definition) and an isomorphism θ of $A^+(V)$ onto $M_s(M_2(F))$ that maps $A^+(L)$ onto $M_s(\mathfrak{O})$, where $s = 2^{r-1}$.*

(ii) *If $t = 3$, then there exist a division quaternion algebra B over F and an isomorphism $\xi : B^\circ \to Z$ such that $\varphi[x\xi] = dxx^\iota$ for every $x \in B^\circ$ with a prime element d of $F^\times$ independent of x, where B° is defined by (1.2). Moreover there exists an isomorphism θ of $A^+(V)$ onto $M_s(B)$ that maps $A^+(L)$ onto $M_s(\mathfrak{O})$, where $s = 2^r$ and $\mathfrak{O}$ is the unique maximal order in B.*

(iii) *In both cases $t = 1$ and $t = 3$ we have $\nu(J) = \mathfrak{g}^\times$, $C = \tau(T)$, $[C : \tau(J)] = 2$, and $T = F^\times(J \cup J\eta)$ with an element η such that $\eta \notin F^\times J$, $\eta^2 \in F^\times$, $J\eta = \eta J$, and $\nu(\eta)\mathfrak{g} = \mathfrak{p}$.*

This will be proved after the proof of Lemma 5.4.

THEOREM 1.9. *In the setting of Theorem 1.3 with even or odd $n > 2$, let t be the core dimension of (V, φ) and let $\sigma : SO^\varphi(V) \to F^\times/F^{\times 2}$ be the spinor norm map of* [S3, (3.7)]. Then $\sigma(C) = \mathfrak{g}^\times F^{\times 2}$ in the following three cases: (i) $t = 0$; (ii) $t = 1$ *and* $\delta(\varphi) \cap \mathfrak{g}^\times \neq \emptyset$; (iii) $t = 2$ *and* $\widetilde{L} = L$. *We have $\sigma(C) = F^\times$ in all the remaining cases.*

Proof. That $\sigma(C) = \mathfrak{g}^\times F^{\times 2}$ in the three cases specified above is shown by Proposition 8.8 (iii) and Theorem 8.9 (i) of [S3]. Therefore we assume that those cases do not apply. If $\delta(\varphi) \cap \mathfrak{g}^\times \neq \emptyset$ or n is even, then Theorem 8.9 (ii) of [S3] shows that $\sigma(C) = F^\times$. If n is odd and $\delta(\varphi) \cap \mathfrak{g}^\times = \emptyset$, then from (iii) of Theorem 1.8 we obtain $\sigma(C) = \nu(T) = F^\times$, which completes the proof.

2. GLOBAL QUADRATIC DIOPHANTINE EQUATIONS

2.1. Let us now turn to the global case. The main theorems of [S3, Sections 11 through 13] were given under a condition stated in [S3, (9.2)]: *if n is odd, then $\delta(\varphi)$, at every nonarchimedean prime, contains a local unit.* By virtue of Theorem 1.3 we can now remove this condition from [S3, Theorem 11.6] and some others, which we will indicate in Remark 2.4 (3), (5), (6), and (7).

Throughout this section we assume that F is an algebraic number field. Given a quadratic space (V, φ) over F, we can define, for each $v \in \mathbf{v}$, the v-localization (V_v, φ_v) of (V, φ) as a quadratic space over F_v by putting $V_v = V \otimes_F F_v$ and extending φ to an F_v-valued quadratic form φ_v on V_v in a natural way.

In [S3, Section 13], we treated the equation $\varphi[h] = q$ not only for a scalar q, but also for a symmetric matrix q of size $n-1$. We now consider a more general case by taking a new approach. Given ${}^t q = q \in GL_m(F)$ and ${}^t\varphi = \varphi \in GL_n(F)$, we consider the solutions $h \in F_n^m$ of the equation $h\varphi \cdot {}^t h = q$. Here and throughout this and the next sections we assume that $n > 2$ and $n > m > 0$. More intrinsically, take (V, φ) as before and take also (X, q) with a nondegenerate quadratic form q on a vector space X over F of dimension m. We put

$$\mathscr{V} = \mathrm{Hom}(X, V), \tag{2.1}$$

and consider $h \in \mathscr{V}$ such that $\varphi[xh] = q[x]$ for every $x \in X$. Since q is nondegenerate, h must be injective. To simplify our notation, for every $k \in \mathscr{V}$ we denote by $\varphi[k]$ the quadratic form on X defined by $\varphi[k][x] = \varphi[xk]$ for every $x \in X$. Then our problem concerns the solutions $h \in \mathscr{V}$ of the equation $\varphi[h] = q$ for a fixed q. If $m = 1$ and $X = F$, then $q \in F^\times$, and an element h of V defines an element of $\mathscr{V}$ that sends c to ch for $c \in F$, and $\mathscr{V}$ consists of all such maps. Thus we can put $\mathscr{V} = V$ if $m = 1$ and the problem about $\varphi[h] = q$ with $q \in F^\times$ is the one-dimensional special case. For $m \geq 1$, if $h \in \mathscr{V}$ and $\det(\varphi[h]) \neq 0$, then

$$\big\{k \in \mathscr{V} \,\big|\, \varphi[k] = \varphi[h]\big\} = h \cdot SO^\varphi(V). \tag{2.2}$$

This is a generalization of (1.1) and follows easily from the Witt theorem; see [S3, Lemma 1.5 (i)].

For a fixed $h \in \mathscr{V}$ such that $\det\big(\varphi[h]\big) \neq 0$, put $W = (Xh)^\perp$, $G = SO^\varphi(V)$, and $H = SO^\varphi(W)$. We identify H with the subgroup of G consisting of the elements that are the identity map on Xh; thus

$$H = \big\{\alpha \in G \,\big|\, h\alpha = h\big\}. \tag{2.3}$$

For $\xi \in G_{\mathbf{A}}$ the symbol $h\xi$ is meaningful as an element of $\mathscr{V}_{\mathbf{A}}$, and so for a subset Ξ of $G_{\mathbf{A}}$ the symbol $h\Xi$ is meaningful as a subset of $\mathscr{V}_{\mathbf{A}}$.

THEOREM 2.2. *Let $D = D_0 G_{\mathbf{a}}$ with an open compact subgroup D_0 of $G_{\mathbf{h}}$. Then the following assertions hold.*

(i) *For $y \in G_{\mathbf{A}}$ we have $H_{\mathbf{A}} \cap GyD \neq \emptyset$ if and only if $\mathscr{V} \cap hDy^{-1} \neq \emptyset$.*

(ii) *Fixing $y \in G_{\mathbf{A}}$, for every $\varepsilon \in H_{\mathbf{A}} \cap GyD$ take $\alpha \in G$ so that $\varepsilon \in \alpha y D$. Then the map $\varepsilon \mapsto h\alpha$ gives a bijection of $H\backslash(H_{\mathbf{A}} \cap GyD)/(H_{\mathbf{A}} \cap D)$ onto $(\mathscr{V} \cap hDy^{-1})/\Gamma^y$, where $\Gamma^y = G \cap yDy^{-1}$.*

(iii) *Take $Y \subset G_{\mathbf{A}}$ so that $G_{\mathbf{A}} = \bigsqcup_{y \in Y} GyD$. Then*

$$\#\big\{H\backslash H_{\mathbf{A}}/(H_{\mathbf{A}} \cap D)\big\} = \sum_{y \in Y} \#\big\{(\mathscr{V} \cap hDy^{-1})/\Gamma^y\big\}. \tag{2.4}$$

(iv) *In particular, suppose $m = 1$ and $n > 2$. With a fixed maximal lattice L in V put $C = \big\{\xi \in G_{\mathbf{A}} \,\big|\, L\xi = L\big\}$, $q = \varphi[h]$, and $\mathfrak{b} = \varphi(h, L)$. Then*

$$V \cap hCy^{-1} = (Ly^{-1})[q, \mathfrak{b}] \quad \textit{for every} \quad y \in G_{\mathbf{A}}. \tag{2.5}$$

Note: Since $H\backslash H_{\mathbf{A}}/(H_{\mathbf{A}} \cap D)$ is a finite set, from (ii) we see that $(\mathscr{V} \cap hDy^{-1})/\Gamma^y$ is a finite set.

Proof. Let y, ε, and α be as in (ii); then clearly $h\alpha \in \mathscr{V} \cap hDy^{-1}$. If $\eta\varepsilon\zeta \in \beta yD$ with $\eta \in H$, $\zeta \in H_{\mathbf{A}} \cap D$, and $\beta \in G$, then $\beta^{-1}\eta\alpha \in G \cap yDy^{-1} = \Gamma^y$, and hence $h\alpha = h\eta\alpha \in h\beta\Gamma^y$. Thus our map is well defined. Next let $k \in \mathscr{V} \cap hDy^{-1}$. Then $k = h\delta y^{-1}$ with $\delta \in D$, and moreover, by (2.2), $k = h\xi$ with $\xi \in G$. Then $h = h\xi y\delta^{-1}$, so that $\xi y\delta^{-1} \in H_{\mathbf{A}}$ by (2.3). Thus $\xi y\delta^{-1} \in H_{\mathbf{A}} \cap GyD$. This shows that k is the image of an element of $H_{\mathbf{A}} \cap GyD$. To prove that the map is injective, suppose $\varepsilon \in \alpha yD \cap H_{\mathbf{A}}$ and $\delta \in \beta yD \cap H_{\mathbf{A}}$ with α, $\beta \in G$, and $h\alpha = h\beta\sigma$ with $\sigma \in \Gamma^y$. Put $\omega = \beta\sigma\alpha^{-1}$. Then $h\omega = h$, so that $\omega \in H$. Since $\sigma \in yDy^{-1}$, we have $\beta yD = \beta\sigma yD = \omega\alpha yD$, and hence $\delta \in \beta yD \cap H_{\mathbf{A}} = \omega\alpha yD \cap H_{\mathbf{A}} = \omega(\alpha yD \cap H_{\mathbf{A}}) = \omega(\varepsilon D \cap H_{\mathbf{A}}) = \omega\varepsilon(D \cap H_{\mathbf{A}}) \subset H\varepsilon(D \cap H_{\mathbf{A}})$. This proves the injectivity, and completes the proof of (ii). At the same time we obtain (i).

Now $H_{\mathbf{A}} = \bigsqcup_{y\in Y}(H_{\mathbf{A}} \cap GyD)$, and so (iii) follows immediately from (ii). As for (iv), clearly $V \cap hC \subset L[q, \mathfrak{b}]$. Conversely, every element of $L[q, \mathfrak{b}]$ belongs to hC by virtue of Theorem 1.3. Thus

$$V \cap hC = L[q, \mathfrak{b}]. \tag{2.6}$$

Let $k \in V \cap hCy^{-1}$ with $y \in G_{\mathbf{A}}$; put $M = Ly^{-1}$. Then $\varphi[k] = q$, $\varphi(k, M) = \varphi(h, L) = \mathfrak{b}$, and $kyCy^{-1} = hCy^{-1}$. Taking k, M, and yCy^{-1} in place of h, L and C in (2.6), we obtain $V \cap hCy^{-1} = V \cap kyCy^{-1} = M[q, \mathfrak{b}] = (Ly^{-1})[q, \mathfrak{b}]$. This proves (2.5) when $V \cap hCy^{-1} \neq \emptyset$. To prove the remaining case, suppose $\ell \in (Ly^{-1})[q, \mathfrak{b}]$; then $\varphi(\ell y_v, L_v) = \mathfrak{b}_v = \varphi(h, L)_v$ for every $v \in \mathbf{h}$. By Theorem 1.3, $\ell y \in hC$, and hence $\ell \in hCy^{-1}$. This shows that if $(Ly^{-1})[q, \mathfrak{b}] \neq \emptyset$, then $V \cap hCy^{-1} \neq \emptyset$, and hence (2.5) holds for every $y \in G_{\mathbf{A}}$. This completes the proof.

If k and ℓ are two elements of $\mathscr{V} \cap hDy^{-1}$, then $k = \ell x_v$ with $x_v \in y_vD_vy_v^{-1}$ for every $v \in \mathbf{h}$. Therefore we are tempted to say that *k and ℓ belong to the same genus*. Then each orbit of $(\mathscr{V} \cap hDy^{-1})/\Gamma^y$ may be called a *class*. Thus (ii) of Theorem 2.2 connects such classes of elements of $\mathscr{V}$ with the classes of H with respect to $H_{\mathbf{A}} \cap D$.

Combining (2.4) with (2.5), we obtain, in the setting of (iv),

$$\#\{H\backslash H_{\mathbf{A}}/(H_{\mathbf{A}} \cap C)\} = \sum_{y\in Y} \#\{(Ly^{-1})[q, \mathfrak{b}]/\Gamma^y\}. \tag{2.7}$$

This was stated in [S3, (11.7)] under the condition mentioned at the beginning of this section, which we can now remove.

The above theorem concerns $SO^{\varphi}\big((Xh)^{\perp}\big)$. Let us now show that we can formulate a result with respect to $SO^q(X)$ instead, when $m = n - 1$. The notation being as above, put $Y = Xh$ and $J = SO^{\varphi}(Y)$. We identify J with $\{\alpha \in G \mid \alpha = \mathrm{id.}\ \text{on}\ W\}$. For every $\delta \in J$ there is a unique element δ' of $SO^q(X)$ such that $\delta' h = h\delta$, and $\delta \mapsto \delta'$ gives an isomorphism of J onto $SO^q(X)$. In this section we employ the symbol δ' always in this sense. We note a simple fact:

(2.8) *If $m = n - 1$ and $h\xi = h\eta$ for ξ, $\eta \in G_{\mathbf{A}}$, then $\xi = \eta$.*

Indeed, for each $v \in \mathbf{v}$ we have $h\xi_v\eta_v^{-1} = h$, and so $\xi_v\eta_v^{-1}$ is the identity map on $X_v h$. Since $m = n-1$, it must be the identity map on the whole V_v. Thus $\xi_v\eta_v^{-1} = 1$ for every $v \in \mathbf{v}$, which proves (2.8).

THEOREM 2.3. *Let $D = D_0 G_\mathbf{a}$ as in Theorem 2.2 and let $E = E_0 J_\mathbf{a}$ with an open compact subgroup E_0 of $J_\mathbf{h}$. Suppose $m = n-1$ and $J_\mathbf{A} \cap D \subset E$. Then for every $z \in J_\mathbf{A}$ and $y \in G_\mathbf{A}$ there exists a bijection*

$$J\backslash(JzE \cap GyD)/(J_\mathbf{A} \cap D) \longrightarrow \Delta'\backslash(\mathscr{V} \cap hzEDy^{-1})/\Gamma, \tag{2.9}$$

where $\Gamma = yDy^{-1} \cap G$ and $\Delta' = \{\delta' \mid \delta \in \Delta\}$, $\Delta = zEz^{-1} \cap J$.

Proof. Given $\sigma \in JzE \cap GyD$, take $\alpha \in G$ and $\beta \in J$ so that $\sigma \in \beta zE \cap \alpha yD$, and put $k = h\beta^{-1}\alpha$. Then $k \in \mathscr{V} \cap hzEDy^{-1}$. If $\sigma \in \beta_1 zE \cap \alpha_1 yD$ with $\alpha_1 \in G$ and $\beta_1 \in J$, then $\beta_1^{-1}\beta \in \Delta$ and $\alpha_1^{-1}\alpha \in \Gamma$, and so $\sigma \to h\beta^{-1}\alpha$ is a well-defined map as in (2.9). To show that it is surjective, take $k \in \mathscr{V} \cap hzEDy^{-1}$; then $\varphi[k] = q$, and so $k = h\alpha$ with $\alpha \in G$ by (2.2). We have also $k = hz\varepsilon\zeta y^{-1}$ with $\varepsilon \in E$ and $\zeta \in D$. By (2.8), $z\varepsilon\zeta y^{-1} = \alpha$, and so $z\varepsilon = \alpha y\zeta^{-1} \in zE \cap \alpha yD$. This shows that k is the image of $z\varepsilon$, which proves the surjectivity. To prove the injectivity, let $\sigma \in \beta zE \cap \alpha yD$ and $\sigma_1 \in \beta_1 zE \cap \alpha_1 yD$ with $\alpha, \alpha_1 \in G$ and $\beta, \beta_1 \in E$. Suppose $h\beta_1^{-1}\alpha_1 = \delta' h\beta^{-1}\alpha\gamma$ with $\delta \in \Delta$ and $\gamma \in \Gamma$. Then $\beta_1^{-1}\alpha_1 = \delta\beta^{-1}\alpha\gamma$ by (2.8). Put $\lambda = \alpha\gamma\alpha_1^{-1}$. Then $\lambda = \beta\delta^{-1}\beta_1^{-1} \in J$. Since $\gamma \in yDy^{-1}$, we have $\sigma \in \alpha yD = \alpha\gamma yD = \lambda\alpha_1 yD = \lambda\sigma_1 D \subset J\sigma_1 D$. Also, since $\sigma, \sigma_1 \in J_\mathbf{A}$, we have $\sigma \in J\sigma_1(J_\mathbf{A} \cap D)$. This proves the injectivity, and completes the proof.

REMARK 2.4. (1) We can take $E = J_\mathbf{A} \cap D$ in the above theorem. Then (2.9) takes a simpler form

$$J\backslash(JzE \cap GyD)/E \longrightarrow \Delta'\backslash(\mathscr{V} \cap hzDy^{-1})/\Gamma, \tag{2.10}$$

(2) Let $G_\mathbf{A} = \bigsqcup_{y \in Y} GyD$ as in Theorem 2.2 and let $J_\mathbf{A} = \bigsqcup_{z \in Z} JzE$ with a finite subset Z of $J_\mathbf{A}$. Then $J_\mathbf{A} = \bigsqcup_{z,y}(JzE \cap GyD)$. For each (z, y) such that $JzE \cap GyD \neq \emptyset$ pick $\xi \in JzE \cap GyD$ and denote by Ξ the set of all such ξ. Then $J_\mathbf{A} = \bigsqcup_{\xi \in \Xi}(J\xi E \cap G\xi D)$, and we obtain

$$\#\{J\backslash J_\mathbf{A}/(J_\mathbf{A} \cap D)\} = \sum_{\xi \in \Xi} \#\{\Delta'_\xi\backslash(\mathscr{V} \cap h\xi ED\xi^{-1})/\Gamma^\xi\}, \tag{2.11}$$

where $\Gamma^\xi = \xi D\xi^{-1} \cap G$ and $\Delta'_\xi = \{\delta' \mid \delta \in J \cap \xi E\xi^{-1}\}$.

(3) For $\ell \in V$ let us denote by $\varphi(h, \ell)$ the element of $\mathrm{Hom}(X, F)$ defined by $x\varphi(h, \ell) = \varphi(xh, \ell)$ for $x \in X$; then for a subset S of V let us put $\varphi(h, S) = \{\varphi(h, s) \mid s \in S\}$. Now, fixing a maximal lattice L in V, put $\Lambda = \widetilde{L}$, $\mathfrak{B} = \varphi(h, \Lambda)$, and $E = \{\varepsilon \in J_\mathbf{A} \mid \varepsilon'\mathfrak{B} = \mathfrak{B}\}$. Since $\mathfrak{B}$ is a $\mathfrak{g}$-lattice in $\mathrm{Hom}(X, F)$, we see that E is a subgroup of $J_\mathbf{A}$ of the type considered in Theorem 2.3. Then we can prove

$$\mathscr{V} \cap hzEDy^{-1} = \{k \in \mathscr{V} \mid \varphi[k] = q,\ \varphi(k, \Lambda y^{-1}) = z'\mathfrak{B}\}. \tag{2.12}$$

This is similar to (2.5), and proved in the proof of [S3, Theorem 13.10]. It should be noted that condition (9.2) imposed in that theorem and also (8.1) in [S3 Theorem 13.8] are unnecessary for the reason explained at the beginning of this section.

(4) Suppose $m = n - 1$ in the setting of Theorem 2.2; then $H = \{1\}$. Therefore $H_{\mathbf{A}} \cap GyD \neq \emptyset$ only when $GyD = GD$. Then (2.4) gives $\#\{(\mathscr{V} \cap hD)/(G \cap D)\} = 1$, but this is an immediate consequence of (2.2).

(5) Let us note some more statements in [S3] from which condition (9.2) of [S3] can be removed. First of all, that is the case with Theorem 9.26 of [S3]. In fact, we can state an improved version of that theorem as follows. Using the notation employed in the theorem, let $\mathfrak{I}'_{\varphi}$ be the subgroup of $\mathfrak{I}$ generated by $\mathfrak{I}_0$ and the prime ideals for which $\sigma(C_v) = F_v^{\times}$. (Such prime ideals are determined by Theorem 1.9.) *Then* $\#\{SO^{\varphi}\backslash SO^{\varphi}_{\mathbf{A}}/C) = [\mathfrak{I} : \mathfrak{I}'_{\varphi}]$ *for* C *of* [S3, (9.15)], which equals C of Theorem 2.2 (iv). The last group index is 1 if $F = \mathbf{Q}$, and so the genus of maximal lattices consists of a single class. Consequently we can state a clear-cut result as follows. Let $\mathfrak{S}_n$ denote the set of symmetric matrices of $GL_n(\mathbf{Q})$ that represent $\mathbf{Z}$-valued quadratic forms on $\mathbf{Z}^n$, and $\mathfrak{S}_n^0$ the subset of $\mathfrak{S}_n$ consisting of the reduced elements of $\mathfrak{S}_n$ in the sense of [S5], that is, the set of $\Phi \in \mathfrak{S}_n$ which cannot be represented nontrivially over $\mathbf{Z}$ by another element of $\mathfrak{S}_n$. For $\Phi \in \mathfrak{S}_n$ put $\sigma(\Phi) = p - q$ where Φ as a real matrix has p positive and q negative eigenvalues. Now the number of genera of $\Phi \in \mathfrak{S}_n^0$ with given $\sigma(\Phi)$ and $\det(2\Phi)$ was determined in [S5, Theorem 6.6]. If $n \geq 3$ and $\sigma(\Phi) \neq n$, then the number of genera of Φ given there equals the number of classes, as each genus of maximal lattices consists of a single class as explained above.

(6) Next, (9.2) is unnecessary in Theorem 12.1 of [S3]. What we need, in addition to Theorem 1.3, is the following fact: *If* $v \in \mathbf{h}$, $\varphi[h]\mathfrak{g}_v = \varphi(h, L_v)^2$, *then* $L_v \cap W_v$ *is maximal and* $C \cap H_v = \{\alpha \in H_v \mid (L_v \cap W_v)\alpha = L_v \cap W_v\}$. Here the notation is the same as in the proof of Theorem 12.1. The statement is valid irrespective of the nature of L_v and t_v. Indeed, replacing h by an element of $F^{\times}h$, we may assume that $\varphi[h]\mathfrak{g}_v = \varphi(h, L_v) = \mathfrak{g}_v$. Then, by Lemma 10.2 (i) of [S3], we have $L_v = \mathfrak{g}_v h \oplus (L_v \cap W_v)$, from which we immediately obtain the expected facts on $L_v \cap W_v$. Consequently, $\mathfrak{g}_v^{\times} \subset \sigma(C \cap H_v)$ by Theorem 1.9, as $n - 1 > 2$, and the original proof of Theorem 12.1 is valid without (9.2).

(7) In [S4] we proved a result of type (2.7) in terms of $G^+(V)$, but imposed the condition that *if* n *is odd, then* $\det(\varphi)\mathfrak{g}$ *is a square in the ideal group of* F; see Theorem 1.6 (iv) of [S4]. We can remove this condition by virtue of Theorem 1.3.

3. NEW MASS FORMULAS

3.1. We first recall the symbols $\mathfrak{m}(G, D)$ and $\nu(\Gamma)$ introduced in [S1] and [S2]. Here $G = SO^{\varphi}(V)$ and D is an open subgroup of $G_{\mathbf{A}}$ containing $G_{\mathbf{a}}$ and such that $D \cap G_{\mathbf{h}}$ is compact. Fixing such a D, we put $\Gamma^a = G \cap aDa^{-1}$ for every $a \in G_{\mathbf{A}}$. Let $C_{\mathbf{a}}$ be a maximal compact subgroup of $G_{\mathbf{a}}$ and let $\mathscr{Z} = G_{\mathbf{a}}/C_{\mathbf{a}}$.

Then $\mathscr{X}$ is a symmetric space on which G acts through its projection into $G_{\mathbf{a}}$. Taking a complete set of representatives $\mathscr{B}$ for $G\backslash G_{\mathbf{A}}/D$, we put

$$\mathfrak{m}(G,\, D) = \mathfrak{m}(D) = \sum_{a\in\mathscr{B}} \nu(\Gamma^a). \tag{3.1}$$

Here $\nu(\Gamma)$ is a quantity defined by

$$\nu(\Gamma) = \begin{cases} [\Gamma : 1]^{-1} & \text{if } G_{\mathbf{a}} \text{ is compact,} \\ \mathrm{vol}(\Gamma\backslash\mathscr{X})/\#(\Gamma \cap \{\pm 1\}) & \text{otherwise.} \end{cases} \tag{3.2}$$

We fix a Haar measure on $G_{\mathbf{a}}$, which determines a unique $G_{\mathbf{a}}$-invariant measure on $\mathscr{X}$ by a well known principle. (In fact, $\mathfrak{m}(D)$ and $\nu(\Gamma)$ are the measure of $D \cap G_{\mathbf{a}}$ and that of $\Gamma\backslash G_{\mathbf{a}}$; see [S6, Theorem 9].) We easily see that $\mathfrak{m}(D)$ does not depend on the choice of $\mathscr{B}$. We call $\mathfrak{m}(G,\, D)$ *the mass of G relative to D*. If D' is a subgroup of $G_{\mathbf{A}}$ of the same type as D and $D' \subset D$, then, as proven in [S1, Lemma 24.2 (1)],

$$\mathfrak{m}(G,\, D') = [D : D']\mathfrak{m}(G,\, D). \tag{3.3}$$

Next, in the setting of §2.1 we consider a subset S of $\mathscr{V}$ of the form $S = \bigsqcup_{\zeta\in Z} h\zeta\Gamma$, where h is as in (2.2), Z is a finite subset of G, and $\Gamma = G \cap D$ with D as above. Then we put

$$\mathfrak{m}(S) = \sum_{\zeta\in Z} \nu(\Delta_\zeta)/\nu(\Gamma), \quad \Delta_\zeta = H \cap \zeta\Gamma\zeta^{-1}, \tag{3.4}$$

and call $\mathfrak{m}(S)$ the *mass* of S. Here, to define $\nu(\Delta_\zeta)$, we need to fix a measure on $H_{\mathbf{a}}$. Thus $\mathfrak{m}(S)$ depends on the choice of measures on $G_{\mathbf{a}}$ and $H_{\mathbf{a}}$, but $\mathfrak{m}(S)$ is independent of the choice of Z and Γ. (Let $\mathfrak{Y} = \{x \in \mathscr{V}_{\mathbf{a}} \,|\, \varphi[x] = q\}$. Since $\mathfrak{Y}$ can be identified with $H_{\mathbf{a}}\backslash G_{\mathbf{a}}$, once a measure on $G_{\mathbf{a}}$ is fixed, a measure on $\mathfrak{Y}$ determines a measure on $H_{\mathbf{a}}$, and vice versa. The replacement of h by an element of hG changes the group H, but that does not change $\mathfrak{m}(S)$ if we start with a fixed measure on $\mathfrak{Y}$. Notice also that an identification of $\mathscr{V}_{\mathbf{a}}$ with a Euclidean space determines a $G_{\mathbf{a}}$-invariant measure on $\mathfrak{Y}$.) Since the left-hand side of (2.4) is finite, we see that $(\mathscr{V} \cap hDy^{-1})/\Gamma^y$ is a finite set for every $y \in G_{\mathbf{A}}$. Thus we can define $\mathfrak{m}(\mathscr{V} \cap hDy^{-1})$ for every $y \in G_{\mathbf{A}}$.

If $G_{\mathbf{a}}$ is compact, we naturally take the measures of $G_{\mathbf{a}}$ and $H_{\mathbf{a}}$ to be 1. Then $\mathfrak{m}(S)$ can be defined in a unique way. We easily see that $S = \bigsqcup_{\zeta\in Z}\bigsqcup_{\gamma\in\Delta'_\zeta\backslash\Gamma} h\zeta\gamma$, where $\Delta'_\zeta = \zeta^{-1}\Delta_\zeta\zeta$, and so $\#S = \sum_{\zeta\in Z}[\Gamma : \Delta'_\zeta] = \mathfrak{m}(S)$ if $G_{\mathbf{a}}$ is compact. Thus we obtain

$$\mathfrak{m}(S) = \#(S) \quad \text{if } G_{\mathbf{a}} \text{ is compact.} \tag{3.5}$$

THEOREM 3.2. *The notation being the same as in Theorem 2.2, we have*

$$\mathfrak{m}(H,\, H_{\mathbf{A}} \cap D) = \sum_{y\in Y} \nu(\Gamma^y)\mathfrak{m}(\mathscr{V} \cap hDy^{-1}). \tag{3.6}$$

In particular if $m = 1$ and the notation is as in (iv) *of Theorem 2.2, then*

$$\mathfrak{m}(H,\, H_{\mathbf{A}} \cap C) = \sum_{y \in Y} \nu(\Gamma^y)\mathfrak{m}\big((Ly^{-1})[q,\, \mathfrak{b}]\big). \tag{3.7}$$

Proof. Let $E_y = H\backslash(H_{\mathbf{A}} \cap GyD)/(H_{\mathbf{A}} \cap D)$. For each $\varepsilon \in E_y$ take $\zeta_\varepsilon \in G$ so that $\varepsilon \in \zeta_\varepsilon yD$. By Theorem 2.2 (ii) we have $\mathscr{V} \cap hDy^{-1} = \bigsqcup_{\varepsilon \in E_y} h\zeta_\varepsilon \Gamma^y$ and $H \cap \varepsilon(H_{\mathbf{A}} \cap D)\varepsilon^{-1} = H \cap H_{\mathbf{A}} \cap \varepsilon D\varepsilon^{-1} = H \cap \zeta_\varepsilon yDy^{-1}\zeta_\varepsilon^{-1} = H \cap \zeta_\varepsilon \Gamma^y \zeta_\varepsilon^{-1}$, so that

$$\nu(\Gamma^y)\mathfrak{m}(\mathscr{V} \cap hDy^{-1}) = \sum_{\varepsilon \in E_y} \nu(H \cap \zeta_\varepsilon \Gamma^y \zeta_\varepsilon^{-1}) = \sum_{\varepsilon \in E_y} \nu(H \cap \varepsilon(H_{\mathbf{A}} \cap D)\varepsilon^{-1}).$$

Since $\bigsqcup_{y \in Y} E_y$ gives $H\backslash H_{\mathbf{A}}/(H_{\mathbf{A}} \cap D)$, we obtain

$$\mathfrak{m}(H,\, H_{\mathbf{A}} \cap D) = \sum_{y \in Y} \sum_{\varepsilon \in E_y} \nu(H \cap \varepsilon(H_{\mathbf{A}} \cap D)\varepsilon^{-1}) = \sum_{y \in Y} \nu(\Gamma^y)\mathfrak{m}(\mathscr{V} \cap hDy^{-1}).$$

This proves (3.6), which combined with (2.5) gives (3.7).

Formula (3.7) was given in [S3, (13.17b)] under the condition mentioned at the beginning of Section 2. That condition can be removed by the above theorem.

It should be noted that (3.6) and (3.7) are different from any of the mass formulas of Siegel. Indeed, the left-hand side of (3.6) concerns an orthogonal group in dimension $n - m$, and the right-hand side concerns the space $\mathscr{V}$, whereas both sides of Siegel's formulas are defined with respect to matrices of the same size. See also [S3, p. 137] for a comment on the connection with the work of Eisenstein and Minkowski on the sums of five squares.

4. CLASSIFICATION AND GENERA OF QUADRATIC FORMS IN TERMS OF MATRICES

4.1. Traditionally the genus and class of a quadratic form over $\mathbf{Q}$ were defined in terms of matrices, but it is easy to see that they are equivalent to those defined in terms of lattices. In the general case, however, there is a standard definition in terms of lattices, but the definition in terms of matrices is nontrivial. Also, we treated in [S5] the classification of quadratic forms over a number field, but gave explicit results in terms of matrices only when $\mathbf{Q}$ is the base field. Let us now discuss how we can handle such problems over an arbitrary number field, as the generalization is far from obvious and requires some new concepts. Before proceeding, we recall two basic facts: Given $(V,\, \varphi)$ over a global F and a lattice L in V, the $O^\varphi(V)$-genus of L is the same as the $SO^\varphi(V)$-genus of L; all the maximal lattices in V form a single $O^\varphi(V)$-genus; see [S3, Lemmas 6.8 and 6.9].

We take a global field F, and denote by F_n^1 the vector space of all n-dimensional row vectors with components in F, and by $\mathfrak{g}_n^1$ the set of elements of F_n^1 with components in $\mathfrak{g}$. Define subgroups E and E_ξ of $GL_n(F)_{\mathbf{A}}$ by

$$E = GL_n(F)_{\mathbf{a}} \prod_{v \in \mathbf{h}} GL_n(\mathfrak{g}_v), \qquad E_\xi = \xi^{-1}E\xi \quad (\xi \in GL_n(F)_{\mathbf{A}}). \tag{4.1}$$

Put $L_0 = \mathfrak{g}_n^1$. An arbitrary $\mathfrak{g}$-lattice L in F_n^1 can be given as $L = L_0\xi$ with $\xi \in GL_n(F)_{\mathbf{A}}$; then $E_\xi = \{y \in GL_n(F)_{\mathbf{A}} \mid Ly = L\}$. It is well known that the map $x \mapsto \det(x)\mathfrak{g}$ gives a bijection of $E\backslash GL_n(F)_{\mathbf{A}}/GL_n(F)$ onto the ideal class group of F; see [S1, Lemma 8.14], for example. Thus the ideal class of $\det(\xi)\mathfrak{g}$ is determined by the $GL_n(F)$-class of $L_0\xi$, and vice versa. Consequently, L is isomorphic as a $\mathfrak{g}$-module to the direct sum of $\mathfrak{g}_{n-1}^1$ and $\det(\xi)\mathfrak{g}$.

To define the generalization of $\mathbf{Z}$-valued quadratic forms, we denote by S_n the set of all symmetric elements of $GL_n(F)$, fix an element ξ of $GL_n(F)_{\mathbf{A}}$, and denote by $S_n(\xi)$ the set of all $T \in S_n$ such that $xT \cdot {}^t x \in \mathfrak{g}$ for every $x \in L_0\xi$. We call such a T *reduced* (relative to ξ) if the following condition is satisfied:

$$T \in S_n(\zeta^{-1}\xi) \text{ with } \zeta \in GL_n(F)_{\mathbf{h}} \cap \textstyle\prod_{v \in \mathbf{h}} M_n(\mathfrak{g}_v) \Longrightarrow \zeta \in E. \tag{4.2}$$

We denote by $S_n^0(\xi)$ the set of all reduced elements of $S_n(\xi)$. These depend essentially on $E\xi GL_n(F)$, as will be seen in Proposition 4.3 (i) below.

4.2. To define the genus and class of an element of S_n, put

$$\Delta_\xi = E_\xi \cap GL_n(F), \qquad \Delta_\xi^1 = E_\xi \cap SL_n(F). \tag{4.3}$$

We say that two elements Φ and Ψ of $S_n(\xi)$ belong to the same *genus* (relative to ξ) if there exists an element ε of E_ξ such that $\varepsilon\Phi \cdot {}^t\varepsilon = \Psi$; they are said to belong to the same *O-class* (resp. *SO-class*) if $\alpha\Phi \cdot {}^t\alpha = \Psi$ for some $\alpha \in \Delta_\xi$ (resp. $\alpha \in \Delta_\xi^1$). These depend on the choice of $L = L_0\xi$, or rather, on the choice of the ideal class of $\det(\xi)\mathfrak{g}$. There is no reason to think that $\xi = 1$ is the most natural choice.

Given $\Phi \in S_n$, put $V = F_n^1$ and $\varphi[x] = x\Phi \cdot {}^t x$ for $x \in V$. Then we obtain a quadratic space (V, φ), which we denote by $[\Phi]$; we put then $O(\Phi) = O^\varphi(V)$ and $SO(\Phi) = SO^\varphi(V)$. Now put $L = L_0\xi$ with $\xi \in GL_n(F)_{\mathbf{A}}$. Clearly L is integral if $\Phi \in S_n(\xi)$, in which case L is maximal if and only if $\Phi \in S_n^0(\xi)$.

PROPOSITION 4.3. (i) *If $\eta \in E\xi\alpha$ with $\xi, \eta \in GL_n(F)_{\mathbf{A}}$ and $\alpha \in GL_n(F)$, then $S_n(\xi) = \alpha S_n(\eta) \cdot {}^t\alpha$ and $S_n^0(\xi) = \alpha S_n^0(\eta) \cdot {}^t\alpha$.*

(ii) *For $\Phi, \Psi \in S_n^0(\xi)$, $\xi \in GL_n(F)_{\mathbf{A}}$, the spaces $[\Phi]$ and $[\Psi]$ are isomorphic if and only if they belong to the same genus.*

(iii) *Let X be a complete set of representatives for $E\backslash GL_n(F)_{\mathbf{A}}/GL_n(F)$ and for each $\xi \in GL_n(F)_{\mathbf{A}}$ let Y_ξ be a complete set of representatives for the genera of the elements of $S_n^0(\xi)$. Then the spaces $[\Phi]$ obtained from $\Phi \in Y_\xi$ for all $\xi \in X$ exhaust all isomorphism classes of n-dimensional quadratic spaces over F without overlapping.*

Proof. Assertion (i) can be verified in a straightforward way. Let Φ and Ψ be elements of $S_n^0(\xi)$ belonging to the same genus. Then there exists an element $\varepsilon \in E_\xi$ such that $\varepsilon\Phi \cdot {}^t\varepsilon = \Psi$, and the Hasse principle guarantees an element α of $GL_n(F)$ such that $\Psi = \alpha\Phi \cdot {}^t\alpha$. Thus $[\Psi]$ is isomorphic to $[\Phi]$. Conversely, suppose $[\Phi]$ and $[\Psi]$ are isomorphic for $\Phi, \Psi \in S_n^0(\xi)$. Then $\Phi = \beta\Psi \cdot {}^t\beta$ for some $\beta \in GL_n(F)$. Now $L_0\xi$ is maximal in both $[\Phi]$ and $[\Psi]$, and $L_0\xi\beta$ is maximal in $[\Psi]$. Thus $L_0\xi\beta = L_0\xi\gamma$ with $\gamma \in O(\Psi)_{\mathbf{A}}$. Put $\zeta = \beta\gamma^{-1}$. Then $\zeta \in E_\xi$, and $\zeta\Psi \cdot {}^t\zeta = \Phi$. Therefore Ψ belongs to the genus of Φ. This proves (ii). As for (iii), clearly every n-dimensional quadratic space is isomorphic to $[\Psi]$ for some $\Psi \in S_n$. Putting $V = F_n^1$, pick a maximal lattice L in V and put $L = L_0\eta$ with $\eta \in GL_n(F)_{\mathbf{A}}$. Then $\eta \in E\xi GL_n(F)$ for some $\xi \in X$, and by (i) we can replace η by ξ. Take a member Φ of Y_ξ belonging to the genus of Ψ. Then by (ii), $[\Psi]$ is isomorphic to $[\Phi]$. Thus every n-dimensional quadratic space is isomorphic to $[\Phi]$ for some $\Phi \in Y_\xi$ with $\xi \in X$. To prove that there is no overlapping, take $\Phi_i \in Y_{\xi_i}$ with $\xi_i \in X$, $i = 1, 2$, and suppose that $[\Phi_1]$ is isomorphic to $[\Phi_2]$. Then $\Phi_1 = \alpha\Phi_2 \cdot {}^t\alpha$ with $\alpha \in GL_n(F)$. Now $L_0\xi_1$ is maximal in $[\Phi_1]$, and so $L_0\xi_1\alpha$ is maximal in $[\Phi_2]$. Therefore $L_0\xi_1\alpha = L_0\xi_2\gamma$ with $\gamma \in O(\Phi_2)_{\mathbf{A}}$. Then $\xi_1\alpha\gamma^{-1}\xi_2^{-1} \in E$, and so $\det(\xi_1\xi_2^{-1}) \in F^\times \det(E)$, which implies that $\xi_1 = \xi_2$, as $\xi_i \in X$. Thus both Φ_1 and Φ_2 belong to Y_{ξ_1}. By (ii) they must belong to the same genus. This completes the proof.

4.4. Take $\Phi \in S_n^0(\xi)$ and suppose another member Ψ of$S_n^0(\xi)$ belongs to the genus of Φ. Put $L = L_0\xi$. Then $\Psi = \varepsilon\Phi \cdot {}^t\varepsilon = \alpha\Phi \cdot {}^t\alpha$ with $\varepsilon \in E_\xi$ and $\alpha \in GL_n(F)$, as observed in the above proof. Clearly $\varepsilon^{-1}\alpha \in O(\Phi)_{\mathbf{A}}$. Since $L\alpha = L\varepsilon^{-1}\alpha$, we see that $L\alpha$ belongs to the genus of L. We associate the $O(\Phi)$-class of $L\alpha$ to the O-class of Ψ. We easily see that the class of $L\alpha$ is determined by the class of Ψ. Moreover, it gives a bijection of the set of O-classes in the genus of Φ onto the set of $O(\Phi)$-classes in the genus of $L_0\xi$. Indeed, let $M = L\sigma$ with $\sigma \in SO(\Phi)_{\mathbf{A}}$. Then $\det(\sigma)\mathfrak{g} = \mathfrak{g}$, and so $\sigma = \tau\beta$ with $\tau \in E_\xi$ and $\beta \in GL_n(F)$. Put $\Psi = \tau^{-1}\Phi \cdot {}^t\tau^{-1}$. Then $\Psi = \beta\Phi \cdot {}^t\beta$ and $M = L\beta$, which corresponds to Ψ. This proves the surjectivity. The injectivity can be easily verified too.

For $\Phi \in S_n^0(\xi)$ we define the *SO-genus* of Φ to be the set of all Ψ in the genus of Φ such that $\det(\Psi) = \det(\Phi)$. For $\Psi = \varepsilon\Phi \cdot {}^t\varepsilon = \alpha\Phi \cdot {}^t\alpha$ as above, suppose $\det(\Psi) = \det(\Phi)$; then $\det(\alpha)^2 = 1$. Since $-1 \in \det\bigl(O(\Phi)\bigr)$, changing α for $\alpha\gamma$ with a suitable $\gamma \in O(\Phi)$, we may assume that $\det(\alpha) = 1$. We then associate the $SO(\Phi)$-class of $L\alpha$ to Ψ We can easily verify that the set of all SO-classes in the SO-genus of Φ contained in $S_n^0(\xi)$ are in one-to-one correspondence with the set of $SO(\Phi)$-classes in the genus of L.

In general, if $\Psi = \varepsilon\Phi{\cdot}{}^t\varepsilon = \alpha\Phi{\cdot}{}^t\alpha$ as above, then $\det(\alpha)^2 = \det(\varepsilon)^2 \in \det(E_\xi)$, and so $\det(\alpha) \in \mathfrak{g}^\times$. Therefore, if $F = \mathbf{Q}$, then $\det(\Psi) = \det(\Phi)$, and the SO-genus of Φ coincides with the genus of Φ.

4.5. Define (V, φ) by $V = F_n^1$ and $\varphi[x] = x\Phi \cdot {}^t x$ as above with any $\Phi \in S_n$. Put $L = L_0\xi$ with $\xi \in GL_n(F)_{\mathbf{A}}$. We easily see that $\widetilde{L} = L_0(2\Phi \cdot {}^t\xi)^{-1}$, and so

$$[\widetilde{L}/L] = \det(2\Phi\xi^2)\mathfrak{g} \quad \text{if} \quad L = L_0\xi. \tag{4.4}$$

In order to state our main results, we need a basic fact on an algebraic extension K of F. Let $\mathfrak{r}$ be the maximal order of K, and $\mathfrak{d}$ the different of K relative to F; let $\mathfrak{d}_0 = N_{K/F}(\mathfrak{d})$, $m = [K : F]$, and $d(K/F) = \det\bigl[(\mathrm{Tr}_{K/F}(e_ie_j))_{i,j=1}^m\bigr]$ with an F-basis $\{e_i\}_{i=1}^m$ of K. Strictly speaking, $d(K/F)$ should be viewed as a coset in $F^\times/F^{\times 2}$ represented by that determinant, but we denote any number in that coset by $d(K/F)$. By the *characteristic ideal class* for the extension K/F we understand the ideal class $\mathfrak{k}$ in F determined by the property that $\mathfrak{r}$ is isomorphic as a $\mathfrak{g}$-module to $\mathfrak{g}^1_{m-1} \oplus \mathfrak{x}$ with an ideal $\mathfrak{x}$ belonging to $\mathfrak{k}$. Then we have

(4.5) *The characteristic ideal class for K/F contains an ideal $\mathfrak{x}$ such that $\mathfrak{d}_0 = d(K/F)\mathfrak{x}^2$.*

To prove this, take $V = K$, $\varphi(x, y) = \mathrm{Tr}_{K/F}(xy)$ for $x, y \in K$, and $L = \mathfrak{r}$ in (4.4). Then $\widetilde{L} = (2\mathfrak{d})^{-1}$, and (4.4) shows that $\mathfrak{d}_0 = d(K/F)\mathfrak{x}^2$ with $\mathfrak{x} = \det(\xi)\mathfrak{g}$, which proves (4.5). This fact was noted, in substance, by Artin in 1949; that $\mathfrak{d}$ defines a square ideal class in K is due to Hecke. We can easily generalize (4.5) to the case of a maximal order $\mathfrak{o}$ in a simple algebra over F. The ideal class which is an obvious analogue of $\mathfrak{k}$ is independent of the choice of $\mathfrak{o}$.

We also need a few more symbols. We denote by $\mathbf{r}$ the set of all real primes in $\mathbf{a}$. For $T \in S_n$ and $v \in \mathbf{r}$ we put $s_v(T) = p_v - q_v$ if T as a real symmetric matrix in $GL_n(F_v) = GL_v(\mathbf{R})$ has p_v positive and q_v negative eigenvalues. Given a quadratic space (V, φ), we denote by $Q(\varphi)$ the characteristic quaternion algebra of (V, φ) in the sense of [S5, §3.1]; if φ is obtained from $\Phi \in S_n$ as above, we put $\delta(\Phi) = \delta(\varphi)$ and $Q(\Phi) = Q(\varphi)$. In [S5, Theorem 4.4] we showed that the isomorphism class of (V, φ) is determined by $\{n, \delta(\varphi), Q(\varphi), \{s_v(\varphi)\}_{v\in\mathbf{r}}\}$. We will now state this fact in terms of the matrices Φ in $S^0_n(\xi)$.

THEOREM 4.6 (The case of even n). *Let the symbols n, $\{\sigma_v\}_{v\in\mathbf{r}}$, δ, K_0, $\mathfrak{e}_0$, $\mathfrak{e}_1$, and ξ be given as follows: $4 \le n \in 2\mathbf{Z}$, $\sigma_v \in 2\mathbf{Z}$, $|\sigma_v| \le n$; $\delta \in F^\times$, $K_0 = F(\sqrt{\delta})$; $\mathfrak{e}_0$ and $\mathfrak{e}_1$ are squarefree integral ideals in F; $\xi \in GL_n(F)_\mathbf{A}$. Let r be the number of prime factors of $\mathfrak{e}_0\mathfrak{e}_1$, and $\mathfrak{d}$ the different of K_0 relative to F; put $\mathfrak{d}_0 = \mathfrak{d}^2 \cap F$. Suppose that $\mathfrak{e}_0$ divides $\mathfrak{d}_0$, $\mathfrak{e}_1$ is prime to $\mathfrak{d}_0$, $(-1)^{\sigma_v/2}\delta > 0$ at each $v \in \mathbf{r}$, $\det(\xi)\mathfrak{e}_1^{-1}$ belongs to the characteristic ideal class for K_0/F, and*

$$r + \#\{v \in \mathbf{r} \mid \sigma_v \equiv 4 \text{ or } 6 \pmod 8\} \in 2\mathbf{Z}. \tag{4.6}$$

Then there exists an element Φ of $S^0_n(\xi)$ such that

$$\delta \in \delta(\Phi), \quad \det(2\Phi\xi^2)\mathfrak{g} = \mathfrak{d}_0\mathfrak{e}_1^2, \quad s_v(\Phi) = \sigma_v \text{ for every } v \in \mathbf{r}, \tag{4.7}$$

and $Q(\Phi)$ is ramified at $v \in \mathbf{h}$ if and only if $v|\mathfrak{e}_0\mathfrak{e}_1$. Moreover, every element of $S^0_n(\xi)$ is of this type, and its genus is determined by $(\{\sigma_v\}_{v\in\mathbf{r}}, \delta, \mathfrak{e}_0, \mathfrak{e}_1)$. These statements are true even for $n = 2$ under the following additional condition on $\mathfrak{e}_1$: if $\mathfrak{e}_1 \ne \mathfrak{g}$, then $K_0 \ne F$ and every prime factor of $\mathfrak{e}_1$ remains prime in K_0.

Proof. Given the symbols as in our theorem, let B be the quaternion algebra over F ramified exactly at the prime factos of $\mathfrak{e}_0\mathfrak{e}_1$ and at those $v \in \mathbf{r}$ for which $\sigma_v \equiv 4$ or $6 \pmod 8$. Such a B exists because of (4.6). By the main theorem of classification [S5, Theorem 4.4] there exists a quadratic space (V, φ) over F such that $B = Q(\varphi)$, $\delta \in \delta(\varphi)$, and $\sigma_v = s_v(\varphi)$ for every $v \in \mathbf{r}$. By Proposition 4.3 (ii) we can take $(V, \varphi) = [\Psi]$ with $\Psi \in S^0_n(\eta)$, $\eta \in GL_n(F)_\mathbf{A}$. Put $L = L_0\eta$. By [S5, Theorem 6.2 (ii)] we have $[\widetilde{L}/L] = \mathfrak{d}_0\mathfrak{e}_1^2$, which combined with (4.4) shows that $\det(2\Psi\eta^2)\mathfrak{g} = \mathfrak{d}_0\mathfrak{e}_1^2$. By our assumption on ξ and (4.5) we have $\mathfrak{d}_0\mathfrak{e}_1^2 = c^2\delta\det(\xi)^2\mathfrak{g}$ with $c \in F^\times$. Thus $\det(2\Psi\eta^2)\mathfrak{g} = c^2\delta\det(\xi)^2\mathfrak{g}$. Since $(-1)^{n/2}\delta = b^2\det(2\Psi)$ with $b \in F^\times$, we see that $\det(\eta)\mathfrak{g} = bc\det(\xi)\mathfrak{g}$, which implies that $\eta \in E\xi\alpha$ with $\alpha \in GL_n(F)$. Then by Proposition 4.3 (i) we can replace η by the given ξ, and we obtain the first part of our theorem.

Next, given $\Phi \in S_n^0(\xi)$, put $L = L_0\xi$. Let $\mathfrak{e}$ be the product of all the prime ideals in F ramified in $Q(\Phi)$ and let $K_0 = F(\sqrt{\delta})$ with $\delta \in \delta(\Phi)$. By [S5, (4.2b)], $Q(\Phi)$ is ramified at $v \in \mathbf{r}$ if and only if $s_v(\Phi) \equiv 4$ or $6 \pmod 8$. Put $\mathfrak{e} = \mathfrak{e}_0\mathfrak{e}_1$, where $\mathfrak{e}_0$ is the product of the prime factors of $\mathfrak{e}$ ramified in K_0. Then in [S5, Theorem 6.2 (ii)] we showed that $[\widetilde{L}/L] = \mathfrak{d}_0\mathfrak{e}_1^2$, where $\mathfrak{d}_0$ is defined for this K_0 as in our theorem. Combining this with (4.4) we obtain $\det(2\Phi\xi^2)\mathfrak{g} = \mathfrak{d}_0\mathfrak{e}_1^2$. We have $\mathfrak{d}_0 = \delta\mathfrak{x}^2$ with an ideal $\mathfrak{x}$ as in (4.5), and so $\det(\xi)\mathfrak{g} = \mathfrak{x}\mathfrak{e}_1$, if we take δ to be $(-1)^{n/2}\det(2\Phi)$. This proves the second part of our theorem. The last part concerning the case $n = 2$ follows from [S5, Theorem 4.4, (4.4b)].

THEOREM 4.7 (The case of odd n). *Let the symbols n, $\{\sigma_v\}_{v\in\mathbf{r}}$, δ, K_0, $\mathfrak{e}$, and ξ be given as follows: $0 < n-1 \in 2\mathbf{Z}$, $\sigma_v - 1 \in 2\mathbf{Z}$, $|\sigma_v| \le n$; $\delta \in F^\times$, $K_0 = F(\sqrt{\delta})$; $\mathfrak{e}$ is a squarefree integral ideal in F; $\xi \in GL_n(F)_\mathbf{A}$. Let r be the number of prime factors of $\mathfrak{e}$; let $\delta\mathfrak{g} = \mathfrak{a}\mathfrak{b}^2$ with a squarefree integral ideal $\mathfrak{a}$ and a fractional ideal $\mathfrak{b}$ in F; put $\mathfrak{e} = \mathfrak{e}_0\mathfrak{e}_1$ with $\mathfrak{e}_1 = \mathfrak{a} + \mathfrak{e}$. Suppose $(-1)^{(\sigma_v-1)/2}\delta > 0$ at each $v \in \mathbf{r}$, $\det(\xi)\mathfrak{b}$ belongs to the ideal class of $\mathfrak{e}_0$, and*

$$r + \#\{v \in \mathbf{r} \mid \sigma_v \equiv \pm 3 \pmod 8\} \in 2\mathbf{Z}. \tag{4.8}$$

Then there exists an element Φ of $S_n^0(\xi)$ such that

$$\delta \in \delta(\Phi), \quad \det(2\Phi\xi^2)\mathfrak{g} = 2\mathfrak{a}\mathfrak{e}_0^2, \quad s_v(\Phi) = \sigma_v \text{ for every } v \in \mathbf{r}, \tag{4.9}$$

and $Q(\Phi)$ is ramified at $v \in \mathbf{h}$ if and only if $v|\mathfrak{e}$. Moreover, every element of $S_n^0(\xi)$ is of this type, and its genus is determined by $(\{\sigma_v\}_{v\in\mathbf{r}}, \delta, \mathfrak{e})$.

Proof. In [S5, Theorem 6.2 (iii)] we showed that $[\widetilde{L}/L] = 2\mathfrak{a}^{-1}\mathfrak{e}^2 \cap 2\mathfrak{a}$ for a maximal lattice L if $\delta(\varphi)\mathfrak{g} = \mathfrak{a}\mathfrak{b}^2$ as above. We easily see that $2\mathfrak{a}^{-1}\mathfrak{e}^2 \cap 2\mathfrak{a} = 2\mathfrak{a}\mathfrak{e}_0^2$. Therefore the proof can be given in exactly the same fashion as for Theorem 4.6.

Remark. (1) Let $\mathfrak{x}$ be the ideal belonging to the characteristic ideal class for K_0/F such that $\mathfrak{d}_0 = \delta\mathfrak{x}^2$. Suppose Φ has the last two properties of (4.7) and $\det(\xi)\mathfrak{e}_1^{-1}$ belongs to the class of $\mathfrak{x}$. Then $b^2\det(2\Phi)\mathfrak{g} = \delta\mathfrak{g}$ with $b \in F^\times$, and so $(-1)^{n/2}\det(2\Phi) = b^{-2}\delta c$ with $c \in \mathfrak{g}^\times$. Since $s_v(\Phi) = \sigma_v$ for every $v \in \mathbf{r}$, we see that $c > 0$ at every $v \in \mathbf{r}$. Suppose an element of $\mathfrak{g}^\times$ positive at every $v \in \mathbf{r}$ is always a square. Then we obtain $\delta \in \delta(\Phi)$. Thus the first property of Φ in (4.7) follows from the assumption on $\det(\xi)\mathfrak{g}$ and the last two properties of (4.7) under that condition on $\mathfrak{g}^\times$. The same comment applies to (4.9).

(2) The last two statements of (1) apply to the case $F = \mathbf{Q}$. Therefore the above two theorems for $F = \mathbf{Q}$ are the same as [S5, Theorems 6.4 and 6.5]. The new features in the general case are that we need $\xi \in GL_n(F)_\mathbf{A}$ and we have to impose a condition on the ideal class of $\det(\xi)\mathfrak{g}$. In other words, the

coset $E\xi GL_n(F)$ is determined by $\{\sigma_v\}_{v\in\mathbf{r}}$, δ, and the characteristic quaternion algebra.

(3) A $\mathfrak{g}$-valued *symmetric* form is represented by an element of S_n with entries in $\mathfrak{g}$. This is different from the notion of a $\mathfrak{g}$-valued *quadratic* form, and so the classification of $\mathfrak{g}$-valued symmetric forms is different from that of $\mathfrak{g}$-valued quadratic forms. We refer the reader to [S7] for the classification of $\mathfrak{g}$-valued symmetric forms and its connection with the classification of quadratic forms.

5. Ternary forms

5.1. Before proceeding further, let us recall several basic facts on quaternion algebras. Let B be a quaternion algebra over a global field F, and $\mathfrak{O}$ an order in B containing $\mathfrak{g}$. For $\xi \in B_{\mathbf{A}}^{\times}$ the principle of §1.2 about the lattice Lx enables us to define $\mathfrak{O}\xi$ as a $\mathfrak{g}$-lattice in B. We call a $\mathfrak{g}$-lattice in B of this form a *proper left* $\mathfrak{O}$*-ideal.* Thus the $B^{\times}$-genus of $\mathfrak{O}$ consists of all proper left $\mathfrak{O}$-ideals, and the genus is stable under right multiplication by the elements of $B^{\times}$, and so a $B^{\times}$-class of proper left $\mathfrak{O}$-ideals is meaningful. Define a subgroup U of $B_{\mathbf{A}}^{\times}$ by $U = B_{\mathbf{a}}^{\times}\prod_{v\in\mathbf{h}}\mathfrak{O}_v^{\times}$. Then $\#(U\backslash B_{\mathbf{A}}^{\times}/B^{\times})$ gives the number of $B^{\times}$-classes in this genus, which we call *the class number of* $\mathfrak{O}$. Next, for $x \in B_{\mathbf{A}}^{\times}$ we denote by $x\mathfrak{O}x^{-1}$ the order $\mathfrak{O}'$ in B such that $\mathfrak{O}'_v = x_v\mathfrak{O}_v x_v^{-1}$ for every $v \in \mathbf{h}$. By *the type number of* $\mathfrak{O}$ we mean $\#S$ for a minimal set S of such orders $\mathfrak{O}'$ with the property that every order of type $x\mathfrak{O}x^{-1}$ can be transformed to a member of S by an inner automorphism of B.

Given an order $\mathfrak{O}$ in B over a local or global F, put $\widetilde{\mathfrak{O}} = \{x \in B \,|\, \mathrm{Tr}_{B/F}(x\mathfrak{O}) \subset \mathfrak{g}\}$. It can be shown that $[\widetilde{\mathfrak{O}}/\mathfrak{O}]$ is a square of an integral ideal $\mathfrak{t}$. We call $\mathfrak{t}$ *the discriminant of* $\mathfrak{O}$. In the local case, if B is a division algebra, then $\mathfrak{O}$ is maximal if and only if the disriminant is the prime ideal of $\mathfrak{g}$. In the global case, if $\mathfrak{O}$ is maximal, then the discriminant of $\mathfrak{O}$ is the product of all the prime ideals where B is ramified. Thus we call it the *discriminant of* B. If $F = \mathbf{Q}$, then writing the discriminant of $\mathfrak{O}$ or of B in the form $t\mathbf{Z}$ with a positive integer t, we call t *the discriminant of* $\mathfrak{O}$ *or of* B. Let us quote here a result due to Eichler [E2, Satz 3]:

(5.1) *If F is local, B is isomorphic to $M_2(F)$, and the discriminant of an order $\mathfrak{O}$ in B is the prime ideal $\mathfrak{p}$ of $\mathfrak{g}$, then there is an isomorphism of B onto $M_2(F)$ that maps $\mathfrak{O}$ onto*

(5.1a) $$\left\{\begin{bmatrix} a & b \\ c & d \end{bmatrix} \in M_2(\mathfrak{g}) \,\middle|\, c \in \mathfrak{p}\right\}.$$

If $\mathfrak{O}$ is the order of (5.1a), then we can easily verify that

(5.1b) $$\widetilde{\mathfrak{O}} = \mathfrak{O}\eta^{-1} = \left\{\begin{bmatrix} a & b \\ c & d \end{bmatrix} \,\middle|\, a,\, c,\, d \in \mathfrak{g},\, b \in \mathfrak{p}^{-1}\right\}, \quad \eta = \begin{bmatrix} 0 & -1 \\ \pi & 0 \end{bmatrix},$$

where π is a prime element of F.

Let $\mathfrak{O}$ be an order in B in the global case, and let $\mathfrak{e}$ be the discriminant of B; suppose the discriminant of $\mathfrak{O}$ is squarefree. Then the discriminant of

$\mathfrak{O}$ is of the form $\mathfrak{a}_0\mathfrak{e}$ with a squarefree integral ideal $\mathfrak{a}_0$ prime to $\mathfrak{e}$. For each $v|\mathfrak{a}_0$ the order $\mathfrak{O}_v$ can be transformed to an order of type (5.1a). We will be considering such an order $\mathfrak{O}$ in the following treatment.

5.2. Let us now consider (V, φ) with $n = 3$ over a local or global F. As shown in [S3, §7.3], we can find a quaternion algebra B over F with which we can put $V = B^\circ\xi$, $A(V) = B + B\xi$, $A^+(V) = B$, $\varphi[x\xi] = dxx^\iota$ and $2\varphi(x\xi, y\xi) = d\mathrm{Tr}_{B/F}(xy^\iota)$ for $x, y \in B^\circ$, where B° is as in (1.2) and ξ is an element such that $\xi^2 = -d \in \delta(\varphi)$; $F + F\xi$ is the center of $A(V)$; $G^+(V) = B^\times$; $\nu(\alpha) = N_{B/F}(\alpha)$ and $x\xi\tau(\alpha) = \alpha^{-1}x\alpha\xi$ for $x \in B^\circ$ and $\alpha \in B^\times$.

Given $h \in V$ such that $\varphi[h] \neq 0$, take $k \in B^\circ$ so that $h = k\xi$. Put $W = (Fh)^\perp$ and $K = F + Fk$. It can easily be seen that $K = \{\alpha \in B \mid \alpha k = k\alpha\}$, and K is either a quadratic extension of F, or isomorphic to $F \times F$. In either case we can find an element ω of B such that

$$B = K + \omega K, \quad \omega^2 \in F^\times, \quad \omega k = -k\omega. \tag{5.2}$$

Then $B^\circ = Fk + \omega K$ and $\mathrm{Tr}_{B/F}(k\omega K) = 0$, and so $W = \omega K\xi = F\omega\xi + F\omega k\xi$, and $\varphi[\omega x\xi] = -d\omega^2 xx^\rho$ for $x \in K$, where ρ is the nontrivial automorphism of K over F. Since $-dk^2 = \varphi[h]$, we have $K = F \times F$ if $-d\varphi[h]$ is a square in F; otherwise $K = F(\kappa^{1/2})$ with $\kappa = -d\varphi[h]$. We easily see that

$$A^+(W) = K, \qquad G^+(W) = K^\times. \tag{5.3}$$

The group $SO^\varphi(W)$ can be identified with $\{b \in K^\times \mid bb^\rho = 1\}$, and the map $\tau : K^\times \to SO^\varphi(W)$ is given by $\tau(a) = a/a^\rho$ for $a \in K^\times$. To be precise, $\omega x\xi\tau(a) = \omega(a/a^\rho)x\xi$ for $x \in K$.

Clearly the isomorphism class of (V, φ) is determined by B and $dF^{\times 2}$, and vice versa. If F is a totally real number field and φ is positive definite at every $v \in \mathbf{a}$, then B is totally definite and d is totally positive.

Assuming F to be global, put $d\mathfrak{g} = \mathfrak{a}\mathfrak{x}^2$ with a squarefree integral ideal $\mathfrak{a}$ and a fractional ideal $\mathfrak{x}$; let $\mathfrak{e}$ be the product of all the prime ideals of F ramified in B. In general the pair $(\mathfrak{e}, \mathfrak{a})$ does not necessarily determine (V, φ), but it does if $F = \mathbf{Q}$ and φ is positive definite.

LEMMA 5.3. *Suppose F is global; let (V, φ), B, $\mathfrak{a}$, and $\mathfrak{e}$ be as above. Put $\mathfrak{e}_1 = \mathfrak{a} + \mathfrak{e}$, $\mathfrak{e} = \mathfrak{e}_0\mathfrak{e}_1$, and $\mathfrak{a} = \mathfrak{a}_0\mathfrak{e}_1$. Let L be an integral lattice in V. Then the following assertions hold:*

(i) *L is maximal if and only if $[\widetilde{L}/L] = 2\mathfrak{a}\mathfrak{e}_0^2$.*

(ii) *If L is maximal, then there is a unique order of B $\bigl(= A^+(V)\bigr)$ of discriminant $\mathfrak{a}_0\mathfrak{e}$ containing $A^+(L)$.*

Proof. Let Λ be a maximal lattice containing L. Since $[\widetilde{L}/L] = [\widetilde{\Lambda}/\Lambda][\Lambda/L]^2$, the "if"-part of (i) follows from the "only if"-part. Clearly the problem can be reduced to the local case. In the local case, we may assume that $d\mathfrak{g}$ equals $\mathfrak{g}$ or the prime ideal $\mathfrak{p}$ of $\mathfrak{g}$. Let t and M be as in §1.4. Then $[\widetilde{L}/L] = [\widetilde{M}/M]$. If $t = 1$, we can easily find $[\widetilde{M}/M]$. Suppose $t = 3$; if $d\mathfrak{g} = \mathfrak{p}$, then the explicit forms of M and $\widetilde{M}$ given in §1.5, (B) show that $[\widetilde{M}/M] = 2\mathfrak{p}$. If $d\mathfrak{g} = \mathfrak{g}$. we have $[\widetilde{M}/M] = 2\mathfrak{p}^2$ as shown in [S3, (7.9)]. This proves (i).

Next, since all the maximal lattices form a genus, it is sufficient to prove (ii) for a special choice of L. We take an order $\mathfrak{O}$ in B of discriminant $\mathfrak{a}\mathfrak{e}_0$, and take the $\mathfrak{g}$-lattice $\mathfrak{M}$ in B such that $\mathfrak{M}_v = \widetilde{\mathfrak{O}}_v$ if $v|\mathfrak{a}$ and $\mathfrak{M}_v = \mathfrak{O}_v$ if $v\nmid\mathfrak{a}$. We then put

$$L = \mathfrak{x}^{-1}(\mathfrak{M} \cap B^\circ)\xi, \tag{5.4}$$

Our task is to show that L is maximal and to prove (ii) for this L. The problems can be reduced to the local case. For simplicity we fix $v \in \mathbf{h}$ and denote F_v and $\mathfrak{g}_v$ by F and $\mathfrak{g}$, suppressing the subscript v. We can take $\mathfrak{x} = \mathfrak{g}$, and $d\mathfrak{g}$ to be $\mathfrak{g}$ or $\mathfrak{p}$. If $B = M_2(F)$ and $d\mathfrak{g} = \mathfrak{g}$, then $\mathfrak{O} = \widetilde{\mathfrak{O}} = M_2(\mathfrak{g})$. If $B = M_2(F)$ and $d\mathfrak{g} = \mathfrak{p}$, then $\mathfrak{O}$ and $\widetilde{\mathfrak{O}}$ can be given by (5.1a, b). Thus, for either type of $d\mathfrak{g}$ we have

$$L = \left\{ \begin{bmatrix} a & b \\ c & -a \end{bmatrix} \,\middle|\, a,\, c \in \mathfrak{g},\, b \in d^{-1}\mathfrak{g} \right\} \cdot \xi, \tag{5.5a}$$

$$\widetilde{L} = \left\{ \begin{bmatrix} a & b \\ c & -a \end{bmatrix} \,\middle|\, a \in (2d)^{-1}\mathfrak{g},\, c \in \mathfrak{g},\, b \in d^{-1}\mathfrak{g} \right\} \cdot \xi, \tag{5.5b}$$

so that $[\widetilde{L}/L] = 2d\mathfrak{g}$, which together with (i) shows that L is maximal. If $v|\mathfrak{e}$, then there is a unique maximal lattice in V given in the form $\{x \in V \mid \varphi[x] \in \mathfrak{g}\}$. It can easily be seen that this coincides with (5.4). Now, in the local case we easily see that $A^+(L) = \mathfrak{O}$ except when B is a division algebra and $d\mathfrak{g} = \mathfrak{g}$, in which case B has a unique order of prime discriminant. This completes the proof.

LEMMA 5.4. *In the setting of Lemma 5.3 let L be a maximal lattice in V and let $\mathfrak{O}$ be the order in B of discriminat $\mathfrak{a}_0\mathfrak{e}$ established in Lemma 5.3 (ii). Put*

$$C = \{x \in SO^\varphi(V)_\mathbf{A} \mid Lx = L\}, \tag{5.6}$$

$$T = \{y \in B_\mathbf{A}^\times \mid y\mathfrak{O} = \mathfrak{O}y\}, \quad T_v = B_v^\times \cap T. \tag{5.7}$$

Then $C = \tau(T)$; moreover, for $v \in \mathbf{h}$ we have $T_v = B_v^\times$ if $v|\mathfrak{e}$, $T_v = F_v^\times\mathfrak{O}_v^\times$ if $v\nmid\mathfrak{a}_0\mathfrak{e}$, and $T_v = F_v^\times(\mathfrak{O}_v^\times \cup \mathfrak{O}_v^\times\eta_v)$ if $v|\mathfrak{a}_0$, where η_v is η of (5.1b). Furthermore,

$$\{\gamma \in SO^\varphi(V) \mid L\gamma = L\} = \tau(\{\alpha \in B^\times \mid \alpha\mathfrak{O} = \mathfrak{O}\alpha\}). \tag{5.8}$$

Proof. The statements about T_v are well known if $\mathfrak{O}_v$ is maximal, that is, if $v\nmid\mathfrak{a}_0$. If $v|\mathfrak{a}_0$, then $\mathfrak{O}_v$ is of type (5.1a), and the desired fact follows from [E2, Satz 5]. To prove $\tau(T) = C$, we first recall the equality $SO^\varphi(V) = \tau(G^+(V))$, which holds over any field. Thus it is sufficient to show that $\tau(T_v) = C_v$ for every $v \in \mathbf{h}$, where $C_v = C \cap SO^\varphi(V)_v$. This is trivial if $v|\mathfrak{e}$, as $C_v = SO^\varphi(V)_v$ and $T_v = B_v^\times$ then. For $v \in \mathbf{h}$ put $D_v = \{\alpha \in B_v^\times \mid \alpha^{-1}L_v\alpha = L_v\}$. Then $\tau(D_v) = C_v$. We have seen in the proof of Lemma 5.3 that $A^+(L_v) = \mathfrak{O}_v$ if $v\nmid\mathfrak{e}$. Therefore if $\alpha \in D_v$ for such a v, then $\alpha^{-1}\mathfrak{O}_v\alpha = \mathfrak{O}_v$, and so $\alpha \in T_v$. Conversely, if $\alpha \in T_v$, then $\alpha\mathfrak{O}_v = \mathfrak{O}_v\alpha$ and we easily see that $\alpha\widetilde{\mathfrak{O}}_v = \widetilde{\mathfrak{O}}_v\alpha$.

We may assume that L is of the form (5.4). Since $\mathfrak{M}_v$ is $\mathfrak{O}_v$ or $\widetilde{\mathfrak{O}}_v$, we have $\alpha^{-1}L_v\alpha = L_v$, and so $\alpha \in D_v$. Thus $T_v = D_v$. Since $\tau(D_v) = C_v$, we obtain $\tau(T) = C$. Since $\mathrm{Ker}(\tau) = F_{\mathbf{A}}^{\times} \subset T$, if $\tau(\alpha) \in C$ with $\alpha \in B^{\times}$, then $\alpha \in T \cap B^{\times}$, from which we obtain (5.8).

Proof of Theorem 1.8. Since $\tau\bigl(G^+(V)\bigr) = SO^{\varphi}(V)$, the equality $\tau(T) = C$ stated in (iii) is clear from the definition of T. To prove the remaining statements, we first consider the case $n = 3$. Then C of §1.7 and Theorem 1.8 is C_v of Lemma 5.4 with a divisor v of $\mathfrak{a}$, and $A^+(L) = \mathfrak{O}_v$ as noted at the end of the proof of Lemma 5.3. Moreover, by Lemma 5.4, we have $F_v^{\times} \subset T_v$ and $\tau(T_v) = C_v$, and so T_v coincides with T of our theorem. Thus $T = B_v^{\times}$ if $v|\mathfrak{e}$ and $T = F_v^{\times}(\mathfrak{O}_v^{\times} \cup \mathfrak{O}_v^{\times}\eta_v)$ if $v|\mathfrak{a}_0$. The last expression for T is valid even for $v|\mathfrak{e}$, if we take η_v to be an element of B_v such that η_v^2 is a prime element of F_v. Clearly $J = \mathfrak{O}_v^{\times}$ and $\nu(J) = \mathfrak{g}_v^{\times}$; also, all the other statements of our theorem can easily be verified.

Next suppose $n > 3$; using the symbols of (1.4a, b, c), put $W = Z$ and $N = M$ if $t = 3$; put $W = Z + Fe_r + Ff_r$ and $N = M + \mathfrak{g}e_r + \mathfrak{g}f_r$ if $t = 1$. Then $V = W + \sum_{i=1}^{u}(Fe_i + Ff_i)$ and $L = N + \sum_{i=1}^{u}(\mathfrak{g}e_i + \mathfrak{g}f_i)$, where $u = r$ if $t = 3$ and $u = r - 1$ if $t = 1$. Observe that N contains an element g such that $\varphi[g] \in \mathfrak{g}^{\times}$. Therefore, as shown in [S3, §8.2], there exists an isomorphism θ of $A^+(V)$ onto $M_2\bigl(A^+(W)\bigr)$ such that $\theta\bigl(A^+(L)\bigr) = M_s\bigl(A^+(N)\bigr)$, where $s = 2^u$. Thus (i) and (ii) of our theorem follow from what we proved in the case $n = 3$; namely, $A^+(L)$ can be identified with $M_s(\mathfrak{O})$ with an order $\mathfrak{O}$ as stated. As for (iii) in the general case, clearly $\nu(J) = \mathfrak{g}^{\times}$. We already found an element η of $G^+(W)$ such that $N\tau(\eta) = N$, $\eta^2 \in F^{\times}$, and $\nu(\eta)\mathfrak{g} = \mathfrak{p}$; also $\eta J = J\eta$ as can easily be seen. Put $T_0 = F^{\times}(J \cup J\eta)$. Then T_0 is a subgroup of $G^+(V)$, $\tau(T_0) \subset C$, and $\nu(T_0) = F^{\times}$. Let $\alpha \in T$. Take $\beta \in T_0$ so that $\nu(\alpha) = \nu(\beta)$. Then $\nu(\beta^{-1}\alpha) = 1$, and so $\beta^{-1}\alpha \in J$. Thus $\alpha \in \beta J \subset T_0$, and we obtain $T = T_0$ and $C = \tau(T_0) = \tau(J) \cup \tau(J\eta)$. If $\tau(\eta) \in \tau(J)$, then $\eta \in F^{\times}J$, which is impossible, as $\nu(\eta)\mathfrak{g} = \mathfrak{p}$. Thus $\tau(\eta) \notin \tau(J)$, and so $[C : \tau(J)] = 2$. This completes the proof.

5.5. We now return to (V, φ) over a global F of an odd dimension $n > 1$, and fix a maximal lattice L in V. For each $v \in \mathbf{h}$ we take an element $\varepsilon_v \in \delta(\varphi_v)$ that is either a unit or a prime element of F_v. Then $\varepsilon_v\mathfrak{g}_v^{\times 2}$ is determined by φ_v, where $\mathfrak{g}_v^{\times 2} = \{a^2 \mid a \in \mathfrak{g}_v^{\times}\}$. Given $h \in V$ such that $\varphi[h] \neq 0$, take an element $\beta_v \in F_v$ such that $\varphi(h, L_v) = \beta_v\mathfrak{g}_v$, and put $r_v(h) = \varepsilon_v^{-1}\varphi[h]\beta_v^{-2}$. Then $r_v(h)$ determines a coset of $F_v^{\times}/\mathfrak{g}_v^{\times 2}$, which depends only on φ_v, L, and $F_v^{\times}hC(L_v)$. Strictly speaking, $r_v(h)$ should be defined as a coset, but for simplicity we view it as an element of $F_v^{\times}$, and write $r_v(h) \in X$ or $r_v(h) \in X$ for any subset X of F_v stable under multiplication by the elements of $\mathfrak{g}_v^{\times 2}$. For example, we can take as X the set

$$\mathfrak{E}_v = \bigl\{u^2 + 4w \bigm| u,\, w \in \mathfrak{g}_v\bigr\}. \tag{5.9}$$

Then the condition $r_v(h) \in \mathfrak{E}_v$ is meaningful. Obviously $\mathfrak{E}_v = \mathfrak{g}_v$ if $v \nmid 2$. We can also define $r_v(h)$ for $h \in V_v$.

The following lemma concerns the local case, that is, v is fixed, and so we write $r(h)$ and $\mathfrak{E}$ for $r_v(h)$ and $\mathfrak{E}_v$.

LEMMA 5.6. *Let* (V, φ), B, ξ, d, *and* B° *be as in* §5.2 *with a local field* F *and* $B = M_2(F)$; *suppose* d *is a prime element of* F; *let* $\mathfrak{O}$ *be the order of* (5.1a) *and* L *be as in* (5.5a); *put* $C = \{\alpha \in SO^\varphi(V) \mid L\alpha = L\}$. *Fixing* $h = k\xi$ *as in* §5.2, *put* $K = F[k]$, $\mathfrak{f} = K \cap \mathfrak{O}$, $\mathfrak{y} = \widetilde{\mathfrak{O}} \cap K$, $W = (Fh)^\perp$, *and*

$$J = \{\alpha \in SO^\varphi(W) \mid (L \cap W)\alpha = L \cap W\}. \tag{5.10}$$

Then the following assertions hold:

(i) $\mathfrak{f}$ *is the order of* K *whose discriminant is* $r(h)\mathfrak{p}^2$.

(ii) *There exists an element* ω *of* $B^\times$ *such that* (5.2) *is satisfied,* $\tau(\omega) \in C$, $W = \omega K\xi$; $L \cap W = \omega\mathfrak{f}\xi$ *if* $r(h) \notin \mathfrak{p}^{-1}$, *and* $L \cap W = \omega\mathfrak{y}\xi$ *if* $r(h) \in \mathfrak{p}^{-1}$.

(iii) *If* $r(h) \notin \mathfrak{p}^{-1}$, *then* $K \cong F \times F$, $\mathfrak{f}$ *is the maximal order of* K, $L \cap W$ *is a maximal lattice in* W, *and* $J = SO^\varphi(W) \cap C = \tau(\mathfrak{f}^\times)$.

(iv) *If* $r(h) \in \mathfrak{E}$, *then* $\mathfrak{y}$ *is the order in* K *whose discriminant is* $r(h)\mathfrak{g}$, $\mathfrak{f}$ *is not maximal,* $J = \tau(\{a \in K^\times \mid a/a^\rho \in \mathfrak{y}^\times\})$, *and* $SO^\varphi(W) \cap C = \tau(\mathfrak{f}^\times)$.

(v) *If* $r(h) \in \mathfrak{p}^{-1}$ *and* $r(h) \notin \mathfrak{E}$, *then* K *is ramified over* F, $\mathfrak{f}$ *is the maximal order of* K, $J = \tau(\{a \in K^\times \mid a/a^\rho \in \mathfrak{f}^\times\})$, *and* $SO^\varphi(W) \cap C = \{x \in K \mid xx^\rho = 1\}$, *which has* $\tau(\mathfrak{f}^\times)$ *as a subgroup of index 2.*

(vi) *Cases* (iii), (iv), *and* (v) *cover all possibilities for* h, *and mutually exclusive.*

Proof. To prove our assertions, we can replace h by any element of $F^\times hC$. Indeed, if we replace h by $ch\tau(\alpha)$ with $c \in F^\times$ and α in the set T_v of (5.7), then K, $\mathfrak{f}$, and other symbols are replaced by their images under the inner asutomorphism $x \mapsto \alpha^{-1}x\alpha$ of $A(V)$. Thus we may assume that $2\varphi(h, L) = \mathfrak{g}$. We can take $\begin{bmatrix} 0 & -d^{-1} \\ 0 & 0 \end{bmatrix}\xi$, $\begin{bmatrix} 0 & 0 \\ 1 & 0 \end{bmatrix}\xi$, and $\begin{bmatrix} 1 & 0 \\ 0 & -1 \end{bmatrix}\xi$ as the elements e_1, f_1, and g of §1.6. Therefore the result there guaratees an element $\gamma \in C$ such that $h\gamma = j\xi$ with j of the following two types: (1) $j = \mathrm{diag}[c, -c]$, $2c\mathfrak{p} = \mathfrak{g}$; (2) $j = \begin{bmatrix} c & b \\ 1 & -c \end{bmatrix}$ with $b \in \mathfrak{p}^{-1}$ and $c \in 2^{-1}\mathfrak{g}$. Thus we may assume that $k = j$ with such a j. Take $\omega = \begin{bmatrix} 0 & d^{-1} \\ -1 & 0 \end{bmatrix}$ for type (1) and $\omega = \begin{bmatrix} -1 & 2c \\ 0 & 1 \end{bmatrix}$ for type (2). Then $\tau(\omega) \in C$ and $\omega^{-1}k\omega = -k$; also $W = \omega K\xi$ and $L \cap W = (\widetilde{\mathfrak{O}} \cap \omega K)\xi$. We now treat our problems according to the type of j.

Type (1). Put $\ell = \mathrm{diag}[2c, 0]$. Then $K = F + F\ell$, as $\ell = j + c$. Clearly K consists of the diagonal matrices, and $\mathfrak{f}$ is its maximal order. Since $2c\mathfrak{p} = \mathfrak{g}$, we have $\mathfrak{f} = \mathfrak{g}[d\ell]$. Now, $\widetilde{\mathfrak{O}} \cap \omega K = \omega(\mathfrak{O} \cap K) = \omega\mathfrak{f}$, so that $L \cap W = \omega\mathfrak{f}\xi$, which is a maximal lattice in W. It can easily be seen that $J = SO^\varphi(W) \cap C = \tau(\mathfrak{f}^\times)$. Also, $r(h)\mathfrak{g} = 4c^2\mathfrak{g} = \mathfrak{p}^{-2}$, so that $r(h) \notin \mathfrak{p}^{-1}$, and $r(h)\mathfrak{p}^2 = \mathfrak{g}$, which is the discriminant of $\mathfrak{f}$. Since $r(h) \in \mathfrak{p}^{-1}$ for type (2) as can easily be seen, we have type (1) if and only if $r(h) \notin \mathfrak{p}^{-1}$.

Type (2). For j of type (2), we have $r(h) \in \mathfrak{p}^{-1}$. We easily see that $r(h) \in \mathfrak{E}$ if $b \in \mathfrak{g}$. Conversely suppose $r(h) \in \mathfrak{E}$; then we have $-4\varphi[h] = d(u^2 + 4w)$

with $u, w \in \mathfrak{g}$. Put $h' = j'\xi$ with $j' = \begin{bmatrix} u/2 & w \\ 1 & -u/2 \end{bmatrix}$. Then $\varphi[h'] = \varphi[h]$ and $2\varphi(h', L) = \mathfrak{g} = 2\varphi(h, L)$, so that $h' \in hC$ by Theorem 1.3. Thus if $r(h) \in \mathfrak{E}$, then we can put $h = j\xi$ with j of the above form such that $b \in \mathfrak{g}$. Without assuming $b \in \mathfrak{g}$, put $\ell = \begin{bmatrix} 2c & b \\ 1 & 0 \end{bmatrix}$. Then $\ell = c + j$ and $K = F + F\ell$. For $y, z \in F$ we have $y + z\ell = \begin{bmatrix} y+2cz & bz \\ z & y \end{bmatrix}$. This belongs to $\mathfrak{O}$ if and only if $y \in \mathfrak{g}$ and $z \in \mathfrak{p}$. Thus $\mathfrak{f} = \mathfrak{g}[d\ell]$. This has discriminant $4(b + c^2)\mathfrak{p}^2$, which equals $r(h)\mathfrak{p}^2$. Since $\omega^2 = 1$ and $\omega \in \mathfrak{O}^\times$, we have $L \cap W = \omega(\widetilde{\mathfrak{O}} \cap K)\xi$. We see that $y + z\ell \in \widetilde{\mathfrak{O}}$ if and only if $y \in \mathfrak{g}$ and $z \in \mathfrak{g}$. Thus $\mathfrak{y} = \mathfrak{g} + \mathfrak{g}\ell$ and $L \cap W = \omega\mathfrak{y}\xi$. For $x \in K$ and $a \in K^\times = G^+(W)$, we have $\omega x\xi\tau(a) = \omega ex\xi$, where $e = a/a^\rho$. Thus $(L \cap W)\tau(a) = L \cap W$ if and only if $(a/a^\rho)\mathfrak{y} = \mathfrak{y}$. Also $\tau(a) \in C$ if and only if $\tau(a) = \tau(s)$ with $s \in \mathfrak{O}^\times \cup \mathfrak{O}^\times\eta^{-1}$, as can be seen from Lemma 5.4. If $s \in \mathfrak{O}^\times$, then $a \in F^\times\mathfrak{O}^\times \cap K^\times = F^\times\mathfrak{f}^\times$, so that $\tau(a) \in \tau(\mathfrak{f}^\times)$. If $s \in \mathfrak{O}^\times\eta^{-1}$, then $s \in \mathfrak{O}^\times\eta^{-1} \cap K \subset \mathfrak{y}$.

Suppose now $r(h) \in \mathfrak{E}$; then we may assume that $b \in \mathfrak{g}$. Since $\ell^2 = 2c\ell + b$, we see that $\mathfrak{y}$ is an order, and so J is as in (iv). If $s \in \mathfrak{O}^\times\eta^{-1}$, then $dss^\iota \in \mathfrak{g}^\times$, which is impossible as $\mathfrak{y}$ is an order. Thus $SO^\varphi(W) \cap C = \tau(\mathfrak{f}^\times)$. Since $\mathfrak{f} = \mathfrak{g}[d\ell] \neq \mathfrak{g}[\ell] = \mathfrak{y}$, $\mathfrak{f}$ is not maximal and the discriminant of $\mathfrak{y}$ is $r(h)\mathfrak{g}$. This proves (iv).

Next suppose $r(h) \notin \mathfrak{E}$; then $b \notin \mathfrak{g}$ and $b\mathfrak{p} = \mathfrak{g}$. We easily see that $\ell\mathfrak{f} = \mathfrak{y}$, and so $e\mathfrak{y} = \mathfrak{y}$ if and only if $e \in \mathfrak{f}^\times$. Thus J is as in (v). Put $\pi = b^{-1}$ and $\sigma = \pi\ell$. Then $K = F[\sigma]$. Since $\sigma^2 = 2c\pi\sigma + \pi$, K is ramified over F and its maximal order is $\mathfrak{g}[\sigma]$, which coincides with $\mathfrak{f}$. Now $SO^\varphi(W) = \{x \in K^\times \mid xx^\rho = 1\}$, which has $\tau(\mathfrak{f}^\times)$ as a subgroup of index 2 by a general principle [S3, Lemma 5.6 (iii)]. Since $\ell \in \mathfrak{O}^\times\eta^{-1}$, we have $\tau(\ell) \in SO^\varphi(W) \cap C$. If $\tau(\ell) \in \tau(\mathfrak{f}^\times)$, then $\ell \in F^\times\mathfrak{f}^\times$, which is impossible. Thus $\tau(\mathfrak{f}^\times) \subsetneqq SO^\varphi(W) \cap C \subset SO^\varphi(W)$, which proves (v). Finally (vi) is clear from the above discussion. This completes the proof, as (i) and (ii) have been proved.

If we apply (2.7) to the present setting, then $H = \{b \in K^\times \mid bb^\rho = 1\}$ with $K = F + Fk$ as noted in §5.2. Thus H is commutative and the left-hand side of (2.7) becomes $[H_{\mathbf{A}} : H(H_{\mathbf{A}} \cap C)]$. We can determine this index explicitly as follows.

THEOREM 5.7. *In the setting of §5.2 with a global F, let L be a maximal lattice in V, and $\mathfrak{e}$ the product of all the prime ideals in F ramified in B; put $d\mathfrak{g} = \mathfrak{a}\mathfrak{x}^2$ with a squarefree integral ideal $\mathfrak{a}$ and a fractional ideal $\mathfrak{x}$. Let $\mathfrak{O}$ be the order in B of discriminant $\mathfrak{a} \cap \mathfrak{e}$ containing $A^+(L)$. (See Lemma 5.3 (ii).) Given $h = k\xi \in V$ with $k \in B^\circ$ such that $\varphi[h] \neq 0$, put $K = F + Fk$; denote by $\mathfrak{r}$ the maximal order of K and by $\mathfrak{d}$ the different of K relative to F. Let $\mathfrak{a}^*$ be the product of the prime factors v of $\mathfrak{a}$ such that $v \nmid \mathfrak{e}$, $r_v(h) \in \mathfrak{p}_v^{-1}$, and $r_v(h) \notin \mathfrak{E}_v$, where $\mathfrak{p}_v$ is the local prime ideal at v. Then the order $\mathfrak{f}$ in K given by $\mathfrak{f} = K \cap \mathfrak{O}$ has conductor $\mathfrak{c}$, which can be determined by the condition that $\mathfrak{c}_v = \mathfrak{g}_v$ if $v|\mathfrak{a}^*\mathfrak{e}$ and $\mathfrak{c}_v^2 N_{K/F}(\mathfrak{d})_v = \mathfrak{a}_v\varphi[h]\varphi(h, L)_v^{-2}$ if $v \nmid \mathfrak{e}$. Moreover,* *put* $H = SO^{\varphi}(W)$ *and identify* H *with* $\{\alpha \in K^\times \mid \alpha\alpha^\iota = 1\}$; *define* C *by (5.6). Suppose* K *is a field; then*

$$(5.11) \qquad [H_{\mathbf{A}} : H(H_{\mathbf{A}} \cap C)] = (c_K/c_F) \cdot 2^{1-\mu-\nu}[\mathfrak{g}^\times : N_{K/F}(\mathfrak{r}^\times)] \cdot [U : U']^{-1}N(\mathfrak{c})\prod_{\mathfrak{p}|\mathfrak{c}}\{1 - [K/F,\, \mathfrak{p}]N(\mathfrak{p})^{-1}\}.$$

Here c_K *resp.* c_F *is the class number of* K *resp.* F; μ *is the number of prime ideals dividing* $\mathfrak{a}^*\mathfrak{e}$ *and ramified in* K; ν *is the number of* $v \in \mathbf{a}$ *ramified in* K;

$$U = \{x \in \mathfrak{r}^\times \mid xx^\rho = 1\} \quad \textit{and} \quad U' = \{x \in U \mid x - 1 \in \mathfrak{c}_v\mathfrak{d}_v \textit{ for every } v \nmid \mathfrak{a}^*\mathfrak{e}\};$$

$\mathfrak{p}$ *runs over all prime factors of* $\mathfrak{c}$; $[K/F,\, \mathfrak{p}]$ *denotes* 1, -1, *or* 0 *according as* $\mathfrak{p}$ *splits in* K, *remains prime in* K, *or is ramified in* K.

Proof. If $d \in \mathfrak{g}^\times$, this is [S3, Theorem 12.3]. Our proof here is a modified version of the proof there. For $v|\mathfrak{e}$ we have $\mathfrak{r}_v \subset \mathfrak{O}_v$, so that $\mathfrak{f}_v = \mathfrak{r}_v$ and $\mathfrak{c}_v = \mathfrak{g}_v$. Next suppose $v \nmid \mathfrak{e}$; then we can put $B_v = M_2(F_v)$, and so [S3, Lemma 11.11] is applicable if $v \nmid \mathfrak{a}$; by (i) of that lemma, $\mathfrak{c}_v^2 N_{K/F}(\mathfrak{d})_v = \varphi[h]\varphi(h,\, L)_v^{-2}$. If $v|\mathfrak{a}$, then we use (i) of Lemma 5.6. If $v|\mathfrak{a}^*$, then $\mathfrak{c}_v = \mathfrak{g}_v$ by (v) of Lamma 5.6. Thus we obtain our assertion concerning $\mathfrak{c}$.

To prove (5.11), suppose K is a field. Then we have

$$[H_{\mathbf{A}} : HE] = (c_K/c_F) \cdot 2^{1-\kappa}[\mathfrak{g}^\times : N_{K/F}(\mathfrak{r}^\times)],$$

where $E = H_{\mathbf{a}}\prod_{v \in \mathbf{h}} E_v$ with $E_v = \mathfrak{r}_v^\times \cap H_v$ and κ is the number of $v \in \mathbf{v}$ ramified in K. Indeed, $[H_{\mathbf{A}} : HE]$ equals the number of classes in the genus of $\mathfrak{g}$-maximal lattices in W, and in [S3, (9.16)] we noted that it is the right-hand side of the above equality. Put $D = H_{\mathbf{A}} \cap C$ and $D_v = D \cap H_v$. If $v|\mathfrak{e}$, then φ_v is anisotropic, so that $C_v = G_v$; thus $D_v = H_v = E_v$ if $v|\mathfrak{e}$. If $v \nmid \mathfrak{a}^*\mathfrak{e}$, then, by [S3, Lemma 11.11 (iv)] and Lemma 5.6 (iii), (iv), we have $D_v = \tau(\mathfrak{f}_v^\times)$. If $v|\mathfrak{a}^*$, then $D_v = E_v$ by Lemma 5.6 (v). Thus $U \cap D = \{x \in U \mid x \in \tau(\mathfrak{f}_v^\times) \text{ for every } v \nmid \mathfrak{a}^*\mathfrak{e}\}$. Applying [S3, Lemma 11.10 (iii)] to $\tau(\mathfrak{f}_v^\times)$, we obtain $U \cap D = U'$. Now we have $U = E \cap H$,

$$[HE : HD] = [E : E \cap HD] = [E : UD] = [E : D]/[UD : D] = [E : D]/[U : U \cap D],$$

$$[E : D] = \textstyle\prod_{v \nmid \mathfrak{a}^*\mathfrak{e}}[E_v : D_v] = \prod_{v \nmid \mathfrak{a}^*\mathfrak{e}}[E_v : \tau(\mathfrak{r}_v^\times)][\tau(\mathfrak{r}_v^\times) : \tau(\mathfrak{f}_v^\times)].$$

By [S3, Lemma 5.6 (iii)], $[E_v : \tau(\mathfrak{r}_v^\times)] = 2$ if v is ramified in K, and $= 1$ otherwise. The index $[\tau(\mathfrak{r}_v^\times) : \tau(\mathfrak{f}_v^\times)]$ is given by [S3, Lemma 11.10 (i), (iv)]. Thus we obtain

$$[E : D] = 2^b N(\mathfrak{c})\textstyle\prod_{\mathfrak{p}|\mathfrak{c}}\{1 - [K/F,\, \mathfrak{p}]N(\mathfrak{p})^{-1}\},$$

where b is the number of the primes $v \nmid \mathfrak{a}^*\mathfrak{e}$ ramified in K. Now $[H_{\mathbf{A}} : HD] = [H_{\mathbf{A}} : HE][HE : HD]$. Combining all these, we obtain (5.11).

COROLLARY 5.8. *The notation and assumption being as in Theorem 5.7, let* $c(\mathfrak{f})$ *denote the class number of* $\mathfrak{f}$ *in the sense of* [S3, §12.5] *and let* $U_{\mathfrak{f}} = U \cap \mathfrak{f}^\times$. *Then* $U' \subset U_{\mathfrak{f}}$ *and*

(5.12) $\big[H_{\mathbf{A}} : H(H_{\mathbf{A}} \cap C)\big] = \big(c(\mathfrak{f})/c_F\big) \cdot 2^{1-\mu-\nu}\big[\mathfrak{g}^\times : N_{K/F}(\mathfrak{f}^\times)\big][U_\mathfrak{f} : U']^{-1}.$

Proof. By [S3, Lemma 11.10 (ii)] we have $U_\mathfrak{f} = \big\{ x \in U \,\big|\, x^\rho - x \in \mathfrak{c}\mathfrak{d} \big\}$. Since $\mathfrak{c}$ is prime to $\mathfrak{a}^*\mathfrak{e}$, for $x \in U$ we have $x^\rho - x \in \mathfrak{c}\mathfrak{d}$ if and only if $x^\rho - x \in \mathfrak{c}_v\mathfrak{d}_v$ for every $v \nmid \mathfrak{a}^*\mathfrak{e}$. We have also $x^\rho - x = (1-x)(1+x^\rho)$, and hence $U' \subset U_\mathfrak{f}$. Now we recall a well known formula (see [S3, (12.3)])

$$c(\mathfrak{f}) = c_K \cdot [\mathfrak{r}^\times : \mathfrak{f}^\times]^{-1} N(\mathfrak{c}) \prod_{\mathfrak{p}|\mathfrak{c}} \{1 - [K/F,\, \mathfrak{p}]N(\mathfrak{p})^{-1}\}. \tag{5.13}$$

As shown in [S3, p. 117, line 7], we have

$$[\mathfrak{r}^\times : \mathfrak{f}^\times] = [U : U_\mathfrak{f}]\big[N_{K/F}(\mathfrak{r}^\times) : N_{K/F}(\mathfrak{f}^\times)\big]. \tag{5.14}$$

Combining (5.11), (5.13), and (5.14) together, we obtain (5.12).

THEOREM 5.9. *In the setting of Lemma 5.4 and Theorem 5.7, the class number of the genus of maximal lattices in the ternary space (V, φ) equals the type number of $\mathfrak{O}$.*

Proof. We easily see that the type number of $\mathfrak{O}$ equals $\#\{B^\times \backslash B_{\mathbf{A}}^\times / T\}$ with T of (5.7). By Lemma 5.4, $\tau(T) = C$ and clearly $F_{\mathbf{A}}^\times \subset T$, and so τ gives a bijection of $B^\times \backslash B_{\mathbf{A}}^\times / T$ onto $SO^\varphi(V) \backslash SO^\varphi(V)_{\mathbf{A}} / C$. This proves our theorem.

This theorem should not be confused with the results in [Pe] (Satz 9 and its corollary), which concern the classes with respect to the group of similitudes.

THEOREM 5.10. *In the setting of Theorem 5.7, suppose the genus of maximal lattices consists of a single class (which is the case if $\mathfrak{O}$ has type number 1). Put $\Gamma(L) = \big\{\gamma \in SO^\varphi(V) \,\big|\, L\gamma = L\big\}$. Then*

$$\#\big\{L[q,\, \mathfrak{b}]/\Gamma(L)\big\} = \big(c(\mathfrak{f})/c_F\big) \cdot 2^{1-\mu-\nu}\big[\mathfrak{g}^\times : N_{K/F}(\mathfrak{f}^\times)\big][U_\mathfrak{f} : U']^{-1}. \tag{5.15a}$$

Moreover, if B is totally definite, then

$$\#L[q,\, \mathfrak{b}]/\#\Gamma(L) = \big(c(\mathfrak{f})/c_F\big) \cdot 2^{1-\mu-\nu}\big[\mathfrak{g}^\times : N_{K/F}(\mathfrak{f}^\times)\big]\#(U_\mathfrak{f})^{-1}. \tag{5.15b}$$

Proof. The left-hand side of (2.7) in the present case is $\big[H_{\mathbf{A}} : H(H_{\mathbf{A}} \cap C)\big]$, as H is commutative; the right-hand side consists of a single term. Thus (2.7) combined with (5.12) gives (5.15a). Since $H \cap C = U \cap D = U'$, we have $\mathfrak{m}(H,\, H_{\mathbf{A}} \cap C) = \big[H_{\mathbf{A}} : H(H_{\mathbf{A}} \cap C)\big]\#(U')^{-1}$, which together with (3.7) and (5.15a) proves (5.15b).

LEMMA 5.11. *Let B be a quaternion algebra over a global field F, and $\mathfrak{e}$ the product of all the prime ideals in F ramified in B; let $\mathfrak{a}_0$ be a squarefree integral ideal prime to $\mathfrak{e}$. Further let K be a quadratic extension of F contained in B and $\mathfrak{f}$ an order in K containing $\mathfrak{g}$ that has conductor $\mathfrak{c}$. Then there exists an order $\mathfrak{O}$ in B of discriminant $\mathfrak{a}_0\mathfrak{e}$ such that $\mathfrak{O} \cap K = \mathfrak{f}$ if and only if $\mathfrak{c} + \mathfrak{e} = \mathfrak{g}$ and every prime factor of $\mathfrak{a}_0$ not dividing $\mathfrak{c}$ does not remain prime in K.*

This is due to Eichler [E2, Satz 6].

LEMMA 5.12. *Let B, $\mathfrak{e}$, $\mathfrak{a}_0$, K, $\mathfrak{f}$ and $\mathfrak{c}$ be as in Lemma 5.11; let $\mathfrak{O}$ be an order in B of discriminant $\mathfrak{a}_0\mathfrak{e}$. Suppose that $\mathfrak{O}$ has type number 1, $\mathfrak{c}+\mathfrak{e}=\mathfrak{g}$ and every prime factor of $\mathfrak{a}_0$ not dividing $\mathfrak{c}$ does not remain prime in K. Then there exists an element α of $B^\times$ such that $\alpha^{-1}\mathfrak{O}\alpha\cap K=\mathfrak{f}$.*

Proof. Since $\mathfrak{O}$ has type number 1, every order in B of discriminant $\mathfrak{a}_0\mathfrak{e}$ is of the form $\alpha^{-1}\mathfrak{O}\alpha$ with $\alpha\in B^\times$. Therefore our assertion follows from Lemma 5.11.

6. POSITIVE DEFINITE TERNARY FORMS OVER $\mathbf{Q}$

6.1. Every positive definite ternary quadratic space $(V,\,\varphi)$ over $\mathbf{Q}$ is obtained by taking B to be a definite quaternion algebra over $\mathbf{Q}$ and d to be a squarefree positive integer in the setting of §5.2. If we represent φ with respect to a $\mathbf{Z}$-basis of an integral $\mathbf{Z}$-lattice L in V, then we obtain a $\mathbf{Z}$-valued ternary form. If L is maximal, then the form is *reduced* in the sense that it cannot be represented nontrivially by another $\mathbf{Z}$-valued ternary form, and vice versa. This definition of a reduced form is different from Eisenstein's terminology for ternary forms.

Define $\mathfrak{a}$ and $\mathfrak{e}$ as in §5.2. Then $\mathfrak{a}=d\mathbf{Z}$ and $\mathfrak{e}=e\mathbf{Z}$ with a squarefree positive integer e. We have $\mathfrak{e}\cap\mathfrak{a}=d_0e\mathbf{Z}$ with a positive divisor d_0 of d prime to e. Thus $\mathfrak{O}$ of Theorem 5.7 is an order of discriminant d_0e. The pair $(e,\,d)$ determines the isomorphism class of $(V,\,\varphi)$ and hence the genus of a reduced $\mathbf{Z}$-valued ternary form. If Φ is a matrix which represents a ternary form belonging to that genus, then $\det(2\Phi)=2d_0^2e^2/d$; this follows from Lemma 5.3 (i). As to the general theory of reduced forms and $\det(2\Phi)$ for an arbitrary n, the reader is referred to [S5]. Taking a maximal lattice L in V, put $C=\{SO^\varphi(V)_\mathbf{A}\,|\,L\alpha=L\}$. Then we have

LEMMA 6.2. $$\mathfrak{m}\big(SO^\varphi(V),\,C\big)=\frac{1}{12}\prod_{p|e}\frac{p-1}{2}\prod_{p|d_0}\frac{p+1}{2}.$$

Proof. Define a quadratic form β on B° by $\beta[x]=xx^\iota$ for $x\in B^\circ$; put $G=SO^\beta(B^\circ)$. Then $x\mapsto x\xi$ for $x\in B^\circ$ gives an isomorphism of $(B^\circ,\,d\beta)$ onto $(V,\,\varphi)$, and also an isomorphism of G onto $SO^\varphi(V)$. Moreover, for $\alpha\in B^\times$ we have $\alpha^{-1}x\alpha\xi=\alpha^{-1}x\xi\alpha$. Thus the symbol $\tau(\alpha)$ is consistent. We can identify C with the subgroup $\{\gamma\in G_\mathbf{A}\,|\,(\mathfrak{M}\cap B^\circ)\gamma=(\mathfrak{M}\cap B^\circ)\}$ of $G_\mathbf{A}$, where $\mathfrak{M}$ is as in (5.4). Let $\mathfrak{O}_0$ be a maximal order in B containing $\mathfrak{O}$; put $M=\mathfrak{O}_0\cap B^\circ$ and $D=\{\gamma\in G_\mathbf{A}\,|\,M\gamma=M\}$. Then M is a maximal lattice with respect to β, and from [S2, Theorem 5.8] we obtain

$$\mathfrak{m}(G,\,D)=\frac{1}{12}\prod_{p|e}\frac{p-1}{2}. \tag{6.1}$$

By (3.3) we have $[C:C\cap D]\mathfrak{m}(G,\,C)=\mathfrak{m}(G,\,C\cap D)=[D:C\cap D]\mathfrak{m}(G,\,D)$. Clearly $D_p=C_p$ if $p\nmid d_0$. Suppose $p|d_0$; then we can put $B_p=M_2(\mathbf{Q}_p)$, $(\mathfrak{O}_0)_p=M_2(\mathbf{Z}_p)$, $D_p=\tau\big(GL_2(\mathbf{Z}_p)\big)$, and $\mathfrak{O}_p$ is of the type (5.1a). From Lemma 5.4 we obtain $[C_p:C_p\cap D_p]=2$ and $[D_p:C_p\cap D_p]=p+1$.

Thus $\mathfrak{m}(G, C) = \mathfrak{m}(G, D)\prod_{p|d_0}\{(p+1)/2\}$. Combining this with (6.1), we obtain our lemma.

6.3. Now Eichler gave a formula for the class number of $\mathfrak{O}$ and also a formula for the type number of $\mathfrak{O}$ in [E2, (64)]. However, his formula for the type number is not completely correct, and correct formulas were given by Peters in [Pe] and by Pizer in [Pi]. There is a table for the type number of $\mathfrak{O}$ for $d_0e \leq 30$ in [Pe, p. 360]; a larger table for $d_0e \leq 210$ is given at the end of [Pi]; in these papers, e and d_0 are denoted by q_1 and q_2. From these tables we see that the type number of $\mathfrak{O}$ for $d_0e \leq 210$ is 1 exactly when (e, d_0) is one of the following 20 pairs:

$$(2, 1), (2, 3), (2, 5), (2, 7), (2, 11), (2, 23), (2, 15), (3, 1), (3, 2), (3, 5),$$
$$(3, 11), (5, 1), (5, 2), (7, 1), (7, 3), (13, 1), (30, 1), (42, 1), (70, 1), (78, 1).$$

Now d is d_0 times a factor of e. Therefore if e has exactly t prime factors, then there are 2^t choices for (e, d) with the same (e, d_0). Consequently there are exactly 64 choices for (e, d) obtained from the above 20 pairs of (e, d_0), and each (e, d) determines (V, φ); $\det(2\Phi) = 2d_0^2e^2/d$ as noted in §6.1. By Theorem 5.9, $\#\{SO^{\varphi}(V)\backslash SO^{\varphi}(V)_{\mathbf{A}}/C\} = 1$ in all these cases. We have actually

THEOREM 6.4. *The spaces (V, φ) obtained from these 64 pairs (e, d) exhaust all positive definite ternary quadratic spaces over $\mathbf{Q}$ for which the genus of maximal lattices has class number 1. In other words, there are exactly 64 genera of positive definite, $\mathbf{Z}$-valued, and reduced ternary forms consisting of a single class.*

Proof. Our task is to show that the class number is not 1 for $d_0e > 210$. If the inverse of the right-hand side of the equality of Lemma 6.2 is not an integer, then the class number cannot be 1. For $d_0e > 210$, the inverse is an integer exactly when (e, d_0) is one of the following six: $(2\cdot 7\cdot 17, 1)$, $(2\cdot 5\cdot 13, 3)$, $(2\cdot 3\cdot 17, 5)$, $(2\cdot 3\cdot 13, 7)$, $(2\cdot 3\cdot 5, 23)$, $(2\cdot 3\cdot 5, 11)$. The verification is easy, as the mass ≤ 1 for relatively few cases of (e, d_0). Among those six, the mass is $1/2$ for the last one; in the other five cases the mass is 1. Now, by the formula in [Pe, p. 361] the type number of $\mathfrak{O}$ (which is the class number in question by virtue of Theorem 5.9) is given by $2^{-\kappa}\sum_t Sp\{P^*(t)\}$, where κ is the number of prime factors of d_0e, t runs over all positive divisors of d_0e, and $Sp\{P^*(t)\}$ is given on the same page. From the formulas there we can easily verify that $Sp\{P^*(1)\} = 2^{\kappa}$ and $Sp\{P^*(d_0e)\} > 0$ in the first five cases, and hence the type number is greater than 1. As for the last case $(2\cdot 3\cdot 5, 11)$, computing $Sp\{P^*(t)\}$ for all $t|d_0e$, we find that the type number is 2. Thus we obtain our theorem.

Remark. In the setting of §6.1, if $\det(2\Phi)/2$ is squarefree, then Φ must be reduced and $\det(2\Phi)/2 = d = d_0e$ with squarefree d_0 and e. If the genus of Φ has class number 1, then in view of the above theorem such a Φ is obtained from one of the pairs listed in §6.3 by taking $d = d_0e$. Thus there are exactly 20 genera of $\mathbf{Z}$-valued positive definite ternary forms with class number 1 such that $\det(2\Phi)/2$ is squarefree. This result was obtained by Watson [W], and we derived it from a stronger result, Theorem 6.4.

6.5. We are going to specialize Theorem 5.7 and (5.15b) to these cases, and give explicit formulas for $\#L[q,\,\mathbf{Z}]$. For simplicity, we consider only the cases in which e is a prime, and d_0 is 1 or a prime; thus we consider only 15 pairs $(e,\,d_0)$, and consequently 30 pairs $(e,\,d)$ with $d=d_0$ or $d=d_0e$. In this setting we have

$$\#\{\Gamma(L)\} = 48/[(d_0+1)(e-1)]. \tag{6.2}$$

This is because $\#\{\Gamma(L)\}$ equals the inverse of the quantity of Lemma 6.2, as the class number is 1.

To study the nature of q for which $L[q,\,\mathbf{Z}]\neq\emptyset$, we consider the localization of B at e. Let $\mathfrak{r}$ denote the maximal order in the unramified quadratic extension of $\mathbf{Q}_e$. We can put $B_e^\circ=\mathbf{Q}_e\sigma+\mathbf{Q}_e\eta$, $\mathfrak{O}_e=\mathfrak{r}+\mathfrak{r}\eta$, and $\widetilde{\mathfrak{O}}_e=\mathfrak{r}+\mathfrak{r}\eta^{-1}$ with σ and η as in (B) of §1.5; we can take $\eta^2=e$, though this is often unnecessary. We have $L=(\mathfrak{M}\cap B^\circ)\xi$, and so

$$L_e=(\mathbf{Z}_e\sigma+\mathfrak{r}\eta)\xi \quad\text{and}\quad \widetilde{L}_e=(2^{-1}\mathbf{Z}_e\sigma+\mathfrak{r}\eta^{-1})\xi \quad\text{if}\quad e\nmid d, \tag{6.3a}$$

$$L_e=(\mathbf{Z}_e\sigma+\mathfrak{r}\eta^{-1})\xi \quad\text{and}\quad \widetilde{L}_e=((2e)^{-1}\mathbf{Z}_e\sigma+\mathfrak{r}\eta^{-1})\xi \quad\text{if}\quad e|d. \tag{6.3b}$$

Here is an easy fact: given an element η_0 of $B_e^\times$ such that η_0^2 is a prime element, we can find an element α of $B_e^\times$ such that (6.3a, b) are true with $(\alpha^{-1}\sigma\alpha,\,\alpha^{-1}\mathfrak{r}\alpha,\,\eta_0)$ in place of $(\sigma,\,\mathfrak{r},\,\eta)$. Indeed, since $\eta_0^2/\eta^2\in\mathbf{Z}_e^\times$, we have $\eta_0^2/\eta^2=aa^\iota$ with $a\in\mathfrak{r}^\times$. Then $\eta_0^2=(a\eta)^2$, and hence there exists an element α of $B_e^\times$ such that $\alpha^{-1}a\eta\alpha=\eta_0$. Since $\alpha^{-1}L_e\alpha=L_e$ and $\alpha^{-1}\widetilde{L}_e\alpha=\widetilde{L}_e$, we obtain the desired result.

THEOREM 6.6. *Given $0<q\in\mathbf{Q}$ in the setting of §6.5, put $K_0=\mathbf{Q}(\sqrt{-dq}\,)$ and denote by δ the discriminant of K_0. Then the following assertions hold:*

(i) *$L[q,\,\mathbf{Z}]\neq\emptyset$ only if q is as follows:*

$$d_0eq=r^2m \quad \textit{when} \quad e\neq 2 \quad \textit{or} \quad e|d, \tag{6.4a}$$

$$d_0q=r^2m \quad \textit{when} \quad e=2 \quad \textit{and} \quad e\nmid d. \tag{6.4b}$$

Here m is a squarefree positive integer such that e does not split in K_0, and also that $m-3\in 8\mathbf{Z}$ if $e=2$ and $2|d$; r is a positive integer prime to e.

(ii) *Moreover, put $a^*=d_0$ in the following two cases:* (A) *$d_0\geq 2$, $d_0|m$, and $d_0\nmid r$;* (B) *$d_0=2$, $r-2\in 2\mathbf{Z}$, and 2 is ramified in K_0; put $a^*=1$ if neither* (A) *nor* (B) *applies. Let $\mathfrak{C}$ be the set of all positive integers c prime to a^*e such that $d_0|c$ if d_0 is a prime that remains prime in K_0. (Thus $\mathfrak{C}$ depends on d, e, and q.) Then $r/2\in\mathfrak{C}$ if $4|\delta$ and $e\neq 2$; $r\in\mathfrak{C}$ otherwise.*

(iii) *Conversely, given r and m satisfying all these conditions, determine q by (6.4a, b). Then $L[q,\,\mathbf{Z}]\neq\emptyset$.*

Notice that $K_0=\mathbf{Q}(\sqrt{-m}\,)$ if $e|d$ or $e=2$, and $K_0=\mathbf{Q}(\sqrt{-em}\,)$ otherwise. Also, $\mathfrak{C}=\emptyset$ if a^* is a prime that remains prime in K_0.

Proof. Suppose $h = k\xi \in L[q, \mathbf{Z}]$ with $k \in B^\circ$; put $K = F[k]$ as in §5.2 and $\mathfrak{f} = K \cap \mathfrak{O}$; let c be the conductor of $\mathfrak{f}$. Then $K \cong K_0$ and $h \in 2\widetilde{L}$. Then from (6.3a, b) we see that $eq \in \mathbf{Z}_e$ if $e \neq 2$ or $e|d$, and $q \in \mathbf{Z}_e$ otherwise. The set $\widetilde{L}_p$ for $p \neq e$ can be given by (5.5b), and so $dq \in \mathbf{Z}_p$. Thus $d_0eq \in \mathbf{Z}$ if $e \neq 2$ or $e|d$; $d_0q \in \mathbf{Z}$ otherwise. Take $0 < r \in \mathbf{Z}$ and a squarefree positive integer m as in (6.4a, b). Define $\mathfrak{a}^*$ as in Theorem 5.7 and put $\mathfrak{a}^* = a^*\mathbf{Z}$ with $0 < a^* \in \mathbf{Z}$. Clearly $a^* = 1$ if $d_0 = 1$. Suppose d_0 is a prime number. To make our exposition easier, denote this prime by s; then $s \neq e$ and $r_s(h) = -d^{-1}q$. We can easily verify that $a^* = s$ exactly in Cases (A) and (B) stated in (ii) above. By Theorem 5.7 and Lemma 5.11, we see that $c \in \mathfrak{C}$. Notice that s is ramified in K_0 if $a^* = s$.

(1) We first consider the case $e|d$. Then $d = d_0e$ and $dq = d_0eq$ and so $K \cong \mathbf{Q}(\sqrt{-m}\,)$. Suppose $e|r$; put $\ell = e^{-1}k$. Then $d^2\ell\ell^\iota = de^{-2}q = (r/e)^2m \in \mathbf{Z}$, and hence $d\ell \in \mathfrak{O}_e \cap B^\circ$. Thus $\ell \in e^{-1}\mathfrak{O}_e \cap B^\circ = e^{-1}\mathbf{Z}_e\sigma + \eta^{-1}\mathfrak{r}$, and so $\ell\xi \in 2\widetilde{L}_e$ by (6.3b). Therefore $h = e\ell\xi \in 2e\widetilde{L}_e$, which implies that $\varphi(h, L)_e \subset e\mathbf{Z}_e$, a contradiction. Thus $e\nmid r$.

(2) Next suppose $e \nmid d$; then $d = d_0$. Assuming $e|r$, put $\ell = e^{-1}k$. Then $d^2\ell\ell^\iota = de^{-2}q \in e^{-1}\mathbf{Z}_e$, and hence $\ell \in \widetilde{\mathfrak{O}}_e \cap B^\circ = \mathbf{Z}_e\sigma + \eta^{-1}\mathfrak{r}$. Thus $\ell\xi \in 2\widetilde{L}_e$ by (6.3a), which leads to a contradiction for the same reason as in Case (1). Therefore $e\nmid r$ in this case too. This reasoning is valid even when $e = 2$.

(3) Since K_0 must be embeddable in B, the prime e cannot split in K_0. Suppose $e = 2$, $e|d$, and $e|m$. Since $kk^\iota = d^{-2}r^2m$, we have $k \in \eta^{-1}\mathfrak{O}_e^\times \cap B^\circ$, so that $k = a\sigma + \eta^{-1}b$ with $a \in \mathbf{Z}_e$ and $b \in \mathfrak{r}$. Since $\eta k \in \mathfrak{O}_e^\times$, we have $b \in \mathfrak{r}^\times$. By (6.3b), $2\varphi(h, L)_e = 2\mathrm{Tr}_{B/F}\big(k(\mathbf{Z}_e\sigma + \eta^{-1}\mathfrak{r})\big) = 4a\mathbf{Z}_e + \mathrm{Tr}_{K/\mathbf{Q}}(b\mathfrak{r}) = \mathbf{Z}_e$, a contradiction. Therefore m must be odd if $e = 2$ and $e|d$. In this case, $k \in 2^{-1}\mathfrak{O}_e^\times \cap B^\circ$, so that $2k = x\sigma + \eta y$ with $x \in \mathbf{Z}_e^\times$ and $y \in \mathfrak{r}$. Thus $2\mathbf{Z}_e = 2\varphi(h, L)_e = 2\mathrm{Tr}_{B/F}\big(k(\mathbf{Z}_e\sigma+\eta^{-1}\mathfrak{r})\big) = 2\mathbf{Z}_e+\mathrm{Tr}_{K/\mathbf{Q}}(y\mathfrak{r})$, so that $y \in 2\mathfrak{r}$. Now $r^2m = dq = d^2kk^\iota = -d_0^2(x^2\sigma^2 + 2yy^\iota)$. We can take $\mathfrak{r} \subset \mathbf{Q}_2(\sqrt{5}\,)$ and $\sigma = \sqrt{5}$. Therefore we see that $m - 3 \in 8\mathbf{Z}$. Thus we obtain the condition on m as stated in our theorem.

(4) By Theorem 5.7, c is prime to a^*e and $c^2\delta\mathbf{Z}_p = d_0q\mathbf{Z}_p = r^2m\mathbf{Z}_p$ for every $p \neq e$. We have seen that r is prime to e. First suppose $4|\delta$ and $e \neq 2$. Then we see that $\delta\mathbf{Z}_p = 4m\mathbf{Z}_p$ for $p \neq e$, and so $4c^2\mathbf{Z}_p = r^2\mathbf{Z}_p$ for $p \neq e$. Since both $2c$ and r are prime to e, we obtain $2c = r$. Similarly we easily find that $c = r$ if $4\nmid\delta$ or $e = 2$. Thus we obtain (ii). This completes the proof of the "if"-part.

(5) Conversely, suppose q is given with r and m as in our theorem; put $c = r/2$ or $c = r$ according as $r/2 \in \mathfrak{C}$ or $r \in \mathfrak{C}$. Let $\mathfrak{o}$ be the order in K_0 whose conductor is c. By Lemma 5.12, our conditions on r and m guarantees an injection θ of K_0 into B such that $\theta(\mathfrak{o}) = \theta(K_0) \cap \mathfrak{O}$. Our task is to find an element h such that $\varphi[h] = q$ and $\varphi(h, L) = \mathbf{Z}$. Put $\mu = \theta(\sqrt{-m}\,)$ if $K_0 = \mathbf{Q}(\sqrt{-m}\,)$.

(5a) First we consider the case $e|d$ and $|\delta| = m$. Then $d = d_0e$, $r = c$, and $K_0 = \mathbf{Q}(\sqrt{-m}\,)$. Put $h = k\xi$ with $k = d^{-1}r\mu$. Then $\varphi[h] = q$ and by Theorem 5.7, $dq\varphi(h, L)_p^{-2} = c^2\delta\mathbf{Z}_p = r^2m\mathbf{Z}_p$ for every $p \neq e$, so that $\varphi(h, L)_p = \mathbf{Z}_p$ for every $p \neq e$. Now $2\varphi(h, L)_e = d\mathrm{Tr}_{B/F}\bigl(k(\mathbf{Z}_e\sigma + \eta^{-1}\mathfrak{r})\bigr) = \mathrm{Tr}_{B/F}\bigl(\mu(\mathbf{Z}_e\sigma + \eta^{-1}\mathfrak{r})\bigr)$ by (6.3b). If $e|m$, then we can take μ as η, changing $\mathfrak{r}$ by an inner automorphism, as noted after (6.3b). (Since $|\delta| = m$, we have $e \neq 2$.) If $e \nmid m$, then we can take $\mathfrak{r} \subset \theta(K_0)_e$ and $\sigma = \mu$. In either case we easily see that $2\varphi(h, L)_e = 2\mathbf{Z}_e$. Thus $\varphi(h, L) = \mathbf{Z}$, and $L[q, \mathbf{Z}] \neq \emptyset$ as expected.

(5b) Next suppose $e|d$ and $|\delta| = 4m$; then $r = 2c$ if $e \neq 2$ and $r = c$ if $e = 2$. We have $K_0 = \mathbf{Q}(\sqrt{-m})$ and we again take $h = k\xi$ with $k = d^{-1}r\mu$. Then $\varphi[h] = q$ and $\varphi(h, L)_p = \mathbf{Z}_p$ for every $p \neq e$ for the same reason as in (5a). If $e \neq 2$, the same argument as in (5a) shows that $\varphi(h, L)_e = \mathbf{Z}_e$, which leads to the desired result. Thus suppose $e = 2$. Then $m-3 \in 8\mathbf{Z}$ as stipulated in our theorem. We can take $\sigma = \sqrt{5}$ as we did in (3). Put $\mu = a\sigma + \eta b$ with $a \in \mathbf{Z}_e$ and $b \in \mathfrak{r}$. Then $a \in \mathbf{Z}_e^\times$ and $-m = a^2\sigma^2 + 2bb^\rho$. Since $m+\sigma^2 \in 8\mathbf{Z}_e$, we see that $b \in 2\mathfrak{r}$. Thus $2\varphi(h, L)_e = \mathrm{Tr}_{B/F}\bigl(\mu(\mathbf{Z}_e\sigma+\eta^{-1}\mathfrak{r})\bigr) = 2\mathbf{Z}_e + \mathrm{Tr}_{K/\mathbf{Q}}(b\mathfrak{r}) = 2\mathbf{Z}_e$, which gives the expected result.

(5c) Suppose $e \nmid d$ and $e = 2$. Then $d = d_0$ and $K_0 = \mathbf{Q}(\sqrt{-m})$; we take $h = k\xi$ with $k = d^{-1}r\mu$. We find that $\varphi[h] = q$ and $\varphi(h, L)_p = \mathbf{Z}_p$ for every $p \neq 2$ in the same manner as in (5a). Now $2\varphi(h, L)_2 = \mathrm{Tr}_{B/F}\bigl(\mu(\mathbf{Z}_2\sigma + \eta\mathfrak{r})\bigr)$ by (6.3a). If $2|m$, then taking μ as η, we obtain $2\varphi(h, L)_2 = 2\mathbf{Z}_2$ as expected. Suppose $2 \nmid m$; then $\mu = a\sigma + \eta b$ with $a \in \mathbf{Z}_2^\times$ and $b \in \mathfrak{r}$, and so $\mathrm{Tr}_{B/F}\bigl(\mu(\mathbf{Z}_2\sigma + \eta\mathfrak{r})\bigr) = 2\mathbf{Z}_2$, which gives the desired result.

(5d) Finally suppose $e \nmid d$ and $e \neq 2$. Then $d = d_0$ and $K_0 = \mathbf{Q}(\sqrt{-em})$; we take $h = k\xi$ with $k = \theta(d^{-1}e^{-1}r\sqrt{-em})$; then $\varphi[h] = q$. For $p \neq e$, $\delta\mathbf{Z}_p$ equals $m\mathbf{Z}_p$ or $4m\mathbf{Z}_p$, and $r = c$ or $r = 2c$ accordingly. Then we easily see that $\varphi(h, L)_p = \mathbf{Z}_p$ for every $p \neq e$ in the same manner as in (5a). Now $2\varphi(h, L)_e = \mathrm{Tr}_{B/F}\bigl(e^{-1}\theta(\sqrt{-em})(\mathbf{Z}_2\sigma + \eta\mathfrak{r})\bigr)$. If $e|m$, then we can put $\theta(\sqrt{-em}) = e(a\sigma + \eta b)$ with $a \in \mathbf{Z}_e^\times$ and $b \in \mathfrak{r}$, and obtain $\varphi(h, L)_e = \mathbf{Z}_e$. If $e \nmid m$. then taking $\eta = \theta(\sqrt{-em})$, we obtain the desired result. This completes the proof.

THEOREM 6.7. *If $L[q, \mathbf{Z}] \neq \emptyset$ in the setting of Theorem 6.6, then*

$$\#L[q, \mathbf{Z}] = \frac{2^{1-\mu} \cdot 48 \cdot \mathbf{c}(K_0)}{(d_0+1)(e-1)w} \cdot c\prod_{p|c}\bigl\{1 - [K_0/\mathbf{Q}, p]p^{-1}\bigr\}.$$

Here μ is the number of prime factors of a^*e ramified in K_0; $\mathbf{c}(K_0)$ is the class number of K_0; w is the number of roots of unity in K; $c = r/2$ if $4|\delta$ and $e \neq 2$; $c = r$ otherwise; $[K_0/\mathbf{Q}, p]$ is defined as in Theorem 5.7.

Proof. Specialize (5.15b) to the present case. Then $[U : 1] = w$ and $\#\Gamma(L)$ is given by (6.2); $c(\mathfrak{f})$ can be connected to $\mathbf{c}(K_0)$ by (5.13) and (5.14). Thus we obtain our formula.

6.8. Before discussing examples, let us insert here a remark applicable to (V, φ) over $\mathbf{Q}$ with an arbitrary n. If $F = \mathbf{Q}$ and $\mathfrak{g} = \mathbf{Z}$, it is natural to consider $L[q, \mathbf{Z}]$, but it is not always best to formulate the result with respect to a $\mathbf{Z}$-basis of L. To see this, first take the standard basis $\{e_i\}_{i=1}^n$ of $\mathbf{Z}^n = L$, and define a matrix Φ by $\Phi = [\varphi(e_i, e_j)]_{i,j=1}^n$; put $f_i = e_i\Phi^{-1}$. Then $2\widetilde{L} = \sum_{i=1}^n \mathbf{Z}f_i$. Let $h \in L[q, \mathbf{Z}]$. Then $h \in 2\widetilde{L}$, and so $h = \sum_{i=1}^n a_i f_i$ with $a_i \in \mathbf{Z}$. We easily see that $\varphi(h, L) = \mathbf{Z}$ if and only if $\sum_{i=1}^n a_i\mathbf{Z} = \mathbf{Z}$, and also that $\Phi^{-1} = [\varphi(f_i, f_j)]_{i,j=1}^n$. This means that

$$(6.5)\quad L[q, \mathbf{Z}] = \left\{ \textstyle\sum_{i=1}^n a_i f_i \;\middle|\; a\Phi^{-1}\cdot {}^t a = q,\ \sum_{i=1}^n a_i\mathbf{Z} = \mathbf{Z} \right\}, \quad a = (a_i)_{i=1}^n.$$

Therefore if we follow the traditional definition of primitivity, we have to use the matrix Φ^{-1} instead of Φ. Recall that $\Gamma(L) = \Gamma(\widetilde{L})$. Thus $\Gamma(L)$ can be represented, with respect to the basis $\{f_i\}_{i=1}^n$, by the group

$$(6.6)\qquad \Gamma' = \left\{ \gamma \in SL_n(\mathbf{Z}) \,\middle|\, \gamma\Phi^{-1}\cdot {}^t\gamma = \Phi^{-1} \right\}.$$

Consequently, $L[q, \mathbf{Z}]/\Gamma(L)$ corrresponds to the vectors a such that $a\Phi^{-1}\cdot {}^t a = q$ and $\sum_{i=1}^n a_i\mathbf{Z} = \mathbf{Z}$, considered modulo Γ'. We note here an easy fact:

$$(6.7)\qquad |\det(2\Phi)| = [\widetilde{L} : L].$$

6.9. Let us now illustrate Theorem 6.6 by considering five examples and formulating the results in terms of the matrix Φ^{-1} of §6.8. For $(e, d) = (2, 1)$ we obtain the result on sums of three squares, which Gauss treated; we do not include this in our examples, as it is easy and too special. Indeed, in this case we have $\Phi = \Phi^{-1}$, and so the result concerns the primitive solutions of $\sum_{i=1}^3 x_i^2 = q$. But in all other cases, $\Phi \neq \Phi^{-1}$, and the results have more interesting features. From Lemma 5.3 (i) and (6.7) we obtain $\det(2\Phi) = 2d_0^2e^2/d$. In each case we state only the condition for $L[q, \mathbf{Z}] \neq \emptyset$. We will dispense with the statement about $\#L[q, \mathbf{Z}]$, as it is merely a specialization of Theorem 6.7.

(1) We first take $e = d = 3$ in Theorem 6.6. Then

$$(6.8)\qquad 2\Phi = \operatorname{diag}\left[2, \begin{bmatrix} 2 & -1 \\ -1 & 2 \end{bmatrix}\right], \quad 3\Phi^{-1} = \operatorname{diag}\left[3, \begin{bmatrix} 4 & 2 \\ 2 & 4 \end{bmatrix}\right].$$

These forms of matrices can be obtained by analyzing L of (5.4) in the present case. Alternatively, since $\det(2\Phi) = 6$ for Φ of (6.8), we can conclude that Φ is the matrix representing φ by the principle explained in [S5]. Theorem 6.6 (or the form of Φ^{-1}) in the present case shows that $L[q, \mathbf{Z}] \neq \emptyset$ only when $3q \in \mathbf{Z}$. Thus with $s = 3q \in \mathbf{Z}$, the principle of §6.8 says that $L[s/3, \mathbf{Z}]$ corresponds to the set of vectors (x, y, z) such that

$$(6.9)\qquad 3x^2 + 4(y^2 + yz + z^2) = s \quad \text{and} \quad x\mathbf{Z} + y\mathbf{Z} + z\mathbf{Z} = \mathbf{Z}.$$

Theorem 6.6 specialized to this case means that given $0 < s \in \mathbf{Q}$, we can find (x, y, z) satisfying (6.9) if and only if $s = r^2m$ with a squarefree positive integer m such that $m+1 \notin 3\mathbf{Z}$ and a positive integer r prime to 3 such that $2|r$ if $m+1 \notin 4\mathbf{Z}$; $K_0 = \mathbf{Q}(\sqrt{-m})$.

(2) Next we take $e = 3$ and $d = 1$. Then

$$(6.10)\qquad 2\Phi = \operatorname{diag}\left[6, \begin{bmatrix} 2 & -1 \\ -1 & 2 \end{bmatrix}\right], \quad 3\Phi^{-1} = \operatorname{diag}\left[1, \begin{bmatrix} 4 & 2 \\ 2 & 4 \end{bmatrix}\right].$$

Therefore $L[s/3, \mathbf{Z}]$ corresponds to the set of vectors (x, y, z) such that

$$(6.11) \qquad x^2 + 4(y^2 + yz + z^2) = s \quad \text{and} \quad x\mathbf{Z} + y\mathbf{Z} + z\mathbf{Z} = \mathbf{Z}.$$

Such vectors exist if and only if $s = r^2 m$ with a squarefree positive integer m such that 3 does not split in $K_0 = \mathbf{Q}(\sqrt{-3m}\,)$ and a positive integer r prime to 3 such that $2|r$ if 4 divides the discriminant of K_0.

In both cases (1) and (2) the group $\Gamma(L)$ has order 12. Represent $(x, y, z) \in \mathbf{Z}^3$ by (x, β) with $\beta = y + z\zeta$, where ζ is a primitive cubic root of unity. Then Γ' of (6.6) consists of the maps $(x, \beta) \mapsto (x, \varepsilon\beta)$ and $(x, \beta) \mapsto (-x, \varepsilon\overline{\beta})$, where ε is a sixth root of unity. Therefore a set of representatives for $L[q, \mathbf{Z}]/\Gamma(L)$ can be found numerically. For instance, take $s = m = 79$ in (6.9); then we easily find that $\#\{L[q, \mathbf{Z}]/\Gamma(L)\} = 5$, which combined with (5.15a) confirms that $\mathbf{Q}(\sqrt{-79}\,)$ has class number 5.

(3) Our third example concerns the case $e = d = 2$. We have

$$(6.12) \qquad 2\Phi = \begin{bmatrix} 2 & -1 & 0 \\ -1 & 2 & -1 \\ 0 & -1 & 2 \end{bmatrix}, \quad 2\Phi^{-1} = \begin{bmatrix} 3 & 2 & 1 \\ 2 & 4 & 2 \\ 1 & 2 & 3 \end{bmatrix}.$$

Then $L[s/2, \mathbf{Z}]$ corresponds to the set of vectors $x = (x_i)_{i=1}^3 \in \mathbf{Z}^3$ such that

$$(6.13) \qquad x \cdot 2\Phi^{-1} \cdot {}^t x = s \quad \text{and} \quad \textstyle\sum_{i=1}^3 x_i\mathbf{Z} = \mathbf{Z}.$$

Such vectors x exist if and only if $s = r^2 m$ with an odd integer r and a squarefree positive integer m such that $m - 3 \in 8\mathbf{Z}$. In this case, $K_0 = \mathbf{Q}(\sqrt{-m}\,)$.

(4) Take $e = d = 7$. We have then

$$(6.14) \qquad 2\Phi = \mathrm{diag}\left[2, \begin{bmatrix} 4 & -1 \\ -1 & 2 \end{bmatrix}\right], \quad 7\Phi^{-1} = \mathrm{diag}\left[7, \begin{bmatrix} 4 & 2 \\ 2 & 8 \end{bmatrix}\right],$$

and $L[s/7, \mathbf{Z}]$ with $s \in \mathbf{Z}$ corresponds to the set of vectors (x, y, z) such that

$$(6.15) \qquad 7x^2 + 4(y^2 + yz + 2z^2) = s \quad \text{and} \quad x\mathbf{Z} + y\mathbf{Z} + z\mathbf{Z} = \mathbf{Z}.$$

Such vectors exist if and only if $s = r^2 m$ with a positive integer r prime to 7 and a squarefree positive integer m such that 7 does not split in $K_0 = \mathbf{Q}(\sqrt{-m}\,)$; $2|r$ if $m + 1 \notin 4\mathbf{Z}$.

(5) Let us finally take $e = 2$ and $d = 22$. We have then

$$(6.16) \qquad 2\Phi = \mathrm{diag}\left[4, \begin{bmatrix} 6 & -1 \\ -1 & 2 \end{bmatrix}\right], \quad 22\Phi^{-1} = \mathrm{diag}\left[11, \begin{bmatrix} 8 & 4 \\ 4 & 24 \end{bmatrix}\right],$$

and the result about $L[s/22, \mathbf{Z}]$ is what we stated in the introduction.

REFERENCES

[E1] M. Eichler, Quadratische Formen und orthogonale Gruppen, Springer, Berlin, 1952, 2nd ed. 1974.

[E2] M. Eichler, Zur Zahlentheorie der Quaternionen-Algebren, J. reine angew. Math. 195 (1956), 127–151.

[Pe] M. Peters, Ternäre und quaternäre quadratische Formen und Quaternionenalgebren, Acta Arithmetica 15 (1969), 329–365.

[Pi] A. Pizer, Type numbers of Eichler orders, J. reine angew. Math. 264 (1973), 76–102.

[S1] G. Shimura, Euler Products and Eisenstein series, CBMS Regional Conference Series in Mathematics, No. 93, Amer. Math. Soc. 1997.

[S2] G. Shimura, An exact mass formula for orthogonal groups, Duke Math. J. 97 (1999), 1–66 (=Collected Papers, IV, 509–574).

[S3] G. Shimura, Arithmetic and analytic theories of quadratic forms and Clifford groups, Mathematical Surveys and Monographs, vol. 109, Amer. Math. Soc. 2004.

[S4] G. Shimura, Quadratic Diophantine equations and orders in quaternion algebras, Amer. J. Math. 128 (2006), 481–518.

[S5] G. Shimura, Classification, construction, and similitudes of quadratic forms, to appear in Amer. J. Math. 128 (2006).

[S6] G. Shimura, Quadratic Diophantine equations, the class number, and the mass formula, Bulletin, Amer. Math. Soc. 43 (2006), 285–304.

[S7] G. Shimura, Classification of integer-valued symmetric forms, to appear in Amer. J. Math. 129 (2007).

[W] G. L. Watson, One-class genera of positive ternary quadratic forms, Mathematika 19 (1972), 96–104.

Goro Shimura
Department of Mathematics
Princeton University
Princeton, New Jersey 08544-1000
U.S.A.
goro@Math.Princeton.EDU

CLASSIFICATION, CONSTRUCTION, AND SIMILITUDES OF QUADRATIC FORMS

By Goro Shimura

Abstract. We present a theory of classifying quadratic forms over an algebraic number field which is a reformulation of Eichler's methods and different from the traditional theory of Hasse. As applications we determine all possible values of the discriminant of an integer-valued quadratic form over $\mathbf{Q}$ with a given dimension and index, and also the number of genera of such forms. We also investigate purely algebraic problems on the Clifford groups and similitudes of quadratic forms.

Introduction. One of the main themes of this paper concerns an integer-valued quadratic form over the rational number field $\mathbf{Q}$, which can be written $x\Phi \cdot {}^t x$ for a variable row vector $x \in \mathbf{Z}^n$ with a symmetric matrix $\Phi = (\varphi_{ij})$ of size n such that $\varphi_{ii} \in \mathbf{Z}$ and $2\varphi_{ij} \in \mathbf{Z}$ for every i and j. To make our exposition simpler, let us assume for the moment that Φ is positive definite. Then $\det(2\Phi)$ is a positive integer. We can define the genus of Φ as usual, and it is classically known that $\det(2\Phi) = 1$ can happen if and only if 8 divides n, and there is a unique genus for each such n. However, no meaningful results seem to have been obtained for the number of genera with a given n and a given value of $\det(2\Phi)$, not even in the case where $\det(2\Phi)$ is a prime, except in a few low-dimensional cases. One of the principal purposes of the present paper is to answer the questions of this type for an arbitrary value of $\det(2\Phi)$ and arbitrary n. Since this is closely related to the classification problem of quadratic forms over an algebraic number field, let us first recall its solution by Hasse.

Given a quadratic form φ over an algebraic number field F, we can take the dimension, the discriminant, and the signatures of φ at the real archimedean primes of F as obvious invariants. In addition, to a form $\varphi = \sum_{i=1}^{n} a_i x_i^2$ and a prime v Hasse associated a number $\varepsilon_v(\varphi)$, which is either 1 or -1, and defined by

$$\varepsilon_v(\varphi) = \prod_{i \le j} \left(\frac{a_i, a_j}{v}\right).$$

He then showed that the equivalence class of φ corresponds to the set consisting of these numbers $\{\varepsilon_v(\varphi)\}$ and the above "obvious invariants" satisfying certain

---

Manuscript received December 14, 2004.
Shimura, Goro. "Classification, construction, and similtudes of quadratic forms." American Journal Journal of Mathematics 128: (2006), 1521–1552.

conditions, and the correspondence is one-to-one; namely he constructed a unique quadratic form for a given set of such data; see [H].

This was followed by a work of Witt [W]. Given a form φ of n variables, he considered the form φ^* of $2n$ variables which is the direct sum of φ and the form $-\sum_{i=1}^{n} y_i^2$, and denoted by $S(\varphi)$ the Clifford algebra of φ^*; in fact he attributed this idea to Artin. Now $S(\varphi)$ is a central simple algebra over F. He showed that $\varepsilon_v(\varphi)$ is actually the invariant of $S(\varphi)$ at v in the standard sense.

Some later authors chose $\varepsilon'_v(\varphi) = \prod_{i<j} \left(\frac{a_i, a_j}{v}\right)$ instead of $\varepsilon_v(\varphi)$, but this together with the discriminant gives an equivalent set of invariants. Though we can develop a reasonable theory of classification of quadratic forms with these invariants $\varepsilon_v(\varphi)$ and $\varepsilon'_v(\varphi)$, it must be pointed out that they have some features which do not look so natural. Indeed, if φ is the hyperbolic form $x_1x_2 + x_3x_4$ of four variables $\{x_i\}_{i=1}^4$ over $\mathbf{Q}$, then both $\varepsilon_v(\varphi)$ and $\varepsilon'_v(\varphi)$, are -1 for $v = 2$. On the other hand, they are 1 for every prime v if φ is the sum of n squares over any number field F. This is a matter of perspective, but it is more natural to assign a *trivial* quantity to a hyperbolic form.

In fact, better invariants were introduced by Eichler in [E]. He associated with φ a set of invariants $\{\chi_v(\varphi)\}$ different from $\{\varepsilon_v(\varphi)\}$ and $\{\varepsilon'_v(\varphi)\}$, and proved a result of the same type as what Hasse proved. This set $\{\chi_v(\varphi)\}$ is conceptually more natural than $\{\varepsilon_v(\varphi)\}$, as $\chi_v(\varphi)$ is more closely connected with the structure of φ at v. Yet I am not aware of any later book or paper, except for a short article [Ch], in which $\{\chi_v(\varphi)\}$ was employed instead of $\{\varepsilon_v(\varphi)\}$ or $\{\varepsilon'_v(\varphi)\}$, which I find strange and unfortunate.

In the present paper we attempt to resurrect Eichler's idea and to show its efficacy. Indeed, there is a straightforward approach to the problem which is equivalent to his methods but never explained in [E]. Namely, we associate a quaternion algebra $Q(\varphi)$ with φ by the following rule: For n even, $Q(\varphi)$ represents the Brauer class of the Clifford algebra of φ as a central simple algebra over F; for n odd, $Q(\varphi)$ represents the Brauer class of the even Clifford algebra of φ as a central simple algebra over F. It is not difficult to show that $Q(\varphi)$ together with the dimension and the discriminant determines φ at every nonarchimedean prime, and therefore $Q(\varphi)$ and the obvious invariants determine φ by virtue of the Hasse principle (Theorem 4.2). Conversely, given a quaternion algebra B over F, we can prove the existence of a quadratic form over F with the prescribed dimension, discriminant, and signatures at the real archimedean primes and such that $B = Q(\varphi)$, provided the prescribed invariants and B satisfy natural conditions of consistency (Theorem 4.4). It turns out that $\chi_v(\varphi)$ is the invariant of $Q(\varphi)$ at v in the standard sense. This is not completely trivial, if not deep. Our proof of the existence theorem is different from and much simpler than Eichler's proof.

Now the point of this approach is that it enables us to answer the question stated at the beginning in a clear-cut form. Though the case of an arbitrary F can be handled by the same methods, let us explain the results here only in the case $F = \mathbf{Q}$, using the matrix Φ. We call Φ reduced if $P^{-1}\Phi \cdot {}^tP^{-1}$ does not represent integer-valued quadratic form for any element P of $GL_n(\mathbf{Q})$ with entries in $\mathbf{Z}$ such that $|\det(P)| > 1$. Clearly Φ is reduced if $\det(2\Phi)$ is squarefree. We put also $s(\Phi) = p - q$ when Φ as a real matrix has p positive and q negative eigenvalues. Let $\mathfrak{S}_n$ denote the set of all Φ of size n that represents a nondegenerate $\mathbf{Z}$-valued quadratic form, and $\mathfrak{S}_n^0$ the set of all reduced elements of $\mathfrak{S}_n$. Then the genus of an element Φ of $\mathfrak{S}_n^0$ determines an isomorphism class of quadratic spaces over $\mathbf{Q}$, and vice versa. Once we have an interpretation of $|\det(2\Phi)|$ in terms of lattices in V, the classification of φ according to $Q(\varphi)$ and the obvious invariants yields a complete list of the genera of $\Phi \in \mathfrak{S}_n^0$ with given values of $|\det(2\Phi)|$ and $s(\Phi)$, parametrized by several positive integers (Theorems 6.4 and 6.5). As a consequence we can determine the number of genera of $\Phi \in \mathfrak{S}_n^0$ with given values of $|\det(2\Phi)|$ and $s(\Phi)$ (Theorem 6.6). To state the results in easy special cases, take an integer σ such that $|\sigma| \leq n$ and $\sigma - n \in 2\mathbf{Z}$. Then we have:

(1) *If $\Phi \in \mathfrak{S}_n$ and n is odd, then* $\det(2\Phi) \in 2\mathbf{Z}$. *If $\Phi \in \mathfrak{S}_n$, n is even, and* $\det(2\Phi) \in 2\mathbf{Z}$, *then* $\det(2\Phi) \in 4\mathbf{Z}$.

(2) *Let f be a squarefree odd positive integer greater than 1 with exactly s prime factors and let* $2 < n \in 2\mathbf{Z}$. *Then $\mathfrak{S}_n$ contains an element Φ such that* $|\det(2\Phi)| = f$ *and $s(\Phi) = \sigma$ if and only if $f - \sigma - 1 \in 4\mathbf{Z}$, and the number of genera of such Φ's is 2^{s-1} for each (n, σ, f).*

(3) *$\mathfrak{S}_n$ contains an element Φ such that $|\det(2\Phi)| = 2$ and $s(\Phi) = \sigma$ if and only if $\sigma \equiv \pm 1 \pmod 8$, and such a Φ determines a unique genus for each (n, σ).*

(4) *Let $1 < n - 1 \in 2\mathbf{Z}$ and let h be a squarefree odd or even positive integer greater than 1 with exactly s prime factors. Then $\mathfrak{S}_n$ contains an element Φ such that $|\det(2\Phi)| = 2h$ and $s(\Phi) = \sigma$, and the number of genera of such Φ's is 2^{s-1} for each (n, σ, h).*

It should be mentioned that in the paper [Ch], using $\{\chi_v(\varphi)\}$, Chang gave a necessary and sufficient condition for the existence of Φ with given $n, s(\Phi)$, and $\det(2\Phi)$, but did not classify all such Φ's.

Though the classification of Φ is our main topic, we have also included in this paper three more topics concerning quadratic spaces and Clifford groups. The first of these is an explicit construction of a quadratic form φ that is positive definite at every real archimedean prime and has a given discriminant and $Q(\varphi)$ (Section 5). The other two are results over an arbitrary field of characteristic different from 2. In Section 7 we give several general results on Clifford algebras and groups which are possibly known, but not stated in the standard literature. In particular we make a detailed examination of the structure of the algebra and group when the dimension n is in the range $4 \leq n \leq 6$. For instance, we shall prove that for $n = 6$ the even Clifford group of a quadratic form φ equals

$$\{\alpha \in A^+(\varphi)^\times \mid \alpha\alpha^* \in F^\times,\ N(\alpha) = (\alpha\alpha^*)^2\},$$

where $A^+(\varphi)$ is the even Clifford algebra of φ, $\alpha \mapsto \alpha^*$ is its canonical involution, and N is the reduced norm map of $A^+(\varphi)$ to its center (Theorem 7.7). In the last section we construct a subgroup $\mathcal{G}$ of $A^+(\varphi)^\times$ and show that the group of similitudes of φ for any even n can be reduced, roughly speaking, to $\mathcal{G}$ and scalar multiplication (Theorems 8.2 and 8.3).

1. Quadratic forms and orthogonal groups.

1.1. For an associative ring A with an identity element we denote by $A^\times$ the group of all invertible elements of A, by $M_n(A)$ for a positive integer n the ring of all $n \times n$-matrices with entries in A, and by 1_n the identity element of $M_n(A)$.

Throughout the paper F denotes a field of characteristic different from 2. Given a finite-dimensional vector space V over F, we denote by $GL(V, F)$ the group of all F-linear automorphisms of V, and by $SL(V, F)$ the subgroup of $GL(V, F)$ consisting of the elements of determinant 1. If α is an F-linear map of V into another vector space and $x \in V$, then we denote by $x\alpha$ the image of x under α. Let φ: $V \times V \to F$ be a nondegenerate F-valued F-bilinear symmetric form. We put $\varphi[x] = \varphi(x, x)$ for $x \in V$, and

$$O(\varphi) = O(V, \varphi) = \{\alpha \in GL(V, F) \mid \varphi[x\alpha] = \varphi[x] \quad \text{for every } x \in V\},$$
$$SO(\varphi) = SO(V, \varphi) = SL(V, F) \cap O(V, \varphi).$$

In this paper we consider only nondegenerate quadratic forms. We call such a pair (V, φ) a *quadratic space* over F. We call (V, φ) or φ *isotropic* if $\varphi[x] = 0$ for some $x \in V, \neq 0$; we call φ *anisotropic* if $\varphi[x] = 0$ only for $x = 0$. Given two quadratic spaces (V, φ) and (V', φ') over F, we denote by $(V, \varphi) \oplus (V', \varphi')$ the quadratic space $(V \oplus V', \varphi \oplus \varphi')$ given by $(\varphi \oplus \varphi')(x+x', y+y') = \varphi(x, y) + \varphi'(x', y')$ for $x, y \in V$ and $x', y' \in V'$. We can define an isomorphism of (V, φ) onto another quadratic space (V', φ') over F in an obvious way.

Put $F^{\times 2} = \{a^2 \mid a \in F^\times\}$. Given (V, φ) of dimension n, let φ_0 be the $(n \times n)$-matrix that represents φ with respect to an F-basis of V. Then the coset $\det(\varphi_0)F^{\times 2}$ in $F^\times/F^{\times 2}$ is completely determined by φ. We denote this coset by $\det(\varphi)$; we shall often denote by $\det(\varphi)$ any number in this coset, when there is no fear of confusion. We now consider a field K_0 given by

$$K_0 = F(\delta^{1/2}), \qquad \delta = (-1)^{n(n-1)/2} \det(\varphi) \tag{1.1}$$

and call K_0 the *discriminant field* of φ, or of (V, φ), This is either F itself or a quadratic extension of F. We then define the *discriminant algebra* K of φ, or of (V, φ), by

$$K = \begin{cases} K_0 & \text{if } K_0 \neq F, \\ F \times F & \text{if } K_0 = F. \end{cases} \tag{1.2}$$

Given a symmetric matrix ψ of $GL_n(F)$, we denote by $\langle F, \psi\rangle$ the n-dimensional quadratic space (V, φ) over F, where V is the vector space of all n-dimensional row vectors with components in F and $\varphi[x] = x\psi \cdot {}^t x$ for $x \in V$. In particular, for $b \in F^\times$ the symbol $\langle F, b\rangle$ denotes the one-dimensional quadratic space Fg generated by an element g for which the quadratic form takes the value b. We call a quadratic space *split* if it is either $\{0\}$ or isomorphic to $\langle F, \eta_m\rangle$, for some positive integer m, where

$$\eta_m = \begin{bmatrix} 0 & (1/2)1_m \\ (1/2)1_m & 0 \end{bmatrix}. \tag{1.3}$$

The factor $1/2$ is meaningful only when we consider such a space arithmetically; otherwise we can ignore it, as we easily see that $\langle F, \eta_m\rangle$ is isomorphic to $\langle F, c\eta_m\rangle$ for every $c \in F^\times$. If it is unnecessary to mention the base field F, we denote any split quadratic space of dimension $2r$ simply by H_{2r}. Then for any quadratic space (V, φ) we have a decomposition

$$(V, \varphi) = H_{2r} \oplus (Z, \zeta) \tag{1.4}$$

with (Z, ζ) that is anisotropic. Moreover, the isomorphism class of (Z, ζ) is completely determined by (V, φ). We call Z a *core subspace* of (V, φ), or simply of V, and call $\dim(Z)$ the *core dimension* of (V, φ).

1.2. Given an associative ring A with an identity element, by an *involution* of A we mean a bijection τ of A onto itself such that $(x+y)^\tau = x^\tau + y^\tau$, $(xy)^\tau = y^\tau x^\tau$, and $(x^\tau)^\tau = x$ for every $x, y \in A$. A central simple algebra B over F is called a *quaternion algebra* over F if B is of dimension 4 over F. Such a B is either a division algebra or isomorphic to $M_2(F)$. Moreover, it has an involution ι which is uniquely determined by the conditions $x + x^\iota \in F$ and $xx^\iota \in F$ for every $x \in F$. We call it *the main involution* of B. We put $\mathrm{Tr}_{B/F}(x) = x + x^\iota$ and $N_{B/F}(x) = xx^\iota$ for $x \in B$. We have $F = \{x \in B \mid x^\iota = x\}$, and so we have a direct sum decomposition

$$B = F \oplus B^\circ \quad \text{with} \quad B^\circ = \{x \in B \mid x^\iota = -x\}. \tag{1.5}$$

Whenever we have a quaternion algebra, we always denote by ι its main involution.

Given B, put $\beta[x] = xx^\iota$. Then (B, β) is a quadratic space of dimension 4. We call β the *norm form* of B. Denoting by β° the restriction of β to B°, we obtain a quadratic space (B°, β°) of dimension 3. We easily see that

(1.6) B is a division algebra $\Longleftrightarrow$ β is anisotropic $\Longleftrightarrow$ β° is anisotropic.

Since $2\beta(x, y) = \beta[x + y] - \beta[x] - \beta[y]$, we easily see that

$$2\beta(x, y) = \mathrm{Tr}_{B/F}(xy^{\iota}) \qquad (x, y \in B). \tag{1.7}$$

1.3. Next, we take a couple (K, ρ) consisting of an F-algebra K of rank 2 and an F-linear automorphism ρ of K belonging to the following two types:

(I) K is a quadratic extension of F and ρ is the generator of $\mathrm{Gal}(K/F)$;

(II) $K = F \times F$ and $(x, y)^{\rho} = (y, x)$ for $(x, y) \in F \times F$.

In both cases we obtain a quadratic space (K, κ) of dimension 2 by putting $\kappa[x] = xx^{\rho}$ for $x \in K$. We call κ the *normform* of K. Clearly κ is anisotropic if and only if K is a field. In our later treatment we will be considering quadratic spaces $(K, c\kappa), (B^{\circ}, c\beta^{\circ})$, and $(B, c\beta)$. with any fixed $c \in F^{\times}$, where $(K, \kappa), (B^{\circ}, \beta^{\circ})$, and (B, β) are defined as above.

Given (K, ρ) as above and $a \in F^{\times}$, define an F-algebra B by

$$B = K \oplus Kj, \quad j^2 = a, \quad jx = x^{\rho}j \quad \text{for every} \quad x \in K. \tag{1.8}$$

Then it can easily be seen that B is a quaternion algebra over F. In this case ι can be given by $(x + yj)^{\iota} = x^{\rho} - yj$ for $x, y \in K$. We denote by $\{K, a\}$ this quaternion algebra B in which K is embedded as in (1.8). Define (K, κ) and (B, β) as above for these K and B. Then the decomposition $B = K + Kj$ gives

$$(B, \beta) = (K, \kappa) \oplus (K, -a\kappa). \tag{1.9}$$

2. Clifford algebras and Clifford groups.

2.1. Given a quadratic space (V, φ), we denote by $A(V, \varphi)$ or simply by $A(\varphi)$ the Clifford algebra of (V, φ). We can define an F-linear automorphism $x \mapsto x'$ and an F-linear involution $x \mapsto x^*$ of $A(\varphi)$ characterized by the property that $v = -v' = v^*$ for every $v \in V$. We call these the *canonical automorphism* and the *canonical involution* of $A(\varphi)$. We then put

$$A^{+}(\varphi) = A^{+}(V, \varphi) = \{\alpha \in A(\varphi) \mid \alpha' = \alpha\},$$
$$A^{-}(\varphi) = A^{-}(V, \varphi) = \{\alpha \in A(\varphi) \mid \alpha' = -\alpha\}.$$

In particular, $A^{+}(\varphi)$ is called the *even Clifford algebra* of (V, φ). We also put

$$G(\varphi) = G(V, \varphi) = \{\alpha \in A(\varphi)^{\times} \mid \alpha^{-1}V\alpha = V\},$$
$$G^{+}(\varphi) = G^{+}(V, \varphi) = G(\varphi) \cap A^{+}(\varphi).$$

The groups $G(\varphi)$ and $G^{+}(\varphi)$ are called the *Clifford group* and the *even Clifford group* of (V, φ). We can define a homomorphism $\nu\colon G^{+}(\varphi) \to F^{\times}$ and a homomorphism τ: $G^+(\varphi) \to SO(\varphi)$ by $\nu(\alpha) = \alpha\alpha^*$ and $x\tau(\alpha) = \alpha^{-1}x\alpha$ for $\alpha \in G^+(\varphi)$ and $x \in V$. The map τ is surjective, and gives an isomorphism of $G^+(\varphi)/F^\times$ onto $SO(\varphi)$. For this and other basic properties of $A(\varphi)$, $G^+(\varphi)$, ν, and τ, etc., the reader is referred to any standard book on this topic such as [C] and [E], and in particular to [S3], as we shall often need some statements in that volume.

2.2. Hereafter in this section we denote by (V, φ) a quadratic space of dimension n over an arbitrary field F of characteristic $\neq 2$, by K_0 resp. K the discriminant field resp. the discriminant algebra of φ, and by ρ the nontrivial automorphism of K over F. Take an orthogonal basis $\{e_i\}_{i=1}^n$ of V, and put $z = e_1 \cdots e_n$. Then $F + Fz$ is the center of $A(\varphi)$ or $A^+(\varphi)$ according as n is odd or even; also $z^2 = (-1)^{(n-1)n/2}e_1^2 \cdots e_n^2$ and $z^* = (-1)^{(n-1)n/2}z$. These are elementary; see [S3, Lemma 2.7, Theorem 2.8], for example. Thus K is isomorphic to $F + Fz$. Therefore in this section we identify K with $F + Fz$, and use the letter z for an element of this type; z can be replaced by any element of $F^\times z$; we put $\delta = z^2$. We have $zx = (-1)^{n-1}xz$ for every $x \in V$, and hence

(2.1) $\quad ax = xa^\rho \quad$ for every $a \in K$ and every $x \in V$ if $n \in 2\mathbf{Z}$.

LEMMA 2.3. (i) *If n is odd, then $A(\varphi) = A^+(\varphi) \otimes_F K$.*

(ii) *If n is even, then there is an F-linear ring-isomorphism of $A(\varphi) \otimes_F K$ onto $M_2(A^+(\varphi))$ that maps $1 \otimes K$ onto the center of $A^+(\varphi)$. In particular, if $K_0 = F$, then there exists a central simple algebra D over F such that $A(\varphi) \cong M_2(D)$ and $A^+(\varphi) \cong D \times D$.*

Assertion (i) is well known; see [S3, Theorem 2.8], for example. The facts stated in (ii) do not seem to have been given before. We shall prove it after the proof of Proposition 7.1.

LEMMA 2.4. *The center of $G^+(\varphi)$ is $F^\times$ or $K^\times \cap G^+(\varphi)$ according as n is odd or even. Moreover, $K^\times \cap G^+(\varphi) = F^\times \cap F^\times z$ if $2 < n \in 2\mathbf{Z}$ and $K^\times \cap G^+(\varphi) = K^\times$ if $n = 2$.*

Proof. The algebra $A^+(\varphi)$ is spanned by the monomials of an even number of members of an orthogonal basis of V. Since these monomials belong to $G^+(\varphi)$, the center of $G^+(\varphi)$ is contained in the center of $A^+(\varphi)$. From this we can easily derive the first assertion. The second assertion is given in [S3, Theorem 3.6]. □

LEMMA 2.5. *For every $c \in F^\times$ there exists an F-linear ring-isomorphism f of $A^+(V, c\varphi)$ onto $A^+(V, \varphi)$ such that $f(G^+(c\varphi)) = G^+(\varphi)$ and $f(\xi)^* = f(\xi^*)$ for every $\xi \in A^+(\varphi)$; further $\tau(\alpha) = \tau(f(\alpha))$ and $\nu(\alpha) = \nu(f(\alpha))$ for every $\alpha \in G^+(\varphi)$.*

Proof. This is [S3, Lemma 3.8 (i)]. We reproduce the proof here, as we need the symbols in it later. Take an algebraic extension $E = F(d)$ of F with an element d such that $d^2 = c$, and let $Y = V \otimes_F E$. We extend φ to an E-bilinear form $Y \times Y \to E$, and denote it again by φ. Clearly $A(Y,\varphi) = A(V,\varphi) \otimes_F E$. Since $\varphi[dx] = c\varphi[x]$ for $x \in Y$, the map $x \mapsto dx$ for $x \in Y$ can be extended to a ring-isomorphism f of $A(Y, c\varphi)$ onto $A(Y,\varphi)$. We easily see that f maps $A^+(V, c\varphi)$ onto $A^+(V,\varphi)$ and $G^+(V, c\varphi)$ onto $G^+(V,\varphi)$. Clearly $SO(V, c\varphi) = SO(V,\varphi)$ and $f(y\beta) = f(y)\beta$ for $\beta \in SO(V,\varphi)$ and $y \in Y$. Now, for $\alpha \in G^+(V, c\varphi)$ we have $f(y)\tau(\alpha) = f(y\tau(\alpha)) = f(\alpha^{-1}y\alpha) = f(\alpha)^{-1}f(y)f(\alpha) = f(y)\tau(f(\alpha))$, and hence $\tau(\alpha) = \tau(f(\alpha))$. The formulas $f(\xi)^* = f(\xi^*)$ and $\nu(\alpha) = \nu(f(\alpha))$ can easily be verified. □

2.6. Let us now discuss $A(\varphi)$, $A^+(\varphi)$, and $G^+(\varphi)$ for $1 \le n \le 4$. If $n = 1$, then $(V,\varphi) = \langle F, \delta \rangle$, $A(\varphi) = K$, $A^+(\varphi) = F$, and $G^+(\varphi) = F^\times$.

Suppose $n = 2$. Then there exists an element $c \in F^\times$ such that $(V,\varphi) \cong (K, c\kappa)$, where κ is the norm form of K. We have $A(\varphi) \cong \{K, c\}$, and $G^+(\varphi) = K^\times$. In particular, $A(\varphi) \cong M_2(F)$ if $c = 1$. The proof of these is easy; see [S3, §7.2].

Suppose $n = 3$. Then there exists a quaternion algebra B such that $(V,\varphi) \cong (B^\circ, -\delta\beta^\circ)$, where (B°, β°) is defined as in §1.2. We have $A(\varphi) = A^+(\varphi) \otimes_F K$, and $A^+(\varphi) \cong B$; $G^+(\varphi)$ can be identified with $B^\times$; (V,φ) is anisotropic if and only if B is a division algebra. These are also well known; see [S3, §7.3], for example.

2.7. Suppose $n = 4$. The structure of (V,φ) in the most general case is not so simple, and will be discussed in Theorem 7.4. Here we treat the case $n = 4$ under the assumption $1 \in \det(\varphi)$. Then there exists a quaternion algebra B over F such that (V,φ) is isomorphic to $(B, d\beta)$ with $d \in F^\times$, where β is the norm form of B; this is well known, see [S3, Theorem 7.5 (i)], for example. By (1.6), φ is anisotropic if and only if B is a division algebra. To find $A(\varphi)$, we consider a map

$$p\colon B \longrightarrow M_2(B) \text{ given by } p(x) = \begin{bmatrix} 0 & dx \\ x^\iota & 0 \end{bmatrix} \quad \text{for } x \in B. \tag{2.2}$$

Observe that $p(x)^2 = dxx^\iota 1_2 = \varphi[x]1_2$. Now we can easily verify that the whole $M_2(B)$ can be generated by $p(B)$ over F. We can thus put $M_2(B) = A(\varphi)$ and

$$V = \left\{ \begin{bmatrix} 0 & dx \\ x^\iota & 0 \end{bmatrix} \mid x \in B \right\}. \tag{2.3}$$

Then $A^+(\varphi) = \{\mathrm{diag}[a,b] \mid a, b \in B\} \cong B \times B$. For $\alpha = \begin{bmatrix} p & q \\ r & s \end{bmatrix} \in M_2(B) = A(\varphi)$, we have $\alpha^* = \begin{bmatrix} p^\iota & dr^\iota \\ d^{-1}q^\iota & s^\iota \end{bmatrix}$, since this is so for every $\alpha \in V$. If we identify

$A^+(\varphi)$ with $B \times B$ in an obvious fashion, then

$$G^+(\varphi) = \{(a,b) \in B^\times \times B^\times \mid aa^\iota = bb^\iota\}, \tag{2.4}$$

$\nu((a,b)) = aa^\iota$ for $(a,b) \in G^+(\varphi)$, and $p(x)\tau(\alpha) = p(a^{-1}xb)$ for $x \in B$ and $\alpha = (a,b) \in G^+(\varphi)$.

LEMMA 2.8. *With (V,φ) and z as in §2.2, let (V',φ') be another quadratic space; define a vector subspace C of $A(\varphi \oplus \varphi')$ by $C = A^+(\varphi') + zA^-(\varphi')$. Then C is a subalgebra of $A(\varphi \oplus \varphi')$ and the following assertions hold:*

(i) *If n is even, then $A(\varphi \oplus \varphi') = A(\varphi) \otimes_F C$ and $C \cong A(\delta\varphi')$.*

(ii) *If n is odd, then $A^+(\varphi \oplus \varphi') = A^+(\varphi) \otimes_F C$ and $C \cong A(-\delta\varphi')$.*

(iii) *For both even and odd n the canonical involution of $A(\varphi \oplus \varphi')$ restricted to $A^+(\varphi')$ coincides with the restriction of the canonical involution of $A(\pm\delta\varphi')$ to $A^+(\pm\delta\varphi')$.*

Proof. Clearly C is a subalgebra of $A(\varphi \oplus \varphi')$. Define p: $V' \to C$ by $p(x) = zx$ for $x \in V'$. Then $p(x)^2 = (-1)^n\delta\varphi'[x]$, and we easily see that F and $p(V')$ generate C, and so we see that $C \cong A((-1)^n\delta\varphi')$. If n is even, $A(\varphi)$ commutes elementwise with C; also we easily see that $A(\varphi)C = A(\varphi \oplus \varphi')$, and so we obtain (i). For n odd, we take $A^+(\varphi \oplus \varphi')$ and $A^+(\varphi)$ in place of $A(\varphi \oplus \varphi')$ and $A(\varphi)$; then we obtain (ii). The last assertion (iii) can easily be verified. □

3. The characteristic quaternion algebra of a quadratic space.

3.1. In this section we assume that F is an algebraic number field or its completion at a prime of F which may be archimedean or nonarchimedean. Let (V,φ) be a quadratic space of dimension n over such an F. Then $A(\varphi)$ or $A^+(\varphi)$ is a central simple algebra over F according as n is even or odd. The canonical involution gives an antiautomorphism of each such algebra onto itself, and so it can be given as $M_p(D)$, where D is a quaternion algebra over F or $D = F$. If $D = F$ and $n > 1$, then p is even, and so $M_p(F) = M_{p/2}(M_2(F))$. Thus we associate a quaternion algebra $Q(\varphi)$ with (V,φ) by the following rule:

$$A(\varphi) \cong M_s(Q(\varphi)) \quad \text{if} \quad 0 < n \in 2\mathbf{Z}, \tag{3.1a}$$

$$A^+(\varphi) \cong M_s(Q(\varphi)) \quad \text{if} \quad 1 < n \notin 2\mathbf{Z}, \tag{3.1b}$$

with $0 < s \in \mathbf{Z}$; we put $Q(\varphi) = M_2(F)$ if $n = 1$. We call $Q(\varphi)$ the *characteristic (quaternion) algebra* of φ. Thus, with a given (V,φ), we associate two algebras: the discriminant algebra and the characteristic algebra. Notice that these algebras depend only on the isomorphism class of core subspaces of (V,φ).

3.2. Let us now determine the discriminant and characteristic algebras when F is a nonarchimedean local field, that is, the completion of an algebraic number field at a nonarchimedean prime. Given (V,φ) over F, denote by t its core dimension, and by K_0 its discriminant field. It is well known that $t \leq 4$. Here is another basic fact: there is only one isomorphism class of division quaternion algebras over a fixed F. Thus $Q(\varphi)$ is either this unique division quaternion algebra, or $M_2(F)$.

We have $n = 2r + t$ and $(V,\varphi) = H_{2r} \oplus (Z,\zeta)$ with a core subspace Z of dimension t as stated in (1.4). Then K_0 is the discriminant field of (Z,ζ), and $Q(\varphi) = Q(\zeta)$.

We first assume $n \in 2\mathbf{Z}$, that is, $t = 0, 2$, or 4. In each case, as is well known, $A(\varphi) \cong M_m(A(\zeta))$ with $m = 2^r$. If $t = 0$, then $A(\zeta) = F$; thus we have $Q(\varphi) = M_2(F)$ and $K_0 = F$.

If $t = 2$, then $(Z,\zeta) \cong (K, s\kappa)$ with $s \in F^\times$ and a quadratic extension K of F, which is the discriminant field of (Z,ζ); $A(\zeta) \cong \{K, s\}$ as explained in §2.6. We have therefore $Q(\varphi) = \{K, s\}$.

Suppose $t = 4$; then it is well known that $(Z,\zeta) \cong (B,\beta)$ with a division quaternion algebra B over F; see [S3, Theorem 7.5 (ii)]. Then $A(\zeta) \cong M_2(B)$ as explained in §2.7. Thus $Q(\varphi) = B$; also we have $K_0 = F$.

Next suppose $n \notin 2\mathbf{Z}$. Then by a well known principle, $A^+(\varphi) \cong M_m(A^+(\zeta))$ with $m = 2^r$; see [S3, Theorem 2.6], for example. If $t = 1$, then $(Z,\zeta) \cong \langle F, c\rangle$ with $c \in F^\times$, and $A^+(\zeta) = F$; thus we have $Q(\varphi) = M_2(F)$ and $K_0 = F(c^{1/2})$.

If $t = 3$, then as we said in §2.6, $A^+(\zeta)$ is a division quaternion algebra B over F, and hence $Q(\varphi) = B$; $(Z,\zeta) \cong (B^\circ, -\delta\beta^\circ)$ if $K_0 = F(\delta^{1/2})$.

LEMMA 3.3. *The isomorphism class of core subspaces of* (V,φ) *over a nonarchimedean local field as above is completely determined by* K_0, $Q(\varphi)$, *and* n (mod 2), *and consequently the isomorphism class of* (V,φ) *is completely determined by* K_0, $Q(\varphi)$, *and* n.

Proof. Let (Z,ζ) be a core subspace of (V,φ). Since the isomorphism class of (V,φ) is determined by (Z,ζ) and n, it is sufficient to show that (Z,ζ) is determined by K_0, $Q(\varphi)$ and $n \pmod 2$. Suppose $n \in 2\mathbf{Z}$. Then $K_0 \neq F$ if and only if $t = 2$, in which case $(Z,\zeta) \cong (K, s\kappa)$ as in §3.2. The isomorphism class of $(K, s\kappa)$ is determined by $sN_{K/F}(K^\times)$. Now $Q(\zeta) = \{K, s\}$, which is $M_2(F)$ if and only if $s \in N_{K/F}(K^\times)$. Suppose $K_0 = F$ still with $n \in 2\mathbf{Z}$. Then $t = 0$ or $t = 4$ according as $Q(\varphi) \cong M_2(F)$ or $Q(\varphi) \ncong M_2(F)$. In the latter case $(Z,\zeta) \cong (B,\beta)$ with $B = Q(\zeta)$ as explained in §3.2. Thus (Z,ζ) is determined by $Q(\zeta)$ and K_0 if $n \in 2\mathbf{Z}$. Next suppose $n \notin 2\mathbf{Z}$. Then $t = 1$ or $t = 3$ according as $Q(\varphi) \cong M_2(F)$ or $Q(\varphi) \ncong M_2(F)$. Then (Z,ζ) is determined by K_0 as explained in §3.2. □

If the base field F is $\mathbf{C}$, then clearly $K_0 = \mathbf{C}$ and $Q(\varphi) = M_2(\mathbf{C})$. The matter is not so simple if $F = \mathbf{R}$. We discuss this case in §4.3.

4. Classification of quadratic spaces over an algebraic number field.

4.1. Throughout this section we assume that F is an algebraic number field. We denote by $\mathbf{a}$ and $\mathbf{h}$ the sets of archimedean primes and nonarchimedean primes of F respectively, and put $\mathbf{v} = \mathbf{a} \cup \mathbf{h}$; we also denote by $\mathbf{r}$ the set of all real primes in $\mathbf{a}$. For each $v \in \mathbf{v}$ we denote by F_v the v-completion of F. Given a quadratic space (V, φ) over F, we can define, for each $v \in \mathbf{v}$, the v-localization (V_v, φ_v) of (V, φ) as a quadratic space over F_v by putting $V_v = V \otimes_F F_v$ and extending φ to an F_v-valued quadratic form φ_v on V_v in a natural way. Then clearly $Q(\varphi_v) = Q(\varphi) \otimes_F F_v$. By a well known principle, $Q(\varphi)$ is determined by the set of all such local quaternion algebras $Q(\varphi_v)$. Similarly, if K is the discriminant algebra of φ, then $K \otimes_F F_v$ is the discriminant algebra of φ_v. For each $v \in \mathbf{r}$ we have a pair of nonnegative integers (p_v, q_v) such that (V_v, φ_v) is isomorphic to $\langle \mathbf{R}, \mathrm{diag}\,[1_{p_v}, -1_{q_v}] \rangle$, where we of course ignore 1_0. We put $s_v(\varphi) = p_v - q_v$ and call $s_v(\varphi)$ the *index* of φ at v. Clearly $|s_v(\varphi)|$ is the core dimension of φ_v, $s_v(\varphi) - n \in 2\mathbf{Z}$, and $|s_v(\varphi)| \le n$; also, n and $s_v(\varphi)$ determine (p_v, q_v), and vice versa.

THEOREM 4.2. (i) *Two quadratic spaces over F of dimension n and n' have isomorphic core subspaces if and only if $n - n' \in 2\mathbf{Z}$ and they have the same discriminant field, the same characteristic algebra, and the same indices for all $v \in \mathbf{r}$.*

(ii) *The isomorphism class of a quadratic space (V, φ) over F is determined by its dimension, discriminant field, characteristic algebra $Q(\varphi)$, and $s_v(\varphi)$ for all $v \in \mathbf{r}$.*

Proof. Once the dimension, the discriminant field, and $Q(\varphi)$ are given, then by Lemma 3.3, the isomorphism class of (V_v, φ_v) for every $v \in \mathbf{h}$ is determined. This together with the index at each $v \in \mathbf{r}$ determines (V, φ) over F by the Hasse principle, which proves (ii). Next, the "only if"-part of (i) is obvious. Let (V, φ) and (V', φ') be quadratic spaces over F of dimension n and n' with the same set of data as in (i); suppose $0 \le n' - n \in 2\mathbf{Z}$. Let $(V'', \varphi'') = (V, \varphi) \oplus H_r$ with $r = (n' - n)/2$. Then, by (ii), (V'', φ'') must be isomorphic to (V', φ'). This proves the "if"-part of (i). □

4.3. We are going to show the existence of (V, φ) with a given set of data consisting of the following objects:

(4.1a) $0 < n \in \mathbf{Z}$; an extension $K_0 = F(\delta^{1/2})$ of F with $\delta \in F^\times$; a quaternion algebra B over F; an integer σ_v such that $|\sigma_v| \le n$ and $\sigma_v - n \in 2\mathbf{Z}$ for each $v \in \mathbf{r}$.

We also put

(4.1b) $$K = K_0 \text{ if } [K_0 : F] = 2 \text{ and } K = F \times F \text{ if } K_0 = F,$$

$K_v = K \otimes_F F_v$, and $B_v = B \otimes_F F_v$ for $v \in \mathbf{v}$.

Our task is to find (V, φ) of dimension n such that K is its discriminant algebra, $B = Q(\varphi)$, and $\sigma_v = s_v(\varphi)$. These data cannot be arbitrary. Indeed, if $B = Q(\varphi)$ and $\sigma_v = s_v(\varphi)$, then denoting by $\mathbf{H}$ the division ring of Hamilton quaternions, we have

$$\text{(4.2a)} \qquad n \text{ odd: } B_v = \begin{cases} M_2(\mathbf{R}) & \text{if } \sigma_v \equiv \pm 1 \pmod 8, \\ \mathbf{H} & \text{if } \sigma_v \equiv \pm 3 \pmod 8, \end{cases}$$

$$\text{(4.2b)} \qquad n \text{ even: } \quad B_v = \begin{cases} M_2(\mathbf{R}) & \text{if } \sigma_v \equiv 0 \text{ or } 2 \pmod 8, \\ \mathbf{H} & \text{if } \sigma_v \equiv 4 \text{ or } 6 \pmod 8. \end{cases}$$

Indeed, in [S3, (A3.6) and (A3.9)] we determined $A^+(\varphi)$ and $A(\varphi)$ when $\varphi = \pm 1_n$. Then we obtain (4.2a,b), since $Q(\varphi)_v$ depends only on $s_v(\varphi)$. Another proof of (4.2a,b) can be given by induction on the dimension by employing Lemma 2.8.

For δ as in (1.1), we have obvious conditions:

$$\text{(4.3a)} \qquad n \text{ odd: } \ (-1)^{(\sigma_v-1)/2}\delta > 0 \quad \text{at each} \quad v \in \mathbf{r},$$

$$\text{(4.3b)} \qquad n \text{ even: } \ (-1)^{\sigma_v/2}\delta > 0 \quad \text{at each} \quad v \in \mathbf{r}.$$

Now the main theorem of the existence of a quadratic space with a given set of data can be stated as follows.

THEOREM 4.4. *Given a set of data as in (4.1a,b) with $n > 2$, there exists a quadratic space (V, φ) over F of dimension n such that K is the discriminant algebra of φ, $B = Q(\varphi)$, and $\sigma_v = s_v(\varphi)$ for every $v \in \mathbf{r}$, if and only if (4.2a,b) and (4.3a,b) are satisfied. Such a (V, φ) exists even for $n = 1$ or 2 if and only if the following condition (4.4a) or (4.4b) is satisfied in addition to (4.2*a,b*) and (4.3*a,b*):*

$$\text{(4.4a)} \qquad B = M_2(F) \text{ if } n = 1;$$

(4.4b) *If $n = 2$, $v \in \mathbf{h}$, and B_v is a division algebra, then K_v is a field.*

We only have to prove the "if" part. We devote §§4.6 and 4.7 to the proof.

4.6. We first prove Theorem 4.4 for $n \leq 4$. If $n = 1$, we have $(V, \varphi) = \langle F, \delta \rangle$ and $Q(\varphi) = M_2(F)$. Clearly (4.2a), (4.3a), and (4.4a) are necessary and sufficient.

Next suppose $n = 2$. The discussion of §2.6 shows that K must be embeddable in $Q(\varphi)$, and so (4.4b) is necessary. Conversely, if $v \in \mathbf{r}$ and $B_v = \mathbf{H}$, then (4.2b) shows that $\sigma_v = -2$, and so $K_v = \mathbf{C}$ by (4.3b). This combined with (4.4b) means that $B = \{K, r\}$ with some $r \in F^\times$. Take $(V, \varphi) = (K, r\kappa)$ with the notation of §1.3. Then K is the discriminant algebra and $Q(\varphi) = B$. From conditions (4.2b) and (4.3b) we see that $\sigma_v = s_v(\varphi)$ for every $v \in \mathbf{r}$.

If $n = 3$, we take $(V, \varphi) = (B^\circ, -\delta\beta^\circ)$. Then $A^+(\varphi) \cong B$ and K_0 is its discriminant field. From (4.2a) and (4.3a) we see that $s_v(\varphi) = \sigma_v$ for every $v \in \mathbf{r}$.

In the case $n = 4$ we determine (V, φ) by

$$(V, h\varphi) = \langle F, 1\rangle \oplus (B^\circ, \delta\beta^\circ). \tag{4.5}$$

Here we take $h \in N_{K/F}(K^\times)$ so that $h < 0$ at every $v \in \mathbf{r}$ such that $\sigma_v = -4$ and $h > 0$ at all other $v \in \mathbf{r}$. Such an h exists, because $K_v = \mathbf{R} \times \mathbf{R}$ if $\sigma_v = -4$ and K is dense in $K \otimes_{\mathbf{Q}} \mathbf{R}$. We can easily verify that $s_v(\varphi) = \sigma_v$ for every $v \in \mathbf{r}$ and K is the discriminant algebra of φ. Moreover, $A(\varphi) \cong M_2(B)$ by a general principle valid for an arbitrary base field, which will be proven in Corollary 7.5(ii).

4.7. Let us now prove the case $6 \le n \in 2\mathbf{Z}$ by induction on n. Take a quadratic extension $K_1 = F(\varepsilon^{1/2})$ of F with $\varepsilon \in F^\times$ such that $\varepsilon < 0$ at every $v \in \mathbf{r}$. Take also $r \in F^\times$ so that $r > 0$ or $r < 0$ at $v \in \mathbf{r}$ according as $\sigma_v \ge 4 - n$ or $\sigma_v < 4 - n$. Put $\tau_v = 2 - \sigma_v$ if $\sigma_v \ge 4 - n$ and $\tau_v = -2 - \sigma_v$ if $\sigma_v < 4 - n$. Let D be the quaternion algebra over F determined by the condition $D \otimes \{K_1, r\} \cong M_2(B)$. Now we observe that the set $(n-2, F((\varepsilon\delta)^{1/2}), D, \{\tau_v\})$ satisfies conditions (4.2b) and (4.3b). By induction we can find (W, ψ) corresponding to this set. Put $(V, \varphi) = (K_1, r\kappa_1) \oplus (W, \varepsilon\psi)$, where κ_1 is the norm form of K_1. Clearly K is the discriminant algebra of φ, and we easily see that $s_v(\varphi) = \sigma_v$ for every $v \in \mathbf{r}$. By Lemma 2.8 (i), $A(\varphi) \cong \{K_1, r\} \otimes A(\psi)$, and hence $Q(\varphi) = B$ as expected.

It remains to prove the case $4 \le n - 1 \in 2\mathbf{Z}$. Since $A^+(c\varphi) \cong A^+(\varphi)$ for every $c \in F^\times$, and conditions (4.2a) and (4.3a) are consistent with the change of φ for $c\varphi$, it is sufficient to prove the existence of φ when $\sigma_v \ne -n$ for every $v \in \mathbf{r}$. Put $\tau_v = 1 - \sigma_v$ for every $v \in \mathbf{r}$ and observe that the set $(n-1, K_0, B, \{\tau_v\})$ satisfies conditions (4.2b) and (4.3b). Since we have proved the case of even n, there exists (W, ψ) corresponding to that set. Put $(V, \varphi) = \langle F, 1\rangle \oplus (W, -\psi)$. Then K_0 is the discriminant field of φ and $s_v(\varphi) = \sigma_v$ for every $v \in \mathbf{r}$. Taking $\langle F, 1\rangle$ as φ of Lemma 2.8 (ii), we obtain $Q(\varphi) = Q(\psi) = B$. This proves the case of odd n and completes the proof.

Remark 4.8. For each $v \in \mathbf{v}$ let $\chi_v(\varphi)$ be the invariant of $Q(\varphi)$ at v; namely $\chi_v(\varphi) = 1$ or -1 according as $Q(\varphi)_v$ is isomorphic to $M_2(F_v)$ or is a division algebra. It can be shown that $\chi_v(\varphi)$ is exactly the invariant Eichler attached to φ in [E]. He defined the invariant in each case according to the type of φ_v, without using Clifford algebras. The fact that it is the invariant of $Q(\varphi)$ at v can easily be seen for $v \in \mathbf{h}$. The fact for $v \in \mathbf{a}$ follows from formulas (4.2a,b), which is nontrivial. In Satz 23.2 of [E] he constructed a space (V', φ') corresponding to $(m, K_0, B, \{\sigma_v\})$ with some m such that $m - n \in 2\mathbf{Z}$, and proved the product formula $\prod_{v\in\mathbf{v}} \chi_v(\varphi) = 1$ at the end of the proof. Then he derived a theorem equivalent to Theorem 4.4 in Satz 23.3 of [E] by establishing the desired (V, φ) as a subspace of (V', φ').

In the present paper we employed $Q(\varphi)$ instead of $\{\chi_v(\varphi)\}$, as we think this approach is clearer and easier. In fact, it allows us to present the desired (V, φ) more directly without using an auxiliary space of an unspecified dimension.

5. Construction of positive definite quadratic spaces.

5.1. The proof of Theorem 4.4 produces (V, φ) only inductively. We can naturally ask whether there is an explicit way of presenting it. In this section we present (V, φ) explicitly when φ is positive definite at every $v \in \mathbf{r}$.

Thus we assume $n = \sigma_v$ for every $v \in \mathbf{r}$. For $a \in F$ we write $a \gg 0$ if a is positive at every $v \in \mathbf{r}$, and $a \ll 0$ if $-a \gg 0$. Now we look for an explicit form of φ with given n, K, and B such that $s_v(\varphi) = n$ for every $v \in \mathbf{r}$. We first introduce a symbol $\mathfrak{r}(C)$. This denotes the set of all primes $v \in \mathbf{v}$ ramified in a quaternion algebra C over F. It is well known that $\mathfrak{r}(C)$ consists of an even number of primes. To clarify the setting, we state a special case of Theorem 4.4 for such a φ, in which we exclude the cases $n \leq 2$ for simplicity.

THEOREM 5.2. *Let a quaternion algebra B over F, an extension K_0 of F, and a positive integer $n > 2$ be given as follows:*

$$\mathfrak{r}(B) \subset \mathbf{h} \quad \textit{if } n \equiv \pm 1, 0, \textit{or } 2 \pmod 8, \tag{5.1a}$$

$$\mathbf{r} \subset \mathfrak{r}(B) \quad \textit{if } n \equiv \pm 3, 4, \textit{or } 6 \pmod 8; \tag{5.1b}$$

$K_0 = F(\delta^{1/2})$ with $\delta \in F^\times$ such that $(-1)^{n/2}\delta \gg 0$ if $n \in 2\mathbf{Z}$ and $(-1)^{(n-1)/2}\delta \gg 0$ if $n \notin 2\mathbf{Z}$. Then there exists a quadratic space (V, φ) of dimension n over F such that $B = Q(\varphi)$, K_0 is the discriminant field, and φ is positive definite at every $v \in \mathbf{r}$. Such a (V, φ) is uique up to isomorphisms.

Notice that (5.1a,b) are satisfied if $\mathbf{r} = \emptyset$.

5.3. We first make an easy observation. Let C be a quaternion algebra over F and γ its norm form. Let $v \in \mathbf{h}$; then we have $\gamma[xy] = \gamma[x]\gamma[y]$ and $\gamma[C_v^\times] = F_v^\times$, and so $(C, a\gamma)_v \cong (C, b\gamma)_v$ for every $a, b \in F_v^\times$. Thus, for every $a \in F^\times$ we have $(C, \gamma)_v \oplus (C, a\gamma)_v \cong (C, \gamma)_v \oplus (C, -\gamma)_v$, which is split. In particular, let $(W, \psi) = (C, \gamma) \oplus (C, \gamma)$ with (C, γ) such that γ is the sum of four squares. Then ψ is an 8-dimensional quadratic form over F that is split at every $v \in \mathbf{h}$ and is positive definite at every $v \in \mathbf{r}$. Thus, if $m = 8k + n$ with $0 \leq k \in \mathbf{Z}$ and $3 \leq n \leq 10$ and (V_0, φ_0) is a quadratic space of dimension n with the prescribed (δ, B), then the direct sum of φ_0 and k copies of ψ gives the desired space for (m, δ, B). Therefore it is sufficient to construct our space only for $3 \leq n \leq 10$. Notice that the 10-dimensional case cannot be reduced to the 2-dimensional case, because of (4.4b).

In the following treatment we use the symbol (K, κ) in the sense of §1.3; namely κ is the norm form of K. Also we need the symbols (B, β) and (B_i, β_i)

for $i = 1, 2$. Here B is always the given quaternion algebra and B_i is a quaternion algebra over F specified in each case; β resp. β_i is the norm form of B resp. B_i. Once such (B, β) and (B_i, β_i) are fixed we shall put $B^\circ = \{x \in B \mid x^\iota = -x\}$ and $B_i^\circ = \{x \in B_i \mid x^\iota = -x\}$; we then consider (B°, β°) and $(B_i^\circ, \beta_i^\circ)$, where β° resp. β_i° is the restriction of β resp. β_i to B° resp. B_i°; see §1.2.

5.4. For each n in the range $3 \leq n \leq 10$ we shall give (V, φ) in formula (Q.n). Our task is to verify that this has the required properties. In this subsection we consider the even-dimensional cases.

The case $n = 4$. In this case $\mathbf{r} \subset \mathfrak{r}(B)$ and $\delta \gg 0$. Put

(Q.4) $$(V, \varphi) = \langle F, 1 \rangle \oplus (B^\circ, \delta\beta^\circ).$$

This is a special case of (4.5), and so has all the required properties.

The case $n = 6$. We have $\mathbf{r} \subset \mathfrak{r}(B)$ and $\delta \ll 0$ in this case; thus $K_v = \mathbf{C}$ for every $v \in \mathbf{r}$. We put

(Q.6) $$(V, \varphi) = (B, \beta) \oplus (K, \kappa).$$

Then φ is positive definite at every $v \in \mathbf{r}$ and has K as its discriminant field. By Lemma 2.8 (i), $A(\varphi) \cong A(\beta) \otimes A(\kappa) \cong M_2(B) \otimes M_2(F)$, and hence $Q(\varphi) = B$.

The case $n = 8$. We have $\delta \gg 0$. Put $\mathfrak{r}(B) = \mathbf{b}$; then $\mathbf{b} \subset \mathbf{h}$. Take a finite subset $\mathbf{c}$ of $\mathbf{h}$ disjoint with $\mathbf{b}$ so that $\#(\mathbf{r} \cup \mathbf{c}) \in 2\mathbf{Z}$. We can find quaternion algebras B_2 and B_1 over F such that $\mathfrak{r}(B_2) = \mathbf{r} \cup \mathbf{b} \cup \mathbf{c}$ and $\mathfrak{r}(B_1) = \mathbf{r} \cup \mathbf{c}$. Put

(Q.8) $$(V, \varphi) = (B_2, \beta_2) \oplus \langle F, 1 \rangle \oplus (B_1^\circ, \delta\beta_1^\circ).$$

Clearly K is the discriminant algebra. Let $(V_1, \varphi_1) = \langle F, 1 \rangle \oplus (B_1^\circ, \delta\beta_1^\circ)$. For the same reason as in the case $n = 4$, we have $Q(\varphi_1) = B_1$, and hence, by Lemma 2.8 (i), $A(\varphi) \cong A(\beta_2) \otimes A(\varphi_1) \cong M_2(B_2) \otimes M_2(B_1) \cong M_4(B_2 \otimes B_1) \cong M_8(B)$. Thus $Q(\varphi) = B$.

The case $n = 10$. We have $\mathfrak{r}(B) \subset \mathbf{h}$, and $K_v = \mathbf{C}$ for every $v \in \mathbf{r}$. Take B_i as in the case $n = 8$, and put

(Q.10) $$(V, \varphi) = (B_2, \beta_2) \oplus (B_1, \beta_1) \oplus (K, \kappa).$$

This has K as the discriminant field. By Lemma 2.8 (i) we can easily verify that $Q(\varphi) = B$.

5.5. We now consider the odd-dimensional cases.

The case $n = 3$. In this case $\delta \ll 0$ and $\mathbf{r} \subset \mathfrak{r}(B)$. Put

(Q.3) $$(V, \varphi) = (B^\circ, -\delta\beta^\circ).$$

Clearly φ has all the required properties.

The case $n = 5$. In this case $\delta \gg 0$ and $\mathbf{r} \subset \mathfrak{r}(B)$. Put

$$(V, \varphi) = (B, \beta) \oplus \langle F, \delta \rangle. \tag{Q.5}$$

By Lemma 2.8 (ii), $A^+(\varphi) \cong A(-\delta\beta)$, which is isomorphic to $M_2(B)$ as shown in §2.7. Thus $Q(\varphi) = B$. Clearly φ has the remaining required properties.

The case $n = 7$. We have $\mathfrak{r}(B) \subset \mathbf{h}$ and $\delta \ll 0$. Take B_i as in the case $n = 8$, and put

$$(V, \varphi) = (B_2, \beta_2) \oplus (B_1^\circ, -\delta\beta_1^\circ). \tag{Q.7}$$

By Lemma 2.8 (ii), $A^+(\varphi) \cong A^+(-\delta\beta_1^\circ) \otimes A(-\delta\beta_2) \cong B_1 \otimes M_2(B_2) \cong M_4(B)$; thus $Q(\varphi) = B$. Clearly φ has the other required properties.

The case $n = 9$. Take B_i as in the case $n = 8$, and put

$$(V, \varphi) = (B_2, \beta_2) \oplus (B_1, \beta_1) \oplus \langle F, \delta \rangle. \tag{Q.9}$$

Then by the same technique as in the cases $n = 5$ and 7, we can easily verify that this has all the required properties.

6. Quadratic forms over the integers.

6.1. So far we considered quadratic forms over a field. Let us now investigate the problems over the ring of integers. We shall eventually show how $\mathfrak{r}(Q(\varphi))$ and δ can be connected to $\det(2\Phi)$ for the matrix Φ representing φ. We fix an algebraic number field F as before, and let $\mathfrak{g}$ denote the ring of all algebraic integers in F and $\mathfrak{d}$ the different of K relative to F; we put $\mathfrak{d}_0 = \mathfrak{d}^2 \cap F$. If $K = F \times F$, then we put $\mathfrak{d} = \mathfrak{g} \times \mathfrak{g}$ and $\mathfrak{d}_0 = \mathfrak{g}$. Given a quadratic space (V, φ) over F, by a $\mathfrak{g}$-*lattice* in V, we mean a finitely generated $\mathfrak{g}$-module in V that spans V over F. Given two $\mathfrak{g}$-lattices L and M in V, we denote by $[L/M]$ the fractional ideal generated by $\det(\alpha)$ for all $\alpha \in GL(V, F)$ such that $L\alpha \subset M$. Thus $[L/L\alpha] = \det(\alpha)\mathfrak{g}$. We denote by $\varphi\{L\}$ the fractional ideal generated by $\varphi[x]$ for all $x \in L$; this was written $\mu(L)$ in [S3], and is different from the set $\{\varphi[x] \mid x \in L\}$.

We call a $\mathfrak{g}$-lattice L in V *integral* if $\varphi[x] \in \mathfrak{g}$ for every $x \in L$. We call an integral lattice L *maximal* if L is the only integral lattice containing L. For $v \in \mathbf{h}$ let $\mathfrak{g}_v$ denote the closure of $\mathfrak{g}$ in F_v. Then we can define the notion of $\mathfrak{g}_v$-lattice in V_v in a similar way; the integrality and maximality can be defined as well.

Fixing now a quadratic space (V, φ) over F, for a maximal lattice L in V, put

$$\widetilde{L} = \{x \in V \mid 2\varphi(x, L) \subset \mathfrak{g}\}. \tag{6.1}$$

Let φ_v denote the matrix representing φ with respect to a $\mathfrak{g}_v$-basis of L_v. If we identify V_v with the set of all n-dimensional row vectors with components in F_v

with respect to that basis, then $\widetilde{L}_v = L_v(2\varphi_v)^{-1}$. Clearly $[\widetilde{L}/L] = \prod_{v\in\mathbf{h}}[\widetilde{L}_v/L_v]$ and $\varphi\{\widetilde{L}\} = \prod_{v\in\mathbf{h}}\varphi_v\{\widetilde{L}_v\}$, because $\varphi_v\{\widetilde{L}_v\} = \varphi\{\widetilde{L}\}_v$ as shown in [S3, Lemma 9.4 (iii)]. These $[\widetilde{L}/L]$ and $\varphi\{\widetilde{L}\}$ do not depend on the choice of L, as all the maximal lattices in V form one genus; see [S3, Lemma 6.9, §9.7]. We call $[\widetilde{L}/L]$ the *discriminant ideal* of φ and $\varphi\{\widetilde{L}\}^{-1}$ the *level ideal* of φ. These are natural invariants of (V, φ).

The determination of $[\widetilde{L}_v/L_v]$ and $\varphi\{\widetilde{L}_v\}$ can be reduced to the anisotropic case as follows. Let (Z, ζ) be a core subspace of (V_v, φ_v), and M a maximal $\mathfrak{g}_v$-lattice in Z; let $\widetilde{M} = \{z \in Z \mid 2\zeta(z, M) \subset \mathfrak{g}_v\}$. We have $V_v = H_r \oplus Z$ with a split space H_r, and so can take $L_v = S \oplus M$ with a maximal lattice S in H_r, as shown in [S3, Lemma 6.3]. Since $\widetilde{S} = S$, we have $[\widetilde{L}_v/L_v] = [\widetilde{M}/M]$ and $\varphi\{\widetilde{L}_v\} = \zeta\{\widetilde{M}\}$, the latter being justified only after showing that $\mathfrak{g}_v \subset \zeta\{\widetilde{M}\}$.

THEOREM 6.2. *Given (V, φ) over F, let $K_0 = F(\delta^{1/2})$ be the discriminant field of (V, φ), and L a maximal lattice in V. Let $\mathfrak{e}$ be the product of the prime ideals ramified in $Q(\varphi)$ and $\mathfrak{e}_1$ the product of the prime factors of $\mathfrak{e}$ that do not ramify in K_0. Then the following assertions hold:*

(i) *$\varphi\{L\} = \mathfrak{g}$ if $n > 2$.*

(ii) *If n is even, then $[\widetilde{L}/L] = \mathfrak{d}_0\mathfrak{e}_1^2$ and $\varphi\{\widetilde{L}\}^{-1} = \mathfrak{d}_0\mathfrak{e}_1$.*

(iii) *Suppose n is odd; let $\delta\mathfrak{g} = \mathfrak{a}\mathfrak{b}^2$ with a squarefree integral ideal $\mathfrak{a}$ and a fractional ideal $\mathfrak{b}$. Then $[\widetilde{L}/L] = 2\mathfrak{a}^{-1}\mathfrak{e}^2 \cap 2\mathfrak{a}$ and $\varphi\{\widetilde{L}\}^{-1} = 4\mathfrak{a} \cap \mathfrak{e}$.*

Proof. For each $v \in \mathbf{h}$ let t_v denote the core dimension of (V_v, φ_v), and $\mathfrak{p}_v$ the maximal ideal of $\mathfrak{g}_v$. Take (Z, ζ), M and $\widetilde{M}$ as in §6.1. Our task is to find $\varphi\{L\}_v, [\widetilde{M}/M]$, and $\zeta\{\widetilde{M}\}$. The case $t_v = 0$ is trivial. If $t_v = 2$, we have $(Z, \zeta) = (K_v, c_v\kappa)$ with $c_v \in F_v^\times$; we can take $c_v \in \mathfrak{g}_v^\times$ if K_v is ramified over F_v; otherwise c_v is a prime element or a unit. Clearly $\varphi\{L\}_v = \mathfrak{g}_v$ if $n > 2$. Let $\mathfrak{r}$ be the maximal order of K; then $M = \mathfrak{r}_v$, and so $\widetilde{M} = c_v^{-1}\mathfrak{d}_v^{-1}$. Since $Q(\varphi)_v = \{K_v, c_v\}$ as observed in §§2.6 and 3.2, we have $c_v\mathfrak{g}_v = \mathfrak{p}_v$ if and only if $v|\mathfrak{e}_1$. Then we easily find $[\widetilde{M}/M]$ and $\zeta\{\widetilde{M}\}$ when $t_v = 2$. Suppose $t_v = 4$; then $(Z, \zeta) = (B, \beta)$ with a division quaternion algebra B over F_v; $Q(\varphi)_v = B$ and $K_v = F_v \times F_v$; M is the unique maximal order in B. Then $\zeta\{M\} = \mathfrak{g}_v$ and we easily find $[\widetilde{M}/M]$, as we already did in [S3, §7.7]. The ideal $\zeta\{\widetilde{M}\}$ can also be determined. Combining these, we obtain our formulas for even n.

To study the case $t_v = 3$, let us recall a few basic facts on a division quaternion algebra B over F_v. For simplicity we fix $v \in \mathbf{h}$ and write simply F and $\mathfrak{g}$ for F_v and $\mathfrak{g}_v$; we denote by $\mathfrak{p}$ the maximal ideal of $\mathfrak{g}$. Now B can be given as $B = J + J\omega$ with an unramified quadratic extension J of F and an element ω such that ω^2 is a prime element of F and $a\omega = \omega a^\sigma$ for every $a \in J$, where σ is the nontrivial automorphism of J over F. Let $\mathfrak{r}$ be the maximal order of J and let $\mathfrak{o} = \mathfrak{r} + \mathfrak{r}\omega$. Then $\mathfrak{o}$ is the unique maximal order of B. Take u so that $\mathfrak{r} = \mathfrak{g} + \mathfrak{g}u$ and put $y = u - u^\sigma$; then $y \in \mathfrak{r}^\times$, as J is unramified over F. Let $(Z, \zeta) = (B^\circ, -\delta\beta^\circ)$ with $\delta \in F^\times$. Clearly $B^\circ = Fy + J\omega$. Let M be a maximal lattice in Z. Then $M = \{x \in B^\circ \mid \delta x x^\iota \in \mathfrak{g}\}$ by a general principle; see [S3, Lemma 6.4]. We have

$\widetilde{M} = \{z \in B^\circ \mid \delta\mathrm{Tr}_{B/F}(zM) \subset \mathfrak{g}\}$. Our task is to find $[\widetilde{M}/M]$, $\zeta\{M\}$, and $\zeta\{\widetilde{M}\}$. We can reduce the problem to the case where $\delta\mathfrak{g}$ is $\mathfrak{g}$ or $\mathfrak{p}$. If $\delta\mathfrak{g} = \mathfrak{g}$, then $M = \mathfrak{g}y + \mathfrak{r}\omega$, $\zeta\{M\} = \mathfrak{g}$, and $\widetilde{M} = 2^{-1}\mathfrak{g}y + \mathfrak{r}\omega^{-1}$. If $\delta\mathfrak{g} = \mathfrak{p}$, then $M = \mathfrak{g}y + \mathfrak{r}\omega^{-1}$, $\zeta\{M\} = \mathfrak{g}$, and $\widetilde{M} = (2\mathfrak{p})^{-1}y + \mathfrak{r}\omega^{-1}$. Thus $[\widetilde{M}/M] = 2\delta^{-1}\mathfrak{p}^2$ and $\zeta\{\widetilde{M}\}^{-1} = 4\delta\mathfrak{g} \cap \mathfrak{p}$ in both cases.

The case $t_v = 1$ is much easier. We can easily verify that $[\widetilde{M}/M] = 2\delta\mathfrak{g}$ and $\zeta\{\widetilde{M}\}^{-1} = 4\delta\mathfrak{g}$ under the assumption that $\delta\mathfrak{g}$ equals $\mathfrak{g}$ or $\mathfrak{p}$. Combining these results for all $v \in \mathbf{h}$, we obtain our formulas when n is odd. □

6.3. By Theorem 4.4 we can enumerate all the isomorphism classes of (V, φ) with given n, K, B, and $\{\sigma_v\}$. We can express the results in connection with $[\widetilde{L}/L]$ and $\varphi\{\widetilde{L}\}$ by virtue of Theorem 6.2. For simplicity let us discuss here only when $F = \mathbf{Q}$, in which case we can give a list according to the nature of a symmetric matrix representing φ with respect to a $\mathbf{Z}$-basis of L. We call a symmetric element $T = (t_{ij})$ of $GL_n(\mathbf{Q})$ *semi-integral* if $2T \in M_n(\mathbf{Z})$ and $t_{ii} \in \mathbf{Z}$ for every i; also we denote by $s(T)$ and $Q(T)$ the index and the characteristic algebra of a quadratic form represented by T. We let $\mathfrak{S}_n$ denote the set of all symmetric elements of $GL_n(\mathbf{Q})$ that are semi-integral. We call an element T of $\mathfrak{S}_n$ *reduced* if the following condition is satisfied:

$$(6.2) \qquad P \in M_n(\mathbf{Z}) \cap GL_n(\mathbf{Q}) \text{ and } P^{-1}T \cdot {}^tP^{-1} \in \mathfrak{S}_n \implies \det(P) = \pm 1.$$

Let N be the smallest positive integer such that $N(4T)^{-1}$ is semi-integral. We put $N = \ell(T)$ and call it *the level of* T. We let $\mathfrak{S}_n^0$ denote the subset of $\mathfrak{S}_n$ consisting of the reduced elements of $\mathfrak{S}_n$. Clearly if $T \in \mathfrak{S}_n$ and $\det(2T)$ is squarefree, then $T \in \mathfrak{S}_n^0$.

Given (V, φ) over $\mathbf{Q}$, take a maximal lattice L in V and let Φ be the matrix that represents φ with respect to a $\mathbf{Z}$-basis of L. Then $\det(2\Phi)\mathbf{Z} = [\widetilde{L}/L]$, and it can easily be verified that $\Phi \in \mathfrak{S}_n^0$. It is also easy to see that conversely, every $\Phi \in \mathfrak{S}_n^0$ can be obtained in this fashion. Since all the maximal lattices form one genus, we see that the genus of Φ in the classical sense is determined by the isomorphism class of (V, φ), and vice versa. (As for the relationship between the genus of matrices and the genus of lattices, the reader is referred to [S3, §9.27].) Also, we can easily verify that $\varphi\{\widetilde{L}\}^{-1} = \ell(\Phi)\mathbf{Z}$. If $n > 2$, then $\varphi\{L\} = \mathbf{Z}$ by Theorem 6.2 (i), and hence the greatest common divisor of $2\varphi_{ij}$ and φ_{ii} for all i,j is 1.

Now the classification of all Φ in $\mathfrak{S}_n^0$ can be stated as follows:

Theorem 6.4 (The case of even n). *Let three integers n, σ, δ, and two squarefree positive integers e_0, e_1 be given as follows: $4 \le n \in 2\mathbf{Z}$, $|\sigma| \le n, \sigma \in 2\mathbf{Z}$; $\delta = 1$ or δ is the discriminant of a quadratic extension K_0 of $\mathbf{Q}$; e_0 divides δ and e_1 is prime to δ. Let r be the number of prime factors of e_0e_1. Suppose that $(-1)^{\sigma/2}\delta > 0$*

and $\sigma - 4r \equiv 0$ *or* $2 \pmod 8$. *Then there exists an element* Φ *of* $\mathfrak{S}_n^0$ *such that*

$$|\det(2\Phi)| = |\delta|e_1^2, \quad s(\Phi) = \sigma, \quad \ell(\Phi) = |\delta|e_1,$$

and $Q(\Phi)$ *is ramified at a prime number* p *if and only if* $p|e_0e_1$. *Moreover, every element of* $\mathfrak{S}_n^0$ *is of this type, and its genus is determined by* $(\sigma, \delta, e_0, e_1)$. *These statements are true even for* $n = 2$ *under the following additional condition on* e_1: *if* $e_1 \neq 1$, *then* $\delta \neq 1$ *and every prime factor of* e_1 *remains prime in* K_0.

Proof. Take $F = \mathbf{Q}$ in Theorems 4.4 and 6.2; put $\mathfrak{e} = e\mathbf{Z}$ with $e = e_0e_1$. From (4.2b) we see that $\sigma - 4r \equiv 0$ or $2 \pmod 8$. Since B is determined by e, and (V, φ) by (n, σ, δ, B), we obtain our theorem. □

THEOREM 6.5 (The case of odd n). *Let two integers* n *and* σ *be given as follows:* $0 < n - 1 \in 2\mathbf{Z}, |\sigma| \leq n$, *and* $\sigma - 1 \in 2\mathbf{Z}$; *also let* m *and* e *be squarefree positive integers; let* r *be the number of prime factors of* e. *Suppose that* $\sigma \equiv 4r \pm 1 \pmod 8$. *Then there exists an element* Φ *of* $\mathfrak{S}_n^0$ *such that*

$$|\det(2\Phi)| = 2(m \cap e)^2/m, \quad s(\Phi) = \sigma, \quad \ell(\Phi) = 4m \cap e,$$

and $Q(\Phi)$ *is ramified at a prime number* p *if and only if* $p|e$, *where* $x \cap y$ *denotes the least common multiple of* x *and* y. *Moreover, every element of* $\mathfrak{S}_n^0$ *is of this type, and its genus is determined by* (σ, m, e).

Proof. Since $F = \mathbf{Q}$, (4.2a) holds if and only if $\sigma \equiv 4r \pm 1 \pmod 8$. Given m and e, take $\delta = (-1)^{(\sigma-1)/2}m$ and $\mathfrak{e} = e\mathbf{Z}$ in Theorems 4.4 and 6.2 (iii). Then $\mathfrak{a} = m\mathbf{Z}$, and we immediately obtain the desired results. □

As the easiest special case we can consider $\Phi \in \mathfrak{S}_n$ such that $|\det(2\Phi)| = 1$. From the above theorems we immediately obtain the classical result that such a Φ exists if and only if $s(\Phi) \in 8\mathbf{Z}$; also, Φ determines a unique genus for each such n.

Let us now state more general results derived from the above two theorems.

THEOREM 6.6. *In the following statements* σ *is an integer such that* $|\sigma| \leq n$ *and* $\sigma - n \in 2\mathbf{Z}$.

(i) *Let* f *and* g *be squarefree positive integers whose greatest common divisor is 2 if both* n *and* f *are even, and 1 otherwise. Let* $\Phi \in \mathfrak{S}_n^0$. *Then* $|\det(2\Phi)| = fg^2$ *with such* f *and* g *if* n *is even, and* $|\det(2\Phi)| = 2fg^2$ *with such* f *and* g *if* n *is odd.*

(ii) *There is no* $\Phi \in \mathfrak{S}_n$ *for even* n *such that* $2^{-1}\det(2\Phi)$ *is an odd integer.*

(iii) *Let* g *be the product of* $t(\geq 0)$ *different prime numbers and let* $2 < n \in 2\mathbf{Z}$. *Then* $\mathfrak{S}_n^0$ *contains an element* Φ *such that* $|\det(2\Phi)| = g^2$ *and* $s(\Phi) = \sigma$ *if and only if* $\sigma - g \in 4\mathbf{Z}$ *or* $\sigma - 4t \in 8\mathbf{Z}$, *and the genus of such a* Φ *is determined by* (σ, g).

(iv) *For* g *as in* (iii) *and* $0 < n - 1 \in 2\mathbf{Z}$ *the set* $\mathfrak{S}_n^0$ *contains an element* Φ *such that* $|\det(2\Phi)| = 2g^2$ *and* $s(\Phi) = \sigma$ *if and only if* $\sigma \equiv 4t \pm 1 \pmod 8$, *and the genus of such a* Φ *is determined by* (σ, g).

(v) *Let f and g be as in* (i) *and let s be the number of prime factors of f; suppose that $s > 0$ and $2 < n \in 2\mathbf{Z}$. Then the number of genera of $\Phi \in \mathfrak{S}_n^0$ such that $|\det(2\Phi)| = fg^2$ and $s(\Phi) = \sigma$ is 2^s iff $-\sigma + 1 \in 4\mathbf{Z}$ and 2^{s-1} otherwise. Moreover, $g \in 2\mathbf{Z}$ iff $-\sigma + 1 \in 4\mathbf{Z}$.*

(vi) *Let f, g, and s be as in* (v). *Then for $1 < n - 1 \in 2\mathbf{Z}$ the number of genera of $\Phi \in \mathfrak{S}_n^0$ such that $|\det(2\Phi)| = 2fg^2$ and $s(\Phi) = \sigma$ is 2^{s-1}.*

(vii) *There exists $\Phi \in \mathfrak{S}_n$ such that $|\det(2\Phi)| = 2$ and $s(\Phi) = \sigma$ if and only if $\sigma \equiv \pm 1 \pmod 8$, and such a Φ determines a unique genus for each (n, σ).*

Proof. We can derive (i) from Theorems 6.4 and 6.5 combined with the well known facts on the discriminant of a quadratic field; (ii) follows easily from (i). Suppose $2 < n \in 2\mathbf{Z}$ and $|\det(2\Phi)| = g^2$ with g as in (iii). Then in Theorem 6.4 we have $(\delta, e_1) = (1, g)$ or $(-4, g/2)$, which is so according as $\sigma \in 4\mathbf{Z}$ or $\sigma - 2 \in 4\mathbf{Z}$. If $\sigma \in 4\mathbf{Z}$, then $t = r$ and so $\sigma - 4t \in 8\mathbf{Z}$; if $\sigma - 2 \in 4\mathbf{Z}$, then e_1 is odd, and so $\sigma - g \in 4\mathbf{Z}$. In the latter case we take $e = e_1$ or $2e_1$ according to the parity of t and $\sigma \pmod 8$. Thus we obtain (iii). The proof of (iv) is similar and simpler.

As for (v), we have $|\delta| = f$ or $4f$, and (σ, δ) determines (σ, f) and vice versa. Now δ has exactly $s + 1$ prime factors if and only if f is odd and $|\delta| = 4f$, which can happen if and only if $f - \sigma + 1 \in 4\mathbf{Z}$. Otherwise δ has exactly s prime factors. Once (σ, δ, e_1) is fixed, the genus of Φ is determined by a factor e_0 of δ, and the number of the prime factors of e_0 must have the parity determined by r and e_1. Since σ determines $r \pmod 2$, we obtain (v). The proof of (vi) is similar and simpler. Finally, (vii) is trivially true if $n = 1$. For $n > 1$, it follows immediately from (ii) and (iv).

Remark. (A) If $n = 2$, we can easily classify all $\Phi \in \mathfrak{S}_2$ without assuming Φ to be reduced. In the higher-dimensional case, however, no clear-cut results seem obtainable without the assumption that Φ is reduced, and therefore we believe that our restriction of Φ to $\mathfrak{S}_n^0$ is more than natural.

(B) When $F \neq \mathbf{Q}$, analogues of the statements of Theorem 6.6 are not so simple. To illustrate this point, let us take a real quadratic field of class number 1 as F. We consider $\Phi = (\varphi_{ij}) \in GL_n(F)$ such that $\varphi_{ii} \in \mathfrak{g}$ and $2\varphi_{ij} \in \mathfrak{g}$ for every i and j. We assume that n is even and look for such a Φ under the condition $\det(2\Phi) \in \mathfrak{g}^\times$. Let $\mathbf{a} = \{v_1, v_2\}$ and $\sigma_i = s_{v_i}(\Phi)$; put also $B = Q(\Phi)$. We have $\mathfrak{d}_0 = \mathfrak{e} = \mathfrak{g}$ in Theorem 6.2, and so $\mathfrak{r}(B) = \mathbf{a}$ or $\mathfrak{r}(B) = \emptyset$. Let ε be a fundamental unit of F.

First suppose that $N_{F/\mathbf{Q}}(\varepsilon) = -1$. Then $K_0 = F$ and $\sigma_i \in 4\mathbf{Z}$. By (4.2b) we have $\sigma_1 \equiv \sigma_2 \equiv 0 \pmod 8$ if $\mathfrak{r}(B) = \emptyset$, and $\sigma_1 \equiv \sigma_2 \equiv 4 \pmod 8$ if $\mathfrak{r}(B) = \mathbf{a}$. Thus there is exactly one genus of Φ for each choice of (n, σ_i).

Next suppose that $N_{F/\mathbf{Q}}(\varepsilon) = 1$. Then there are exactly two choices for $K_0 : K_0$ can be F or a quadratic extension of F ramified exactly at v_1 and v_2. If $K_0 = F$, then we obtain the same result as in the case $N_{F/\mathbf{Q}}(\varepsilon) = -1$. Suppose $K_0 \neq F$; then $\sigma_i - 2 \in 4\mathbf{Z}$. By (4.2b) we have $\sigma_1 \equiv \sigma_2 \equiv 2 \pmod 8$ if $\mathfrak{r}(B) = \emptyset$, and

$\sigma_1 \equiv \sigma_2 \equiv 6 \pmod 8$ if $\mathfrak{r}(B) = \mathbf{a}$. There is exactly one genus of Φ for each choice of (n, K_0, σ_i).

6.7. Let us now explain the meaning of $\ell(\Phi)$ in terms of the theta series associated with Φ, which is given by

$$\theta(z, \Phi) = \sum_g \exp(2\pi i z \cdot g\Phi \cdot {}^t g). \tag{6.3}$$

Here Φ is a positive definite element of $\mathfrak{S}_n$, g runs over the n-dimensional row vectors with components in $\mathbf{Z}$, and z is a variable on the half plane $H = \{z \in \mathbf{C} \mid \mathrm{Im}(z) > 0\}$. This is a modular form of weight $n/2$ and of level $\ell(\Phi)$, which is why we call $\ell(\Phi)$ the level of Φ. To be precise, put $N = \ell(\Phi)$. Then for every $\gamma = \begin{bmatrix} a & b \\ c & d \end{bmatrix} \in SL_2(\mathbf{Z})$ such that $c \in N\mathbf{Z}$ we have

$$\theta(\gamma(z), \Phi) = \left(\frac{\Delta}{d}\right)(cz + d)^{n/2}\theta(z, \Phi) \qquad \text{if } n \in 2\mathbf{Z}, \tag{6.4a}$$

$$\theta(\gamma(z), \Phi) = \left(\frac{\Delta}{d}\right) h_\gamma(z)(cz + d)^{(n-1)/2}\theta(z, \Phi) \qquad \text{if } n \notin 2\mathbf{Z}. \tag{6.4b}$$

Here $\gamma(z) = (az + b)(cz + d)^{-1}$, $\Delta = (-1)^{n/2}\det(2\Phi)$ if $n \in 2\mathbf{Z}$, and $\Delta = (-1)^{(n-1)/2}2\det(2\Phi)$ if $n \notin 2\mathbf{Z}$; $h_\gamma(z)$ is the standard factor of automorphy of weight $1/2$ which was written $j(\gamma, z)$ in [S1]. Clearly Δ can be replaced by the discriminant of the discriminant field of the quadratic form represented by Φ.

To obtain (6.4a,b), we consider the theta series $\theta(z; h, A, N, P)$ of [S1, (2.0)] with $h = 0, A = 2\Phi$, and $P = 1$. Then it gives our series of (6.3), and so we obtain (6.4a,b) as special cases of Proposition 2.1 of [S1].

We can similarly associate a theta series in the Hilbert modular case with an integral lattice L in (V, φ) when F is totally real and φ is totally positive. Then we can derive transformation formulas similar to (6.4a,b) from [S2, Theorem 3.5, Proposition 3b.2]. The ideal $\varphi\{\widetilde{L}\}^{-1}$ is the level of the series in the same sense.

7. Canonical forms of quadratic spaces and Clifford groups for dimension ≤ 6. In this section we denote by F an arbitrary field of characteristic different from 2, and use the symbols n, (V, φ), K_0, K, z, ρ, and δ in the sense of §2.2. Our principal aim is to find a "canonical way" of constructing a quadratic space for $n = 4, 5$, or 6, under a mild condition in each case that is always satisfied when F is an algebraic number field or its completion at an archimedean or a nonarchimedean prime. We begin with the case of an arbitrary n.

PROPOSITION 7.1. *Suppose n is even; let ζ be an element of $A^-(V)$ such that $\zeta^2 \in F^\times$ and $\zeta^* = \pm\zeta$. Put $\zeta^2 = h, x^\sigma = \zeta x \zeta^{-1}$ for $x \in A^+(\varphi)$, and $D = \{x \in$*

$A^+(\varphi) \mid x^\sigma = x\}$. *Then the following assertions hold:*

(i) *The map σ gives an automorphism of $A^+(\varphi)$ of order 2 and $\sigma = \rho$ on K. Moreovere, D is a central simple algebra over F stable under the canonical involution $x \mapsto x^*$, and $A^+(\varphi) = D \otimes_F K$.*

(ii) *$A(\varphi) \cong D \otimes_F \{K, h\}$; in particular, $A(\varphi) \cong M_2(D)$ if $h \in N_{K/F}(K^\times)$.*

(iii) *We can take any element e of V such that $\varphi[e] \neq 0$ as ζ; then $h = \varphi[e]$.*

(iv) *For every $c \in F^\times$ the algebra $A(c\varphi)$ belongs to the same Brauer algebra class as $A(\varphi) \otimes_F \{K, c\}$. Consequently $A(c\varphi) \cong A(d\varphi)$ for $c, d \in F^\times$ if and only if $c/d \in N_{K/F}(K^\times)$.*

Proof. We have $zu = -uz$ for every $u \in V$ as noted in (2.1). Thus $zyz^{-1} = -y$ for every $y \in A^-(\varphi)$, so that $z\zeta = -\zeta z$. Then we can easily verify the first part of (i) and $A^+(\varphi) = D \otimes_F K$. Since $A^+(\varphi)$ is central simple over K or the direct sum of two central simple algebras over F, we see that D is central simple over F. Since $\zeta^* = \pm\zeta$ and $\zeta^2 \in F$, we have $(x^\sigma)^* = (x^*)^\sigma$, and so $D^* = D$. Now $A(\varphi) = A^+(\varphi) + A^+(\varphi)\zeta$. Clearly $K + K\zeta$ form a quaternion algebra $\{K, h\}$, which commutes elementwise with D. Thus we obtain (ii). Assertion (iii) is obvious. To prove (iv), pick e as in (iii) and put $D_1 = \{x \in A^+(V, c\varphi) \mid xe = ex\}$. Let f be as in the proof of Lemma 2.5. Then clearly $f(D_1) = D$. By (ii), $A(c\varphi) \cong D_1 \otimes_F \{K, ch\}$. Therefore we obtain (iv). $\square$

Let us now prove (ii) of Lemma 2.3. With D as above, we have $\{K, h\} \otimes_F K \cong M_2(K)$, and so $A(\varphi) \otimes_F K \cong M_2(D \otimes_F K) = M_2(A^+(\varphi))$, and the isomorphism maps $1 \otimes k$ to $k1_2$ for $k \in K$. Thus we obtain the desired fact.

Remark. If $K_0 = F$, then the isomorphism class of D is determined by (V, φ), as can be seen from (ii). If $K_0 \neq F$, however, (V, φ) alone cannot determine D; once a number h of $F^\times \cap \varphi[V]$ is fixed, then D is determined by (V, φ) and the coset $h \cdot N_{K/F}(K^\times)$. In particular, if $N_{K/F}(K^\times) \cap \varphi[V] \neq \emptyset$, then taking $\zeta = e \in V$ so that $\varphi[e] \in N_{K/F}(K^\times)$, we obtain $A(\varphi) \cong M_2(D)$. If F is an algebraic number field and $n \in 2\mathbf{Z}$, then $A(\varphi) \cong M_s(B)$ with $s = 2^{(n-2)/2}$ and a quaternion algebra B over F, and so $A^+(\varphi) \cong M_{s/2}(B \otimes_F K)$ by Lemma 2.3 (ii). Therefore whether or not $N_{K/F}(K^\times) \cap \varphi[V] \neq \emptyset$ may be a problem of minor interest for such an F. Still, an explicit construction of D by means of ζ as above is significant, as will be explained by Proposition 7.3 below. For that reason we note:

LEMMA 7.2. *Suppose F is an algebraic number field and $4 \leq n \in 2\mathbf{Z}$. Then $N_{K/F}(K^\times) \cap (c\varphi)[V] \neq \emptyset$ for some $c \in F^\times$. We can take $c = 1$ except when $n - 2 \in 4\mathbf{Z}$ and there exists a real archimedean prime of F for which φ is negative definite.*

Proof. We are going to find an element g of $K^\times$ such that $N_{K/F}(g) \in \varphi[V_v]$ for every real $v \in \mathbf{a}$. Such a g does not exist if there is a real $v \in \mathbf{a}$ such that φ is negative definite at v and $K_v = \mathbf{C}$, which can happen only when $n - 2 \in 4\mathbf{Z}$.

If that does not occur, then we can find a desired element g of K, since K is dense in $\prod_{v\in\mathbf{a}} K_v$. Now $\varphi[V_v] = F_v$ for every $v \in \mathbf{h}$, since $n \geq 4$. Therefore, by the Hasse principle, $N_{K/F}(g) \in \varphi[V]$ as expected, provided $n \in 4\mathbf{Z}$ or there is no real $v \in \mathbf{a}$ for which φ is negative definite. In the general case that condition is satisfied by $c\varphi$ for a suitable choice of $c \in F^\times$. Therefore we obtain our Lemma. □

We now consider the homomorphsim τ: $G^+(\varphi) \to SO(\varphi)$ defined by $x\tau(\alpha) = \alpha^{-1}x\alpha$ for $x \in V$ and $\alpha \in G^+(\varphi)$. Since the space V is contained in $A^-(\varphi)$ and not in $A^+(\varphi)$, in general this action of $G^+(\varphi)$ on V is not transparant. However, we can embed V into $A^+(\varphi)$ and describe the action of $G^+(\varphi)$ on V within the algebra $A^+(\varphi)$ as follows.

PROPOSITION 7.3. *For both even and odd n let ζ be an element of $A^-(V)$ such that $\zeta^2 \in F^\times$ and $\zeta^* = \pm\zeta$. Put $\alpha^\sigma = \zeta\alpha\zeta^{-1}$ for $\alpha \in A^+(\varphi)$, $Y = V\zeta$, and $\psi[y] = yy^*$ for $y \in Y$. Then the following assertions hold:*

(i) *$Y \subset A^+(\varphi)$, ψ is a quadratic form on Y, and the map $x \mapsto x\zeta$ gives an isomorphism of $(V, \zeta\zeta^*\varphi)$ onto (Y, ψ).*

(ii) *For every $\alpha \in G^+(\varphi)$ the set Y is stable under the map $y \mapsto \alpha^{-1}y\alpha^\sigma$, and moreover, $(x\tau(\alpha))\zeta = \alpha^{-1}x\zeta\alpha^\sigma$ for every $x \in V$. Thus the action of α on V can be transferred to the map $y \mapsto \alpha^{-1}y\alpha^\sigma$ on Y.*

(iii) *If n is odd and $\zeta = z$, then $\xi^\sigma = \xi$ for every $\xi \in A^+(\varphi)$ and $y^* = (-1)^{(n-1)/2}y$ for every $y \in Y$.*

(iv) *If n is even and $\zeta \in V$, then $y^\sigma = y^*$ for every $y \in Y$.*

All these can be verified in a straightforward way.

THEOREM 7.4. *Suppose $n = 4$; let $h = \varphi[e] \neq 0$ with $e \in V$; define an automorphism σ of $A^+(\varphi)$ by $x^\sigma = exe^{-1}$ for $x \in A^+(\varphi)$ and put $B = \{x \in A^+(\varphi) \mid x^\sigma = x\}$. Then the following assertions hold:*

(i) *B is a quaternion algebra over F, $A(\varphi) \cong B \otimes_F \{K, h\}$ and $A^+(\varphi) = B \otimes_F K$; in particular, $A(\varphi) \cong M_2(B)$ if and only if $h \in N_{K/F}(K^\times)$.*

(ii) *The canonical involution $x \mapsto x^*$ gives the main involution of B, and $\alpha\alpha^* \in K$ for every $\alpha \in A^+(\varphi)$.*

(iii) *$G^+(\varphi) = \{\alpha \in A^+(\varphi)^\times \mid \alpha\alpha^* \in F^\times\}$.*

(iv) *Put $Y = Ve$ and $\psi[y] = yy^*$ for $y \in Y$. Then ψ is a quadratic form on Y, and the map $x \mapsto xe$ for $x \in V$ gives an isomorphism of $(V, h\varphi)$ onto (Y, ψ). Moreover, $(x\tau(\alpha))e = \alpha^{-1}xe\alpha^\sigma$ for every $x \in V$,*

$$Y = \{y \in A^+(\varphi) \mid y^\sigma = y^*\} = F \oplus zB^\circ, \tag{7.1}$$

and the decomposition $Y = F \oplus zB^\circ$ gives

$$(Y, \psi) = \langle F, 1\rangle \oplus (B^\circ, \delta\beta^\circ), \tag{7.2}$$

where (B°, β°) *is as in* §*1.2.*

(v) *For* $c, d \in F^\times$ *we have* $(V, c\varphi) \cong (V, d\varphi)$ *if and only if* $c/d \in \mathcal{W}$, *where* $\mathcal{W} = N(B^\times)$ *if* $K_0 = F$, *and* $\mathcal{W} = N_{K/F}(F^\times N(B_K^\times))$ *otherwise, where* $N(\xi) = \xi\xi^*$ *for* $\xi \in B_K = B \otimes_F K$.

Proof. Take $\zeta = e$ and define σ and D as in Proposition 7.1; write B for D. Then we obtain (i) as special cases of Proposition 7.1 (i) and (ii); clearly B must be a quaternion algebra over F; also B is stable under $*$. Therefore we can find an element d of $B^\times$ such that $x^* = dx^\iota d^{-1}$ for every $x \in B$ and $d^\iota = \pm d$. Suppose $d^\iota = -d$. Then we have $\{x \in B \mid x^* = x\} = dB^\circ$, which combined with (i) shows that $\{x \in A^+(\varphi) \mid x^* = x\}$ has dimension 6 over F, which is a contradiction, as the last set equals K. Thus $d^\iota = d$, and so $d \in F^\times$, and $x^* = x^\iota$ for $x \in B$. Since $z^* = z$, we obtain (ii). Now taking an F-basis of $A(\varphi)$ consisting of monomials of the members of an orthogonal basis of V, we easily see that

$$V = \{x \in A^-(\varphi) \mid x^* = x\} \quad \text{if} \quad n = 4. \tag{7.3}$$

If $\alpha \in A^+(\varphi)$ and $\alpha\alpha^* \in F^\times$, then for $x \in V$ we see that $\alpha^* x\alpha$ belongs to the right-hand side of (7.3), and hence $\alpha \in G^+(\varphi)$. From this we obtain (iii). With (Y, ψ) as in (iv), we can easily verify (7.1) and (7.2). The first part of (iv) is a special case of Proposition 7.3 (i). Assertion (v) will be proven at the end of §8.5. □

The above theorem can be summarized as follows: For every (V, φ) with $n = 4$, $G^+(\varphi)$ can be identified with $\mathcal{G} = \{\alpha \in B_K^\times \mid \alpha\alpha^* \in F^\times\}$, and the map $\tau\colon G^+(\varphi) \to SO(\varphi)$ can be transferred to the map $\tau'\colon \mathcal{G} \to SO(Y, \psi)$ with a space Y of hermitian elements y in the sense $y^{\sigma *} = y$, and $y\tau'(\alpha) = \alpha^{-1} y \alpha^\sigma$.

In general the quaternion algebra B depends on the choice of e as already remarked after the proof of Proposition 7.1. If F is an algebraic number field, however, B can be determined, and so (V, φ) can be given canonically as follows.

COROLLARY 7.5. (i) *Suppose* $n = 4$ *and* $N_{K/F}(K^\times) \cap \varphi[V] \neq \emptyset$. *Then* $A(\varphi) \cong M_2(B)$ *and* $A^+(\varphi) \cong B \otimes K$ *with a quaternion algebra* B *over* F, *and*

$$(V, h\varphi) \cong \langle F, 1\rangle \oplus (B^\circ, \delta\beta^\circ) \tag{7.4}$$

for every $h \in N_{K/F}(K^\times) \cap \varphi[V]$. *In particular,* $N_{K/F}(K^\times) \cap \varphi[V] \neq \emptyset$ *if* F *is an algebraic number field.*

(ii) *Conversely, given a quaternion algebra* B *over* F, *a field* $K_0 = F(\delta^{1/2})$ *with* $\delta \in F^\times$, *and an arbitrary* $h \in F^\times$, *define* (V, φ) *so that (7.4) holds; define also* K *by (1.2). Then* K_0 *is the discriminant field of* (V, φ), $A(\varphi) \cong B \otimes_F \{K, h\}$, *and* $A^+(\varphi) \cong B \otimes_F K$; *in particular,* $A(\varphi) \cong M_2(B)$ *if* $h \in N_{K/F}(K^\times)$.

Proof. Assertion (i) follows immediately from Lemma 7.2 and Theorem 7.4. Let the symbols be as in (ii). Clearly K_0 is the discriminant field of (V, φ). We see that $1 \in h\varphi[V]$, and hence $h = \varphi[e]$ with some $e \in V$. Let B_1 be the quaternion algebra defined in Theorem 7.4 for these (V, φ) and e. Then $(V, h\varphi) \cong \langle F, 1\rangle \oplus (B_1^\circ, \delta\beta_1^\circ)$ with $(B_1^\circ, \beta_1^\circ)$ defined with respect to B_1. Thus $(B^\circ, \delta\beta^\circ) \cong (B_1^\circ, \delta\beta_1^\circ)$, and hence $B \cong A^+(\delta\beta^\circ) \cong A^+(\delta\beta_1^\circ) \cong B_1$. This together with Theorem 7.4 proves (ii). □

THEOREM 7.6. *If $n = 5$, then*

$$G^+(\varphi) = \{\alpha \in A^+(\varphi)^\times \mid \alpha\alpha^* \in F^\times\}, \tag{7.5a}$$

$$Vz = \{x \in A^+(\varphi) \mid x^* = x,\ \mathrm{tr}(x) = 0\}, \tag{7.5b}$$

where tr *is the reduced trace map of $A^+(\varphi)$ to F. Moreover, suppose $\delta \in \varphi[V]$. (See Remark below.) Then there exists a quaternion algebra B over F with the following properties:*

(i) *$A^+(\varphi) \cong M_2(B)$ and $A(\varphi) \cong M_2(B) \otimes_F K$.*

(ii) *$(V, \delta\varphi) \cong \langle F, 1\rangle \oplus (B, d\beta)$ with some $d \in F^\times$.*

(iii) *Put ${}^t\xi^\iota = \begin{bmatrix} r^\iota & u^\iota \\ s^\iota & v^\iota \end{bmatrix}$ for $\xi = \begin{bmatrix} r & s \\ u & v \end{bmatrix} \in M_2(B)$ and $\varepsilon = \mathrm{diag}[d, 1]$ with d of* (ii). *Then there exists an F-linear isomorphism g of $A^+(\varphi)$ onto $M_2(B)$ such that $g(\alpha^*) = \varepsilon \cdot {}^t g(\alpha)^\iota \varepsilon^{-1}$ for every $\alpha \in A^+(\varphi)$, so that*

$$g(Vz) = \{y \in M_2(B) \mid \varepsilon \cdot {}^t y^\iota \varepsilon^{-1} = y,\ \mathrm{tr}(y) = 0\}, \tag{7.6}$$

where tr *is the reduced trace map of $M_2(B)$ to F. Furthermore,*

$$g(G^+(\varphi)) = \{\xi \in GL_2(B) \mid \xi\varepsilon \cdot {}^t\xi^\iota = \mu(\xi)\varepsilon \ \ \text{with}\ \ \mu(\xi) \in F^\times\} \tag{7.7}$$

and $\nu(\alpha) = \mu(g(\alpha))$ for every $\alpha \in G^+(\varphi)$.

Remark. For $n = 5$, the condition $\delta \in \varphi[V]$ is always satisfied if F is an algebraic number field or its completion at an archimedean or a nonarchimedean prime.

Proof. Taking an F-basis of $A(\varphi)$ consisting of monomials of the members of an orthogonal basis of V, we find that $\{x \in A^-(\varphi) \mid x^* = x\} = V + Fz$. To prove (7.5a), suppose $\alpha \in A^+(\varphi)^\times, \alpha\alpha^* \in F^\times$, and $v = \alpha^{-1}u\alpha$ with $0 \neq u \in V$. Our task is to show that $v \in V$. Since $v \in A^-(V)$ and $v^* = v$, we have $v = w + cz$ with $w \in V$ and $c \in F$. If $w = 0$, then $u = \alpha cz\alpha^{-1} = cz$, a contradiction. Thus $w \neq 0$, and since $v^2 \in F$, we obtain $czw \in F$, which can happen only when $c = 0$, that is, $v \in V$ as desired. Thus $\alpha^{-1}V\alpha \subset V$ if $\alpha\alpha^* \in F^\times$, and so we obtain (7.5a). To prove the remaining part of the theorem, we first assume that $\delta \in \varphi[V]$, take an element e of V such that $\varphi[e] = \delta$, and let $W = \{x \in V \mid xe = -ex\}$. Then $\delta\varphi[\delta^{-1}e] = 1$, $V = Fe \oplus W$, and $(V, \delta\varphi) = \langle F, 1\rangle \oplus (W, \psi)$, where ψ is the restriction of $\delta\varphi$ to W. Clearly $1 \in \det(\psi)$, and so as noted in §2.7, $(W, \psi) \cong (B, d\beta)$ with a quaternion algebra B over F and some $d \in F^\times$. Identifying V with $F \oplus B$, define maps q: $V \longrightarrow M_2(B)$ and p: $V \longrightarrow M_2(B) \times M_2(B)$ by

$$q(a, b) = \begin{bmatrix} a & db \\ b^\iota & -a \end{bmatrix} \qquad (a \in F, b \in B), \tag{7.8a}$$

$$p(x) = (q(x), -q(x)) \qquad (x \in V). \tag{7.8b}$$

We have clearly $p(x)^2 = \delta\varphi[x]$ for every $x \in V$. The elements $q(x)$ for all $x \in V$ generate $M_2(B)$. Indeed, we have seen in [S3, §7.4, (B)] that the elements $q(0, b)$ for all $b \in B$ generate $M_2(B)$. By the definition of $A(\delta\varphi)$, there exists an F-linear ring-homomorphism h of $A(\delta\varphi)$ into $M_2(B)\times M_2(B)$ such that $h(x) = p(x)$ for every $x \in V$. Now h maps $A^+(\delta\varphi)$ into $\{(y, y) \mid y \in M_2(B)\}$. Since $A^+(\delta\varphi)$ is a simple algebra, h is either 0 or an injection. Since $h \neq 0$ and $\dim(A^+(\delta\varphi)) = 2^4$, we have $h(A^+(\delta\varphi)) = \{(y, y) \mid y \in M_2(B)\}$. From (7.8b) we see that $h(A(\delta\varphi))$ contains an element of the form $(w, -w)$ with $w \in B^\times$, and hence we see that both (y, y) and $(y, -y)$ belong to $h(A(\delta\varphi))$ for every $y \in B$. Therefore, identifying V with $h(V)$, we have $A(\delta\varphi) = M_2(B) \times M_2(B)$ and $A^+(\delta\varphi) = \{(y, y) \mid y \in M_2(B)\}$. Put $\alpha^* = \varepsilon \cdot {}^t\alpha^\iota\varepsilon^{-1}$ for $\alpha \in M_2(B)$ with the notation of (iii). Then $(\alpha, \beta)^* = (\alpha^*, \beta^*)$ for $(\alpha, \beta) \in A(\delta\varphi)$, since $q(x)^* = q(x)$ for every $x \in V$. By Lemma 2.4 there exists an isomorphism f of $A^+(\varphi)$ onto $A^+(\delta\varphi)$ with the properties described there. Thus $A^+(\varphi) \cong M_2(B)$, so that $A(\varphi) \cong M_2(B) \otimes K$ by Lemma 2.3 (i). Define g: $A^+(\varphi) \longrightarrow M_2(B)$ by $f(\alpha) = (g(\alpha), g(\alpha))$ for $\alpha \in A^+(\varphi)$. Then $g(\alpha^*) = g(\alpha)^*$, and we obtain (7.7) from (7.5a). Clearly (7.6) is equivalent to (7.5b). To prove (7.5b), we first observe that

$$q(V) = \{x \in M_2(B) \mid x^* = x,\ \mathrm{tr}(x) = 0\}. \tag{7.9}$$

Now assume that F is algebraically closed. Then $K = F \times F$, and we can put $\delta = 1, A(\delta\varphi) = A(\varphi)$, and $h(z) = (1, -1)$, and so $g(Vz) = q(V)$, which proves (7.6), and hence (7.5b) too, under the assumption that F is algebraically closed. In the general case, we have (7.5b) over an algebraic extension of F. Since both sides of (7.5b) are vector spaces over F, (7.5b) must hold over the original F. This completes the proof. □

THEOREM 7.7. *Suppose $n = 6$; then*

$$G^+(\varphi) = \{\alpha \in A^+(\varphi)^\times \mid \alpha\alpha^* \in F^\times,\ N(\alpha) = (\alpha\alpha^*)^2\}, \tag{7.10}$$

where N denotes the reduced norm map from $A^+(\varphi)$ to its center K. (For the definition, see below.)

If K is a field, then the reduced norm map is well defined. If $K = F \times F$, then by Lemma 2.3 (ii), $A^+(\varphi) = D \times D$ with a central simple algebra D over F. We then define the reduced norm map $N\colon D \times D \to K$ by $N(a,b) = (N_0(a), N_0(b))$ for $a, b \in D$, where N_0 is the reduced norm map from D to F.

Proof. We first consider a special case in which (V, φ) is given by

$$V = \{h \in M_2(B) \mid {}^t h^\iota = h\}, \tag{7.11a}$$

$$\varphi[h] = ad - bb^\iota \quad \text{for} \quad h = \begin{bmatrix} a & b \\ b^\iota & d \end{bmatrix} \in V, \tag{7.11b}$$

where B is a quaternion algebra over F and ${}^t h^\iota$ is defined as in Theorem 7.6 (iii). (Notice that every split six-dimensional quadratic space can be given in this form with $B = M_2(F)$.) Now for $h = \begin{bmatrix} a & b \\ b^\iota & d \end{bmatrix} \in V$ we put

$$\widehat{h} = h^\wedge = \begin{bmatrix} d & -b \\ -b^\iota & a \end{bmatrix}. \tag{7.12}$$

Then $h\widehat{h} = \varphi[h]1_2$. To make our formulas short, let us hereafter put $x^* = {}^t x^\iota$ for $x \in M_2(B)$, and let N denote the reduced norm map of $M_2(B)$ to F. Then

$$N(h) = \varphi[h]^2 \text{ and } \varphi[\alpha^* h \alpha] = N(\alpha)\varphi[h] \text{ if } h \in V \text{ and } \alpha \in M_2(B). \tag{7.13}$$

Indeed, if $B = M_2(F)$ and $J = \operatorname{diag}[j, j]$ with $j = \begin{bmatrix} 0 & -1 \\ 1 & 0 \end{bmatrix}$, we easily see that $\varphi[h]$ is the Pfaffian of the alternating matrix hJ of size 4. Thus (7.13) can be shown by taking the algebraic closure of F.

Now we define a map $p\colon V \to M_4(B)$ by

$$p(h) = \begin{bmatrix} 0 & h \\ \widehat{h} & 0 \end{bmatrix}. \tag{7.14}$$

Then $p(h)^2 = \varphi[h]1_4$. It is not difficult to show that $p(V)$ generates $M_4(B)$, and so we can identify V with $p(V)$ and $A(\varphi)$ with $M_4(B)$. Then we see that

$$A^+(\varphi) = \{\operatorname{diag}[x, y] \mid x, y \in M_2(B)\} \cong M_2(B) \times M_2(B). \tag{7.15}$$

Also we have

$$\operatorname{diag}[x, y]^* = \operatorname{diag}[y^*, x^*] \quad \text{for} \quad \operatorname{diag}[x, y] \in A^+(\varphi), \tag{7.16}$$

where $*$ on the left-hand side is the canonical involution of $A^+(\varphi)$. This is because we can verify (7.16) when $\operatorname{diag}[x, y] = p(h)p(k)$ with $h, k \in V$.

Suppose $c = \xi\xi^* \in F^\times$ with $\xi = \mathrm{diag}[\alpha, \beta] \in A^+(\varphi)^\times$. Then by (7.16) we have $\alpha\beta^* = c$. We now examine whether $\xi^{-1}p(h)\xi \in V$ for every $h \in V$. We have

$$\xi^{-1}p(h)\xi = \begin{bmatrix} 0 & \alpha^{-1}h\beta \\ \beta^{-1}\widehat{h}\alpha & 0 \end{bmatrix}.$$

Now $\alpha^{-1}h\beta = c^{-1}\beta^*h\beta \in V$. Since $\widehat{h} = \varphi[h]h^{-1}$, we have, by (7.13),

$$(\alpha^{-1}h\beta)^\wedge = \varphi[c^{-1}\beta^*h\beta](\alpha^{-1}h\beta)^{-1} = c^{-2}N(\beta)\varphi[h]\beta^{-1}h^{-1}\alpha = c^{-2}N(\beta)\beta^{-1}\widehat{h}\alpha.$$

Therefore $\xi^{-1}p(h)\xi = p(k)$ with $k \in V$ if and only if $N(\beta) = c^2$, which means that $\xi \in G^+(\varphi)$ if and only if $N(\beta) = c^2$, in which case $N(\alpha) = c^2$, as $N(\alpha)N(\beta) = c^4$. This proves (7.10) for (V, φ) as in (7.11a, b).

To prove (7.10) in the general case, given (V, φ), let $V_1 = V \otimes_F F_1$ with an algebraic extension F_1 of F and extend φ to φ_1: $V_1 \to F_1$. Take F_1 so that (V_1, φ_1) is split. Let $\alpha \in G^+(\varphi)$; then $\alpha \in G^+(\varphi_1)$ and $\alpha\alpha^* \in F^\times$. Since (7.10) holds for $G^+(\varphi_1)$, we have $N(\alpha) = (\alpha\alpha^*)^2$. Conversely, if $\alpha \in A^+(\varphi)^\times, \alpha\alpha^* \in F^\times$, and $N(\alpha) = (\alpha\alpha^*)^2$, then $\alpha \in G^+(\varphi_1)$, so that $\alpha^{-1}V_1\alpha = V_1$. Now $A(\varphi_1) = A(\varphi)\otimes_F F_1$, and so $\alpha^{-1}V\alpha \subset V_1 \cap A(\varphi) = V$; thus $\alpha \in G^+(\varphi)$. This completes the proof. □

7.8. The above theorem does not give an explicit form of (V, φ) for $n = 6$. We shall now present such under the following assumption. *If K is the discriminant algebra of (V, φ), then (V, φ) represents $(K, s\kappa)$ for some $s \in F^\times$, that is, (V, φ) is the direct sum of $(K, s\kappa)$ and a quadratic space (W, ψ) of dimension 4.* (We can easily show that this is always the case if F is an algebraic number field or its completion at an archimedean or a nonarchimedean prime.) Then the discriminant field of ψ is F, and hence (W, ψ) is of the form $(B, d\beta)$ with a quaternion algebra B over F and $d \in F^\times$; see [S3, Theorem 7.5 (i)]. Thus

$$(V, \varphi) \cong (K, s\kappa) \oplus (B, d\beta). \tag{7.17}$$

Then by Lemma 2.8 (i), $A(\varphi) \cong M_2(B) \otimes A(s\kappa) \cong M_2(B) \otimes \{K, s\}$. Clearly $s \in \varphi[V]$, and so, taking s as h in Proposition 7.1, we see that D there is isomorphic to $M_2(B)$. Thus $A^+(\varphi) \cong M_2(B) \otimes K$ by Proposition 7.1 (i).

8. The groups of similitudes.

8.1. Throughout this section we use the notation of §2.2 with an arbitrary field F of characteristic $\neq 2$. By a *similitude* of φ we mean an element γ of $GL(V, F)$ such that $\varphi[x\gamma] = c\varphi[x]$ for every $x \in V$ with an element c of $F^\times$ independent of x; we put then $\mu(\gamma) = c$, and denote by $\mathscr{S}(\varphi)$ the group of all similitudes of φ. Every similitude γ of φ can be extended naturally to a ring-isomorphism of $A(V, c\varphi)$ onto $A(V, \varphi)$. Combining this with the isomorphism f^{-1}: $A^+(\varphi) \to A^+(c\varphi)$ with f as in Lemma 2.5, we obtain a ring-automorphism $[\gamma]$ of $A^+(\varphi)$ given by

$$(8.1) \qquad (x_1 \cdots x_{2m})^{[\gamma]} = \mu(\gamma)^{-m}(x_1\gamma) \cdots (x_{2m}\gamma)$$

for $x_i \in V$. We have clearly $\det(\gamma)^2 = \mu(\gamma)^n$, and

$$(8.2) \qquad z^{[\gamma]} = \det(\gamma)\mu(\gamma)^{-n/2}z \quad \text{if} \quad n \in 2\mathbf{Z}$$

for a generator z of K as in §2.2, as Fz is independent of the choice of $\{e_i\}_{i=1}^n$. Thus, for n even, $[\gamma]$ gives an automorphism of K, which is the identity map if and only if $\det(\gamma) = \mu(\gamma)^{n/2}$.

Assuming n to be even, we denote by $\mathscr{S}^+(\varphi)$ the subgroup of $\mathscr{S}(\varphi)$ consisting of the elements γ such that $\det(\gamma) = \mu(\gamma)^{n/2}$. By (8.2), a similitude γ belongs to $\mathscr{S}^+(\varphi)$ if and only if $[\gamma]$ is the identity map on the center K of $A^+(\varphi)$, in which case $[\gamma]$ can be given as an inner automorphism of $A^+(\varphi)$. Now for every element e of V such that $e^2 \neq 0$, we have $\tau(e) \in O(\varphi)$ and $\det(\tau(e)) = -1$, so that $\tau(e) \notin \mathscr{S}^+(\varphi)$ and hence $\mathscr{S}(\varphi) = \mathscr{S}^+(\varphi) \sqcup \mathscr{S}^+(\varphi)\tau(e)$. Thus $[\mathscr{S}(\varphi) : \mathscr{S}^+(\varphi)] = 2$, and so in order to determine $\mathscr{S}(\varphi)$, it is sufficient to determine $\mathscr{S}^+(\varphi)$. Since $\mu(\tau(e)) = 1$, we have

$$(8.3) \qquad \mu(\mathscr{S}(\varphi)) = \mu(\mathscr{S}^+(\varphi)).$$

Suppose n is odd. Given $\gamma \in \mathscr{S}(\varphi)$, put $c = \mu(\gamma)$ and $d = c^{(1-n)/2}\det(\gamma)$. Then $d^2 = c$, and we easily see that $d^{-1}\gamma \in SO(\varphi)$. Thus $\mathscr{S}(\varphi) = F^\times SO(\varphi)$, and therefore the question about $\mathscr{S}(\varphi)$ is nontrivial only for even n.

THEOREM 8.2. *Suppose $n \in 4\mathbf{Z}$; define a subgroup $\mathcal{G}$ of $A^+(\varphi)^\times$ by*

$$(8.4) \qquad \mathcal{G} = \{\xi \in A^+(\varphi)^\times \mid \xi\xi^* \in K^\times, \xi^* V\xi = V\}.$$

Given $(c, \xi) \in F^\times \times \mathcal{G}$, define $\gamma \in GL(V, F)$ by $x\gamma = c\xi^* x\xi$ for $x \in V$. Then $\gamma \in \mathscr{S}^+(\varphi)$ and $\mu(\gamma) = c^2(\xi\xi^*)(\xi\xi^*)^\rho$. Moreover, $\mathscr{S}^+(\varphi)$ consists of all such maps γ obtained from the elements of $F^\times \times \mathcal{G}$. Furthermore, let $\mathcal{G}_0$ be the subgroup of $F^\times \times \mathcal{G}$ consisting of the elements (dd^ρ, d^{-1}) with $d \in K^\times$. Then the map $(c, \xi) \mapsto \gamma$ gives an isomorphism of $(F^\times \times \mathcal{G})/\mathcal{G}_0$ onto $\mathscr{S}^+(\varphi)$. In particular we have

$$(8.5) \qquad \mathcal{G} = A^+(\varphi)^\times \quad \textit{if} \quad n = 4.$$

THEOREM 8.3. *Suppose $n - 2 \in 4\mathbf{Z}$; put*

$$(8.6) \qquad \mathcal{G} = \{\xi \in A^+(\varphi)^\times \mid \xi\xi^* \in F^\times, \xi^{-1}V\xi \subset V + zV\}.$$

(i) *If* $\xi \in \mathcal{G}$, *then there exists an element* b *of* $K^\times$ *such that* $\xi^{-1}bV\xi = V$. *Conversely, if* $\xi \in A^+(\varphi)^\times, b \in K^\times$, *and* $\xi^{-1}bV\xi = V$, *then* $\xi \in \mathcal{G}$. *Consequently,* $\mathcal{G}$ *is a subgroup of* $A^+(\varphi)^\times$. *Moreover,* bF *is determined by* ξ *if* $n > 2$.

(ii) *Given* $(b, \xi) \in K^\times \times \mathcal{G}$ *as in* (i), *define* $\gamma \in GL(V, F)$ *by* $x\gamma = \xi^{-1}bx\xi$ *for* $x \in V$. *Then* $\gamma \in \mathscr{S}^+(\varphi)$ *and* $\mu(\gamma) = bb^\rho$. *Moreover,* $\mathscr{S}^+(\varphi)$ *consists of all such maps* γ *obtained from the elements of* $K^\times \times \mathcal{G}$ *as in* (i).

(iii) *Let* $\widetilde{\mathcal{G}}$ *be the subgroup of* $K^\times \times \mathcal{G}$ *consisting of all* (b, ξ) *as in* (i), *and* $\widetilde{\mathcal{G}}_0$ *the subgroup of* $\widetilde{\mathcal{G}}$ *consisting of the elements* $(d(d^\rho)^{-1}, d)$ *with* $d \in K^\times$. *Then the map* $(b, \xi) \mapsto \gamma$ *gives an isomorphism of* $\widetilde{\mathcal{G}}/\widetilde{\mathcal{G}}_0$ *onto* $\mathscr{S}^+(\varphi)$.

(iv) *In particular,* $\mathcal{G} = G^+(V) = K^\times$ *if* $n = 2$, *and*

$$\mathcal{G} = \{\xi \in A^+(\varphi)^\times \mid \xi\xi^* \in F^\times\} \quad \text{if} \quad n = 6. \tag{8.7}$$

The condition $\xi^{-1}V\xi \subset V + zV$ in (8.6) can be replaced by $\xi^{-1}(V + zV)\xi = V + zV$, as can easily be seen. We devote the following two subsections to the proof of these theorems.

8.4. We first recall two basic facts: $ax = xa^\rho$ for every $a \in K$ and $x \in V$ as noted in (2.1); also $a^* = a$ or $a^* = a^\rho$ for every $a \in K$ according as $n \in 4\mathbf{Z}$ or $n - 2 \in 4\mathbf{Z}$.

Let us now treat the case $n = 2$. Then $A^+(\varphi) = K, V = zV$, and $\mathcal{G} = G^+(V) = K^\times$. Let $b \in K^\times$. Then $bV = V$ and $(bx)(by) = bb^\rho xy$ for every $x, y \in V$. Thus the map $x \mapsto bx$ defines an element γ of $\mathscr{S}(\varphi)$ such that $\mu(\gamma) = bb^\rho$ and $[\gamma] = 1$. Now let $\alpha \in \mathscr{S}^+(\varphi)$. Then α gives an isomorphism of $(V, \mu(\alpha)\varphi)$ onto (V, φ), and hence $\mu(\alpha) \in N_{K/F}(K^\times)$ by Proposition 7.1 (iv). Take $b \in K^\times$ so that $\mu(\alpha) = bb^\rho$. Then the map $x \mapsto b^{-1}x\alpha$ is an element of $SO(\varphi)$, and so there is an element $\xi \in G^+(\varphi) = K^\times$ such that $b^{-1}x\alpha = \xi^{-1}x\xi$ for every $x \in V$. This proves the last part of (ii) for $n = 2$. The remaining portion of (i), (ii), and (iii) for $n = 2$ can easily be verified. This completes the proof in the case $n = 2$.

Suppose now $0 < n - 2 \in 4\mathbf{Z}$. To prove the first part of Theorem 8.3 (i), take $\xi \in \mathcal{G}$. Then for $x \in V$ we have $\xi^{-1}x\xi = u + zv$ with $u, v \in V$, and so $x^2 = u^2 - z^2v^2 + z(vu - uv)$, so that $z(vu - uv) \in F$, which can happen only when $vu = uv$. Since $uv + vu \in F$, we have $uv \in F$, and hence u and v are not linearly independent over F. In such a case, $u + zv$ can be written dw with $d \in K$ and $w \in V$. If $x^2 \neq 0$, then $d \in K^\times$. We are going to show that d can be chosen independently of x. For that purpose take an orthogonal basis $\{e_i\}_{i=1}^n$ of V. For each i we have $d_i \in K^\times$ and $f_i \in V$ such that $\xi^{-1}e_i\xi = d_if_i$. Then $\xi^{-1}(\sum_{i=1}^n Ke_i)\xi = \sum_{i=1}^n Kf_i$. Since $\sum_{i=1}^n Ke_i$ has dimension $2n$ over F, so does $\sum_{i=1}^n Kf_i$, and hence $\{f_i\}_{i=1}^n$ is an F-basis of V. Now $\xi^{-1}(\sum_{i=1}^n e_i)\xi = hg$ with $h \in K$ and $g \in V$. Put $g = \sum_{i=1}^n s_if_i$ with $s_i \in F$. Then $\sum_{i=1}^n d_if_i = \sum_{i=1}^n hs_if_i$, and hence $d_i = hs_i$, that is, $d_i \in Fh$ for every i. Thus $h \in K^\times$ and $\xi^{-1}V\xi \subset hV$, which proves the first part of Theorem 8.3 (i).

Next suppose $\xi \in A^+(\varphi)^\times, b \in K^\times$, and $\xi^{-1}bV\xi = V$. Then for $x \in V$ we have $\xi^{-1}bx\xi = (\xi^{-1}bx\xi)^* = \xi^* xb^*(\xi^{-1})^* = \xi^* bx(\xi^{-1})^*$, so that $\xi\xi^* bx = bx\xi\xi^*$. Now bV generates $A(\varphi)$, and hence we obtain $\xi\xi^* \in F^\times$. Therefore $\xi \in \mathcal{G}$. If $d \in K^\times, 0 \neq y \in V$, and $dy \in V$, then clearly $d \in F$. Thus $bF^\times$ is determined by ξ. This completes the proof of Theorem 8.3 (i).

Next, define γ as in Theorem 8.3 (ii). Then for $x, y \in V$ we have

$$(8.8)\quad (x\gamma)(y\gamma) = \xi^{-1}bx\xi \cdot \xi^{-1}by\xi = \xi^{-1}bxby\xi = \xi^{-1}bb^\rho xy\xi = bb^\rho \xi^{-1}xy\xi,$$

and in particular, $\varphi[x\gamma] = bb^\rho \varphi[x]$. Thus $\gamma \in \mathscr{S}(\varphi)$ and $\mu(\gamma) = bb^\rho$. Also, from (8.8) we see that $\alpha^{[\gamma]} = \xi^{-1}\alpha\xi$ for every $\alpha \in A^+(\varphi)$, and so $\gamma \in \mathscr{S}^+(\varphi)$. This proves the first part of Theorem 8.3 (ii).

Before proving the second part of Theorem 8.3 (ii), we consider the case $n \in 4\mathbf{Z}$, and for $(c, \xi) \in F^\times \times \mathcal{G}$ define γ as in Theorem 8.2. Put $b = c\xi\xi^*$. Then $b \in K^\times$ and $x\gamma = \xi^{-1}bx\xi$ for every $x \in V$. We again obtain (8.8) and see that $\gamma \in \mathscr{S}^+(\varphi)$ and $\mu(\gamma) = bb^\rho = c^2(\xi\xi^*)(\xi\xi^*)^\rho$ for the same reason. This proves the direct part of Theorem 8.2.

8.5. It remains to prove the converse part of Theorem 8.2 and the converse part of Theorem 8.3 (ii). Let $\gamma \in \mathscr{S}^+(\varphi)$ in either case. Then γ gives an isomorphism of $(V, \mu(\gamma)\varphi)$ onto (V, φ), and so $\mu(\gamma) \in N_{K/F}(K^\times)$ by Proposition 7.1 (iv). Take $d \in K^\times$ so that $dd^\rho = \mu(\gamma)^{-1}$. Then $(d(x\gamma))^2 = dd^\rho(x\gamma)^2 = x^2$ for every $x \in V$. Therefore the map $x \mapsto d(x\gamma)$ can be extended to an automorphism of $A(\varphi)$. Since $A(\varphi)$ is central simple over F, there exists an element ω of $A(\varphi)^\times$ such that $d(x\gamma) = \omega^{-1}x\omega$ for every $x \in V$. Thus, for $x, y \in V$ we have

$$\omega^{-1}xy\omega = \omega^{-1}x\omega \cdot \omega^{-1}y\omega = d(x\gamma)d(y\gamma) = dd^\rho(x\gamma)(y\gamma) = (xy)^{[\gamma]}$$

with $[\gamma]$ as in (8.1), and so $\omega^{-1}\alpha\omega = \alpha^{[\gamma]}$ for every $\alpha \in A^+(\varphi)$. Since $\gamma \in \mathscr{S}^+(\varphi)$, we have $\alpha^{[\gamma]} = \xi^{-1}\alpha\xi$ for every $\alpha \in A^+(\varphi)$ with some $\xi \in A^+(\varphi)^\times$. Then $\omega\xi^{-1}$ commutes with every element of $A^+(\varphi)$, and so by Proposition 7.1 (i), (ii), $\omega = r\xi$ with $r \in K^\times$. Put $b = d^{-1}r^{-1}r^\rho$. Then $b \in K^\times$ and $x\gamma = d^{-1}\xi^{-1}r^{-1}xr\xi = \xi^{-1}bx\xi$ for every $x \in V$.

Suppose $n - 2 \in 4\mathbf{Z}$; then by Theorem 8.3 (i) we have $\xi \in \mathcal{G}$, and so the proof of Theorem 8.3 (ii) is complete. If $\gamma = 1$, then $\xi^{-1}\alpha\xi = \alpha$ for every $\alpha \in A^+(\varphi)$, and so $\xi \in K^\times$, and (iii) can easily be verified. If $n = 6$, we easily see that

$$(8.9)\qquad \{y \in A^-(V) \mid y^* = y\} = V + zV.$$

Therefore $\xi^* V\xi \subset V + zV$ for every $\xi \in A^+(\varphi)$. If $\xi\xi^* \in F^\times$, then $\xi^{-1}V\xi = \xi^* V\xi$, and hence $\mathcal{G}$ can be given as in (8.7).

Next suppose $n \in 4\mathbf{Z}$. Applying the canonical involution to the element $\xi^{-1}bx\xi$ of V, we obtain $\xi^{-1}bx\xi = (\xi^{-1}bx\xi)^* = \xi^* xb(\xi^{-1})^*$, so that $x\xi\xi^* b^{-1} =$

$\xi\xi^* b^{-1}x$ for every $x \in V$, as b commutes with $\xi\xi^*$. Thus $\xi\xi^* b^{-1}$ commutes with every element of $A(\varphi)$, and so $\xi\xi^* b^{-1} \in F^\times$. Thus $\xi\xi^* \in K^\times$ and $\xi^* \in Fb\xi^{-1}$. Therefore $\xi^* V\xi = \xi^{-1}bV\xi = V\gamma = V$, and so $\xi \in \mathcal{G}$. If $\gamma = 1$, then $\xi \in K^\times$ for the same reason as in the case $n - 2 \in 4\mathbf{Z}$, and so we can easily verify that $(F^\times \times \mathcal{G})/\mathcal{G}_0 \cong \mathscr{S}^+(\varphi)$. To prove (8.5), suppose $n = 4$; then from (7.3) we easily see that $\xi^* V\xi = V$ for every $\xi \in A^+(\varphi)^\times$. This fact combined with Theorem 7.4 (ii) proves (8.5). This completes the proof

Let us now prove Theorem 7.4 (v). If $n = 4$, we see from (8.3), (8.5), and Theorem 8.2 that $\mu(\mathscr{S}(\varphi))$ consists of $c^2 N_{K/F}(N(\xi))$ for all $c \in F^\times$ and $\xi \in A^+(\varphi)^\times$. Therefore we obtain the desired fact. Notice that if $K_0 = F$, then $N(B_K^\times) = N(B^\times) \times N(B^\times)$, which is mapped onto $N(B^\times)$ under $N_{K/F}$.

Department of Mathematics, Princeton University, Princeton, NJ

REFERENCES

[C] C. Chevalley, *The algebraic Theory of Spinors*, Columbia University Press, New York, 1954.

[Ch] K.-S. Chang, Diskriminanten und Signaturen gerader quadratisher Formen, *Arch. Math.* **21** (1970), 59–65.

[E] M. Eichler, *Quadratische Formen und orthogonale Gruppen*, Springer, Berlin, 1952 (2nd ed., 1974).

[H] H. Hasse, Äquivalenz quadratischer Formen in einem beliebigen algebraischen Zahlkörper, *J. Reine Angew. Math.* **153** (1924), 158–162.

[S1] G. Shimura, On modular forms of half integral weight, *Ann. of Math.* **97** (1973), 440–481 (= *Collected Papers*, II, pp. 532–573).

[S2] ———, On the transformation formulas of theta series, *Amer. J. Math.* **115** (1993), 1011–1052 (= *Collected Papers*, IV, pp. 191–232).

[S3] ———, *Arithmetic and Analytic Theories of Quadratic Forms and Clifford Groups*, *Math. Surveys Monogr.*, vol. 109, Amer. Math. Soc., Providence, RI, 2004.

[W] E. Witt, Theorie der quadratischen Formen in beliebigen Körpern, *J. Reine Angew. Math.* **176** (1937), 31–44.

CLASSIFICATION OF INTEGER-VALUED SYMMETRIC FORMS

By Goro Shimura

Abstract. We classify all $\mathbf{Z}$-valued symmetric forms that are reduced in the sense that they cannot be represented nontrivially by another $\mathbf{Z}$-valued symmetric form. We determine the possible values of the discriminant and signature of each such form and also the number of genera of the forms with the given discriminant and signature.

Introduction. Given an associative ring R with identity element, we denote by $R^\times$ the group of invertible elements of R; also, for positive integers m and n we denote by R_n^m the set of all $m \times n$-matrices with components in R. We put $R_n^n = M_n(R)$ when we view the set as a ring, denote by 1_n the identity element of $M_n(R)$, and put $GL_n(R) = M_n(R)^\times$. If R is an integral domain, we denote by $S_n(R)$ the set of all nonsingular symmetric elements of $M_n(R)$. Given square matrices $\alpha, \beta, \ldots$, we denote by $\mathrm{diag}[\alpha, \beta, \ldots]$ the square matrix with $\alpha, \beta, \ldots$ in the diagonal square blocks and zeros in all other blocks.

Let T_n denote the set of all elements $\gamma = (c_{ij})$ of $S_n(\mathbf{Z})$ such that $c_{ii} \in 2\mathbf{Z}$ for every i. Then a $\mathbf{Z}$-valued quadratic form is represented by an element of $2^{-1}T_n$. Now by a $\mathbf{Z}$-valued symmetric form we mean a symmetric bilinear form represented by an element of $S_n(\mathbf{Z})$. Thus, in terms of matrices, the former includes the latter, as $S_n(\mathbf{Z}) \subset 2^{-1}T_n$, but we can view the former as a special case of the latter, as $T_n \subset S_n(\mathbf{Z})$. When we discuss the classification problems of these, however, the distinction is essential.

We first introduce two types of reducibilities of such matrices. We call an element ζ of $2^{-1}T_n$ *q-reduced* (reduced as a quadratic form) if it satisfies

$$(0.1) \qquad \zeta = \alpha\omega \cdot {}^t\alpha, \quad \alpha \in M_n(\mathbf{Z}), \quad \text{and} \quad \omega \in 2^{-1}T_n \Longrightarrow |\det(\alpha)| = 1.$$

We call an element φ of $S_n(\mathbf{Z})$ *s-reduced* (reduced as a symmetric form) if it satisfies

$$(0.2) \qquad \varphi = \alpha\psi \cdot {}^t\alpha, \quad \alpha \in M_n(\mathbf{Z}), \quad \text{and} \quad \psi \in S_n(\mathbf{Z}) \Longrightarrow |\det(\alpha)| = 1.$$

For instance, if $\varphi \in S_n(\mathbf{Z})$ and $\det(\varphi)$ is squarefree, then φ is s-reduced.

Manuscript received March 14, 2006; revised May 11, 2006.
Shimura, Goro. "Classification of integer-valued symmetric forms." American Journal of Mathematics 130:3 (2008), 685–711.

Now, instead of trying to classify all $\mathbf{Z}$-valued quadratic forms or symmetric forms, we restrict ourselves to the q-reduced or s-reduced elements. This idea seems perfectly reasonable for the following reason. In algebraic number theory, without considering an arbitrary ring of algebraic integers, we take the maximal order, and discuss everything in relation with it. Though the parallelism does not go too far, one thing is clear: if we want to classify symmetric forms, we must at least be able to classify such reduced forms. It should also be noted that there are few previous investigations on classification for an arbitrary value of the discriminant of the form beyond the unimodular case.

In the present paper we classify all the genera of s-reduced elements of $S_n(\mathbf{Z})$ and determine the number of genera with given discriminant and signature. The same types of problems for q-reduced matrices were investigated in our previous paper [S4]. To make our exposition smooth, let us write $\varphi \sim \psi$ for $\varphi,\ \psi \in S_n(\mathbf{Q})$ if $\varphi = \alpha\psi \cdot {}^t\alpha$ with $\alpha \in GL_n(\mathbf{Q})$ and $\varphi \approx \psi$ if such an α can be taken from $GL_n(\mathbf{Z})$. We also note that every element of T_n as a quadratic form restricted on $\mathbf{Z}_n^1$ represents only even numbers, and an element of $S_n(\mathbf{Z})$ not contained in T_n represents some odd integers. We say that the matrices of the latter kind are of type 1, and those of the former kind are of type 2; we then denote by $S_n^\lambda(\mathbf{Z})$ the set of s-reduced elements of $S_n(\mathbf{Z})$ of type λ. Clearly if $\varphi \approx \psi$ for $\varphi,\ \psi \in S_n(\mathbf{Z})$, then they are of the same type.

Here are two basic facts we prove in the present paper:

(I) *For $\varphi,\ \psi \in S_n^\lambda(\mathbf{Z})$ we have $\varphi \sim \psi$ if and only if they belong to the same genus* (§3.6).

(II) *The set $S_n^2(\mathbf{Z})$ consists of the elements τ of T_n such that $2^{-1}\tau$ is q-reduced and* $\det(\tau) \notin 4\mathbf{Z}$ (Lemma 3.7 (iii)).

The "if"-part of (I) is trivial, and so the essential point of (I) is that if we consider s-reduced elements of the same type, then the isomorphism class of quadratic spaces determines the genus. Next, statement (II) reduces the classification of the elements of $S_n^2(\mathbf{Z})$ to that of q-reduced elements of $2^{-1}T_n$, which were already done in [S4]. Therefore we state here only the results about $S_n^1(\mathbf{Z})$, though we must note that the notion of a type over an algebraic number field is more complicated.

In order to state the results, for $\varphi \in S_n(\mathbf{Q})$ we put $s(\varphi) = p - q$ if φ as a real matrix has p positive and q negative eigenvalues. Then as one of the main results of this paper we shall prove:

(III) *The set of genera of elements of $S_n^1(\mathbf{Z})$, if $n > 2$, can be parametrized by four integers n, σ, m, e. Here σ is an integer such that $|\sigma| \le n$ and $n - \sigma \in 2\mathbf{Z}$; m and e are squarefree positive integers such that $\sigma - 4r \equiv 0$ or $2 \pmod 8$ if n is even and $\sigma - 4r \equiv \pm 1 \pmod 8$ if n is odd, where r is the number of prime factors of e. For ψ in the genus determined by (n, σ, m, e) we have $s(\psi) = \sigma$ and $|\det(\psi)| = mf^2$, where f is the product of the prime factors of e prime to $2m$* (Theorem 3.8).

From this combined with (I) we can derive:

(IV) *Let* $n \in \mathbf{Z}$, > 2, *and let* σ *be an integer such that* $|\sigma| \leq n$ *and* $n - \sigma \in 2\mathbf{Z}$; *let* m *and* f *be two squarefree positive integers such that* f *is prime to* $2m$, *and let* ν *be the number of different prime factors of* $2m$. *Then there exist exactly* $2^{\nu-1}$ *genera of* $\psi \in S_n^1(\mathbf{Z})$ *such that* $|\det(\psi)| = mf^2$ *and* $s(\psi) = \sigma$ (Theorem 3.9).

We investigate the same types of problems over an algebraic number field. As in the discussion of any number-theoretical problem, we have to study the problems over the completion of a number field at a nonarchimedean prime. Indeed, we can define q-reduced and s-reduced matrices and also their types over such a local field in an obvious way. If the prime does not divide 2, then each s-reduced matrix corresponds to a maximal lattice in a local quadratic space, whose structure is already known. Therefore our principal technical task is to investigate the matrices over a dyadic local field. Thus, let F be an unramified algebraic extension of $\mathbf{Q}_2$ of finite degree, and $\mathfrak{g}$ the ring of local integers in F. We can define $S_n^\lambda(\mathfrak{g})$ as an obvious analogue of $S_n^\lambda(\mathbf{Z})$, as well as the symbols $\sim$ and $\approx$. Then we have a crucial local result on which the whole theory hinges:

(V) *If* $\varphi, \psi \in S_n^\lambda(\mathfrak{g})$ *and* $\varphi \sim \psi$, *then* $\varphi \approx \psi$ (Theorem 2.4(ii)).

It should be noted that this is false if F is ramified over $\mathbf{Q}_2$, as will be shown in §3.11. We devote Section 2 to the proof of (V), which is not so simple even in the case $F = \mathbf{Q}_2$.

In the present paper we are mainly concerned with the *genera* of symmetric forms, as the discussion of *classes* is a problem of a completely different nature. In the indefinite case, however, we can prove a clear-cut result as follows:

(VI) *Let* φ *and* ψ *be indefinite elements of* $S_n^\lambda(\mathbf{Z})$ *with* $n \geq 3$ *such that* $\varphi \sim \psi$. *Then* $\varphi \approx \psi$ (Theorem 3.10(i)).

Combining (IV) and (VI), we obtain

(VII) *Let* n, σ, m, *and* ν *be as in* (IV). *If* $|\sigma| \neq n$, *then the elements* φ *of* $S_n(\mathbf{Z})$ *of type 1 such that* $|\det(\varphi)| = m$ *and* $s(\varphi) = \sigma$ *form exactly* $2^{\nu-1}$ *classes.*

This is because φ is s-reduced if $\det(\varphi)$ is squarefree. In fact, (VI) and (VII) are included in [S3, Theorem 9.26] as special cases, except that we have to verify the conditions under which that theorem holds, which is nontrivial. If $m = 1$, (VII) was proved by Milnor in [M]. We also mention [V] by van der Blij, who discussed the relation between n, $s(\varphi)$, and $\det(\varphi)$ modulo 4 or 8 for $\varphi \in S_n(\mathbf{Z})$; a few earlier papers on similar topics are mentioned there.

We end the paper with some applications to quadratic Diophantine equations in the framework introduced in [S3]. As a final remark we note that the characterization of a quadratic space by means of its discriminant, signatures, and characteristic quaternion algebra given in [S4] is essential in our theory.

1. Basic symbols and terminology.

1.1. In this paper a *global field* means an algebraic number field of finite degree, and a *local field* the completion of a global field at a nonarchimedean prime. For a global field F we denote by $\mathfrak{g}$ the ring of algebraic integers in F; for a local F we denote by $\mathfrak{g}$ the ring of local integers in F in the standard sense. Whenever F is local, we denote by $\mathfrak{p}$ the maximal ideal of $\mathfrak{g}$. An archimedean completion of a global field will not be called a local field, though such will be needed at some point.

Let V be a vector space of dimension n over a local or global field F, and $\varphi : V \times V \to F$ a nondegenerate F-bilinear symmetric form. We put $\varphi[x] = \varphi(x, x)$ for $x \in V$. We then call the structure (V, φ) a *quadratic space* over F of dimension n. We call φ *isotropic* if $\varphi[x] = 0$ for some $x \in V$, $\neq 0$, and call φ *anisotropic* otherwise. We denote by $O^{\varphi}(V)$ and $SO^{\varphi}(V)$ the orthogonal group and special orthogonal group of φ defined as usual. We let every F-linear endomorphism α of V act on the right, so that $x\alpha$ is the image of $x \in V$ under α.

Put $F^{\times 2} = \{a^2 \mid a \in F^{\times}\}$. Given (V, φ) of dimension n as above, let Φ be the $(n \times n)$-matrix that represents φ with respect to an F-basis of V. Then we denote by $\delta_0(\varphi)$ the coset in $F^{\times}/F^{\times 2}$ represented by $(-1)^{n(n-1)/2} \det(\Phi)$. We now consider a field K_0 given by

$$K_0 = F(\delta^{1/2}), \qquad \delta \in \delta_0(\varphi) \tag{1.1}$$

and call K_0 the *discriminant field* of φ, or of (V, φ). This is either F itself or a quadratic extension of F; clearly it does not depend on the choice of δ. We then define the *discriminant algebra* K of φ, or of (V, φ), by

$$K = \begin{cases} K_0 & \text{if } K_0 \neq F, \\ F \times F & \text{if } K_0 = F. \end{cases} \tag{1.2}$$

For such a K, we always denote by ρ the nontrivial automorphism of K over F; in particular, $(x, y)^{\rho} = (y, x)$ for $(x, y) \in F \times F$. Given $c \in F^{\times}$ and an algebra K of type (1.2), we denote by $\{K, c\}$ the quaternion algebra B given by

$$B = K \oplus Kj, \quad j^2 = c, \quad jx = x^{\rho} j \text{ for every } x \in K. \tag{1.3}$$

Whenever we have a quaternion algebra B over F, we denote by ι the main involution of B and put

$$B^{\circ} = \{x \in B \mid x^{\iota} = -x\}. \tag{1.4}$$

We also put $\mathrm{Tr}_{B/F}(x) = x + x^{\iota}$ and $N_{B/F}(x) = xx^{\iota}$ for $x \in B$. Then we have quadratic spaces (B, β) and $(B^{\circ}, \beta^{\circ})$ over F of dimension 4 and 3, where $\beta(x, y) = 2^{-1} \mathrm{Tr}_{B/F}(xy^{\iota})$ for $x, y \in B$ and β° is the restriction of β to $B^{\circ} \times B^{\circ}$.

1.2. We associate a quaternion algebra $Q(\varphi)$ with φ by the following rule: For n even, $Q(\varphi)$ represents the Brauer class of the Clifford algebra of φ as a central simple algebra over F; for n odd, $Q(\varphi)$ represents the Brauer class of the even Clifford algebra of φ (that is, the subalgebra of the Clifford algebra of φ consisting of the even elements) as a central simple algebra over F. We call $Q(\varphi)$ *the characteristic quaternion algebra* of φ. In [S4] we showed that if F is global, then the isomorphism class of (V, φ) is determined by $n, \delta_0(\varphi)$, $Q(\varphi)$, and the signature of φ at each real archimedean prime of F. If F is local, only $n, \delta_0(\varphi)$, and $Q(\varphi)$ are sufficient; see (1.12) below.

We define the set $S_n(R)$ as in the introduction and put, in both local and global cases, for $0 < n \in \mathbf{Z}$,

$$(1.5) \qquad E_n = GL_n(\mathfrak{g}), \qquad T_n = \{\tau = (\tau_{ij}) \in S_n(\mathfrak{g}) \mid \tau_{ii} \in 2\mathfrak{g} \text{ for every } i\}.$$

We shall simply put $S_n = S_n(F)$ if F is clear from the context. For every $\varphi \in S_n$ we can define a quadratic space (V, φ) by taking $V = F_n^1$ and $\varphi[x] = x\varphi \cdot {}^t x$ for $x \in V$. We denote this space by $\langle\varphi\rangle$, and the groups $O^{\varphi}(V)$ and $SO^{\varphi}(V)$ defined as subgroups of $GL_n(F)$ by $O(\varphi)$ and $SO(\varphi)$. Then we can naturally use the symbols $\delta_0(\varphi)$ and $Q(\varphi)$ for $\varphi \in S_n$. Clearly $\varphi \in T_n$ if and only if $\varphi[x] \in 2\mathfrak{g}$ for every $x \in \mathfrak{g}_n^1$. For $\varphi, \psi \in S_n$ we write $\varphi \sim \psi$ if $\alpha\varphi \cdot {}^t\alpha = \psi$ for some $\alpha \in GL_n(F)$, and $\varphi \approx \psi$ if such an α can be taken from E_n. Clearly we have $\varphi \sim \psi$ if and only if $\langle\varphi\rangle$ is isomorphic to $\langle\psi\rangle$. If $\varphi \approx \psi$ we say that φ and ψ are *equivalent over* $\mathfrak{g}$. For $\varphi \in S_n$ and $c \in \mathfrak{g}$ we say that φ *represents* c if $\varphi[x] = c$ for some $x \in \mathfrak{g}_n^1$.

We now recall that Witt's theorem guarantees an element τ of S_n such that

$$(1.6) \qquad \varphi \sim \tau = \operatorname{diag}[\eta_r, \zeta], \qquad \eta_r = \begin{bmatrix} 0 & 1_r \\ 1_r & 0 \end{bmatrix},$$

with an anisotropic element ζ of S_t, $0 \le t = n - 2r$. We call t the *core dimension of* φ. We have $t \le 4$ if F is local. Notice that $\delta_0(\varphi) = \delta_0(\tau) = \delta_0(\zeta)$ and $Q(\varphi) = Q(\tau) = Q(\zeta)$.

We call an element φ of $2^{-1}T_n$ *q-reduced* (that is, reduced as a quadratic form) if it satisfies the following condition:

$$(1.7) \qquad \varphi = \alpha\psi \cdot {}^t\alpha, \quad \alpha \in M_n(\mathfrak{g}), \quad \text{and} \quad \psi \in 2^{-1}T_n \Longrightarrow \alpha \in E_n.$$

Similarly we call an element φ of $S_n(\mathfrak{g})$ *s-reduced* (reduced as a symmetric form) if it satisfies

$$(1.8) \qquad \varphi = \alpha\psi \cdot {}^t\alpha, \quad \alpha \in M_n(\mathfrak{g}), \quad \text{and} \quad \psi \in S_n(\mathfrak{g}) \Longrightarrow \alpha \in E_n.$$

If an element φ of $S_n(\mathfrak{g})$ is q-reduced, then it is s-reduced, but the converse is false. For example, η_r is s-reduced, but not q-reduced if $2 \notin \mathfrak{g}^{\times}$, as $\eta_r = \alpha(2^{-1}\eta_r) \cdot {}^t\alpha$ with $\alpha = \operatorname{diag}[1_r, 2 \cdot 1_r]$; $2^{-1}\eta_r$ is q-reduced.

1.3. Let (V, φ) be a quadratic space over a local or global F. We shall call a $\mathfrak{g}$-lattice in V simply a *lattice* in V. We call a lattice L in V *φ-integral* if $\varphi[x] \in \mathfrak{g}$ for every $x \in L$, and call a φ-integral lattice *φ-maximal* if it is the only φ-integral lattice containing it. Similarly we call L *φ^*-integral* if $\varphi(x, y) \in \mathfrak{g}$ for every $x, y \in L$, and call a φ^*-integral lattice *φ^*-maximal* if it is the only φ^*-integral lattice containing it. Given (V, φ) and a lattice L in V, put

$$L^{\varphi} = \{x \in V \mid \varphi(x, L) \subset \mathfrak{g}\}. \tag{1.9}$$

Clearly L^{φ} is a lattice in V. If L is φ^*-integral and M is a φ^*-integral lattice containing L, then $M \subset L^{\varphi}$, and so there is a φ^*-maximal lattice containing L. Similarly every φ-integral lattice is contained in a φ-maximal lattice. Suppose $\mathfrak{g}$ is a principal ideal domain; let ψ be the matrix representing φ with respect to a $\mathfrak{g}$-basis of a lattice L in V. Then we easily see that ψ is q-reduced (resp. s-reduced) if and only if L is φ-maximal (resp. φ^*-maximal).

When F is a global field, we denote by $\mathbf{a}$ and $\mathbf{h}$ the sets of archimedean primes and nonarchimedean primes of F respectively, and put $\mathbf{v} = \mathbf{a} \cup \mathbf{h}$. For each $v \in \mathbf{v}$ we denote by F_v the v-completion of F. Given an algebraic group G defined over F, we define an algebraic group G_v over F_v for each $v \in \mathbf{v}$ and the adelization $G_{\mathbf{A}}$ as usual, and view G and G_v as subgroups of $G_{\mathbf{A}}$. We then denote by $G_{\mathbf{a}}$ and $G_{\mathbf{h}}$ the archimedean and nonarchimedean factors of $G_{\mathbf{A}}$, respectively. Given a quadratic space (V, φ) over F, we can define, for each $v \in \mathbf{v}$, the v-localization (V_v, φ_v) of (V, φ) as a quadratic space over F_v by putting $V_v = V \otimes_F F_v$ and extending φ to an F_v-valued quadratic form φ_v on V_v in a natural way. For a lattice L in V and $x \in G_{\mathbf{A}}$ we define a lattice Lx and also the G-genus and G-class of L as usual; also for a lattice M contained in L we define a $\mathfrak{g}$-ideal $[L/M]$ so that $[L/M] = \prod_{i=1}^{m} \mathfrak{a}_i$ if L/M is isomorphic as a $\mathfrak{g}$-module to $\oplus_{i=1}^{m} \mathfrak{g}/\mathfrak{a}_i$ with ideals $\mathfrak{a}_i$ of $\mathfrak{g}$.

1.4. Let us now recall some elementary facts on the structure of (V, φ) and $Q(\varphi)$. The reader is referred to [S3, Section 7] and [S4, §§2.6 and 2.7] for more details. We denote by t the core dimension, and define K and ρ as in §1.1. We first discuss the cases $n \leq 3$ with a local or global F.

Case $n = 1$. We have $(V, \varphi) = \langle \delta \rangle$ with $\delta \in \delta_0(\varphi)$, and $Q(\varphi) = M_2(F)$.

Case $n = 2$. Put $\kappa(x, y) = 2^{-1}\,\mathrm{Tr}_{K/F}(xy^{\rho})$ for $x, y \in K$. This is a symmetric form on $K \times K$, and $\kappa[x] = xx^{\rho}$. Then (V, φ) is isomorphic to $(K, c\kappa)$ with $c \in F^{\times}$. The coset $c \cdot N_{K/F}(K^{\times})$ is determined by the condition $Q(\varphi) = \{K, c\}$. In particular,

$$Q(\operatorname{diag}[a, 1]) = M_2(F) \quad \text{for every} \quad a \in F^{\times}. \tag{1.10}$$

Case $n = 3$. Let A be the Clifford algebra of (V, φ) and B the even Clifford algebra of (V, φ). Then B is a quaternion algebra isomorphic to $Q(\varphi)$ and $A = B + B\zeta$ with an element ζ of the center of A such that $\zeta^2 = \delta$ with $\delta \in \delta_0(\varphi)$. Moreover, V as a subset of A coincides with $B^{\circ}\zeta$ and $(x\zeta)^2 = -\delta xx^{\iota}$ for every $x \in B^{\circ}$. In other words, (V, φ) is isomorphic to $(B^{\circ}, -\delta\beta^{\circ})$.

Case $n = 4$. The core dimension t is 0, 2, or 4. Instead of (V, φ) we consider $\langle\varphi\rangle$ with $\varphi \in S_4$. If $t = 0$, then $\varphi \sim \eta_2, 1 \in \delta_0(\varphi)$, and $Q(\varphi) = M_2(F)$. If $t = 2$, then $\varphi \sim \mathrm{diag}[\eta_1, \zeta]$ with $\zeta \in S_2$, and the structure of $\langle\zeta\rangle$ has been determined in Case $n = 2$. Thus $Q(\varphi) = Q(\zeta) = \{K, c\}$ as described there. Since $\delta_0(\varphi) = \delta_0(\zeta)$, the field K is the discriminant field of φ.

If $t = 4$ and F is global, the structure of $\langle\varphi\rangle$ and $Q(\varphi)$ is somewhat involved. We do not need such in the present paper; for details, the reader is referred to [S3, Theorem 7.4c] and [S4, Theorem 7.4 and Corollary 7.5]. If F is local and $t = 4$, then $1 \in \delta_0(\varphi)$ and $Q(\varphi)$ is a division algebra. If we denote it by B, then (V, φ) is isomorphic to (B, β).

1.5. In this section, assuming F to be local, we recall a few basic facts on quaternion algebras over F. (All the facts stated here are either well known or can easily be verified, except for some concerning the maximality, which can be found in [S3, §7.7].) First of all, there is a division algebra B over F, unique up to isomorphisms, which can be given as $B = K + Kj = \{K, \pi\}$ as in (1.3), where K is an unramified quadratic extension of F, and π is a prime element of F. Let $\mathfrak{O} = \{x \in B \mid xx^\iota \in \mathfrak{g}\}$ and let $\mathfrak{r}$ be the maximal order in K. Then $\mathfrak{O}$ is a maximal order in B and $\mathfrak{O} = \mathfrak{r} + \mathfrak{r}j$. Define β and β° as in §1.1 and let γ be the matrix representing β with respect to a $\mathfrak{g}$-basis of $\mathfrak{O}$. Then $\mathfrak{O}$ is β-maximal, and so γ is q-reduced and represents every element of $\mathfrak{g}$. However, $\mathfrak{O}$ is β^*-integral if and only if $2 \notin \mathfrak{p}$.

Next we consider (B°, β°). Put $M = \mathfrak{O} \cap B^\circ$. Take $\xi \in \mathfrak{r}$ so that $\mathfrak{r} = \mathfrak{g}[\xi]$ and put $y = \xi - \xi^\rho$. Then $M = \mathfrak{g}y + \mathfrak{r}j$, M is β°-maximal, and $-\beta^\circ[ay + bj] = a^2y^2 + \pi bb^\rho$ for $a \in F$ and $b \in K$. Let γ° be the matrix representing β° with respect to a $\mathfrak{g}$-basis of M. Then γ° is q-reduced, and

(1.11) $\quad \gamma^\circ$ represents every element of $\pi\mathfrak{g}^\times \cup \{a \in \mathfrak{g}^\times \mid -a \notin F^{\times 2}\}$.

Indeed, since $N_{K/F}(\mathfrak{r}^\times) = \mathfrak{g}^\times$, γ° represents the elements of $\pi\mathfrak{g}^\times$. Next, suppose $a \in \mathfrak{g}^\times$ and $-a \notin F^{\times 2}$. Every quadratic extension of F is embeddable in B, and so there exists an element x of B such that $x^2 = -a$. Then $x \in \mathfrak{O}$. Since $x \notin F$, we have $x^\iota = -x$, and so $x \in M$, and $\beta^\circ[x] = xx^\iota = a$. This proves (1.11).

Now we see that M is $(\beta^\circ)^*$-integral if and only if $2 \notin \mathfrak{p}$ or $\mathfrak{p} = 2\mathfrak{g}$, in which case γ° belongs to $S_3(\mathfrak{g})$, and is s-reduced.

Hereafter whenever we take F to be local, we denote by $\mathscr{D}$ this unique division quaternion algebra over F. Thus $Q(\varphi)$ defined in §1.2 is either $M_2(F)$ or $\mathscr{D}$ if F is local. We also use the letter B for an algebra that is $M_2(F)$ or $\mathscr{D}$. We shall often write $B_1 = B_2$ for such algebras even when $B_1 \cong B_2$ is more appropriate, as no confusion is expected. For instance, the characterization of (V, φ) given in [S4, Lemma 3.3] can be stated as follows:

(1.12) $\quad F$ is local, $\varphi, \psi \in S_n$, $\delta_0(\varphi) = \delta_0(\psi)$, and $Q(\varphi) = Q(\psi) \Longrightarrow \varphi \sim \psi$.

LEMMA 1.6. *Suppose F is local and $n > 2$; then for each $\delta \in F^\times$ and a quaternion algebra B over F, there exists a quadratic space (V, φ) over F such that $\delta \in \delta_0(\varphi)$ and $Q(\varphi) = B$. The same holds for $n = 2$ under the assumption that $B = M_2(F)$ if $\delta \in F^{\times 2}$.*

Proof. First suppose n is odd; put $n - 3 = 2r$ and $\varphi = \mathrm{diag}[\eta_r, \zeta]$, where ζ is a matrix representing $-\delta\beta^\circ$. As shown in §1.4, Case $n = 3$, we have $\delta \in \delta_0(\zeta)$ and $B = Q(\zeta)$, which settles the question, as $\delta_0(\zeta) = \delta_0(\varphi)$ and $Q(\varphi) = Q(\zeta)$. Next suppose $0 < n \in 2\mathbf{Z}$; define K by (1.1) and (1.2) for the given δ. If $K \neq F \times F$, then put $\varphi = \mathrm{diag}[\eta_r, c\kappa]$ with $r = (n-2)/2$ and κ as in §1.4, Case $n = 2$; we take $c \in F^\times$ so that $B = \{K, c\}$, which is feasible. Then $\delta \in \delta_0(c\kappa) = \delta_0(\varphi)$ and $Q(\varphi) = Q(c\kappa) = \{K, c\} = B$. Suppose $K = F \times F$. If $n = 2$, then $\langle \eta_1 \rangle$ gives the desired space. Assuming $n > 2$, put $\varphi = \mathrm{diag}[\eta_s, \zeta]$, where $s = (n-4)/2$ and ζ is a matrix representing the norm form β of B. Then $1 \in \delta_0(\zeta) = \delta_0(\varphi)$ and $Q(\varphi) = Q(\zeta) = B$ as expected. This completes the proof. □

1.7. Let (V, φ) be a quadratic space over a global field F, and let L be a lattice in V. It can easily be seen that L is φ-maximal if and only if L_v is φ_v-maximal for every $v \in \mathbf{h}$; see [S3, Lemma 9.4 (iii)]. Since $\varphi[x+y] - \varphi[x] - \varphi[y] = 2\varphi(x, y)$, we see that L_v is φ_v-maximal if and only if it is φ_v^*-maximal, provided $v \nmid 2$. We already treated φ_v-maximal lattices in [S1], [S3], and [S4], and we need nothing new about φ_v-maximal lattices beyond what was done there. If $v|2$, however, the φ_v^*-maximality is different from, and in fact. more complex than, the φ_v-maximality. We devote the next section to the discussion of φ_v^*-maximal lattices for the primes $v|2$. Before doing it, let us recall some basic facts on φ_v-maximal lattices. Instead of using the symbol F_v, we simply take F to be local and use the letters $\mathfrak{g}$ and $\mathfrak{p}$ as we already defined in the local case. We also take (V, φ) over F and define t and K as in §§1.1 and 1.4.

First of all, we note a basic fact that all φ-maximal lattices in V form a single orbit under $SO^\varphi(V)$; see [S3, Lemma 6.9]. More explicitly, the result of [S3, Lemma 6.5] can be stated in terms of matrices as follows: given a q-reduced element ψ of $2^{-1}T_n$, we have $\psi \approx \mathrm{diag}[2^{-1}\eta_r, \omega]$ with an element $\omega \in 2^{-1}T_t$ that is anisotropic and q-reduced. Moreover, for $M = \mathfrak{g}_n^1$ we have $[M^{2\psi}/M] = \det(2\psi)\mathfrak{g} = \det(2\omega)\mathfrak{g}$, and this can be given as follows:

$$(1.13)\qquad [M^{2\psi}/M] = \begin{cases} 2\delta\mathfrak{g} & \text{if } t = 1, \\ 2\delta^{-1}\mathfrak{p}^2 & \text{if } t = 3, \\ \mathfrak{p}^2 & \text{if } t \in 2\mathbf{Z},\ \mathfrak{d} = \mathfrak{r}, \text{ and } Q(\varphi) = \mathscr{D}, \\ N_{K/F}(\mathfrak{d}) & \text{otherwise,} \end{cases}$$

where $\mathfrak{r}$ is the maximal order in K, $\mathfrak{d}$ is the different of K relative to F, and δ is an element of $\delta_0(\varphi)$ such that $\mathfrak{p} \subset \delta\mathfrak{g} \subset \mathfrak{g}$; we understand that $\mathfrak{d} = \mathfrak{r}$ and $N_{K/F}(\mathfrak{d}) = \mathfrak{g}$ if $K = F \times F$. This was shown in [S3, §7.7] when n is even or

$\delta \in \mathfrak{g}^\times$. The case in which n is odd and $\delta\mathfrak{g} = \mathfrak{p}$ is proved in the proof of [S4, Theorem 6.2]. If L is a lattice contained in M, then

$$[L^{2\psi}/L] = [M/L]^2[M^{2\psi}/M]. \tag{1.14}$$

We take F to be local in the following two lemmas.

LEMMA 1.8. *Given $\tau \in S_n$, we have $\tau \approx \mathrm{diag}\,[\tau_1, \dots, \tau_m]$ with τ_i of size 1 or 2, where each τ_i of size 2 is of the form* $\tau_i = \begin{bmatrix} a & b \\ b & d \end{bmatrix}$, $a\mathfrak{g} + d\mathfrak{g} \subsetneqq b\mathfrak{g} \neq \{0\}$.

This is elementary and well known; for the proof, see [S1, Lemma 13.3 (2)], which includes a similar statement in the hermitian case.

LEMMA 1.9. *If $\varphi \in S_n(\mathfrak{g})$, $\varepsilon \in E_m$, $m < n$, and $\varepsilon = \xi\varphi \cdot {}^t\xi$ with $\xi \in \mathfrak{g}_n^m$, then $\varphi \approx \mathrm{diag}\,[\varepsilon, \psi]$ with $\psi \in S_{n-m}(\mathfrak{g})$. In particular, if φ represents an element e of $\mathfrak{g}^\times$, then $\varphi \approx \mathrm{diag}\,[e, \psi]$ with $\psi \in S_{n-1}(\mathfrak{g})$.*

Proof. We have $\mathfrak{g}_1^m = \varepsilon\mathfrak{g}_1^m \subset \xi\mathfrak{g}_1^n \subset \mathfrak{g}_1^m$, so that $\xi\mathfrak{g}_1^n = \mathfrak{g}_1^m$. Therefore we can find an element α of E_n of the form $\alpha = \begin{bmatrix} \xi \\ \eta \end{bmatrix}$ with $\eta \in \mathfrak{g}_n^{n-m}$. Then we can put $\alpha\varphi \cdot {}^t\alpha = \begin{bmatrix} \varepsilon & {}^t\beta \\ \beta & \omega \end{bmatrix}$ with $\beta \in \mathfrak{g}_m^{n-m}$ and $\omega \in \mathfrak{g}_{n-m}^{n-m}$. Put $\gamma = \begin{bmatrix} 1 & 0 \\ -\beta\varepsilon^{-1} & 1 \end{bmatrix} \alpha$. Then $\alpha\varphi \cdot {}^t\alpha = \mathrm{diag}\,[\varepsilon, \psi]$ with $\psi \in S_{n-m}(\mathfrak{g})$. □

LEMMA 1.10. *Given $\varphi \in S_n$, let $A(\varphi)$ denote the Clifford algebra of $\langle\varphi\rangle$ and $A^+(\varphi)$ the even Clifford algebra of $\langle\varphi\rangle$; let $\delta \in \delta_0(\varphi)$ and $\psi \in S_m$. Then the following assertions hold.* (i) *If n is even, then $A(\mathrm{diag}\,[\varphi, \psi]) \cong A(\varphi) \otimes_F A(\delta\psi)$;* (ii) *If n is odd, then $A^+(\mathrm{diag}\,[\varphi, \psi]) \cong A^+(\varphi) \otimes_F A(-\delta\psi)$.*

This holds for every field F; see [S4, Lemma 2.8] for the proof.

2. Classification of s-reduced matrices in the dyadic local case.

2.1. In this section we assume that F is local and $2 \in \mathfrak{p}$. We put $\mathfrak{g}^{\times 2} = \{a^2 \mid a \in \mathfrak{g}^\times\}$ and $1 + \mathfrak{a} = \{1 + a \mid a \in \mathfrak{a}\}$ for an ideal $\mathfrak{a}$ of $\mathfrak{g}$. Here are some useful facts:

$$\mathfrak{g}^\times = \{u^2 + x \mid u \in \mathfrak{g}^\times, x \in \mathfrak{p}\} = \mathfrak{g}^{\times 2}(1 + \mathfrak{p}); \tag{2.1}$$

$$\mathrm{diag}\,[1, \eta_1] \approx \mathrm{diag}\,[1, 1, -1]; \tag{2.2}$$

$$c\eta_r \approx \eta_r \text{ for every } c \in \mathfrak{g}^\times; \tag{2.3}$$

(2.4) $\quad \mathrm{diag}[g,h] \approx c\cdot \mathrm{diag}[g,h]$ if $a,b,g,h \in \mathfrak{g}$ and $c = g(a^2g+b^2h) \in \mathfrak{g}^\times$;

(2.5) $\quad \mathrm{diag}[1,-1] \approx c\cdot \mathrm{diag}[1,-1]$ for every $c \in \mathfrak{g}^\times$, provided $\mathfrak{p} = 2\mathfrak{g}$;

(2.6) $\quad \mathrm{diag}[1,-d] \approx c\cdot \mathrm{diag}[1,-d]$ for every $c \in \mathfrak{g}^\times$, if $d \in \mathfrak{g}^\times$, $F(d^{1/2})$ is an unramified quadratic extension of F, and $\mathfrak{p} = 2\mathfrak{g}$.

Indeed, since $(\mathfrak{g}/\mathfrak{p})^\times$ is of an odd order, given $c \in \mathfrak{g}^\times$, we can find $u \in \mathfrak{g}^\times$ such that $c - u^2 \in \mathfrak{p}$, which gives (2.1). As for (2.2) which is true even when $2 \notin \mathfrak{p}$, we have $\alpha \cdot \mathrm{diag}[1,\eta_1]\cdot {}^t\alpha = \mathrm{diag}[1,1,-1]$ with

$$\alpha = \begin{bmatrix} 1 & 0 & -1 \\ 1 & 1 & 0 \\ 1 & 1 & -1 \end{bmatrix}.$$

Put $\alpha = \mathrm{diag}[1_r, c1_r]$. Then ${}^t\alpha\eta_r\alpha = c\eta_r$, which gives (2.3). Next, in the setting of (2.4) put $\alpha = \begin{bmatrix} ga & -bh \\ gb & ga \end{bmatrix}$. Then $\det(\alpha) = c$ and ${}^t\alpha\cdot\mathrm{diag}[g,h]\alpha = c\cdot\mathrm{diag}[g,h]$, which proves (2.4). As for (2.5), given $c \in \mathfrak{g}^\times$, by (2.1) we can put $c = u^2+2y$ with $u \in \mathfrak{g}^\times$ and $y \in \mathfrak{g}$. Put $b = yu^{-1}$ and $a = u+b$. Then $a^2-b^2 = c$, and so (2.5) follows from (2.4). To prove (2.6), put $K = F(d^{1/2})$ and denote by $\mathfrak{r}$ the maximal order of K. Then $1+\mathfrak{p}^m = N_{K/F}(1+\mathfrak{p}^m\mathfrak{r})$ for every $m \in \mathbf{Z}, > 0$. This is elementary; see [S1, Lemma 17.6 (1)] for the proof. Since the discriminant ideal of the order $\mathfrak{g}[d^{1/2}]$ relative to $\mathfrak{g}$ is $4\mathfrak{g}$, we have $\mathfrak{g}[d^{1/2}] = \mathfrak{g}+2\mathfrak{r}$ by the general principle about such an order in K; see [S3, §11.8]. Given $c \in \mathfrak{g}^\times$, by (2.1), $c = u^2+x$ with $u \in \mathfrak{g}^\times$ and $x \in \mathfrak{p}$. Then $1+u^{-2}x = N_{K/F}(y)$ with $y \in 1+\mathfrak{p}\mathfrak{r}$, and so $c = N_{K/F}(uy)$, and $uy \in \mathfrak{g}+\mathfrak{p}\mathfrak{r} = \mathfrak{g}[d^{1/2}]$ as $\mathfrak{p} = 2\mathfrak{g}$. Put $uy = a+bd^{1/2}$ with $a,b \in \mathfrak{g}$. Then $c = a^2-b^2d$ and we obtain (2.6) from (2.4).

It should be noted that the condition $\mathfrak{p} = 2\mathfrak{g}$ in (2.5) and (2.6) is indispensable. We shall discuss this point in more detail in §3.11.

LEMMA 2.2 (i) *Let* $\tau = \mathrm{diag}[\tau_1,\dots,\tau_m] \in S_n(\mathfrak{g})$ *with* τ_i *as in Lemma 1.8. Suppose* τ *is s-reduced. Then every* τ_i *of size 2 belongs to* E_2 *and there is at most one* τ_i *of size 1 not contained in* $\mathfrak{g}^\times$*; such a* τ_i *not contained in* $\mathfrak{g}^\times$ *is a prime element of* F*. Moreover, every* τ_i *of size 2 belongs to* T_2 *if* $\mathfrak{p} = 2\mathfrak{g}$.

(ii) *An element* τ *of* $S_n(\mathfrak{g})$ *is s-reduced if and only if* $\tau \in E_n$ *or* $\det(\tau)\mathfrak{g} = \mathfrak{p}$.

Proof. Suppose $\tau_i = \begin{bmatrix} a & b \\ b & d \end{bmatrix}$ and $a\mathfrak{g}+d\mathfrak{g} \subsetneq b\mathfrak{g} \neq \{0\}$. If $b \in \mathfrak{p}$, then $\alpha\tau_i\cdot{}^t\alpha \in S_n(\mathfrak{g})$ for $\alpha = \mathrm{diag}[1,\pi^{-1}]$ with a prime element π, and so τ cannot be s-reduced. Therefore if τ is s-reduced, then $b \in \mathfrak{g}^\times$, and so $\tau_i \in E_2$; clearly $\tau_i \in T_2$ if $\mathfrak{p} = 2\mathfrak{g}$. Similarly τ cannot be s-reduced if $\tau_i \in \mathfrak{p}^2$ for some i. Thus every τ_i of size 1 belongs to $\mathfrak{g}^\times$ or is a prime element of F. Suppose τ_1 and τ_2 are prime elements.

By (2.1) we have $-\tau_2/\tau_1 = u^2 - \tau_1 a$ with some $u, a \in \mathfrak{g}$. Put $\alpha = \begin{bmatrix} 1 & 0 \\ -u & \tau_1 \end{bmatrix}$ and $\psi = \begin{bmatrix} \tau_1 & u \\ u & a \end{bmatrix}$. Then $\alpha\psi \cdot {}^t\alpha = \mathrm{diag}[\tau_1, \tau_2]$, which cannot happen if τ is s-reduced. Thus we obtain (i), which immediately implies the "only if"-part of (ii). The "if"-part of (ii) follows from (1.8). □

2.3. We now put

$$J_n = S_n \cap E_n, \qquad D_n = \{\psi \in S_n(\mathfrak{g}) \mid \det(\psi)\mathfrak{g} = \mathfrak{p}\}, \tag{2.7a}$$

$$J_n^2 = J_n \cap T_n, \qquad J_n^1 = \{\psi \in J_n \mid \psi \notin J_n^2\}, \tag{2.7b}$$

$$D_n^2 = D_n \cap T_n, \qquad D_n^1 = \{\psi \in D_n \mid \psi \notin D_n^2\}. \tag{2.7c}$$

By Lemma 2.2 (ii), $J_n \cup D_n$ is the set of s-reduced elements of $S_n(\mathfrak{g})$. Clearly $J_n = J_n^1 \sqcup J_n^2$, $D_n = D_n^1 \sqcup D_n^2$; also J_n^λ and D_n^λ are stable under the maps $\psi \mapsto \alpha\psi \cdot {}^t\alpha$ for all $\alpha \in E_n$. From Lemmas 1.8 and 2.2 we see that $J_n^2 = \emptyset$ if n is odd and $D_n^2 = \emptyset$ if n is even. For $\varphi \in S_n$ we put

$$J_n(\varphi) = \{\psi \in J_n \mid \psi \sim \varphi\}, \quad J_n^\lambda(\varphi) = J_n^\lambda \cap J_n(\varphi) \quad (\lambda = 1, 2), \tag{2.8a}$$

$$D_n(\varphi) = \{\psi \in D_n \mid \psi \sim \varphi\}, \quad D_n^\lambda(\varphi) = D_n^\lambda \cap D_n(\varphi) \quad (\lambda = 1, 2). \tag{2.8b}$$

In the rest of this section we assume

$$\mathfrak{p} = 2\mathfrak{g}. \tag{2.9}$$

This means that F is unramified over $\mathbf{Q}_2$.

Let (V, φ) be a quadratic space over F. We say that a φ^*-integral lattice L in V is of type 1 (with respect to φ) if $\varphi[x] \in \mathfrak{g}^\times$ for some $x \in L$; otherwise we say that L is of type 2. Under our assumption (2.9), L is of type 2 if and only if L is $(2^{-1}\varphi)$-integral. Such an L is φ^*-maximal only if it is $(2^{-1}\varphi)$-maximal. However, a $(2^{-1}\varphi)$-maximal lattice is not necessarily φ^*-maximal. The lattice $\mathfrak{g}_n^1$ in $\langle\varphi\rangle$ for $\varphi \in S_n(\mathfrak{g})$ is of type 2 if and only if $\varphi \in T_n$.

The principal aim of this section is to prove:

THEOREM 2.4. *The following assertions hold if F is local and $\mathfrak{p} = 2\mathfrak{g}$:*

(i) *Let $\varphi \in S_n$. Then $J_n(\varphi) \neq \emptyset$ if and only if $\delta_0(\varphi) \cap \mathfrak{g}^\times \neq \emptyset$, and $D_n(\varphi) \neq \emptyset$ if and only if $\delta_0(\varphi) \cap \mathfrak{g}^\times = \emptyset$.*

(ii) *Each of the sets $J_n^\lambda(\varphi)$ and $D_n^\lambda(\varphi)$ for a fixed $\varphi \in S_n(\mathfrak{g})$, if nonempty, consists of a single equivalence class over $\mathfrak{g}$.*

(iii) *Let* (V, φ) *be a quadratic space over* F; *let* L *and* M *be* φ^**-maximal lattices in* V *of the same type. Then* $L = M\omega$ *for some* $\omega \in SO^{\varphi}(V)$.

Proof. The "only if"-parts of (i) are obvious. Conversely, given $\varphi \in S_n$, take a φ^*-maximal lattice L in $\langle\varphi\rangle$ and denote by ψ the matrix representing φ with respect to a $\mathfrak{g}$-basis of L. Then ψ is s-reduced as observed in §1.3, and so $\psi \in J_n \cup D_n$, from which we can derive the "if"-parts of (i).

Next, let us prove that (iii) follows from (ii). For that purpose we take (V, φ) to be $\langle\varphi\rangle$ with $\varphi \in S_n$. Given L and M as in (iii), we can put $L = \mathfrak{g}_n^1\alpha$ and $M = \mathfrak{g}_n^1\beta$ with $\alpha, \beta \in GL_n(F)$. Put $\gamma = \alpha\varphi \cdot {}^t\alpha$ and $\varepsilon = \beta\varphi \cdot {}^t\beta$. Then γ and ε are s-reduced, as L and M are φ^*-maximal. If $\delta_0(\varphi) \cap \mathfrak{g}^\times \neq \emptyset$, then $\gamma, \varepsilon \in J_n(\varphi)$; moreover, $\gamma, \varepsilon \in J_n^\lambda(\varphi)$ if L and M are of type λ. Similarly $\gamma, \varepsilon \in D_n^\lambda(\varphi)$ if $\delta_0(\varphi) \cap \mathfrak{g}^\times = \emptyset$. Therefore, by (ii), $\xi\gamma \cdot {}^t\xi = \varepsilon$ with some $\xi \in E_n$. Then $\beta^{-1}\xi\alpha \in O(\varphi)$, and $\det(\beta^{-1}\xi\alpha) = \pm 1$. By [S3, Lemma 6.8] there exists an element ζ of $O(\varphi)$ such that $M\zeta = M$ and $\det(\zeta) = \det(\beta^{-1}\xi\alpha)$. Put $\omega = \zeta\beta^{-1}\xi\alpha$. Then $\omega \in SO(\varphi)$ and $M\omega = L$. Thus we obtain (iii).

The proof of (ii) will be completed in §2.10 for J_n^λ and in §2.14 for D_n^λ. □

It should be noted that Theorem 2.4(ii) is false if F is ramified over $\mathbf{Q}_2$, as will be shown in §3.11.

Hereafter until the end of this section, for simplicity, we write $[a, b, \ldots]$ for $\mathrm{diag}[a, b, \ldots]$, though we use the latter in Lemmas and displayed formulas.

LEMMA 2.5. *For* $\varphi \in S_n$ *let* t *be the core dimension of* φ *and let* $n = 2r + t$. *Then the following assertions hold:*

(i) *If* $\tau \in J_n^2(\varphi)$ *or* $\tau \in D_n^2(\varphi)$, *then* $\mathfrak{g}_n^1$ *is* $(2^{-1}\tau)$*-maximal,* $2^{-1}\tau$ *is* q*-reduced, and* $\tau \approx [\eta_r, \gamma]$ *with* $\gamma \in J_t^2$ *or* $\gamma \in D_t^2$ *accordingly.*

(ii) *If* $J_n^2(\varphi) \neq \emptyset$, *then it consists of a single equivalence class over* $\mathfrak{g}$.

(iii) *We have* $D_n^2(\varphi) \neq \emptyset$ *if and only if* $t = 1$ *and* $\mathfrak{g}^\times \cap \delta_0(\varphi) = \emptyset$, *in which case* $D_n^2(\varphi) = \{\psi \in S_n(\mathfrak{g}) \mid \psi \approx \mathrm{diag}[\pi, \eta_r]\}$, *where* π *is a prime element contained in* $\delta_0(\varphi)$ *and* $r = (n-1)/2$.

Proof. Given τ as in (i), put $\psi = 2^{-1}\tau, V = F_n^1$, and $L = \mathfrak{g}_n^1$. Then L is ψ-maximal as observed in §2.3. Thus ψ is q-reduced, and as noted a few lines above (1.13), $\psi \approx \mathrm{diag}[2^{-1}\eta_r, \omega]$ with a q-reduced ω. Put $\gamma = 2\omega$; then $\det(\gamma)\mathfrak{g} = \det(\tau)\mathfrak{g}$, and so we obtain (i). To prove (ii), let $\sigma, \tau \in J_n^2(\varphi)$ and let $\psi = 2^{-1}\tau$ as above. Then $\sigma = \alpha\tau \cdot {}^t\alpha$ with $\alpha \in GL_n(F)$. Since L is $(2^{-1}\sigma)$-maximal, $L\alpha$ is ψ-maximal. Therefore $L\alpha\beta = L$ with $\beta \in SO(\psi)$ for the reason explained in §1.7. Put $\alpha\beta = \varepsilon$. Then $\varepsilon \in E_n$ and $\sigma = \varepsilon\tau \cdot {}^t\varepsilon$, which proves (ii). Next suppose $\tau \in D_n^2(\varphi)$. Then $\delta_0(\varphi)$ contains a prime element, say π, and $[L^\tau/L] = \mathfrak{p}$. From (1.13) and (1.14) we see that this can happen only if $t = 1$. (If n is odd, then $\delta_0(\psi) = 2\delta_0(\tau)$, and so $\delta_0(\psi) \cap \mathfrak{g}^\times \neq \emptyset$. Also, if $t = 2$, we have $N_{K/F}(\mathfrak{d}) = \mathfrak{g}$ or $N_{K/F}(\mathfrak{d}) \subset \mathfrak{p}^2$, as $2 \in \mathfrak{p}$.) Then $\tau \approx [\pi, \eta_r]$ by (i). Conversely, if $t = 1$ and $\delta_0(\varphi)$ contains a prime element π, then $[\pi, \eta_r] \in D_n^2(\varphi)$. This proves (iii). □

2.6. Theorem 2.4(ii) is trivial if $n = 1$. Let us now determine $J_n^\lambda(\varphi)$ when $n = 2$. Fixing $\varphi \in S_2$, let t and K be as in §1.4. In Case $n = 2$ there we have seen that $\langle\varphi\rangle$ is isomorphic to $(K, c\kappa)$ with $c \in F^\times$. First suppose $t = 0$; then $K = F \times F$ and $Q(\varphi) = M_2(F)$. Let $\psi \in J_2^1(\varphi)$ in such a case; then by Lemma 1.8, $\psi \approx [a, b]$ with $a, b \in \mathfrak{g}^\times$. Since $t = 0$, we can put $-b = ac^2$ with $c \in \mathfrak{g}^\times$. Therefore $\psi \approx \varepsilon_a$ with $\varepsilon_a = [a, -a]$. Since $\varepsilon_a \approx \varepsilon_1$ by (2.5), we see that $J_2^1(\varphi)$ is represented by $[1, -1]$. Clearly $\eta_1 \in J_2^2(\varphi)$, and by Lemma 2.5 (ii), $J_2^2(\varphi)$ is represented by η_1.

Next suppose $t = 2$; then K is a quadratic extension of F. Let $\mathfrak{r}$ be the maximal order in K. We assume that $J_2(\varphi) \neq \emptyset$, and so $K = F(\sqrt{\delta})$ with $\delta \in \mathfrak{g}^\times \cap \delta_0(\varphi)$. Then $Q(\varphi) = \{K, c\}$. If $\psi \in J_2(\varphi)$, then ψ corresponds to a $(c\kappa)^*$-maximal lattice L in K. Thus $cxx^\rho \in \mathfrak{g}$ for every $x \in L$. The constant c can be replaced by any element of $cN_{K/F}(K^\times)$.

Case 1: K is unramified over F and $Q(\varphi) = \mathscr{D}$. Then we may assume that $c = 2$. Thus $2xx^\rho \in \mathfrak{g}$ for every $x \in L$, and so $L \subset \mathfrak{r}$. Since $c\kappa(\mathfrak{r}, \mathfrak{r}) = \mathrm{Tr}_{K/F}(\mathfrak{r}) = \mathfrak{g}$ and L is $(c\kappa)^*$-maximal, we have $L = \mathfrak{r}$. Therefore ψ is the matrix representing $c\kappa$ with respect to a $\mathfrak{g}$-basis of $\mathfrak{r}$, and $\psi \in J_2^2$. Thus $J_2^2(\varphi)$ consists of a unique equivalence class over $\mathfrak{g}$ (which is a special case of Lemma 2.5 (ii)), and $J_2^1(\varphi) = \emptyset$.

Case 2: K is unramified over F and $Q(\varphi) = M_2(F)$. Then we can take $c = 1$. If L is of type 2, then $L \subset \{x \in K \mid xx^\rho \in 2\mathfrak{g}\} = 2\mathfrak{r}$, but the matrix representing κ on $2\mathfrak{r}$ does not belong to E_2, a contradiction. Thus $J_2^2(\varphi) = \emptyset$. Let $\psi \in J_2^1(\varphi)$. By Lemma 2.2, we have $\psi \approx e[1, -d]$ with $e, d \in \mathfrak{g}^\times$. Then $d \in \delta_0(\varphi)$, $K = F(d^{1/2})$, and $d\mathfrak{g}^{\times 2}$ is uniquely determined. Thus $J_2(\varphi) = J_2^1(\varphi)$, which consists of a single class over $\mathfrak{g}$ represented by $[1, -d]$, as $e[1, -d] \approx [1, -d]$ by (2.6).

Case 3: K is ramified over F. Then we may assume that $\mathfrak{r} = \mathfrak{g} + \mathfrak{g}\sqrt{\delta}$ with $\delta \in \mathfrak{g}^\times \cap \delta_0(\varphi)$ and $c \in \mathfrak{g}^\times$, and so $L \subset \mathfrak{r}$. Thus $c\kappa$ can be represented by $c[1, -\delta]$ with respect to the $\mathfrak{g}$-basis $\{1, \sqrt{\delta}\}$ of $\mathfrak{r}$. If $\psi \in J_2(\varphi)$, then L is ψ^*-maximal, and so $L = \mathfrak{r}$ and $\psi \approx c[1, -\delta]$. Now $cN_{K/F}(K^\times)$ is determined by $Q(\varphi)$ and $F^\times / N_{K/F}(K^\times) \cong \mathfrak{g}^\times / N_{K/F}(\mathfrak{r}^\times)$. Thus, in view of (2.4), we see that $J_2^1(\varphi)$ consists of a single class over $\mathfrak{g}$, and $J_2^2(\varphi) = \emptyset$.

LEMMA 2.7. *Given $\varphi \in J_3$, put $B = Q(\varphi)$. Then $\varphi \approx -e\gamma^\circ$ if $B = \mathscr{D}$, and $\varphi \approx e \cdot \mathrm{diag}[1_2, -1]$ if $B = M_2(F)$, where e is an element of $\mathfrak{g}^\times \cap \delta_0(\varphi)$, and γ° is the matrix of §1.5.*

Proof. As observed in §1.4, Case $n = 3$, there is an F-linear isomorphism f of B° onto F_3^1 such that $\varphi[xf] = -exx^\iota$ for $x \in B^\circ$ with $e \in \delta_0(\varphi)$. Since $\varphi \in J_3$, we may assume that $e \in \mathfrak{g}^\times$. Put $L = \mathfrak{g}_3^1$. Then L is contained in a φ-maximal lattice M, and Mf^{-1} is β°-maximal. Suppose $B = \mathscr{D}$; then applying (1.13) to Mf^{-1}, we obtain $[M^{2\varphi}/M] = \mathfrak{p}^3$, as $\mathfrak{p} = 2\mathfrak{g}$. Also, $[L^{2\varphi}/L] = \det(2\varphi)\mathfrak{g} = \mathfrak{p}^3$. Thus (1.14) shows that $L = M$. This proves the desired fact when $B = \mathscr{D}$. Next suppose $B = M_2(F)$. By Lemma 1.8 we may assume that $\varphi = [b, \psi]$ with $b \in \mathfrak{g}^\times$

and $\psi \in J_2$. Now, by (1.12), $[b, \psi] \sim [b, -b, e]$ with $e \in \delta_0(\varphi) \cap \mathfrak{g}^\times$, and so $\psi \sim [-b, e]$. This means that $J_2^1(\psi) \neq \emptyset$. The results of §2.6 show that such can happen only if $\psi \in J_2^1$ or $\psi \approx \eta_1$. If $\psi \in J_2^1$, then the same results show that $\psi \approx [-b, e]$, as $[-b, e] \in J_2^1$. Thus $[b, \psi] \approx [b, -b, e] \approx e[1_2, -1]$ by (2.5). Suppose $\psi \approx \eta_1$; then $\varphi \approx [b, b\eta_1] \approx b[1_2, -1]$ by (2.3) and (2.2). This completes the proof. □

LEMMA 2.8. (i) *Every element of* $\mathfrak{g}^\times$ *can be represented by any fixed element of* J_n^1 *at least in the following two cases*: (a) $n > 3$; (b) $n = 3$ *and* φ *is isotropic.*

(ii) *Every element of* $2\mathfrak{g}$ *can be represented by any fixed element of* J_n^2, $n \geq 4$.

(iii) *Every element of* $\mathfrak{g}^\times$ *can be represented by any fixed element of* D_n^1, $n \geq 3$.

Proof. To prove (ii), let $\varphi \in J_n^2$ with $n \geq 4$. Then $\mathfrak{g}_n^1$ is $(2^{-1}\varphi)$-maximal by Lemma 2.5 (i). If φ is anisotropic, then $n = 4$ and $2^{-1}\varphi$ is γ of §1.5, which represents every element of $\mathfrak{g}$. If φ is isotropic, then $\varphi \approx [\eta_1, \zeta]$ with $\zeta \in J_{n-2}$ by Lemma 2.5 (i). Clearly η_1 represents every element of $2\mathfrak{g}$. Thus we obtain (ii).

To prove (i), first take an isotropic element φ of J_3^1. By Lemma 2.7 and (2.2), $\varphi \approx e[1, \eta_1]$ with $e \in \mathfrak{g}^\times$. Given $a \in \mathfrak{g}^\times$, by (2.1), $a/e = u^2 + 2x$ with $u, x \in \mathfrak{g}$. Thus $e[1, \eta_1]$ represents a. Next, let $\varphi \in J_4^1$. By Lemma 2.2, $\varphi \approx [a, \tau]$ with $a \in \mathfrak{g}^\times$ and $\tau \in J_3$. The matter is settled if τ is isotropic, and so we assume that τ is anisotropic. Then $\tau \approx c\gamma^\circ$ with $c \in \mathfrak{g}^\times$ by Lemma 2.7. Replacing φ by $c^{-1}\varphi$, we may assume that $\varphi = [b, \gamma^\circ]$ with $b \in \mathfrak{g}^\times$. Suppose φ is isotropic. Then $-b = \gamma^\circ[x]$ with some $x \in F_3^1$. Since $b \in \mathfrak{g}$, we have $x \in \mathfrak{g}_3^1$, as $\mathfrak{g}_3^1$ corresponds to $M = \mathfrak{O} \cap B^\circ$ of §1.5. Thus γ° represents $-b$, and so $\gamma^\circ \approx [-b, \varepsilon]$ with $\varepsilon \in J_2$. Consequently $\varphi \approx [b, -b, \varepsilon]$, which represents every element of $\mathfrak{g}^\times$ by (2.5). Suppose φ is anisotropic; then γ° does not represent $-b$, and so $b \in F^{\times 2}$ by (1.11). Thus $b \in \mathfrak{g}^{\times 2}$, and $\varphi \approx [1, \gamma^\circ]$. Let $e \in \mathfrak{g}^\times$. If $-e \notin \mathfrak{g}^{\times 2}$, then γ° represents e by (1.11); so assume that $-e = u^2$ with $u \in \mathfrak{g}^\times$. Then $e = -u^2 = u^2 - 2u^2$. By (1.11), γ° represents $-2u^2$, and so $[1, \gamma^\circ]$ represents e. This proves (i) for $n = 4$.

We now prove (i) by induction on n. Let $\varphi \in J_n^1$ with $n > 4$. By Lemma 2.2, $\varphi \approx [b, \psi]$ with $b \in \mathfrak{g}^\times$ and $\psi \in J_{n-1}$. If $\psi \in J_{n-1}^1$, then the problem is settled by induction. Suppose $\psi \in J_{n-1}^2$. Given $c \in \mathfrak{g}^\times$, we can put $c/b = u^2 + 2x$ with $u, x \in \mathfrak{g}$ by (2.1). Then $c = bu^2 + 2bx$. Since $n - 1 \geq 4$, ψ represents $2bx$ by (ii), and so φ represents c as expected. This proves (i). The proof of (iii) will be given in §2.14. □

LEMMA 2.9. *Given* $\varphi \in S_n$, *let* t *be the core dimension of* φ *and* K *the discriminant algebra of* φ. *Then the following assertions hold.*

(i) *We have* $J_n(\varphi) \neq \emptyset$ *if and only if* $\delta_0(\varphi) \cap \mathfrak{g}^\times \neq \emptyset$, *in which case* $J_n^1(\varphi) \neq \emptyset$ *with the following exception:* $J_n^1(\varphi) = \emptyset$ *if* $n = t = 2$, K *is unramified over* F, *and* $Q(\varphi) = \mathscr{D}$.

(ii) *We have* $J_n^2(\varphi) \neq \emptyset$ *exactly in the following two cases*: (a) $t = 0$; (b) $t = 2$, K *is unramified over* F, *and* $Q(\varphi) = \mathscr{D}$.

Proof. The first part of (i) is included in Theorem 2.4(i). Now suppose $\delta_0(\varphi) \cap \mathfrak{g}^\times \neq \emptyset$; put $n = t + 2r$. If $t = 0$, then $\eta_r \in J_n^2(\varphi)$ and $[1_r, -1_r] \in J_n^1(\varphi)$. Assuming $t > 0$, take τ and ζ as in (1.6). Let ω be the matrix representing ζ with respect to a $\mathfrak{g}$-basis of a ζ^*-maximal lattice in $\langle \zeta \rangle$. Then $\omega \in J_t(\zeta)$. Put $\psi = [1_r, -1_r, \omega]$. Then $\psi \in J_n^1(\varphi)$ if $r > 0$ or t is odd. If $\gamma \in J_t^2(\zeta)$, then $[\eta_r, \gamma] \in J_n^2(\varphi)$. Conversely, if $\alpha \in J_n^2(\varphi)$, then by Lemma 2.5 (i), $\alpha \approx [\eta_r, \beta]$ with $\beta \in J_t^2(\zeta)$. Thus $J_n^2(\varphi) \neq \emptyset$ if and only if $J_t^2(\zeta) \neq \emptyset$. Let $\gamma \in J_t^2(\zeta)$. Then $2^{-1}\gamma$ is q-reduced, $\delta_0(2^{-1}\gamma) = 2^t \delta_0(\gamma)$, and $\det(2 \cdot 2^{-1}\gamma) \in \mathfrak{g}^\times$. From (1.13) we see that such is possible only if $t = 0$ or $t = 2$. Thus $J_n^2(\varphi) = \emptyset$ if $t > 0, \neq 2$, and $J_n^1(\varphi) \neq \emptyset$ then. For $t = 2$, in §2.6 we have seen that $J_2^2(\zeta) \neq \emptyset$ if and only if K is unramified over F and $Q(\zeta) = \mathscr{D}$, in which case $J_2^1(\zeta) = \emptyset$. This completes the proof. □

2.10. *Proof of Theorem 2.4(ii) for J_n^λ.* The case $\lambda = 2$ is proved in Lemma 2.5 (ii), and so we take $\lambda = 1$. The matter is easy if $n = 1$; also the cases $n = 2$ and $n = 3$ are settled in §2.6 and Lemma 2.7. Let $\sigma, \tau \in J_n^1$ with $n \geq 4$. Suppose $\sigma \sim \tau$. By Lemmas 2.8 (i) and 1.9 we may assume that $\sigma = [1, \varphi]$ and $\tau = [1, \psi]$ with $\varphi, \psi \in J_{n-1}$. We have $\varphi \sim \psi$ by Witt's theorem. If n is even, then both φ and ψ belong to J_{n-1}^1, and the matter is settled by induction. Thus suppose n is odd, ≥ 5. The only case we have to check is when $\varphi \in J_{n-1}^1$ and $\psi \in J_{n-1}^2$, as induction applies otherwise. By Lemma 2.9 (ii), ψ must be isotropic. Therefore $\psi \approx [\eta_1, \omega]$ with $\omega \in J_{n-3}$ by Lemma 2.5 (i). Thus $[1, \varphi] \approx \sigma \sim \tau \approx [1, \eta_1, \omega] \approx [1, 1, -1, \omega]$ by (2.2). Then $\varphi \sim [1, -1, \omega]$. Since $\varphi \in J_{n-1}^1$, induction shows that $\varphi \approx [1, -1, \omega]$, and so $\sigma \approx \tau$. This completes the proof.

2.11. We already determined $D_n^2(\varphi)$ in Lemma 2.5 (iii). Our next task is to investigate $D_n^1(\varphi)$. We first examine the case $n = 2$. Given $\varphi \in D_2$, define K and κ as in §1.4. By Theorem 2.4(i) and Lemma 2.5 (iii), $D_2^1(\varphi) \neq \emptyset$. Let $\psi \in D_2^1(\varphi)$. By Lemma 2.2, $\psi \approx [c, \pi]$ with $c \in \mathfrak{g}^\times$ and a prime element π. Putting $\delta = -\pi/c$, we see that $\psi \approx c[1, -\delta], \delta \in \delta_0(\varphi), K = F(\delta^{1/2})$, and $\mathfrak{g} + \mathfrak{g}\delta^{1/2}$ is the maximal order in K. Since κ is represented by $[1, -\delta]$, we have $Q(\varphi) = \{K, c\}$. Therefore, for the same reason as in Case 3 of §2.6, $D_2^1(\varphi)$ consists of a single class over $\mathfrak{g}$.

LEMMA 2.12. *Given $\psi \in D_3^1$, put $Q(\psi) = B$. Then $\psi \approx \operatorname{diag}[\pi, 1, -1]$ if $B = M_2(F)$ and $\psi \approx \operatorname{diag}[\pi, 1, -d]$ if $B = \mathscr{D}$, where d is an element of $\mathfrak{g}^\times$ uniquely determined modulo $\mathfrak{g}^{\times 2}$ such that $F(d^{1/2})$ is an unramified quadratic extension of F, and π is a prime element of F determined modulo $\mathfrak{g}^{\times 2}$ by $\delta_0(\psi)$. Consequently, if $\varphi \in S_3$ and $\delta_0(\varphi) \cap \mathfrak{g}^\times = \emptyset$, then $D_3^1(\varphi)$ is nonempty and consists of a single equivalence class over $\mathfrak{g}$.*

Proof. By Lemmas 1.8 and 2.2 we have $\psi \approx [\pi, \tau]$ with a prime element π and $\tau \in J_2^1$. As shown in §2.6, we may assume that τ has one of the following forms: $[1, -1], [1, -d], c[1, -e]$. Here d is as in our lemma and $c, e \in \mathfrak{g}^\times$; $F(e^{1/2})$

is ramified over F. We have $B = M_2(F)$ if $\tau = [1, -1]$ and $B = \mathscr{D}$ if $\tau = [1, -d]$, as no prime element can be written $x^2 - dy^2$ with $x, y \in F$. Both types of B can happen for $\tau = c[1, -e]$. Fix $e \in \mathfrak{g}^\times$ and assume that $Q(\psi) = M_2(F)$ for $\psi = [\pi, c, -ce]$. By (1.12), $[\pi, c, -ce] \sim [e\pi, c, -c]$. By Witt's theorem, $[\pi, -ce] \sim [e\pi, -c]$. By the result of §2.11, we have $[\pi, -ce] \approx [e\pi, -c]$, and so $[\pi, c, -ce] \approx [e\pi, c, -c] \approx [e\pi, 1, -1]$ by (2.5). This proves our lemma when $B = M_2(F)$.

Next, suppose $B = \mathscr{D}$ and $\psi = [\pi, c, -ce]$. Take d as above. We have $[c, -cd] \approx [1, -d]$ by (2.6) and $Q([de\pi, c, -cd]) = \mathscr{D}$, as no prime element can be written $cx^2 - cdy^2$ with $x, y \in F$. Then $[\pi, c, -ce] \sim [de\pi, c, -cd]$ by (1.12), and so $[\pi, -ce] \sim [de\pi, -cd]$. Thus, by the result of §2.11, $[\pi, -ce] \approx [de\pi, -cd]$, and so $[\pi, c, -ce] \approx [de\pi, c, -cd] \approx [de\pi, 1, -d]$ by (2.6). This proves our lemma when $B = \mathscr{D}$ and completes the proof. □

LEMMA 2.13. *For $\varphi \in S_4$ the set $D_4^1(\varphi)$ is nonempty if and only if $\delta_0(\varphi)$ contains a prime element, in which case $D_4^1(\varphi)$ consists of a single equivalence class over $\mathfrak{g}$ represented by* $\mathrm{diag}[\pi, 1_2, -1]$ *if $B = M_2(F)$ and by* $\mathrm{diag}[\pi, 1_2, -d]$ *if $B = \mathscr{D}$, where d is as in Lemma 2.12 and π is a prime element of F determined modulo $\mathfrak{g}^{\times 2}$ by $\delta_0(\varphi)$.*

Proof. Let $\psi \in D_4^1(\varphi)$. We first show that ψ represents 1. By Lemmas 1.8 and 2.2, we have $\psi \approx [a, \tau]$ with $a \in \mathfrak{g}^\times$ and $\tau \in D_3$. If $\tau \in D_3^1$, then τ represents 1 by Lemma 2.12; so suppose $\tau \in D_3^2$. Then, by Lemma 2.5 (iii), $\tau \approx [\pi, \eta_1]$ with a prime lement π. Thus $\psi \approx [\pi, \omega]$ with $\omega = [a, \eta_1]$. Then ω is isotropic and $\omega \in J_3^1$. By Lemma 2.8 (i), ω represents 1. Therefore by Lemma 1.9, $\psi \approx [1, \zeta]$ with $\zeta \in D_3$. If $\zeta \in D_3^2$, then by Lemma 2.5 (iii), $\zeta \approx [\pi, \eta_1]$ with a prime lement π, and so $\psi \approx [\pi, \eta_1, 1] \approx [\pi, 1_2, -1]$ by (2.2). Suppose $\zeta \in D_3^1$. Applying Lemma 2.12 to ζ, we have $\psi \approx [\pi, 1_2, -1]$ or $\psi \approx [\pi, 1_2, -d]$ with a prime element π. In the first case, $\psi \approx [\pi, 1, \eta_1]$ by (2.2) and $Q(\psi) = Q([\pi, 1]) = M_2(F)$. In the second case, by Lemma 1.10 (i), $A([\pi, 1_2, -d]) \cong A([\pi, 1]) \otimes A(-\pi[1, -d]) \cong M_2(F) \otimes \{F(d^{1/2}), -\pi\} \cong M_2(\mathscr{D})$. Thus $Q([\pi, 1_2, -d]) = \mathscr{D}$. This completes the proof. □

2.14. *Proof of Theorem 2.4(ii) for D_n^λ and Lemma 2.8(iii).* Since the case $\lambda = 2$ was proved in Lemma 2.5 (iii), we assume that $\lambda = 1$. As the cases $n \le 4$ have been settled, we prove the desired fact by induction on n, assuming that $n > 4$. Let $\psi \in D_n^1(\varphi)$. By Lemma 2.2, $\psi \approx [\pi, \zeta]$ with a prime element π and $\zeta \in J_{n-1}^1$. By Lemma 2.8 (i), ζ represents 1, and so we have $\psi \approx [1, \alpha]$ with $\alpha \in D_{n-1}$. This together with Lemmas 2.12 and 2.13 shows that every element of $D_n^1, n > 2$, represents 1, which proves Lemma 2.8 (iii), as D_n^1 is stable under multiplication by the elements of $\mathfrak{g}^\times$. Take another $\psi' \in D_n^1(\varphi)$; then $\psi' \approx [1, \beta]$ with $\beta \in D_{n-1}$ for the same reason, and clearly $\alpha \sim \beta$. If both α and β belong to D_{n-1}^λ with the same λ, then $\alpha \approx \beta$ by induction, and so $\psi \approx \psi'$. Thus we have to check the case in which $\alpha \in D^1_{n-1}$ and $\beta \in D^2_{n-1}$. By Lemma 2.5 (iii), $\beta \approx [\eta_r, \pi]$ with a prime element π and $r = (n-2)/2$. Put $\gamma = [\eta_{r-1}, \pi]$. Then $\beta \approx [\eta_1, \gamma]$, and so $[1, \beta] \approx [1_2, -1, \gamma]$ by (2.2). Put $\varepsilon = [1, -1, \gamma]$. Then $[1, \alpha] \sim [1, \varepsilon]$, and so $\alpha \sim \varepsilon$. Thus we have $\alpha \approx \varepsilon$ by induction, as $\alpha, \varepsilon \in D^1_{n-1}$, from which we obtain $\psi \approx \psi'$ as expected. This completes the proof.

LEMMA 2.15. *For $\varphi \in S_n$ we have $D_n(\varphi) \neq \emptyset$ if and only if $\delta_0(\varphi) \cap \mathfrak{g}^\times = \emptyset$, in which case $D^1_n(\varphi) \neq \emptyset$ except when $n = 1$.* (As for $D^2_n(\varphi)$, see Lemma 2.5(iii).)

Proof. The first part is included in Theorem 2.4(i). If n is even, then $D^2_n = \emptyset$ by Lemma 2.5 (iii), and so $D^1_n(\varphi) \neq \emptyset$. Suppose n is odd and > 1; put $r = (n-3)/2$. By Lemma 2.12 we can find $\psi \in D^1_3$ such that $\delta_0(\psi) = \delta_0(\varphi)$ and $Q(\psi) = Q(\varphi)$. Then $\varphi \sim [\psi, \eta_r] \in D^1$. This proves our lemma. □

LEMMA 2.16. *Given an s-reduced element φ of $S_n(\mathfrak{g})$ with $n > 2$, put $C(\varphi) = E_n \cap SO(\varphi)$; let $\sigma : SO(\varphi) \to F^\times/F^{\times 2}$ be the spinor norm map of* [S3, (3.7)]. *Then $\mathfrak{g}^\times \subset \sigma(C(\varphi))$.*

Proof. We first note that $SO(\varphi) = SO(b\varphi)$ for every $b \in F^\times$ and the map σ is the same for φ and $b\varphi$, as can be seen from [S3, Lemma 3.8]. Suppose $\varphi \in T_n$; put $\psi = 2^{-1}\varphi$. Then $\mathfrak{g}^1_n$ is ψ-maximal by Lemma 2.5 (i), and so the question is settled in [S5, Theorem 1.9] in that case. Thus we may assume that $\varphi \in J^1_n \cup D^1_n$. Though we assume $n > 2$ in our lemma, we first consider the cases $\varphi = [1, -1]$ and $\varphi = [1, -d]$, where d is as in (2.6). The discriminant algebra of φ can be written $F + F\varepsilon$ with an element ε such that $\varepsilon^2 = 1$ or $= d$. Then $\langle\varphi\rangle$ is isomorphic to (K, κ) with κ as in §1.4, Case $n = 2$. Let $\mathfrak{o} = \mathfrak{g} + \mathfrak{g}\varepsilon$. Then φ is the matrix representing κ with respect to the basis $\{1, \varepsilon\}$. For $x \in \mathfrak{o}^\times$, multiplication by x^ρ/x on K gives an element of $C(\varphi)$, and $\sigma(x^\rho/x) = xx^\rho$ for every $x \in \mathfrak{o}^\times$. This follows easily from the structure of the Clifford group of (K, κ); see [S3, §7.2]. Now $\mathfrak{g}^\times \subset N_{K/F}(\mathfrak{o}^\times)$ as shown in the proof of (2.5) and (2.6). Therefore $\mathfrak{g}^\times \subset \sigma(C(\varphi))$ for such φ.

We now prove our lemma for $\varphi \in J^1_n \cup D^1_n$ with $n > 2$ by induction on n. We first note that $\sigma(C(\psi)) \subset \sigma(C(\varphi))$ if $\varphi = [\psi, \zeta]$, and $\sigma(C(\varphi)) = \sigma(C(a\varphi))$ for every $a \in \mathfrak{g}^\times$, for the reason explained at the beginning. If $\varphi \in D^1_3$, Lemma 2.12 reduces the problem to the case $[1, -1]$ or $[1, -d]$. If $\varphi \in J^1_3$, then by Lemma 2.7 we may assume that $\varphi = c\gamma^\circ$ or $\varphi = c[1_2, -1]$ with $c \in \mathfrak{g}^\times$. Since γ° is q-reduced, [S5, Theorem 1.9] is applicable to it, and so $\mathfrak{g}^\times \subset \sigma(C(\gamma^\circ)) = \sigma(C(c\gamma^\circ))$. As for the other case, $\mathfrak{g}^\times \subset \sigma(C([1, -1])) \subset \sigma(C(c[1_2, -1]))$. Next let $\varphi \in J^1_n$ with $n > 3$. We have $\varphi \approx [1, \psi]$ with $\psi \in J_{n-1}$ as shown in §2.10. Then, by induction, $\mathfrak{g}^\times \subset \sigma(C(\psi)) \subset \sigma(C(\varphi))$. Finally let $\varphi \in D^1_n$ with $n > 3$. By Lemma 2.2, $\varphi \approx [\pi, \zeta]$ with a prime element π and $\zeta \in J^1_{n-1}$, and we obtain the desired result by applying induction to ζ. This completes the proof. □

3. Classification of s-reduced matrices over a number field.

3.1. We now take F to be an algebraic number field, and assume:

(3.1) *Every prime factor of 2 in F is unramified over* $\mathbf{Q}$.

This means that condition (2.9) is satisfied by F_v for every $v|2$. Let (V, φ) be a quadratic space over F, and L a φ^*-integral lattice in V. Then L is φ^*-maximal if and only if L_v is φ_v^*-maximal for every $v \in \mathbf{h}$. The "if"-part is obvious. Conversely, suppose L_u is not φ_u^*-maximal for some $u \in \mathbf{h}$. Then L_u is contained in a φ_u^*-integral lattice $M, \neq L_u$. Then there exists a $\mathfrak{g}$-lattice L' in V such that $L'_u = M$ and $L'_v = L_v$ for every $v \neq u$. Then $L \subsetneq L'$ and L' is φ^*-integral. This proves the "only if"-part.

We now consider a set of integers $\lambda = (\lambda_v)_{v|2}$, where λ_v is 1 or 2 for each prime factor v of 2. We say that L is of *type* λ if L_v is of type λ_v in the sense of §2.3 for every $v|2$.

THEOREM 3.2. *Given a quadratic space* (V, φ) *of dimension* $n > 2$ *over a number field F satisfying (3.1), let K be the discriminant algebra of* φ, *and* $\mathfrak{e}$ *the product of the prime ideals ramified in the characteristic quaternion algebra* $Q(\varphi)$. *Then for* $\lambda = (\lambda_v)_{v|2}$ *as above, the following assertions hold:*

(i) *The set of all* φ^**-maximal lattices of type* λ, *if nonempty, consists of a single genus with respect to* $SO^{\varphi}(V)$.

(ii) *There exists a* φ^**-maximal lattice in V of type* λ *if and only if the following conditions are satisfied*:

(3.2a) $$v|2, \delta_0(\varphi_v) \cap \mathfrak{g}_v^{\times} = \emptyset, \lambda_v = 2 \Longrightarrow n \notin 2\mathbf{Z} \quad \text{and} \quad v \nmid \mathfrak{e}.$$

(3.2b) $$v|2, \delta_0(\varphi_v) \cap \mathfrak{g}_v^{\times} \neq \emptyset, \lambda_v = 2 \Longrightarrow n \in 2\mathbf{Z};\ 1 \in \delta_0(\varphi_v) \text{ if } v \nmid \mathfrak{e} \text{ and } v \text{ remains prime in } K \text{ if } v|\mathfrak{e}.$$

Proof. We can always find a φ-maximal lattice L' in V. For each $v|2$, let M_v be a φ_v^*-maximal lattice in V_v. Then we can find a lattice L in V such that $L_v = M_v$ for $v|2$ and $L_v = L'_v$ for $v \nmid 2$. Thus a φ^*-maximal lattice of type λ in V exists if and only if a φ_v^*-maximal lattice of type λ_v in V_v exists for every $v|2$. Therefore (ii) follows from Lemmas 2.5 (iii), 2.9, and 2.14. As for (i), if L and M are φ^*-maximal lattices of type λ, then for each $v|2$, by Theorem 2.4 (iii) we can find an element α_v of $SO^{\varphi}(V)_v$ such that $L_v\alpha_v = M_v$. For $v \nmid 2$ the φ_v^*-maximality is the same as the φ_v-maximality, and so such an α_v is guaranteed by [S3, Lemma 6.9]. Thus we obtain (i). □

THEOREM 3.3. *In the setting of Theorem 3.2 let L be a* φ^**-maximal lattice in V. For any* $\delta \in \delta_0(\varphi)$ *put* $\delta\mathfrak{g} = \mathfrak{a}\mathfrak{b}^2$ *with a squarefree integral ideal* $\mathfrak{a}$ *and a fractional ideal* $\mathfrak{b}$; *let* $\mathfrak{f}$ *be the product of the prime factors of* $\mathfrak{e}$ *prime to* $2\mathfrak{a}$. *Then* $[L^{\varphi}/L] = \mathfrak{a}\mathfrak{f}^2$.

Proof. We have $[L^\varphi/L] = \prod_{v\in\mathbf{h}}[L_v^\varphi/L_v]$. If $v\nmid 2$, then L_v is φ_v-maximal and $L_v^\varphi = L_v^{2\varphi}$. Therefore $[L_v^\varphi/L_v]$ can be given by (1.13), and it can be written $\mathfrak{a}_v\mathfrak{f}_v^2$; see also [S4, Theorem 6.2]. For $v|2$, let τ_v be the matrix representing φ with respect to a $\mathfrak{g}_v$-basis of L_v. Then τ_v is s-reduced and $[L_v^\varphi/L_v] = \det(\tau_v)\mathfrak{g}_v = \mathfrak{a}_v$ by Lemma 2.2 (ii). Therefore we obtain $[L^\varphi/L]$ as stated in our theorem. □

THEOREM 3.4. *Given (V,φ) as in Theorem 3.2, suppose that $O^\varphi(V)_\mathbf{a}$ is not compact. Let $\mathfrak{I}$ denote the ideal group of F and $\mathfrak{I}_0$ the subgroup of $\mathfrak{I}$ generated by all the squares of ideals and the principal ideals $a\mathfrak{g}$ with the elements $a \in F^\times$ that are positive at every archimedean prime v for which $O^\varphi(V)_v$ is compact. Then the class number of the genus of φ^*-maximal lattices of type λ with a fixed λ as in §3.1 is not greater than $[\mathfrak{I} : \mathfrak{I}_0]$.*

Proof. Let $C_v = \{x \in SO^\varphi(V)_v \mid L_vx = L_v\}$ for each $v \in \mathbf{h}$ with a φ^*-maximal lattice L in V, and let $\sigma_v : SO^\varphi(V)_v \to F_v^\times/F_v^{\times 2}$ be the spinor norm map. Then $\mathfrak{g}_v^\times \subset \sigma_v(C_v)$ for $v|2$ by Lemma 2.16, and also for $v\nmid 2$ by [S5, Theorem 1.9]. Therefore we obtain our theorem as a special case of [S3, Theorem 9.26]. See also [S5, Remark 2.4, (5)]. □

3.5. In the rest of this section except in §3.11 we take $F = \mathbf{Q}$ and $\mathfrak{g} = \mathbf{Z}$; we use the symbols $\sim$ and $\approx$ of §1.2. Then $S_n(\mathbf{Z})$ is the set of nondegenerate symmetric matrices with entries in $\mathbf{Z}$, and we can define s-reduced elements of $S_n(\mathbf{Z})$ by (1.8) with $\mathfrak{g} = \mathbf{Z}$. Given a quadratic space (V,φ) over $\mathbf{Q}$, we can speak of a φ^*-maximal lattice in V of type λ for $\lambda = 1$ or 2. Let ψ be an s-reduced element of $S_n(\mathbf{Z})$. We say that ψ is of *type* 1 or *type* 2 according as ψ represents an odd integer or ψ represents only even integers. Then $\mathbf{Z}_n^1$ is a ψ^*-maximal lattice in $\langle\psi\rangle$, which is of type 1 or 2 according as ψ is of type 1 or 2. Clearly ψ is of type 2 if and only if ψ belongs to the set T_n of (1.5).

Given $\varphi \in S_n$, we put $s(\varphi) = p - q$, where p (resp. q) is the number of positive (resp. negative) eigenvalues of φ as a real matrix. Given $n \in \mathbf{Z}, > 2$, $\delta \in \mathbf{Q}^\times, \sigma \in \mathbf{Z}$, and a quaternion algebra B over $\mathbf{Q}$, there exists $\varphi \in S_n$ such that

$$\delta \in \delta_0(\varphi), \quad \sigma = s(\varphi), \quad \text{and} \quad Q(\varphi) = B, \tag{3.3}$$

if and only if the following conditions are satisfied:

$$|\sigma| \le n,\ \sigma - n \in 2\mathbf{Z}, \quad (-1)^{[\sigma/2]}\delta > 0, \tag{3.4a}$$

$$B \otimes_\mathbf{Q} \mathbf{R} \cong M_2(\mathbf{R}) \Longleftrightarrow \sigma \equiv 0, \pm 1, \text{ or } 2 \pmod 8, \tag{3.4b}$$

where $[\sigma/2]$ denotes $\sigma/2$ if n is even and $(\sigma - 1)/2$ if n is odd. Moreover, the isomorphism class of $\langle\varphi\rangle$ is determined by n, σ, δ, and B. These are special cases of [S4, Theorems 4.2 and 4.4], which concern similar results over an arbitrary number field, including the case $n = 2$. Here we have assumed $n > 2$ for simplicity.

3.6. We say that two elements φ and ψ of S_n belong to the same *genus* if $s(\varphi) = s(\psi)$ and $\alpha_p\varphi \cdot {}^t\alpha_p = \psi$ with some $\alpha_p \in GL_n(\mathbf{Z}_p)$ for every prime p. Then the Hasse principle guarantees an element β of $GL_n(\mathbf{Q})$ such that $\beta\varphi \cdot {}^t\beta = \psi$. We see that $\det(\psi)/\det(\varphi)$ is positive and belongs to $\mathbf{Z}_p^\times$ for every p. Thus $\det(\varphi) = \det(\psi)$, and so $\det(\alpha_p) = \pm 1$. Also, if φ is of type λ, then we easily see that ψ must be of tye λ too.

For $\lambda = 1$ or 2, let us now denote by $S_n^\lambda(\mathbf{Z})$ the set of all s-reduced elements of $S_n(\mathbf{Z})$ of type λ. By Lemma 2.2 (ii), for an s-reduced $\varphi \in S_n(\mathbf{Z})$ we see that $\det(\varphi)$ is either odd or twice an odd integer. Both cases of $\det(\varphi)$ can occur for each type of λ, and so there are four combinations of λ and the parity of $\det(\varphi)$.

Given $\varphi, \psi \in S_n^\lambda(\mathbf{Z})$ with a fixed λ, we have

$$\varphi \sim \psi \Longleftrightarrow \varphi \text{ and } \psi \text{ belong to the same genus.} \tag{3.5}$$

Indeed, we have seen $\Leftarrow$. Conversely suppose $\varphi \sim \psi$. Then ψ, being an s-reduced element of $S_n(\mathbf{Z}_2)$, must belong to the set J_n or D_n of (2.7a) defined over $\mathbf{Q}_2$. Thus $\psi \in J_n(\varphi)$ or $\psi \in D_n(\varphi)$ accordingly. If φ and ψ are of type λ, then $\psi \in J_n^\lambda(\varphi)$ or $\psi \in D_n^\lambda(\varphi)$, and so ψ is equivalent to φ over $\mathbf{Z}_2$ by virtue of Theorem 2.4(ii). If $p \neq 2$, then φ and ψ are q-reduced, and they must be equivalent over $\mathbf{Z}_p$, as all the φ_p-maximal lattices in $\langle\varphi\rangle_p$ form a single orbit under $SO^\varphi(V)_p$. This completes the proof of (3.5).

Thus the genus of an element φ of $S_n^\lambda(\mathbf{Z})$ is determined by the isomorphism class of the quadratic space $\langle\varphi\rangle$, and vice versa. Consequently, the genus of φ is determined by λ, n, $\delta_0(\varphi)$, $s(\varphi)$, and $Q(\varphi)$.

Let $L = \mathbf{Z}_n^1$ and $\varphi \in S_n^\lambda(\mathbf{Z})$. Then L is a φ^*-maximal lattice of type λ. Suppose $\psi \in S_n^\lambda(\mathbf{Z})$ and $\psi \sim \varphi$. Take $\beta \in GL_n(\mathbf{Q})$ so that $\beta\varphi \cdot {}^t\beta = \psi$. Then L is a ψ^*-maximal lattice of type λ, and so $L\beta$ is φ^*-maximal and of type λ. Therefore, by Theorem 3.2(i), $L\beta$ belongs to the genus of L with respect to $SO(\varphi)$. Now assigning the $SO(\varphi)$-class of $L\beta$ to ψ, we easily see that the set of equivalence classes of such ψ over $\mathbf{Z}$ correspond bijectively to the $SO(\varphi)$-classes of lattices in the $SO(\varphi)$-genus of L.

We are going to enumerate the genera of such elements φ of $S_n^\lambda(\mathbf{Z})$ with prescribed $\det(\varphi)$ and $s(\varphi)$. First we note:

LEMMA 3.7. *Given $\varphi \in S_n(\mathbf{Q})$ with $n > 2$, let K be the discriminant algebra of φ, and e the product of the prime numbers ramified in $Q(\varphi)$; let a be a squarefree integer belonging to $\delta_0(\varphi)$. Then the following assertions hold:*

(i) *There always exists an element ψ of $S_n^1(\mathbf{Z})$ such that $\psi \sim \varphi$. Moreover, the genus of such a ψ is uniquely determined by φ.*

(ii) *There exists an element ψ of $S_n^2(\mathbf{Z})$ such that $\psi \sim \varphi$ exactly in the following three cases*: (a) $2 \nmid en$ *and* $2|a$; (b) $2|n, 2 \nmid ea$, *and* $K \otimes_\mathbf{Q} \mathbf{Q}_2 \cong \mathbf{Q}_2 \times \mathbf{Q}_2$; (c) $2|n, 2|e, 2 \nmid a$, *and $K \otimes_\mathbf{Q} \mathbf{Q}_2$ is a field unramified over $\mathbf{Q}_2$. Moreover, the genus of such a ψ is uniquely determined by φ.*

(iii) *The set* $S_n^2(\mathbf{Z})$ *consists of the elements* τ *of the set* T_n *of the introduction such that* $2^{-1}\tau$ *is q-reduced and* $\det(\tau) \notin 4\mathbf{Z}$.

Proof. Given $\varphi \in S_n$, we can find, by Theorem 3.2, a φ^*-maximal lattice M of type 1 in $\langle\varphi\rangle$. Let ψ be the matrix representing φ with respect to a $\mathbf{Z}$-basis of M. Then ψ gives the element of (i), and its genus is determined by the principle of (3.5). Assertion (ii) can be proved in a similar way; we find the three cases of (ii) by checking (3.2a, b). As for (iii), take $\tau \in S_n^2(\mathbf{Z})$. By Lemma 2.2 (ii), $\det(\tau) \notin 4\mathbf{Z}$. From (1.7) and (1.8) we easily see that $2^{-1}\tau$ is q-reduced. Conversely, suppose $\tau \in T_n$ and $2^{-1}\tau$ is q-reduced. Put $L = \mathbf{Z}_n^1$. Then L is $(2^{-1}\tau)$-maximal, and so L_p is τ^*-maximal for every $p \neq 2$. Suppose now $\det(\tau) \notin 4\mathbf{Z}$. Then by Lemma 2.2 (ii), τ is s-reduced at the prime 2. Thus L_2 is τ^*-maximal, and so L is τ^*-maximal. Consequently τ is s-reduced. This completes the proof. □

By means of (iii) of this lemma, the classification of the elements of $S_n^2(\mathbf{Z})$ and their genera can be reduced to the classification of the reduced $\mathbf{Z}$-valued quadratic forms represented by the elements Φ of $2^{-1}T_n$, which was settled in [S4]. We obtain the desired results on $S_n^2(\mathbf{Z})$ by simply imposing an additional condition that $\det(2\Phi) \notin 4\mathbf{Z}$ in [S4, Theorems 6.4, 6.5, and 6.6]. For this reason, we restrict our results on the classification of s-reduced elements to those of type 1.

THEOREM 3.8. *Let two integers n and* σ *be given as follows:* $3 \leq n \in \mathbf{Z}$, $n - \sigma \in 2\mathbf{Z}$, $|\sigma| \leq n$; *also let m and e be squarefree positive integers. Let r be the number of prime factors of e, and f the product of the prime factors of e prime to* $2m$. *Suppose that* $\sigma - 4r \equiv 0$ *or* $2 \pmod 8$ *if n is even and* $\sigma - 4r \equiv \pm 1 \pmod 8$ *if n is odd. Then there exists an element* ψ *of* $S_n^1(\mathbf{Z})$ *such that*

$$|\det(\psi)| = mf^2, \qquad s(\psi) = \sigma,$$

and $Q(\psi)$ *is ramified at a prime number p if and only if* $p|e$. *Moreover, every element of* $S_n^1(\mathbf{Z})$ *is of this type, and its genus is determined by* (n, σ, m, e).

Proof. Let $\delta = (-1)^{[\sigma/2]}m$ and let B be a quarternion algebra over $\mathbf{Q}$ that is ramified exactly at the prime factors of e, and also at the archimedean prime if r is odd. The condition on $\sigma - 4r$ implies (3.4b). Then as we said in §3.5, there exists $\varphi \in S_n$ satisfying (3.3). By Lemma 3.7 (i) we can find $\psi \in S_n^1(\mathbf{Z})$ such that $\psi \sim \varphi$. Now for $L = \mathbf{Z}^n$ we have $\det(\psi)\mathbf{Z} = [L^\psi/L]$, which can be given by Theorem 3.3. We have $\mathfrak{a} = m\mathbf{Z}$ and $\mathfrak{f} = f\mathbf{Z}$, and so $|\det(\psi)| = mf^2$. This proves the first part of our theorem. Next, given $\tau \in S_n^1(\mathbf{Z})$, let e be the product of the prime numbers ramified in $Q(\tau)$; take a squarefree positive integer m so that $\pm m \in \delta_0(\tau)$. Then we find that τ is of the type described in our theorem. Now the genus of τ is determined by n, $s(\tau)$, $\delta_0(\tau)$, and $Q(\tau)$; also, $Q(\tau)$ is determined by e. Therefore we obtain our assertion about the genus. □

THEOREM 3.9. *Let $n \in \mathbf{Z}, > 2$, and let σ be an integer such that $|\sigma| \leq n$ and $\sigma - n \in 2\mathbf{Z}$. Let m and f be two squarefree positive integers such that f is prime to $2m$, and let ν be the number of different prime factors of $2m$. Then there exist exactly $2^{\nu-1}$ genera of $\psi \in S_n^1(\mathbf{Z})$ such that $|\det(\psi)| = mf^2$ and $s(\psi) = \sigma$.*

Proof. Since the genus of ψ is determined by (n, σ, m, e) as stated in Theorem 3.8, our task is to count the number of possible e's. Clearly $e = fg$ with a squarefree factor g of $2m$. Let μ be the number of prime factors of g. Then $\mu \leq \nu$ and $\sigma - 4r$ modulo 8 must be determined as in Theorem 3.8. We easily see that each (σ, f) determines μ modulo 2, and so we have exactly $2^{\nu-1}$ choices for g. This proves our theorem. □

THEOREM 3.10. *Let φ and ψ be indefinite elements of $S_n^\lambda(\mathbf{Z})$ with $n \geq 3$. Then the following assertions hold:*

(i) *If $\varphi \sim \psi$, then $\varphi \approx \psi$.*

(ii) *If $\lambda = 1$, $s(\varphi) = s(\psi)$, and $|\det(\varphi)| = |\det(\psi)| = f^2$ or $= 2f^2$ with a squarefree odd integer f, then $\varphi \approx \psi$. Such a φ exists for both values f^2 and $2f^2$ and also for a given $s(\varphi)$, provided $|s(\varphi)| \leq n$ and $s(\varphi) - n \in 2\mathbf{Z}$.*

Proof. If $F = \mathbf{Q}$, we have $\mathfrak{I} = \mathfrak{I}_0$ in Theorem 3.4. As noted in §3.6, the class number of φ^*-maximal lattices of type λ is the same as the number of equivalence classes over $\mathbf{Z}$ of $\psi \in S_n^\lambda(\mathbf{Z})$ such that $\psi \sim \varphi$. Therefore we obtain (i) from Theorem 3.4. Next, if $\varphi \in S_n^1(\mathbf{Z})$ and $\det(\varphi)$ is as in (ii), then the genus of φ is determined by $\det(\varphi)$ and $s(\varphi)$, by virtue of Theorem 3.9. Therefore (ii) follows from (i). □

Remark. Lemma 3.7 reduces the problem for $S_n^2(\mathbf{Z})$ to q-reduced matrices, for which we determined the number of genera in [S4, Theorem 6.6]. For example, suppose $\varphi \in S_n(\mathbf{Z})$, φ is indefinite, and $|\det(\varphi)| = 2$. Then our definition (1.8) shows that φ is s-reduced. Therefore if it is of type 1, then (ii) above shows that there is exactly one equivalence class over $\mathbf{Z}$ for each choice of $s(\varphi)$. If φ is of type 2, then n must be odd, and there is a unique genus with a given value of $s(\varphi)$, provided $s(\varphi) \equiv \pm 1 \pmod 8$; see [S4, Theorem 6.6(vii)]. That genus consists of a single class by (i) of Theorem 3.10.

3.11. Let us now explain why we need condition (3.1) or its local version (2.9). In fact, many statements of Section 2 are false without (2.9). To be precise, take F to be local and $2\mathfrak{g} = \mathfrak{p}^\kappa$ with $0 < \kappa \in \mathbf{Z}$. Put $q = [\mathfrak{g} : \mathfrak{p}]$ and $\mathfrak{h} = \mathfrak{g}^{\times 2}(1+2\mathfrak{g})$. This is a subgroup of $\mathfrak{g}^\times$. We first note:

$$[\mathfrak{g}^\times : \mathfrak{h}] = q^{[\kappa/2]}; \tag{3.6}$$

(3.7a) $\mathfrak{h}$ is the set of the elements of $\mathfrak{g}^\times$ that $\mathrm{diag}[1, -1]$ represents;

(3.7b) $\mathfrak{h}$ is the set of the elements of $\mathfrak{g}^\times$ that $\mathrm{diag}[1, -d]$ represents, if $d \in \mathfrak{g}^\times$ and $F(d^{1/2})$ is an unramified quadratic extension of F.

Equality (3.6) is a special case of [S2, Lemma 3.5 (1)]. To prove (3.7a), let $c = x^2 - y^2 \in \mathfrak{g}^\times$ with $x, y \in \mathfrak{g}$. Put $x + y = a$. Then $c = a(a - 2y)$, and so $a \in \mathfrak{g}^\times$. Thus $c = a^2(1 - 2a^{-1}y) \in \mathfrak{h}$. Conversely, if $c = a^2(1 + 2z)$ with $a \in \mathfrak{g}^\times$ and $z \in \mathfrak{g}$, then $c = (a + az)^2 - (az)^2$. This proves (3.7a). As for (3.7b), put $K = F(d^{1/2})$ and denote by $\mathfrak{r}$ the maximal order of K. Then $\mathfrak{g}[d^{1/2}] = \mathfrak{g} + 2\mathfrak{r}$ and $1 + 2\mathfrak{g} = N_{K/F}(1 + 2\mathfrak{r})$ as shown in the proof of (2.6). Therefore we see that $\mathfrak{h} \subset N_{K/F}(\mathfrak{g}[d^{1/2}])$, and so $\mathrm{diag}[1, -d]$ represents every element of $\mathfrak{h}$. Conversely, suppose $c = x^2 - dy^2 \in \mathfrak{g}^\times$ with $x, y \in \mathfrak{g}$; then we can put $x + d^{1/2}y = a + 2b$ with $a \in \mathfrak{g}$ and $b \in \mathfrak{r}$. Clearly $a \in \mathfrak{g}^\times$ and $c = N_{K/F}(a + 2b) \in a^2N_{K/F}(1 + 2\mathfrak{r}) \subset \mathfrak{h}$. This proves (3.7b).

Let $\tau = \mathrm{diag}[1, -1]$ or $\tau = \mathrm{diag}[1, -d]$ with d as in (3.7b). Then by (1.12), $c\tau \sim \tau$ for every $c \in \mathfrak{g}^\times$. Combining (3.7a, b) with (2.4), we see that $a\tau \approx b\tau$ for $a, b \in \mathfrak{g}^\times$ if and only if $a/b \in \mathfrak{h}$. Similarly, comparing the representable numbers, we find that $\mathrm{diag}[a\tau, \eta_r] \approx \mathrm{diag}[b\tau, \eta_r]$ if and only if $a/b \in \mathfrak{h}$. Thus $J_n^1(\mathrm{diag}[\tau, \eta_r])$ for $n = 2 + 2r$ contains at least $q^{[\kappa/2]}$ equivalence classes over $\mathfrak{g}$. There are some other unpleasant (or interesting) phenomena which make the classification in the case $\kappa > 1$ more complicated, but we will not discuss them here.

The classification of lattices in a quadratic space over a dyadic local field was investigated by O'Meara in [O], but the results are very involved, and it is impossible to formulate clear-cut results in the global case by using them. Therefore, instead of considering arbitrary lattices, we have introduced the notion of s-reduced matrices and imposed the condition that the prime 2 is unramified in F. This approach has enabled us to state the results in a transparent and satisfactory way.

4. Quadratic Diophantine equations.

4.1. Let (V, φ) be a quadratic space of dimension n over a local or global field F. For a $\mathfrak{g}$-lattice L in V, an element q in $F^\times$, and a fractional ideal $\mathfrak{b}$ in F, we put

$$L[q, \mathfrak{b}] = \{x \in V \mid \varphi[x] = q, \varphi(x, L) = \mathfrak{b}\}, \tag{4.1}$$

$$\Gamma(L) = \{\alpha \in SO^\varphi(V) \mid L\alpha = L\}. \tag{4.2}$$

Clearly $L[q, \mathfrak{b}]$ is stable under right multiplication by the elements of $\Gamma(L)$. In [S3] and [S5] we developed a theory of quadratic Diophantine equation $\varphi[x] = q$

by associating the set $L[q, \mathfrak{b}]/\Gamma(L)$ with the classes in the group $SO^{\varphi}(W)$ with a subspace W of V of dimension $n-1$.

In particular, we proved as a crucial fact that $\#\{L[q, \mathfrak{b}]/\Gamma(L)\} \leq 1$ if F is local, L is φ-maximal, and $n > 2$; see [S3, Theorem 10.5] and [S5, Theorem 1.3]. Our aim of this section is to examine the case where L is φ^*-maximal in the local case, and its consequences in the global case. The φ^*-maximality is the same as the φ-maximality if $2 \notin \mathfrak{p}$, and so our local question concerns only the case $2 \in \mathfrak{p}$. For the reason explained in §3.11, we assume that $\mathfrak{p} = 2\mathfrak{g}$, as we did in Section 2. Now, if L is φ^*-maximal and of type 2, then L is $(2^{-1}\varphi)$-maximal, and the problem can be reduced to the cases studied in [S3] and [S5]. Therefore we consider the case where L is of type 1; also we restrict our discussion to the case of $L[q, \mathfrak{g}]$ with $q \in \mathfrak{g}^{\times}$. Notice that $L[q, \mathfrak{g}] = \{x \in L \mid \varphi[x] = q\}$ if $q \in \mathfrak{g}^{\times}$ and L is φ^*-maximal. Indeed, suppose $x \in L$ and $\varphi[x] \in \mathfrak{g}^{\times}$; then $\mathfrak{g} = \varphi[x]\mathfrak{g} \subset \varphi(x, L) \subset \mathfrak{g}$, and so $\varphi(x, L) = \mathfrak{g}$. Conversely, suppose $\varphi(x, L) = \mathfrak{g}$ and $\varphi[x] = q \in \mathfrak{g}^{\times}$. Then $L + \mathfrak{g}x$ is φ^*-integral, and so $x \in L$, as L is φ^*-maximal. Thus we obtain the desired equality.

THEOREM 4.2. *Let (V, φ) be a quadratic space of dimension $n > 2$ over a local field F, and L a φ^*-maximal lattice of type 1 in V; let $e \in \mathfrak{g}^{\times}$. Suppose $\mathfrak{p} = 2\mathfrak{g}$. Then $\#\{L[e, \mathfrak{g}]/\Gamma(L)\} = 2$ in the following three cases:* (A) *$n \notin 2\mathbf{Z}, e \in \delta_0(\varphi)$, and $Q(\varphi) = M_2(F)$;* (B) *$2 < n-1 \in 2\mathbf{Z}$, the discriminant field of $e\varphi$ is an unramified quadratic extension of F, and $Q(\varphi) = \mathscr{D}$;* (C) *$n \in 2\mathbf{Z}, \delta_0(\varphi) \cap \mathfrak{g}^{\times} = \emptyset$, and $Q(\varphi) = M_2(F)$. Also, $\#\{L[e, \mathfrak{g}]/\Gamma(L)\} \leq 1$ if $n = 3$, $\delta_0(\varphi) \cap \mathfrak{g}^{\times} \neq \emptyset$, and φ is anisotropic. We have $\#\{L[e, \mathfrak{g}]/\Gamma(L)\} = 1$ in all the other cases.*

Proof. We can take L to be $\mathfrak{g}_n^1$ considered in $\langle\varphi\rangle$ with $\varphi \in J_n^1 \cup D_n^1$. Replacing φ by $e^{-1}\varphi$, we may assume that $e = 1$. For simplicity put $N = \#\{L[e, \mathfrak{g}]/\Gamma(L)\}$; also we write $[a, b, \ldots]$ for $\mathrm{diag}\,[a, b, \ldots]$ as we did in Section 2. We first treat the case where φ is anisotropic and belongs to J_3^1. By Lemma 2.7, $\varphi \approx c\gamma^{\circ}$ with $c \in \mathfrak{g}^{\times}$. As noted in §1.5, γ° is q-reduced, and so L is φ-maximal. Therefore the question is settled by [S3, Theorem 10.5] as we said in §4.1; thus $N \leq 1$. We hereafter exclude that case. Now let $\varphi \in J_n^1$. By Lemma 2.8 (i), $L[1, \mathfrak{g}] \neq \emptyset$. Suppose $h \in L[1, \mathfrak{g}]$; then the standard technique of orthogonal projection shows that $L = \mathfrak{g}h + M$ with $M = \{x \in L \mid \varphi(h, x) = 0\}$. This means $\varphi \approx [1, \psi]$, where ψ represents φ on M. Similarly if $h' \in L[1, \mathfrak{g}]$, then $L = \mathfrak{g}h' + M'$ with $M' = \{x \in L \mid \varphi(h', x) = 0\}$, and $\varphi \approx [1, \psi']$, where ψ' represents φ on M'. Then $\psi, \psi' \in J_{n-1}$, and $\psi \sim \psi'$. If n is even, then both ψ and ψ' belong to J_{n-1}^1, and so $\psi \approx \psi'$ by Theorem 2.4(ii). This means that there exists a $\mathfrak{g}$-linear map β of M onto M' such that $\psi'[x\beta] = \psi[x]$ for every $x \in M$. Extend β to an element γ of $GL_n(F)$ by putting $h\gamma = h'$. Then $\gamma \in O(\varphi)$. By [S3, Lemma 6.8] there exists an element ζ of $O(\psi)$ such that $M\zeta = M$ and $\det(\zeta) = \det(\gamma)$. Modifying γ on M by ζ, we may assume that $\gamma \in SO(\varphi)$. Then $\gamma \in \Gamma(L)$, and so $N = 1$ if $n \in 2\mathbf{Z}$.

Suppose next $n \notin 2\mathbf{Z}$. We first note that $Q(c\varphi) = Q(\varphi)$ and $\delta_0(c\varphi) = c\delta_0(\varphi)$ for every $c \in F^\times$. We repeat the above argument up to the point where $\varphi \approx [1, \psi] \approx [1, \psi']$ with $\psi, \psi' \in J_{n-1}$. If ψ and ψ' are of the same type, then $\psi \approx \psi'$. Therefore $N = 2$ if it can happen that $\psi \in J_{n-1}^2$ and $\psi' \in J_{n-1}^1$; otherwise $N = 1$. Thus our task is to examine when the first case occurs. Suppose $\varphi \approx [1, \psi]$ with $\psi \in J_{n-1}^2$. By Lemma 2.9 (ii), ψ must belong to Cases (a) and (b) there. In Case (a) we have $\psi \sim \eta_r$ with $r = (n-1)/2$, and so $\psi \approx \eta_r$ as $\eta_r \in J_{n-1}^2(\psi)$. We have $Q(\varphi) = M_2(F)$ and $1 \in \delta_0(\varphi)$. Now $\eta_r \sim [1_r, -1_r]$ and $[1, 1_r, -1_r] \in J_n^1(\varphi)$, and so $N = 2$. Next suppose ψ belongs to Case (b). Then $-\psi$ is also of the same type. We have $\delta_0(\varphi) = \delta_0(\psi)$ and $Q(\varphi) = Q(-\psi)$ by Lemma 1.10 (ii). If $n = 3$, then $J_{n-1}^1(\psi) = \emptyset$ by Lemma 2.9 (i), and so $N = 1$. If $n > 3$, then φ belongs to Case (B), and the same lemma guarantees an element ψ' of $J_{n-1}^1(\psi)$. Then $\varphi \approx [1, \psi']$ and so $N = 2$. This completes the proof for $\varphi \in J_n^1$.

Next, let $\varphi \in D_n^1$. By Lemma 2.8 (iii), $\varphi \approx [1, \psi]$ with $\psi \in D_{n-1}$. Repeating the above argument for J_n^1, we see that N is 1 or 2, and $N = 2$ only if $\varphi \approx [1, \psi]$ with $\psi \in D_{n-1}^2$. Suppose that is so; then by Lemma 2.5 (iii), n is even and $\psi \approx [\pi, \eta_r]$ with a prime element π and $r = (n-2)/2$. Thus $\varphi \approx [1, \pi, \eta_r] \sim [1, \psi']$ with $\psi' = [\pi, 1_r, -1_r]$. Then $\psi' \in D_{n-1}^1$ and $\varphi \approx [1, \psi']$, as $[1, \psi'] \in D_n^1$; thus $N = 2$. We have $Q(\varphi) = Q([1, \pi]) = M_2(F)$ then. Conversely suppose $Q(\varphi) = M_2(F)$ and n is even; take a prime element π' in $\delta_0(\varphi)$. Then $\varphi \sim [1, -\pi', \eta_r]$ by (1.12), and so $\varphi \approx [1, -\pi', \eta_r]$, which is the case we have just discussed. This shows that for $\varphi \in D_n^1$ with $n > 2$ we have $N = 2$ exactly in Case (C). This completes the proof. □

For instance, take $F = \mathbf{Q}_2, \varphi = 1_n$, and $L = (\mathbf{Z}_2)_n^1$ with $n > 3$. Then φ is s-reduced and L is φ^*-maximal. From the above theorem we can easily derive that for an odd integer e,

$$\#\{L[e, \mathbf{Z}_2]/\Gamma(L)\} = \begin{cases} 2 & \text{if } n - e \in 8\mathbf{Z}, \\ 1 & \text{otherwise.} \end{cases} \tag{4.3}$$

As for the case $n = 3$, the reader is referred to [S3, §12.7].

4.3. Let us now take a quadratic space (V, φ) of dimension $n > 2$ over a global field F. Given a lattice L in V and an element h of V such that $\varphi[h] \neq 0$, put

$$G = SO^\varphi(V), \qquad H = \{\alpha \in G \mid h\alpha = h\},$$
$$C = \{x \in G_\mathbf{A} \mid Lx = L\}.$$

Take $Y \subset G_\mathbf{A}$ so that $G_\mathbf{A} = \bigsqcup_{y \in Y} GyC$. In [S5, (2.4), (2.5)] we proved

$$\#\{H\backslash H_\mathbf{A}/(H_\mathbf{A} \cap C)\} = \sum_{y \in Y} \#\{(V \cap hCy^{-1})/\Gamma^y\}, \tag{4.4}$$

where $\Gamma^y = G \cap yCy^{-1}$; moreover, if L is φ-maximal and $h \in L[q, \mathfrak{b}]$, then

$$V \cap hCy^{-1} = (Ly^{-1})[q, \mathfrak{b}] \text{ for every } y \in G_{\mathbf{A}}. \tag{4.5}$$

Now the question is what happens if we take L to be φ^*-maximal and of type λ with λ as in §3.1. The answer in the general case is rather involved, but we can handle the problem under some conditions. First of all, we note that $L[q, \mathfrak{b}] = \bigsqcup_{\zeta \in Z} (V \cap h\zeta C)$ with a finite subset Z of $G_{\mathbf{h}}$, as $\#\{L_v[q, \mathfrak{b}_v]/C_v\}$ is always finite and equal to 1 for almost all $v \in \mathbf{h}$; see [S3, Theorem 10.3]. Suppose the prime 2 is unramified in F. If $v|2$ and $\lambda_v = 2$, then L_v is $(2^{-1}\varphi_v)$-maximal, and so $L_v[q, \mathfrak{b}_v] = hC_v$ by the results in [S3] and [S5] mentioned in §4.1; the same holds if $v \nmid 2$. Thus if q is prime to 2 and $\mathfrak{b} = \mathfrak{g}$, then by Theorem 4.2 we can take Z so that $\#Z \leq 2^\nu$, where ν is the number of the prime factors v of 2 such that $\lambda_v = 1$ and one of the three cases (A), (B), and (C) of that theorem is applicable. If $k \in V \cap h\zeta C$, then $\varphi[k] = \varphi[h]$, and so $k = h\xi$ with $\xi \in G$ by [S3, Lemma 1.5 (ii)]. For such a ξ we have $h\zeta C = h\xi C$, so that we can put $L[q, \mathfrak{g}] = \bigsqcup_{\xi \in X} (V \cap h\xi C)$ with a finite subset X of G such that $\#(X) \leq \#(Z)$.

We have $\#(X) = \#(Z)$ if $V \cap h\zeta C \neq \emptyset$ for every $\zeta \in Z$, which is so if and only if $G \cap H_{\mathbf{A}}\zeta C \neq \emptyset$ for every $\zeta \in Z$. This condition is satisfied, for example, if $G_{\mathbf{A}} = GC$, that is, if the genus of L consists of a single class. Assuming that to be the case, put $H^\xi = \xi^{-1}H\xi$. Then from (4.4) we obtain

$$\sum_{\xi \in X} \#\{H^\xi \backslash H^\xi_{\mathbf{A}}/(H^\xi_{\mathbf{A}} \cap C)\} = \#\{L[q, \mathfrak{g}]/\Gamma^1\}. \tag{4.6}$$

In [S3, Theorem 13.12] and [S5, (3.7)] we proved certain mass formulas that connect the mass $\mathfrak{m}(H, H_{\mathbf{A}} \cap C)$ to the mass of the set $L[q, \mathfrak{b}]$. We can certainly discuss such for a φ^*-maximal lattice L. For instance, in the above setting with $G_{\mathbf{A}} = GC$, suppose F is totally real and φ is definite at every archimedean prime of F. Then from [S5, (3.6)] we obtain

$$\sum_{\xi \in X} \mathfrak{m}(H^\xi, H^\xi_{\mathbf{A}} \cap C) = \#L[q, \mathfrak{g}]/\#\Gamma^1. \tag{4.7}$$

Department of Mathematics, Princeton University, Princeton, NJ

REFERENCES

[E] M. Eichler, *Quadratische Formen und orthogonale Gruppen*, Springer-Verlag, Berlin, 1952, 2nd ed., 1974.

[M] J. Milnor, On simply connected 4-manifolds, *Symposium Internacional Topologia Algebraica*, Mexico, 1958, pp. 122–128.

[O] O. T. O'Meara, *Introduction to Quadratic Forms*, Springer-Verlag, NY, 1963.
[S1] G. Shimura, *Euler Products and Eisenstein Series*, *CBMS Regional Conference Series in Mathematics*, No. 93, Amer. Math. Soc., Providence, RI, 1997.
[S2] ________, The number of representations of an integer by a quadratic form, *Duke Math. J.* **100** (1999), 59–92.
[S3] ________, *Arithmetic and Analytic Theories of Quadratic Forms and Clifford Groups*, *Math. Surv. Monogr.*, vol. 109, Amer. Math. Soc., Providence, RI, 2004.
[S4] ________, Classification, construction, and similitudes of quadratic forms, *Amer. J. Math.* **128** (2006), 1521–1552.
[S5] ________, Integer-valued quadratic forms and quadratic Diophantine equations, *Document. Math.* **11** (2006), 333–367.
[V] F. van der Blij, An invariant of quadratic forms mod 8, *Indag. Math.* **21** (1959), 291–293.

ARITHMETIC OF HERMITIAN FORMS

GORO SHIMURA

Received: May 27, 2008

Communicated by Don Blasius

ABSTRACT. We investigate the following two problems on a hermitian form Φ over an algebraic number field: (1) classification of Φ over the ring of algebraic integers; (2) hermitian Diophantine equations. The same types of problems for quadratic forms were treated in the author's previous articles. Here we discuss the hermitian case. Problem (2) concerns an equation $\xi\Phi \cdot {}^t\xi^\rho = \Psi$, where Φ and Ψ represent hermitian forms. We connect the number of such ξ modulo a group of units with the class number and mass of the unitary group of a form Θ such that $\Phi \approx \Psi \oplus \Theta$.

2000 Mathematics Subject Classification: 11E39 (primary), 11E41, 11D09 (secondary)

INTRODUCTION

To explain Problems (1) and (2) of the abstract, we take a quadratic extension K of an algebraic number field F, a vector space V over K of dimension n, and a nondegenerate hermitian form $\varphi : V \times V \to K$ with respect to the nontrivial automorphism ρ of K over F. We denote by $d_0(\varphi)$ the coset of $F^\times/N_{K/F}(K^\times)$ represented by $(-1)^{n(n-1)/2}\det(\varphi)$. It is classically known that n, $d_0(\varphi)$, and the indices of φ at certain archimedean primes of F, satisfying a natural consistency condition, determine the isomorphism class of (V, φ), and vice versa. This classification does not answer, however, the question of classification over the ring of integers. To be precise, let $\mathfrak{r}$ denote the ring of algebraic integers in K and $\mathfrak{g} = F \cap \mathfrak{r}$; let $\mathfrak{d}$ be the different of K relative to F. We put

$$(0.1) \qquad \mathfrak{H}_n = \{\Phi \in GL_n(K) \,|\, \Phi = {}^t\Phi^\rho\}.$$

We call a matrix $\Phi = (\varphi_{ij}) \in \mathfrak{H}_n$ *semi-integral* if $\varphi_{ij} \in \mathfrak{d}^{-1}$ and $\varphi_{ii} \in \mathfrak{g}$ for every i and j, which means that $\sum_{i,j} \varphi_{ij}x_i x_j^\rho \in \mathfrak{g}$ for every $(x_i)_{i=1}^n \in \mathfrak{r}^n$. Further we call a semi-integral Φ *reduced* if the following condition is satisfied:

Shimura, Goro. "Arithmetic of Hermitian forms." Documenta Mathematica 13 (2008), 739–774.

(R) *If* $\Phi = P\Psi \cdot {}^tP^\rho$ *with a semi-integral* Ψ *and* $P = (p_{ij}) \in GL_n(K)$, $p_{ij} \in \mathfrak{r}$, *then* $\det(P) \in \mathfrak{r}^\times$.

These definitions are natural, but cover only a special class of (V, φ), as an $\mathfrak{r}$-lattice in V may not be isomorphic to $\mathfrak{r}^n$. In order to classify all $\mathfrak{g}$-valued hermitian forms, we have to define the genus of a form relative to an isomorphism class of lattices, and study its connection with the isomorphism class of (V, φ). These are nontrivial, and will be treated in §§2.4 and 2.5. We are then able to classify all the genera of $\mathfrak{g}$-valued hermitian forms in terms of matrices (Theorems 2.10 and 2.11). The results can be presented in simpler forms if K is a real or an imaginary quadratic field of odd class number, in which case the above definitions cover all hermitian spaces. Let d be the discriminant of such a K; then $K = \mathbf{Q}(\sqrt{d})$. For a semi-integral Φ with entries in K, put $s(\Phi) = p - q$ when $d < 0$ and Φ as a complex hermitian matrix has p positive and q negative eigenvalues; we do not define $s(\Phi)$ if $d > 0$. Let $\mathfrak{H}_n^0$ be the set of all reduced semi-integral elements of $\mathfrak{H}_n$. Then we can prove (Theorem 2.14):

(A) *Let three integers* n, σ, *and* e *be given as follows:* $0 < n \in 2\mathbf{Z}$, $\sigma \in 2\mathbf{Z}$, $|\sigma| \le n$; $\sigma = 0$ *if* $d > 0$; e *is positive and squarefree. Let* r *be the number of prime factors of* e. *Suppose that* $\sigma - 2r \in 4\mathbf{Z}$ *and no prime factor of* e *splits in* K. *Then there exists an element* Φ *of* $\mathfrak{H}_n^0$ *such that*

$$\det(\sqrt{d}\,\Phi) = (-1)^{\sigma/2}e \quad \textit{and} \quad s(\Phi) = \sigma \qquad \textit{if } d < 0,$$
$$\det(\sqrt{d}\,\Phi) = \tau e \quad \textit{with} \quad \tau = 1 \textit{ or } -1 \qquad \textit{if } d > 0.$$

Moreover, every element of $\mathfrak{H}_n^0$ *is of this type. Its genus is determined by* (σ, e) *if* $d < 0$, *and by* e *if* $d > 0$. *If* $d > 0$ *and* $-1 \in N_{K/\mathbf{Q}}(K^\times)$, *then both* e *and* $-e$ *can occur as* $\det(\sqrt{d}\,\Phi)$ *for* Φ *in the same genus. If* $d > 0$ *and* $-1 \notin N_{K/\mathbf{Q}}(K^\times)$, *then* τ *is uniquely determined by the condition that a prime number* p *divides* e *if and only if* $\tau e \notin N_{K/\mathbf{Q}}(K_p^\times)$, *where* $K_p = K \otimes_\mathbf{Q} \mathbf{Q}_p$.

This concerns the case of even n. We have similar but somewhat different results for odd n (Theorem 2.15). In fact we discussed in [S5] and [S6] semi-integral and reduced quadratic forms and obtained results of the same type. If K is imaginary, the hermitian case is almost parallel to the case of quadratic forms over $\mathbf{Q}$, but the theory for real K is more subtle, as can be seen from the above statement.

Let us now turn to the second problem. Before explaining the principal results, let us first discuss natural problems which are more basic and which must be settled before investigating the main question. Given (V, φ) as before, let $U^\varphi(V)$ and $SU^\varphi(V)$ denote the unitary group and the special unitary group of φ, defined as subgroups of $GL(V, K)$. Take an $\mathfrak{r}$-lattice L in V and put

(0.2) $$\Gamma(L) = \{\alpha \in U^\varphi(V) \mid L\alpha = L\}, \qquad \Gamma^1(L) = \Gamma(L) \cap SU^\varphi(V).$$

Then we ask, for a fixed $q \in F^\times$, whether the set $\{h \in V \mid \varphi[h] = q\}$ modulo $\Gamma^1(L)$ is a finite set. A similar question can be asked by replacing F, K, (V, φ) and L by their localizations at a nonarchimedean prime and by defining an obvious analogue of $\Gamma^1(L)$. We will prove that the answer is affirmative in both global and local cases, provided $n > 1$ (Theorems 3.3 and 4.2). The same is true for the problem about the solutions ξ of the equation $\xi\Phi \cdot {}^t\xi^\rho = \Psi$, where Ψ is of size m, and ξ belongs to an $\mathfrak{r}$-lattice in the space of $(m \times n)$-matrices with entries in K, where m is a positive integer $< n$ (Theorems 5.2 and 5.3). We already proved in [S3] the analogues of these facts for quadratic forms and orthogonal groups.

In order to go beyond the mere finiteness, we consider the adelizations $U^\varphi(V)_\mathbf{A}$ and $SU^\varphi(V)_\mathbf{A}$, and define their open subgroups C and C^1 by

$$C = \{\gamma \in U^\varphi(V)_\mathbf{A} \mid L\gamma = L\}, \quad C^1 = C \cap SU^\varphi(V)_\mathbf{A}, \tag{0.3}$$

where L is a fixed $\mathfrak{r}$-lattice in V. Given two solutions ξ_0 and ξ_1 of the equation $\xi\Phi \cdot {}^t\xi^\rho = \Psi$, we say that they belong to the same *genus* (with respect to C) if $\xi_0\gamma_v = \xi_1$ for every nonarchimedean prime v with an element $(\gamma_v)_v \in C$. Naturally they are said to belong to the same *class* if $\xi_0\gamma = \xi_1$ with $\gamma \in \Gamma(L)$. Now to explain our principal ideas in the simplest case, put $G = U^\varphi(V)$ and $H = \{\alpha \in G \mid \xi_0\alpha = \xi_0\}$; also assume for the moment that $G_\mathbf{A} = GC$. Then there is a bijection of $H\backslash H_\mathbf{A}/(H_\mathbf{A} \cap C)$ onto the set of classes in the genus of ξ_0, and so

$$\#\{H\backslash H_\mathbf{A}/(H_\mathbf{A} \cap C)\} = \text{the number of classes in the genus of } \xi_0. \tag{B}$$

Here and henceforth $\#\{X\}$ denotes the number of elements in a set X. If $G_\mathbf{A} \neq GC$, the right-hand side becomes a finite sum of the class numbers of several genera (Theorem 5.4). Since the left-hand side is *the class number* of the unitary group H with respect to $H_\mathbf{A} \cap C$, equality (B) connects it to the solutions ξ of $\xi\Phi \cdot {}^t\xi^\rho = \Psi$.

If $m = 1$, the results can be stated in a more transparent way. Returning to the hermitian form $\varphi : V \times V \to K$, put $\varphi[h] = \varphi(h, h)$ for $h \in V$. Then the equation $\xi\Phi \cdot {}^t\xi^\rho = \Psi$ can be written $\varphi[h] = q$ with $h \in V$ and $q \in F^\times$; thus h and q replace ξ and Ψ. Given a fractional ideal $\mathfrak{b}$ in K, put

$$L[q, \mathfrak{b}] = \{h \in V \mid \varphi[h] = q, \varphi(h, L) = \mathfrak{b}\}. \tag{0.4}$$

We call L *integral* if $\varphi[x] \in \mathfrak{g}$ for every $x \in L$ and call L *maximal* if it is maximal among the integral $\mathfrak{r}$-lattices. The point of considering $L[q, \mathfrak{b}]$ is that $L[q, \mathfrak{b}]$, *if nonempty, consists of a single genus with respect to* C in the above sense. This is clearly a result of local nature; unfortunately, its proof given in Section 3 is not short. In this case $H = U^\varphi(W)$, where W is the orthogonal complement of Kh in V. Now we can prove (Theorem 4.4, Corollary 5.8):

(C) *For every* $y \in G_{\mathbf{A}}$, *there is a bijection of* $H\backslash(H_{\mathbf{A}} \cap GyC)/(H_{\mathbf{A}} \cap C)$ *onto* $(Ly^{-1})[q, \mathfrak{b}]/\Delta_y$, *where* $\Delta_y = G \cap yCy^{-1}$.

(D) *Take* $\{y_i\}_{i \in I} \subset G_{\mathbf{A}}$ *so that* $G_{\mathbf{A}} = \bigsqcup_{i \in I} Gy_iC$ *and put* $L_i = Ly_i^{-1}$. *Then*

$$\sum_{i \in I} \#\{L_i[q, \mathfrak{b}]/\Gamma(L_i)\} = \#\{H\backslash H_{\mathbf{A}}/(H_{\mathbf{A}} \cap C)\}.$$

(E) *If the archimedean factor of* $G_{\mathbf{A}}$ *is compact, then*

$$\sum_{i \in I} \#\{L_i[q, \mathfrak{b}]\}/\#\{\Gamma_i\} = \mathfrak{m}(H, H_{\mathbf{A}} \cap C),$$

where the right-hand side is the mass of H *with respect to the open subgroup* $H_{\mathbf{A}} \cap C$ *of* $H_{\mathbf{A}}$ *in the sense of* [S2].

These assertions are true with $SU^{\varphi}(V)$, $SU^{\varphi}(W)$ and C^1 in place of G, H, and C, if $\dim(V) > 2$ and we impose a certain condition on $(q, \mathfrak{b})$. Notice that (E) gives the mass of H by means of the number of solutions h of $\varphi[h] = q$ under the condition $\varphi(h, L) = \mathfrak{b}$, while (D) gives the class number of H.

In our recent book [S3, Chapter III] we developed a theory of a Diophantine equation $\varphi[h] = q$ for a quadratic form φ defined on a vector space over an algebraic number field. The principal result is that to each "primitive solution" h of this equation for a fixed q, considered modulo the group of units Γ, one can associate a "class" of lattices with respect to the orthogonal group H of the restriction of φ to a subspace of codimension 1. Consequently the class number of H equals the number of such h modulo Γ. This includes as a special case the result of Gauss that the number of primitive representations of q as the sum of three squares equals an elementary factor times the class number of primitive binary quadratic forms of discriminant $-q$. Also, formulas of type (B) and (E) were proved in [S3] and [S6] for quadratic forms. The reader is referred to [S7] for some more historical and technical comments on this subject. Now (B), (C), (D), and (E) are hermitian analogues of these results. In order to develop the theory for hermitian forms, we are naturally guided by the formulation in the case of quadratic forms, but we need new ideas and technique, and it is wrong to say that we can do things "in the same way." This is especially so when we consider the problem with respect to the special unitary group instead of the unitary group. Thus there are two theories with respect to these two types of groups, and one, that for the special unitary group, is more complex than the other, and in a sense more interesting.

1. GENERALITIES ON HERMITIAN FORMS AND UNITARY GROUPS

1.1. For an associative ring A with an identity element we denote by $A^{\times}$ the group of all invertible elements of A, and for positive integers m and n we denote by A_n^m the A-module of all $(m \times n)$-matrices with entries in A. We put $M_n(A) = A_n^n$ when we view it as a ring, and denote by 1_n its identity element.

We take a basic field F and a couple (K, ρ) consisting of an F-algebra K and a nontrivial F-linear automorphism ρ of K belonging to the following two types:

(I) K is a separable quadratic extension of F and ρ generates $\mathrm{Gal}(K/F)$;
(II) $K = F \times F$ and $(x, y)^\rho = (y, x)$ for $(x, y) \in F \times F$.

In our later discussion, K of type (II) will appear as a localization of a global K of type (I). For a matrix $\alpha = (a_{ij})$ with entries in K we denote by ${}^t\alpha$ the transpose of α, and put $\alpha^\rho = (a_{ij}^\rho)$ and $\alpha^* = {}^t\alpha^\rho$; we put also $\alpha^{-\rho} = (\alpha^\rho)^{-1}$ when α is invertible. For a subring S of K we write $\alpha \prec S$ if all the entries of α are contained in S. Given a left K-module V, we denote by $\mathrm{End}(V, K)$ the ring of all K-linear endomorphisms of V and put $GL(V, K) = \mathrm{End}(V, K)^\times$. We let $\mathrm{End}(V, K)$ act on V on the right; namely we denote by $w\alpha$ the image of $w \in V$ under $\alpha \in \mathrm{End}(V, K)$.

Let V be a left K-module isomorphic to K_n^1; we put then $n = \dim(V)$. By a hermitian space we mean a structure (V, φ), where φ is a hermitian form on V, that is, an F-bilinear map $\varphi : V \times V \to K$ such that

$$\varphi(x, y)^\rho = \varphi(y, x), \tag{1.1}$$

$$\varphi(ax, by) = ab^\rho\varphi(x, y) \quad \text{for every} \quad a, b \in K. \tag{1.2}$$

Whenever we speak of a hermitian space (V, φ), we assume that φ is nondegenerate, and put $\varphi[x] = \varphi(x, x)$ for $x \in V$. We define groups $U^\varphi(V)$ and $SU^\varphi(V)$ by

$$U^\varphi = U^\varphi(V) = \{\alpha \in GL(V, K) \mid \varphi[x\alpha] = \varphi[x] \ \text{ for every } x \in V\}, \tag{1.3a}$$

$$SU^\varphi = SU^\varphi(V) = \{\alpha \in U^\varphi(V) \mid \det(\alpha) = 1\}. \tag{1.3b}$$

For every free K-submodule X of V on which φ is nondegenerate, we put

$$X^\perp = \{y \in V \mid \varphi(y, X) = 0\}, \tag{1.4}$$

and define $U^\varphi(X)$ and $SU^\varphi(X)$ by (1.3a, b) with X in place of V; namely we use φ for its restriction to X. We always identify $U^\varphi(X)$ with the subgroup of $U^\varphi(V)$ consisting of the elements α such that $y\alpha = y$ for every $y \in X^\perp$. Similarly we view $SU^\varphi(X)$ as a subgroup of $SU^\varphi(V)$.

Let h be an element of V such that $\varphi[h] \neq 0$. Then

$$\{x \in V \mid \varphi[x] = \varphi[h]\} = \begin{cases} h \cdot U^\varphi & \text{if} \quad \dim(V) = 1, \\ h \cdot SU^\varphi & \text{if} \quad \dim(V) > 1. \end{cases} \tag{1.5}$$

This follows easily from the generalized Witt theorem; see [S2, Lemma 1.3], for example. The case $K = F \times F$ is not included in that theorem, but the structure of (V, φ) for such a K is determined by $\dim(V)$ as shown in [S2, §2.13], and so the fact corresponding to the Witt theorem is trivially true; see also §1.8 below.

1.2. Let φ_0 be the matrix that represents φ with respect to a K-basis of V; then we denote by $d_0(V, \varphi)$ the element of $F^\times/N_{K/F}(K^\times)$ represented by $(-1)^{n/2}\det(\varphi_0)$ or $(-1)^{(n-1)/2}\det(\varphi_0)$ according as n is even or odd. This does not depend on the choice of a K-basis of V. We denote $d_0(V, \varphi)$ simply by $d_0(V)$ or $d_0(\varphi)$ when there is no fear of confusion.

Given $s \in F^\times$, we denote by $\{K, s\}$ the quaternion algebra B over F in which K is embedded and which is given by

$$(1.6)\qquad B = K + K\omega, \quad \omega^2 = s, \quad \omega a = a^\rho\omega \text{ for every } a \in K.$$

Since B is determined by K and $sN_{K/F}(K^\times)$, for a coset $\varepsilon \in F^\times/N_{K/F}(K^\times)$ we denote by $\{K, \varepsilon\}$ the algebra $\{K, s\}$ with any $s \in \varepsilon$. In particular, we can associate with (V, φ) a quaternion algebra $\{K, d_0(\varphi)\}$.

LEMMA 1.3. *Given (V, φ) as in §1.1, suppose* $\dim(V) = 2$ *and put* $B = \{K, d_0(\varphi)\}$. *Then there is a ring-injection j of B into* $\mathrm{End}(V, K)$ *and an element ℓ of V such that $\varphi[\ell] \neq 0$, $\ell j(B) = V$, $\ell j(a) = a\ell$ for every $a \in K$, and $\varphi[\ell j(\xi)] = \varphi[\ell]\xi\xi^\iota$ for every $\xi \in B$, where ι is the main involution of B. Moreover,* $\mathrm{Tr}_{K/F}(\varphi(\ell\alpha, \ell\beta)) = \varphi[\ell]\mathrm{Tr}_{B/F}(\alpha\beta^\iota)$ *for every $\alpha, \beta \in B$, and*

$$(1.7a)\qquad U^\varphi(V) = \{z^{-1}\alpha \mid z \in K^\times, \alpha \in B^\times, zz^\rho = \alpha\alpha^\iota\},$$

$$(1.7b)\qquad SU^\varphi(V) = \{\alpha \in B^\times \mid \alpha\alpha^\iota = 1\},$$

where we identify α with $j(\alpha)$ for $\alpha \in B$.

Proof. Identify V with K_2^1 so that $\varphi(x, y) = x\varphi_0 y^*$ for $x, y \in K_2^1$ with $\varphi_0 = \mathrm{diag}[c, -cs]$, where $c, s \in F^\times$. Then $d_0(\varphi) = sN_{K/F}(K^\times)$, and so B is given by (1.6). Define $j : B \to M_2(K)$ by $j(a+b\omega) = \begin{bmatrix} a & b \\ sb^\rho & a^\rho \end{bmatrix}$ and put $\ell = (1, 0)$. Then it is an easy exercise to verify all the statements; cf. [S2, Lemmas 4.3, 4.4, and (4.3.2)]. Notice that $j(B) = \{\alpha \in M_2(K) \mid \alpha^\iota\varphi_0 = \varphi_0\alpha^*\}$.

1.4. When K is a field, by a *weak Witt decomposition of* V we mean a direct sum decomposition of V with $2r$ elements e_i, f_i, and a subspace Z of V such that

$$(1.8a)\qquad V = \textstyle\sum_{i=1}^r (Ke_i + Kf_i) + Z, \qquad Z = \big(\sum_{i=1}^r (Ke_i + Kf_i)\big)^\perp,$$

$$(1.8b)\qquad \varphi(e_i, e_j) = \varphi(f_i, f_j) = 0, \quad \varphi(e_i, f_j) = \delta_{ij} \text{ for every } i \text{ and } j.$$

Clearly $\sum_{i=1}^r (Ke_i + Kf_i)$ is a subspace of dimension $2r$. We call this a *Witt decomposition* if $\varphi[x] \neq 0$ for every $x \in Z, \neq 0$, in which case we call Z a *core subspace* of (V, φ) and $\dim(Z)$ the *core dimension* of (V, φ). If ζ is the restriction of φ to Z, then clearly $d_0(\varphi) = d_0(\zeta)$.

1.5. In this paper a *global field* means an algebraic number field of finite degree, and a *local field* the completion of a global field at a nonarchimedean prime. For a global field F we denote by $\mathfrak{g}$ the ring of algebraic integers in F;

for a local F we denote by $\mathfrak{g}$ the ring of local integers in F in the standard sense. An archimedean completion of a global field will not be called a local field. In both local and global cases, by a $\mathfrak{g}$-*lattice* in a finite-dimensional vector space V over F, we mean a finitely generated $\mathfrak{g}$-submodule of V that spans V over F.

Let (K, ρ) be as in §1.1 with a local or global F. We then denote by $\mathfrak{r}$ the ring of all elements of K integral over $\mathfrak{g}$, and by $\mathfrak{d}$ the different of K relative to F. We have $\mathfrak{r} = \mathfrak{d} = \mathfrak{g} \times \mathfrak{g}$ if $K = F \times F$. By a $\mathfrak{g}$-*ideal* we mean a fractional ideal in F, and similarly by an $\mathfrak{r}$-*ideal* we mean a fractional ideal in K if K is a field. If $K = F \times F$, an $\mathfrak{r}$-*ideal* means a subset of K of the form $\mathfrak{a} \times \mathfrak{b}$ with $\mathfrak{g}$-ideals $\mathfrak{a}$ and $\mathfrak{b}$. In both local and global cases, by an $\mathfrak{r}$-*lattice* in a K-module V as in §1.1 we mean a $\mathfrak{g}$-lattice in V stable under multiplication by the elements of $\mathfrak{r}$. Given two $\mathfrak{r}$-lattices L and M in V, we denote by $[L/M]$ the $\mathfrak{r}$-ideal generated by $\det(\alpha)$ for all $\alpha \in GL(V)$ such that $L\alpha \subset M$. Thus $[L/L\alpha] = \det(\alpha)\mathfrak{r}$. In particular, if $\alpha \in U^{\varphi}$ and K is a field in the local case, then $[L/L\alpha] = \mathfrak{r}$. If $K = F \times F$, however, $[L/L\alpha] = \mathfrak{a} \times \mathfrak{a}^{-1}$ with a $\mathfrak{g}$-ideal $\mathfrak{a}$ for $\alpha \in U^{\varphi}$.

LEMMA 1.6. *Two hermitian spaces* (V, φ) *and* (V', φ') *in the local case are isomorphic if and only if* $\dim(V) = \dim(V')$ *and* $d_0(\varphi) = d_0(\varphi')$.

This is well known. For the proof, see [S2, Proposition 5.3], for example.

1.7. Let (V, φ) be defined with a local or global F. For a $\mathfrak{g}$-lattice L in V we denote by $\mu(L)$ the $\mathfrak{g}$-ideal generated by $\varphi[x]$ for all $x \in L$. We call a $\mathfrak{g}$-lattice L in V *integral* if $\mu(L) \subset \mathfrak{g}$; we call an $\mathfrak{r}$-lattice L *maximal* if L is maximal among the integral $\mathfrak{r}$-lattices. (This is what we called $\mathfrak{g}$-maximal in [S2].) For basic properties of maximal lattices in V the reader is referred to [S1] or Sections 4 and 5 of [S2]. For example, we note ([S2, (4.7.1)])

$$\varphi(L, L) \subset \mathfrak{d}^{-1} \text{ if } L \text{ is integral.} \tag{1.9}$$

If $n > 1$ and L is maximal, then $\mu(L) = \mathfrak{g}$. This fact in the global case follows from the local case, which can be seen from [S2, Lemmas 5.4 and 5.6].

Given an $\mathfrak{r}$-lattice L in V, $q \in F^{\times}$, and an $\mathfrak{r}$-ideal $\mathfrak{b}$, we put

$$\widehat{L} = \{x \in V \mid \varphi(x, L) \subset \mathfrak{d}^{-1}\}, \tag{1.10a}$$

$$L[q] = \{x \in L \mid \varphi[x] = q\}, \tag{1.10b}$$

$$L[q, \mathfrak{b}] = \{x \in V \mid \varphi[x] = q,\, \varphi(x, L) = \mathfrak{b}\}. \tag{1.10c}$$

By (1.9), we have $L \subset \widehat{L}$ if L is integral. The set $L[q, \mathfrak{b}]$ is not necessarily contained in $L[q]$. If M is another $\mathfrak{r}$-lattice in V, then we easily see that $[L/M]^{\rho} = [\widehat{M}/\widehat{L}]$. If $L_1 = L\alpha$ with $\alpha \in U^{\varphi}$, then $\widehat{L}_1 = \widehat{L}\alpha$, and so $[\widehat{L}_1/L_1] = [\widehat{L}/L]$.

The notation being as in (1.8a, b), take a maximal $\mathfrak{r}$-lattice M in Z and put

(1.11) $$L = \sum_{i=1}^{r}(\mathfrak{r}e_i + \mathfrak{d}^{-1}f_i) + M.$$

Then L is maximal; see [S2, Lemma 4.9 (2)]. We can easily verify that

(1.12) $$\widehat{L} = \sum_{i=1}^{r}(\mathfrak{r}e_i + \mathfrak{d}^{-1}f_i) + \widehat{M}, \quad \widehat{M} = \{x \in Z \mid \varphi(x, M) \subset \mathfrak{d}^{-1}\}.$$

1.8. Let us now consider the case $K = F \times F$; then we define the core dimension of (V, φ) to be 0. We can write an element of $V = K_n^1$ in the form (x, y) with $x, y \in F_n^1$. Taking a suitable coordinate system, we can assume that

(1.13) $$\varphi((x, y), (z, w)) = (x \cdot {}^t w, y \cdot {}^t z) \qquad (x, y, z, w \in F_n^1).$$

This is shown in [S2, §2.13]. We have then $\varphi[(x, y)] = x \cdot {}^t y$ and

(1.14) $$U^\varphi = \{(\xi, \widetilde{\xi}) \mid \xi \in GL_n(F)\}, \quad SU^\varphi = \{(\xi, \widetilde{\xi}) \mid \xi \in SL_n(F)\},$$

where $\widetilde{\xi} = {}^t\xi^{-1}$; also $\mathfrak{g}_n^1 \times \mathfrak{g}_n^1$ is a maximal lattice. It should be noted that if $h \in V$ and $\varphi[h] \neq 0$, then Kh is isomorphic to K.

LEMMA 1.9. *Let L be a maximal lattice in V and t the core dimension of (V, φ); suppose K is a local field; put*

$$E = \{e \in \mathfrak{r}^\times \mid ee^\rho = 1\}, \quad E_0 = \{e^\rho/e \mid e \in \mathfrak{r}^\times\}, \quad E_L = \det(C(L)),$$

where $C(L) = \{\alpha \in U^\varphi(V) \mid L\alpha = L\}$. Then the following assertions hold:

(i) *$[E : E_0] = 1$ or 2 according as K is unramified or ramified over F.*

(ii) *$E_L = E_0$ if $t = 0$; $E_L = E$ if $t > 0$.*

This is included in [S1, Lemma 4.16 and Proposition 4.18] and [S2, Lemma 5.11].

Suppose K is a local field ramified over F; let L and M be maximal lattices in V. Then there exists an element $\alpha \in U^\varphi$ such that $M = L\alpha$ as shown in (i) of the following lemma. We then denote by $e(L/M)$ the element of E/E_0 represented by $\det(\alpha)$. This is well defined in view of (ii) above.

LEMMA 1.10. *Let L and M be maximal lattices in V in the local case, and let t be the core dimension of (V, φ). Then the following assertions hold:*

(i) *There exists an element α of U^φ such that $M = L\alpha$.*

(ii) *We can take such an α from SU^φ if $t > 0$ or K is a field unramified over F.*

(iii) *Suppose K is ramified over F and $t = 0$; then $M = L\alpha$ with $\alpha \in SU^\varphi$ if and only if $e(L/M) = 1$.*

(iv) *Suppose $K = F \times F$; then $M = L\alpha$ with $\alpha \in SU^\varphi$ if and only if $[L/M] = \mathfrak{r}$.*

Proof. The first assertion is included in [S1, Propositions 3.3 and 4.13] and also in [S2, Lemmas 4.12 and 5.9]. Next, suppose K is a field; given M, take

$\alpha \in U^\varphi$ so that $L\alpha = M$ and put $c = \det(\alpha)$. Then $c \in E$. If K is unramified over F or $t > 0$, then Lemma 1.9 guarantees an element β of U^φ such that $L\beta = L$ and $\det(\beta) = c^{-1}$. Then $\beta\alpha \in SU^\varphi$ and $L\beta\alpha = M$. This proves (ii). Assertion (iii) can be proved in a similar way. Assertion (iv) is included in [S1, Proposition 3.3].

1.11. Let us now consider the case with $n = 2$ and a local F. Using the symbols B, j, and ℓ of Lemma 1.3, we first observe that

$$B \cong M_2(F) \iff t = 0 \iff 1 \in d_0(\varphi). \tag{1.15}$$

Let $\mathfrak{O}$ be a maximal order in B containing $\mathfrak{r}$. We can find an element $\gamma \in B^\times$ such that $\varphi[\ell]^{-1} = \gamma\gamma^\iota$. Put $M = \ell j(\mathfrak{O}\gamma)$ and $\widehat{\mathfrak{O}} = \{\alpha \in B \,|\, \mathrm{Tr}_{B/F}(\alpha\mathfrak{O}) \subset \mathfrak{g}\}$. Identifying B with $j(B)$, for $\alpha \in B$ we see that

$$\ell\alpha\gamma \in \widehat{M} \iff \varphi(\ell\alpha\gamma,\, M) \subset \mathfrak{d}^{-1} \iff \mathrm{Tr}_{K/F}\big(\mathfrak{r}\varphi(\ell\alpha\gamma,\, M)\big) \subset \mathfrak{g}$$
$$\iff \mathrm{Tr}_{B/F}(\alpha\mathfrak{O}) \subset \mathfrak{g} \iff \alpha \in \widehat{\mathfrak{O}}.$$

Thus $\widehat{M} = \ell j(\widehat{\mathfrak{O}}\gamma)$. If B is not a division algebra, then $\widehat{\mathfrak{O}} = \mathfrak{O}$, so that $\widehat{M} = M$, which means that M is maximal. Suppose B is a division algebra; then $\mathfrak{O} = \{\alpha \in B \,|\, \alpha\alpha^\iota \in \mathfrak{g}\}$ as noted in [S3, Theorem 5.13], and so $M = \{x \in V \,|\, \varphi[x] \in \mathfrak{g}\}$, which is a unique maximal lattice in V by [S2, Lemma 5.4]. Since $\widehat{\mathfrak{O}} = \mathfrak{P}^{-1}$ with the maximal ideal $\mathfrak{P}$ of $\mathfrak{O}$, we have $\widehat{M} = \ell j(\mathfrak{P}^{-1}\gamma)$. Thus $[\widehat{M}/M] = \mathfrak{p}\mathfrak{r}$ with the maximal ideal $\mathfrak{p}$ of $\mathfrak{g}$.

2. Classification of hermitian forms over a global field

2.1. Throughout this section we assume that F is a global field and K is a quadratic extension of F. We denote by $\mathbf{a}$ and $\mathbf{h}$ the sets of archimedean primes and nonarchimedean primes of F respectively, and put $\mathbf{v} = \mathbf{a} \cup \mathbf{h}$. Given an algebraic group G defined over F, we define G_v for each $v \in \mathbf{v}$ and the adelization $G_\mathbf{A}$ as usual, and view G and G_v as subgroups of $G_\mathbf{A}$. We then denote by $G_\mathbf{a}$ and $G_\mathbf{h}$ the archimedean and nonarchimedean factors of $G_\mathbf{A}$, respectively. In particular, the adelization of the multiplicative group $F^\times$ is denoted by $F_\mathbf{A}^\times$, which is the idele group of F. For $x \in G_\mathbf{A}$ and $v \in \mathbf{v}$ we denote by x_v the v-component of x.

Given (V, φ) over F, for each $v \in \mathbf{v}$ we can define the v-localization $(V, \varphi)_v = (V_v, \varphi_v)$ with $\varphi_v : V_v \times V_v \to K_v$ in a natural way. For $v \in \mathbf{h}$ let t_v be the core dimension of $(V, \varphi)_v$. Since $x \mapsto \varphi_v[x]$ for $x \in V_v$ can be viewed as an F_v-valued quadratic form, we have $2t_v \le 4$ by a well known principle, and so $t_v \le 2$. Let $\mathbf{r}_0$ denote the set of all real archimedean primes of F that do not split in K. If $v \in \mathbf{a}$ and $v \notin \mathbf{r}_0$, then there is only one isomorphism class of $(V, \varphi)_v$ for each n. For each fixed $v \in \mathbf{r}_0$ we have a pair of nonnegative integers (p_v, q_v) such that φ_v is represented by $\mathrm{diag}[1_{p_v}, -1_{q_v}]$ when F_v and K_v are identified with $\mathbf{R}$ and $\mathbf{C}$. We put then $s_v(\varphi) = p_v - q_v$, and call $s_v(\varphi)$ the *index* of φ at v. Clearly $|s_v(\varphi)|$ is the core dimension of φ_v, $s_v(\varphi) - n \in 2\mathbf{Z}$, and $|s_v(\varphi)| \le n$; also, n and $s_v(\varphi)$ determine (p_v, q_v), and vice versa.

For an $\mathfrak{r}$-lattice L in V and $v \in \mathbf{h}$ we denote by L_v the $\mathfrak{r}_v$-linear span of L in V_v. Also, for $\xi \in GL(V, K)_{\mathbf{A}}$ we denote by $L\xi$ the lattice in V such that $(L\xi)_v = L_v\xi_v$ for every $v \in \mathbf{h}$. By the $U^\varphi(V)$-genus (resp. $SU^\varphi(V)$*-genus*) of L we understand the set of all lattices of the form $L\xi$ with $\xi \in U^\varphi(V)_{\mathbf{A}}$ (resp. $\xi \in SU^\varphi(V)_{\mathbf{A}}$). Also, by the $U^\varphi(V)$-class (resp. $SU^\varphi(V)$*-class*) of L we understand the set of all lattices of the form $L\alpha$ with $\alpha \in U^\varphi(V)$ (resp. $\alpha \in SU^\varphi(V)$).

The classification of (V, φ) over a number field was done by Landherr in [L]. We formulate the results in the form that suits our later purposes, and give a proof for the reader's convenience. To be precise, we are going to show that the isomorphism classes of hermitian spaces correspond bijectively to the sets of data consisting of the following objects:

(2.1a) $0 < n \in \mathbf{Z}$; $\varepsilon \in F^\times$; *an integer* σ_v, *given for each* $v \in \mathbf{r}_0$, *such that* $|\sigma_v| \le n$ *and* $\sigma_v - n \in 2\mathbf{Z}$.

We look for (V, φ) such that $\dim(V) = n$, $d_0(\varphi)$ is represented by ε, and $s_v(\varphi) = \sigma_v$ for every $v \in \mathbf{r}_0$. Clearly the following condition is necessary:

(2.1b) $(-1)^{\sigma_v/2}\varepsilon > 0$ *for every* $v \in \mathbf{r}_0$ *if* $n \in 2\mathbf{Z}$ *and* $(-1)^{(\sigma_v-1)/2}\varepsilon > 0$ *for every* $v \in \mathbf{r}_0$ *if* $n - 1 \in 2\mathbf{Z}$.

THEOREM 2.2. (i) *The isomorphism class of* (V, φ) *is determined by* n, $\{\sigma_v\}$, *and* $d_0(\varphi)$.

(ii) *Given* n, ε, *and* $\{\sigma_v\}$ *satisfying* (2.1a, b), *there exists a hermitian space* (V, φ) *such that* $\dim(V) = n$, $\varepsilon \in d_0(\varphi)$, *and* $s_v(\varphi) = \sigma_v$ *for every* $v \in \mathbf{r}_0$.

Proof. Clearly n and $\{\sigma_v\}$ determine $(V, \varphi)_v$ for every $v \in \mathbf{a}$, and n and $d_0(\varphi)$ determine $(V, \varphi)_v$ for every $v \in \mathbf{h}$ by Lemma 1.6. Therefore we obtain (i) in view of the Hasse principle. We prove (ii) by induction on n. The case $n = 1$ is trivial, and so we assume $n > 1$. We first prove the case in which $\sigma_v \ge 0$ for every $v \in \mathbf{r}_0$. Let $\tau_v = \sigma_v - 1$. Then the set $\bigl(n - 1, (-1)^{n-1}\varepsilon, \{\tau_v\}\bigr)$ satisfies (2.1a, b), and therefore by induction we can find a hermitian space (W, ψ) such that $\dim(W) = n - 1$, $\tau_v = s_v(\psi)$ for every $v \in \mathbf{r}_0$, and $(-1)^{n-1}\varepsilon \in d_0(\psi)$. Put $V = K \oplus W$ and define φ on V by $\varphi[a \oplus y] = aa^\rho + \psi[y]$ for $a \in K$ and $y \in W$. Then clearly $\varepsilon \in d_0(\varphi)$ and $s_v(\varphi) = \sigma_v$ for every $v \in \mathbf{r}_0$.

Now, given $\{\sigma_v\}$ with possibly negative σ_v, take $c \in F^\times$ so that $c < 0$ or $c > 0$ at $v \in \mathbf{r}_0$ according as $\sigma_v < 0$ or $\sigma_v \ge 0$. Then the set $\bigl(n, c^n\varepsilon, \{|\sigma_v|\}\bigr)$ satisfies (2.1a, b). Therefore we can find a hermitian space (V_1, φ_1) such that $\dim(V_1) = n$, $c^n\varepsilon \in d_0(\varphi_1)$, and $s_v(\varphi_1) = |\sigma_v|$ for every $v \in \mathbf{r}_0$. Put $\varphi = c\varphi_1$. Then $\varepsilon \in d_0(\varphi)$ and $s_v(\varphi) = \sigma_v$ for every $v \in \mathbf{r}_0$. This completes the proof.

THEOREM 2.3. *Given* (V, φ), *put* $B = \{K, d_0(\varphi)\}$ *using the notation of* §1.2. *Let* $\mathfrak{e}$ *be the product of the prime ideals of* F *ramified in* B; *also let* L *be a maximal lattice in* V. *Then the following assertions hold.*

(i) $[\widehat{L}/L] = \mathfrak{e}\mathfrak{r}$ *if* n *is even.*

(ii) *When* n *is odd, put* $d_0(\varphi)\mathfrak{g} = \mathfrak{a}N_{K/F}(\mathfrak{b})$ *with an* $\mathfrak{r}$-*ideal* $\mathfrak{b}$ *and a squarefree integral* $\mathfrak{g}$-*ideal* $\mathfrak{a}$ *whose prime factors remain prime in* K. *Then* $[\widehat{L}/L] = \mathfrak{a}\mathfrak{d}$, *where* $\mathfrak{d}$ *is the different of* K *relative to* F.

Proof. For $v \in \mathbf{h}$ let t_v be the core dimension of $(V, \varphi)_v$. Suppose $n \in 2\mathbf{Z}$; then $t_v = 0$ if and only if $d_0(\varphi)$ is represented by an element of $N_{K/F}(K_v^\times)$, that is, if and only if v does not divide $\mathfrak{e}$. If n is odd, the isomorphism class of $(V, \varphi)_v$ depends on $\mathfrak{a}_v$ and $\mathfrak{d}_v$. Thus our assertions can be reduced to the question about $[\widehat{L}_v/L_v]$ for $v \in \mathbf{h}$. In fact, suppressing the subscript v, we have, in the local case,

(2.2) $[\widehat{L}/L] = \mathfrak{r}$ *if* $t = 0$; $[\widehat{L}/L] = \mathfrak{p}\mathfrak{r}$ *if* $t = 2$; $[\widehat{L}/L] = \mathfrak{d}$ *if* $t = 1$ *and* $d_0(\varphi) \cap \mathfrak{g}^\times \neq \emptyset$; $[\widehat{L}/L] = \mathfrak{p}\mathfrak{d}$ *if* $t = 1$ *and* $d_0(\varphi) \cap \mathfrak{g}^\times = \emptyset$.

Here $\mathfrak{p}$ is the maximal ideal of $\mathfrak{g}$. In view of Lemma 1.10 (i), it is sufficient to prove this for a special choice of L. If $K = F \times F$, then we can put $L = \mathfrak{g}_n^1 \times \mathfrak{g}_n^1$ as noted in §1.8, and so $\widehat{L} = L$. Thus we assume that K is a field. By (1.12), $L = \widehat{L}$ if $t = 0$. Let $M = \{x \in Z \mid \varphi[x] \in \mathfrak{g}\}$. By (1.12), $[\widehat{L}/L] = [\widehat{M}/M]$. We have seen that $[\widehat{M}/M] = \mathfrak{p}\mathfrak{r}$ in §1.11 if $t = 2$. If $t = 1$, then $M = \mathfrak{r}\ell$ with an element ℓ such that $\varphi[\ell]\mathfrak{g}$ is $\mathfrak{g}$ or $\mathfrak{p}\mathfrak{g}$. Thus $\widehat{M} = \mathfrak{d}^{-1}\varphi[\ell]^{-1}\ell$, and so $[\widehat{L}/L] = [\widehat{M}/M] = \varphi[\ell]\mathfrak{d}$, which completes the proof of (2.2). Combining the results on $[\widehat{L}_v/L_v]$ for all $v \in \mathbf{h}$, we obtain our theorem.

2.4. To illustrate Theorem 2.3 in terms of matrices, we have to define the genus and class of a hermitian matrix. We put

$$\mathfrak{G} = GL_n(K), \qquad \mathfrak{H}_n = \{\Phi \in \mathfrak{G} \mid \Phi^* = \Phi\}, \qquad L_0 = \mathfrak{r}_n^1, \tag{2.3}$$

$$E = \mathfrak{G}_{\mathbf{a}} \prod_{v \in \mathbf{h}} GL_n(\mathfrak{r}_v), \qquad E_\xi = \xi^{-1}E\xi \quad (\xi \in \mathfrak{G}_{\mathbf{A}}), \tag{2.4}$$

$$\Delta_\xi = E_\xi \cap \mathfrak{G}, \qquad \Delta_\xi^1 = E_\xi \cap SL_n(K). \tag{2.5}$$

Every $\mathfrak{r}$-lattice L in K_n^1 can be given as $L = L_0\xi$ with $\xi \in \mathfrak{G}_{\mathbf{A}}$, and $E_\xi = \{y \in \mathfrak{G}_{\mathbf{A}} \mid Ly = L\}$. We denote by $\mathfrak{H}_n(\xi)$ the set of all $\Phi \in \mathfrak{H}_n$ such that $x\Phi x^* \in \mathfrak{g}$ for every $x \in L_0\xi$. We call such a Φ *reduced* (relative to ξ) if the following condition is satisfied:

$$\Phi \in \mathfrak{H}_n(\zeta^{-1}\xi) \text{ with } \zeta \in \mathfrak{G}_{\mathbf{h}} \cap \prod_{v \in \mathbf{h}} M_n(\mathfrak{r}_v) \Longrightarrow \zeta \in E. \tag{2.6}$$

We denote by $\mathfrak{H}_n^0(\xi)$ the set of all reduced elements of $\mathfrak{H}_n(\xi)$.

We say that two elements Φ and Ψ of $\mathfrak{H}_n(\xi)$ belong to the same *genus* (relative to ξ) if there exists an element ε of E_ξ such that $\varepsilon\Phi\varepsilon^* = \Psi$; they are said to belong to the same *U-class* (resp. *SU-class*) if $\alpha\Phi\alpha^* = \Psi$ for some $\alpha \in \Delta_\xi$ (resp. $\alpha \in \Delta_\xi^1$). These depend on the choice of $L = L_0\xi$.

Given $\Phi \in \mathfrak{H}_n$, put $V = K_n^1$ and $\varphi[x] = x\Phi x^*$ for $x \in V$. Then we obtain a hermitian space (V, φ), which we denote by $[\Phi]$, and we write $U(\Phi)$ and $SU(\Phi)$ for $U^\varphi(V)$ and $SU^\varphi(V)$ as subgroups of $\mathfrak{G}$. Put $L = L_0\xi$ with $\xi \in \mathfrak{G}_\mathbf{A}$. Clearly L is integral if $\Phi \in \mathfrak{H}_n(\xi)$, in which case L is maximal if and only if $\Phi \in \mathfrak{H}_n^0(\xi)$. Thus an element of $\mathfrak{H}_n^0(\xi)$ determines a hermitian space and a maximal lattice.

2.5. To parametrize all genera of $\Phi \in \mathfrak{H}_n^0(\xi)$, we need a few more symbols:

$$\mathfrak{t} = K_\mathbf{A}^\times \cap \Bigl(K_\mathbf{a}^\times \prod_{v\in\mathbf{h}} \mathfrak{t}_v\Bigr), \qquad \mathfrak{t}_v = \bigl\{y \in K_v^\times \bigm| yy^\rho \in \mathfrak{g}_v^\times\bigr\}, \tag{2.7}$$

$$T = \bigl\{x \in \mathfrak{G}_\mathbf{A} \bigm| \det(x) \in \mathfrak{t}\bigr\}. \tag{2.8}$$

Notice that $\mathfrak{t}_v = \mathfrak{r}_v^\times \cdot \bigl\{y \in K_v^\times \bigm| yy^\rho = 1\bigr\}$ for every $v \in \mathbf{h}$, and $\mathfrak{t}_v = \mathfrak{r}_v^\times$ if v does not split in K. Let $\mathfrak{I}_K$ denote the ideal group of K and $\mathfrak{I}_{K/F}^0$ the subgroup of $\mathfrak{I}_K$ generated by the ideals $\mathfrak{a}$ such that $N_{K/F}(\mathfrak{a}) = \mathfrak{g}$ and the principal ideals. Now there is a sequence of isomorphisms:

$$\mathfrak{G}_\mathbf{A}/T\mathfrak{G} \;\cong\; K_\mathbf{A}^\times/K^\times\mathfrak{t} \;\cong\; \mathfrak{I}_K/\mathfrak{I}_{K/F}^0. \tag{2.9}$$

The last isomorphism can be obtained by the map $y \mapsto y\mathfrak{r}$ for $y \in K_\mathbf{A}^\times$. As for the first isomorphism, we first note, for every $\xi \in \mathfrak{G}_\mathbf{A}$ and $\Phi \in \mathfrak{H}_n$,

$$T\mathfrak{G}\xi = E\xi U(\Phi)_\mathbf{A}\mathfrak{G} = \bigl\{x \in \mathfrak{G}_\mathbf{A} \bigm| \det(\xi^{-1}x) \in K^\times\mathfrak{t}\bigr\}. \tag{2.10}$$

Clearly the last set contains the second set. Conversely, suppose $x \in \mathfrak{G}_\mathbf{A}$ and $\det(\xi^{-1}x) = by$ with $b \in K^\times$ and $y \in \mathfrak{t}$. We can find $z, w \in K_\mathbf{A}^\times$ such that $z_v \in \mathfrak{r}_v^\times$ and $w_v w_v^\rho = 1$ for every $v \in \mathbf{h}$ and $y = zw$. We can find $\varepsilon \in E$, $\alpha \in \mathfrak{G}$, and $\gamma \in U(\Phi)_\mathbf{A}$ such that $\det(\varepsilon) = z$, $\det(\alpha) = b$, and $\det(\gamma) = w$. Then $\det(x^{-1}\varepsilon\xi\gamma\alpha) = 1$. By strong approximation in $SL_n(K)$ we see that $x^{-1}\varepsilon\xi\gamma\alpha \in x^{-1}ExSL_n(K)$, and so $x^{-1}\varepsilon\xi\gamma\alpha = x^{-1}\varepsilon' x\beta$ with $\varepsilon' \in E$ and $\beta \in SL_n(K)$. Then $x = (\varepsilon')^{-1}\varepsilon\xi\gamma\alpha\beta^{-1} \in E\xi U(\Phi)_\mathbf{A}\mathfrak{G}$, which proves the last equality of (2.10). That $T\mathfrak{G}$ equals the last set of (2.10) for $\xi = 1$ can be proved in the same way. Thus $T\mathfrak{G}$ is the inverse image of $K^\times\mathfrak{t}$ under the map $x \mapsto \det(x)$, and so $T\mathfrak{G}$ is a normal subgroup of $\mathfrak{G}_\mathbf{A}$. Then we obtain the first isomorphism of (2.9) and also the first equality of (2.10) for every $\xi \in \mathfrak{G}_\mathbf{A}$.

PROPOSITION 2.6. (i) *For $\Phi, \Psi \in \mathfrak{H}_n^0(\xi)$, $\xi \in \mathfrak{G}_\mathbf{A}$, the spaces $[\Phi]$ and $[\Psi]$ are isomorphic if and only if they belong to the same genus.*

(ii) *Let X be a complete set of representatives for $\mathfrak{G}_\mathbf{A}/T\mathfrak{G}$, and for each $\xi \in \mathfrak{G}_\mathbf{A}$ let Y_ξ be a complete set of representatives for the genera of the elements of $\mathfrak{H}_n^0(\xi)$. Then the hermitian spaces $[\Phi]$ obtained from $\Phi \in Y_\xi$ for all $\xi \in X$ exhaust all isomorphism classes of n-dimensional hermitian spaces without overlapping.*

Proof. Let Φ and Ψ be elements of $\mathfrak{H}_n^0(\xi)$ belonging to the same genus. Then there exists an element $\varepsilon \in E_\xi$ such that $\varepsilon\Phi\varepsilon^* = \Psi$, and the Hasse principle guarantees an element α of $\mathfrak{G}$ such that $\Psi = \alpha\Phi\alpha^*$. Thus $[\Psi]$ is isomorphic to $[\Phi]$. Conversely, suppose $[\Phi]$ and $[\Psi]$ are isomorphic for $\Phi, \Psi \in \mathfrak{H}_n^0(\xi)$. Then $\Phi = \beta\Psi\beta^*$ for some $\beta \in \mathfrak{G}$. Now $L_0\xi$ is maximal in both $[\Phi]$ and $[\Psi]$, and $L_0\xi\beta$ is maximal in $[\Psi]$. Thus $L_0\xi\beta = L_0\xi\gamma$ with $\gamma \in U(\Psi)_\mathbf{A}$ by Lemma 1.10. Put $\zeta = \beta\gamma^{-1}$. Then $\zeta \in E_\xi$, and $\zeta\Psi\zeta^* = \Phi$. Therefore Ψ belongs to the genus of Φ. This proves (i). Clearly every n-dimensional hermitian space is isomorphic to $[\Psi]$ for some $\Psi \in \mathfrak{H}_n$. Take a maximal lattice L in $K_n^1 = V$ and put $L = L_0\eta$ with $\eta \in \mathfrak{G}_\mathbf{A}$. We have then $\eta \in T\mathfrak{G}\xi$ with some $\xi \in X$. By (2.10) we can put $\xi = \varepsilon\eta\gamma\alpha^{-1}$ with $\varepsilon \in E$, $\gamma \in U(\Psi)_\mathbf{A}$, and $\alpha \in \mathfrak{G}$. Put $\Phi = \alpha\Psi\alpha^*$. Then α gives an isomorphism of $[\Phi]$ onto $[\Psi]$. Now $L_0\eta\gamma$ is a maximal lattice in $[\Psi]$ and $L_0\xi\alpha = L_0\eta\gamma$, and so $L_0\xi$ is a maximal lattice in $[\Phi]$. Thus $\Phi \in \mathfrak{H}_n^0(\xi)$. By (i), Φ can be replaced by a member of Y_ξ. This shows that every (V, φ) can be obtained as described in (ii). Now suppose $[\Phi_1]$ and $[\Phi_2]$ are isomorphic for $\Phi_i \in Y_{\xi_i}$ with $\xi_1, \xi_2 \in X$. Then $\Phi_1 = \alpha\Phi_2\alpha^*$ with $\alpha \in \mathfrak{G}$. Since $L_0\xi_i$ is maximal, we have $L_0\xi_1\alpha = L_0\xi_2\zeta$ with $\zeta \in U(\Phi_2)_\mathbf{A}$. Then $\xi_1\alpha\zeta^{-1}\xi_2^{-1} \in E$, and so $\det(\xi_1\xi_2^{-1}) \in K^\times\mathfrak{t}$. We are taking the ξ_i from X, and therefore $\xi_1 = \xi_2$ by (2.10). By (i), Φ_2 belongs to the genus of Φ_1, and so $\Phi_2 = \Phi_1$. This completes the proof.

2.7. The connection of a class of hermitian matrices with a class of lattices is not so simple in general. Given $\Phi \in \mathfrak{H}_n^0(\xi)$ with a fixed ξ, in order to exhaust all classes in the $U(\Phi)$-genus of $L_0\xi$, we have to consider the genera of elements in $\mathfrak{H}_n^0(\xi\zeta)$ for all $\zeta \in T\mathfrak{G}/E_\xi\mathfrak{G}$. Thus $\mathfrak{H}_n^0(\xi)$ is sufficient if and only if $K^\times\mathfrak{t} = K^\times\det(E)$. We will not go into details, as we do not need the result in our later treatment.

The case of SU-class is simpler. Fix $\xi \in \mathfrak{G}$ and put $L = L_0\xi$. For $\Phi \in \mathfrak{H}_n(\xi)$ we define the *SU-genus* (relative to ξ) of Φ to be the set of all $\Psi \in \mathfrak{H}_n^0(\xi)$ such that $\Psi = \varepsilon\Phi\varepsilon^*$ with $\varepsilon \in E_\xi$ such that $\det(\varepsilon) = 1$. Clearly $\det(\Psi) = \det(\Phi)$. Given such Ψ and ε, the Hasse principle guarantees an element $\alpha \in \mathfrak{G}$ such that $\Psi = \alpha\Phi\alpha^*$. Then $\det(\alpha)\det(\alpha)^\rho = 1$. Changing α for $\alpha\gamma$ with a suitable $\gamma \in U(\Phi)$, we may assume that $\det(\alpha) = 1$. Since $L\alpha = L\varepsilon^{-1}\alpha$ and $\varepsilon^{-1}\alpha \in SU(\Phi)_\mathbf{A}$, we see that $L\alpha$ belongs to the $SU(\Phi)$-genus of L. We then associate the $SU(\Phi)$-class of $L\alpha$ to Ψ. We can easily verify that the set of all SU-classes in the SU-genus of Φ contained in $\mathfrak{H}_n(\xi)$ are in one-to-one correspondence with the set of $SU(\Phi)$-classes in the $SU(\Phi)$-genus of L.

2.8. Define (V, φ) by $V = K_n^1$ and $\varphi[x] = x\Phi x^*$ as above with any $\Phi \in \mathfrak{H}_n$. Put $L = L_0\xi$ with $\xi \in \mathfrak{G}_\mathbf{A}$. We easily see that $\widehat{L} = \mathfrak{d}^{-1}L_0(\Phi\xi^*)^{-1}$, and so

$$(2.11) \qquad [\widehat{L}/L] = \det(\Phi\xi\xi^\rho)\mathfrak{d}^n \quad \text{if} \quad L = L_0\xi.$$

We need a few more symbols. First, we put $\mathfrak{d}_0 = \mathfrak{d}^2 \cap F$. For $v \in \mathbf{r}_0$ and (V, φ)

isomorphic to $[\Phi]$ with $\Phi \in \mathfrak{H}_n$ we put $s_v(\Phi) = s_v(\varphi)$ and $d_0(\Phi) = d_0(\varphi)$.

LEMMA 2.9. *Let B be a quaternion algebra over F and K a quadratic extension of F contained in B; let $\mathfrak{r}$ be the maximal order of K and $\mathfrak{O}$ a maximal order in B containing $\mathfrak{r}$; further let $\mathfrak{e}$ be the product of the prime ideals in F ramified in B and $\mathfrak{d}$ the different of K relative to F. Then there exists a $\mathfrak{g}$-ideal $\mathfrak{a}$ such that $\mathfrak{O}$ is isomorphic as a left $\mathfrak{r}$-module to $\mathfrak{r}\oplus\mathfrak{a}$ and $N_{K/F}(\mathfrak{d}\mathfrak{a}) = s\mathfrak{e}$ with an element s such that B is isomorphic to $\{K, s\}$. Moreover, the coset $\mathfrak{a}\mathfrak{I}^0_{K/F}$ is independent of the choice of $\mathfrak{O}$, and $\mathfrak{O}$ is isomorphic as a right $\mathfrak{r}$-module to $\mathfrak{r} \oplus \mathfrak{a}^\iota$, where ι is the main involution of B.*

Proof. Take $\varepsilon \in F^\times$ so that $B = \{K, \varepsilon\}$, and consider $(V, \varphi) = [\Phi]$ over K with $V = K_2^1$ and φ such that $\varepsilon \in d_0(\varphi)$. Using the symbols y, j of Lemma 1.3, identify $j(\alpha)$ with α for $\alpha \in B$, and put $M = y\mathfrak{O}\gamma$ with $\gamma \in B_{\mathbf{h}}^\times$ such that $\varphi[y]^{-1} = \gamma_v\gamma_v^\iota$ for every $v \in \mathbf{h}$. Applying the local result of §1.11 to M_v, we see that M is maximal and $[\widehat{M}/M] = \mathfrak{e}\mathfrak{r}$. Put $M = \mathfrak{r}_2^1\xi$ with some $\xi \in GL_2(K)_{\mathbf{A}}$ and $\mathfrak{a} = \det(\xi)\mathfrak{r}$. Then by a well-known principle M is $\mathfrak{r}$-isomorphic to $\mathfrak{r} \oplus \mathfrak{a}$. By (2.11) we have $\det(\Phi)N_{K/F}(\mathfrak{d}\mathfrak{a}) = \mathfrak{e}$. Then we obtain the first assertion of our lemma by taking $s = -\det(\Phi)^{-1}$. Let $\mathfrak{O}'$ be another maximal order in B containing $\mathfrak{r}$. By the Chevalley-Hasse-Noether theorem (see [E, Satz 7]) there exists an $\mathfrak{r}$-ideal $\mathfrak{b}$ such that $\mathfrak{b}\mathfrak{O}' = \mathfrak{O}\mathfrak{b}$. Take $c \in K_{\mathbf{A}}^\times$ so that $\mathfrak{b} = c\mathfrak{r}$. Then for each $v \in \mathbf{h}$ we can find $\eta_v \in GL_2(K_v)$ such that $yc_v^{-1}xc_v = yx\eta_v$ for every $x \in B_v$. Then $y\mathfrak{O}'_v = y\mathfrak{O}_v\eta_v$, and so $y\mathfrak{O}' = M\gamma^{-1}\eta$ with $\eta = (\eta_v)_{v\in\mathbf{h}}$. Using the map j in the proof of lemma 1.3, we find that $\eta_v = \mathrm{diag}[1, c_v^\rho/c_v]$, and so $\det(\gamma^{-1}\eta)\mathfrak{r} = \varphi[y]\mathfrak{b}^{-1}\mathfrak{b}^\rho \in \mathfrak{I}^0_{K/F}$. Thus $\det(\xi\gamma^{-1}\eta)\mathfrak{r} \in \mathfrak{a}\mathfrak{I}^0_{K/F}$, which proves the second assertion. We can put $\mathfrak{O} = \mathfrak{r}z + \mathfrak{a}w$ with elements z and w. Applying ι to this, we obtain the last assertion.

We call the coset $\mathfrak{a}\mathfrak{I}^0_{K/F}$ in the above lemma *the characteristic coset of K relative to B*. Using this notion, we now reformulate Theorems 2.2 and 2.3 in terms of the matrices Φ in $\mathfrak{H}_n^0(\xi)$.

THEOREM 2.10 (The case of even n). *Let the symbols n, $\{\sigma_v\}_{v\in\mathbf{r}_0}$, ε, and ξ be given as follows: $0 < n \in 2\mathbf{Z}$, $\sigma_v \in 2\mathbf{Z}$, $|\sigma_v| \le n$; $\varepsilon \in F^\times$, $\xi \in \mathfrak{G}_{\mathbf{A}}$. Let $B = \{K, \varepsilon\}$; let $\mathfrak{e}$ be the product of the prime ideals in F ramified in B, and $\mathfrak{k}$ the characteristic coset of K relative to B. Suppose that $(-1)^{\sigma_v/2}\varepsilon > 0$ at each $v \in \mathbf{r}_0$ and*

$$\det(\xi)\mathfrak{d}^{(n-2)/2} \in \mathfrak{k}. \tag{2.12}$$

Then there exists an element Φ of $\mathfrak{H}_n^0(\xi)$ such that

$$\varepsilon \in d_0(\Phi), \quad \det(\Phi\xi\xi^\rho)\mathfrak{d}_0^{n/2} = \mathfrak{e}, \quad s_v(\Phi) = \sigma_v \textit{ for every } v \in \mathbf{r}_0. \tag{2.13}$$

Moreover, every element of $\mathfrak{H}_n^0(\xi)$ is of this type, and the coset $T\mathfrak{G}\xi$ and the genus of Φ are determined by $(\varepsilon N_{K/F}(K^\times), \{\sigma_v\}_{v\in\mathbf{r}_0})$.

Proof. Let the symbols n, $\{\sigma_v\}_{v\in\mathbf{r}_0}$, ε, and ξ be given as in our theorem. Then Theorem 2.2 combined with Proposition 2.6 guarantees an element Ψ of $\mathfrak{H}_n^0(\eta)$ with some $\eta \in \mathfrak{G}_\mathbf{A}$ such that $\varepsilon \in d_0(\Psi)$ and $\sigma_v = s_v(\Psi)$ for every $v \in \mathbf{r}_0$. Put $\mathfrak{y} = \det(\eta)\mathfrak{r}$. By Theorem 2.3 (i), (2.11), and Lemma 2.9 we see that $\mathfrak{y}\mathfrak{d}^{(n-2)/2} \in \mathfrak{k}$. Combining this with our condition (2.12), we see that $\mathfrak{y} \in \det(\xi)\mathfrak{I}_{K/F}^0$, and so $\det(\eta^{-1}\xi) \in K^\times\mathfrak{t}$, which implies $\eta \in T\mathfrak{G}\xi$; see (2.9). By Proposition 2.6 (ii), $[\Psi]$ is isomorphic to $[\Phi]$ with some $\Phi \in \mathfrak{H}_n^0(\xi)$. Replacing (Ψ, η) by (Φ, ξ), we obtain (2.13). This proves the first part of our theorem. Conversely, given $\Phi \in \mathfrak{H}_n^0(\xi)$, put $L = L_0\xi$. Let $\mathfrak{e}$ be the product of the prime ideals in F ramified in $\{K, d_0(\varphi)\}$. By Theorem 2.2 (i) and (2.11), $\det(\Phi\xi\xi^\rho)\mathfrak{d}_0^{n/2} = \mathfrak{e}$, which together with Lemma 2.9 implies condition (2.12). This proves the second part. The last part follows from Proposition 2.6.

THEOREM 2.11 (The case of odd n). *Let the symbols n, $\{\sigma_v\}_{v\in\mathbf{r}}$, ε, and ξ be given as follows: $0 < n-1 \in 2\mathbf{Z}$, $\sigma_v - 1 \in 2\mathbf{Z}$, $|\sigma_v| \le n$; $\varepsilon \in F^\times$ and $\xi \in \mathfrak{G}_\mathbf{A}$. Let $\varepsilon\mathfrak{g} = \mathfrak{a}N_{K/F}(\mathfrak{b})$ with an $\mathfrak{r}$-ideal $\mathfrak{b}$ and a squarefree integral $\mathfrak{g}$-ideal $\mathfrak{a}$ whose prime factors remain prime in K. Suppose $(-1)^{(\sigma_v-1)/2}\varepsilon > 0$ at each $v \in \mathbf{r}_0$ and*

$$\det(\xi)\mathfrak{d}^{(n-1)/2}\mathfrak{b} \in \mathfrak{I}_{K/F}^0. \tag{2.14}$$

Then there exists an element Φ of $\mathfrak{H}_n^0(\xi)$ such that

$$\varepsilon \in d_0(\Phi), \quad \det(\Phi\xi\xi^\rho)\mathfrak{d}_0^{(n-1)/2} = \mathfrak{a}, \quad s_v(\Phi) = \sigma_v \text{ for every } v \in \mathbf{r}_0. \tag{2.15}$$

Moreover, every element of $\mathfrak{H}_n^0(\xi)$ is of this type, and the coset $T\mathfrak{G}\xi$ and the genus of Φ are determined by $(\varepsilon N_{K/F}(K^\times), \{\sigma_v\}_{v\in\mathbf{r}_0})$.

This can be proved in exactly the same fashion as for Theorem 2.10.

LEMMA 2.12. *Suppose F has class number 1. Then the class number of K is odd if and only if $\mathfrak{I}_K = \mathfrak{I}_{K/F}^0$, in which case $-1 \in N_{K/\mathbf{Q}}(K^\times)$ if and only if $-1 \in N_{K/\mathbf{Q}}(\mathfrak{r}^\times)$.*

Proof. Suppose the class number of K is odd. Then every $\mathfrak{r}$-ideal $\mathfrak{a}$ is of the form $\mathfrak{a} = c\mathfrak{b}^2$ with an $\mathfrak{r}$-ideal $\mathfrak{b}$ and $c \in K^\times$. Thus $\mathfrak{a} = c\mathfrak{b}\mathfrak{b}^\rho\mathfrak{b}(\mathfrak{b}^\rho)^{-1} \in \mathfrak{I}_{K/F}^0$ as $\mathfrak{b}\mathfrak{b}^\rho$ is principal, and so $\mathfrak{I}_K = \mathfrak{I}_{K/F}^0$. Suppose the class number of K is even. Then there exists an $\mathfrak{r}$-ideal $\mathfrak{x}$ whose ideal class is not a square. Suppose $\mathfrak{x} \in \mathfrak{I}_{K/F}^0$. Then $\mathfrak{x} = z\mathfrak{y}^{-1}\mathfrak{y}^\rho$ with $z \in K^\times$ and an $\mathfrak{r}$-ideal $\mathfrak{y}$. Thus $\mathfrak{x} = z\mathfrak{y}\mathfrak{y}^\rho\mathfrak{y}^{-2}$, a contradiction, as $\mathfrak{y}\mathfrak{y}^\rho$ is principal. This proves the first part. To prove the second part, suppose $-1 = \alpha\alpha^\rho$ with $\alpha \in K^\times$; put $\alpha\mathfrak{r} = \mathfrak{b}\mathfrak{c}^{-1}$ with integral $\mathfrak{r}$-deals $\mathfrak{b}$ and $\mathfrak{c}$ that are relatively prime. Then $\mathfrak{b}\mathfrak{b}^\rho = \mathfrak{c}\mathfrak{c}^\rho$ and we easily see that $\mathfrak{b} = \mathfrak{c}^\rho$, and so $\mathfrak{c}^2 = \alpha^{-1}\mathfrak{c}\mathfrak{c}^\rho$, which is principal. If the class number of K is odd, then $\mathfrak{c} = c\mathfrak{r}$ with $c \in \mathfrak{r}$. Thus $\alpha c = \varepsilon c^\rho$ with $\varepsilon \in \mathfrak{r}^\times$. Then $\varepsilon\varepsilon^\rho = \alpha\alpha^\rho = -1$. This completes the proof.

The last statement of the above lemma is false if the class number of K is even. For example, let $K = \mathbf{Q}(\sqrt{34})$. Then the class number is 2 and $-1 = \alpha\alpha^\rho$ with $\alpha = (3+\sqrt{34})/5$, but $-1 \notin N_{K/\mathbf{Q}}(\mathfrak{r}^\times)$.

2.13. Let us now take K to be a real or an imaginary quadratic field whose class number is odd. We denote by d the discriminant of K. Thus $F = \mathbf{Q}$ and $K = \mathbf{Q}(\sqrt{d})$. By Lemma 2.12 we have $\mathfrak{I}_K = \mathfrak{I}^0_{K/F}$, and so $\mathfrak{G}_\mathbf{A} = T\mathfrak{G}$ by (2.9). Therefore by Proposition 2.6, every hermitian space over K is isomorphic to $[\Phi]$ with $\Phi \in \mathfrak{H}^0_n(1_n)$, and $\mathfrak{r}^1_n$ is a maximal lattice in it. For simplicity we put $\mathfrak{H}^1_n = \mathfrak{H}_n(1_n)$ and $\mathfrak{H}^0_n = \mathfrak{H}^0_n(1_n)$. Then $\mathfrak{H}^1_n$ consists of all $\Phi = (c_{ij}) \in \mathfrak{H}_n$ such that $\sqrt{d}\,c_{ij} \in \mathfrak{r}$ and $c_{ii} \in \mathbf{Z}$ for every i and j; $\mathfrak{H}^0_n$ consists of all $\Phi \in \mathfrak{H}^1_n$ satisfying the following condition:

(2.16) *If* $\Phi = P\Psi P^*$, $\Psi \in \mathfrak{H}^1_n$, *and* $P \in GL_n(K) \cap M_n(\mathfrak{r})$, *then* $\det(P) \in \mathfrak{r}^\times$.

For $\Phi \in \mathfrak{H}_n$ we put $s(\Phi) = p - q$ if K is imaginary and Φ as a complex hermitian matrix has p positive and q negative eigenvalues; we put $s(\Phi) = 0$ if $d > 0$ and $n \in 2\mathbf{Z}$; we do not define $s(\Phi)$ if $d > 0$ and $n \notin 2\mathbf{Z}$, and so the symbol $s(\Phi)$ in that case must be ignored. Clearly two elements Φ_1 and Φ_2 of $\mathfrak{H}^1_n$ belong to the same genus if $s(\Phi_1) = s(\Phi_2)$ and $\Phi_1 = P_v\Phi_2P_v^*$ with some $P_v \in GL_n(\mathfrak{r}_v)$ for every $v \in \mathbf{h}$. Now, for $L = \mathfrak{r}^1_n$ we have $[\widehat{L}/L] = \det(\sqrt{d}\,\Phi)\mathfrak{r}$ by (2.11). For $d > 0$ we fix an embedding of K into $\mathbf{R}$, and take $\sqrt{d} > 0$.

THEOREM 2.14 (The case of even n). *Let* $K = \mathbf{Q}(\sqrt{d})$ *as in §2.13, and let three integers* n, σ, *and* e *be given as follows:* $0 < n \in 2\mathbf{Z}$, $\sigma \in 2\mathbf{Z}$, $|\sigma| \le n$; $\sigma = 0$ *if* $d > 0$; e *is positive and squarefree. Let* r *be the number of prime factors of* e. *Suppose that* $\sigma - 2r \in 4\mathbf{Z}$ *and no prime factor of* e *splits in* K. *Then there exists an element* Φ *of* $\mathfrak{H}^0_n$ *such that*

(2.17a) $\det(\sqrt{d}\,\Phi) = (-1)^{\sigma/2}e$ *and* $s(\Phi) = \sigma$ *if* $d < 0$,

(2.17b) $\det(\sqrt{d}\,\Phi) = \tau e$ *with* $\tau = 1$ *or* -1 *if* $d > 0$.

Moreover, every element of $\mathfrak{H}^0_n$ *is of this type. Its genus is determined by* (σ, e) *if* $d < 0$, *and by* e *if* $d > 0$. *If* $d > 0$ *and* $-1 \in N_{K/\mathbf{Q}}(K^\times)$, *then both* e *and* $-e$ *can occur as* $\det(\sqrt{d}\,\Phi)$ *for* Φ *in the same genus. If* $d > 0$ *and* $-1 \notin N_{K/\mathbf{Q}}(K^\times)$, *then* τ *is uniquely determined by the condition that a prime number* p *divides* e *if and only if* $\tau e \notin N_{K/\mathbf{Q}}(K_p^\times)$.

Proof. Given (n, σ, e) as in our theorem, we can find a quaternion algebra B over $\mathbf{Q}$ which is ramified at p if and only if $p|e$. Then B is definite if and only if r is odd. Since $\sigma - 2r \in 4\mathbf{Z}$, we see that $d < 0$ if r is odd. Our assumption on the prime factors of e allows us to put $B = \{K, \varepsilon\}$ with $\varepsilon \in \mathbf{Q}^\times$. Then $(-1)^{\sigma/2}\varepsilon > 0$ if $d < 0$. By Theorem 2.10, we obtain $\Phi \in \mathfrak{H}^0_n$ satisfying (2.13) with $\xi = 1_n$, as (2.12) can be ignored. Then $\mathfrak{e} = e\mathbf{Z}$ and $\det(\sqrt{d}\,\Phi) = \tau e$ with $\tau = \pm 1$. Since $s(\Phi) = \sigma$, we see that $\tau = (-1)^{\sigma/2}$ if $d < 0$. The same theorem says that every element of $\mathfrak{H}_n^0$ is of this type, and its genus is determined by $\varepsilon N_{K/F}(K^\times)$ and σ. We easily see that (e, σ) determines $(\varepsilon N_{K/F}(K^\times), \sigma)$, and vice versa. If $-1 \in N_{K/\mathbf{Q}}(K^\times)$, then by Lemma 2.12, $\mathfrak{r}^\times$ contains an element ζ such that $\zeta\zeta^\rho = -1$. Then $\det(P\Phi P^*) = -\det(\Phi)$ for $P = \mathrm{diag}[\zeta, 1_{n-1}]$. Thus both e and $-e$ can happen. Suppose $d > 0$ and $-1 \notin N_{K/\mathbf{Q}}(K^\times)$. Then $\{K, e\}$ is not isomorphic to $\{K, -e\}$. Since $-d \in N_{K/\mathbf{Q}}(K^\times)$, $d_0(\varphi)$ is represented by $\det(\sqrt{d}\Phi)$. If $\det(\sqrt{d}\Phi) = \tau e$, then $B = \{K, \tau e\}$. Thus τ is uniquely determined by the condition that a prime number p divides e if and only if $\tau e \notin N_{K/\mathbf{Q}}(K_p^\times)$.

THEOREM 2.15 (The case of odd n). *Let $K = \mathbf{Q}(\sqrt{d})$ as in §2.13 and let four integers n, σ, τ, and e be given as follows: $0 < n-1 \in 2\mathbf{Z}$; σ is necessary only if $d < 0$, $\sigma - 1 \in 2\mathbf{Z}$, and $|\sigma| \le n$; τ is necessary only if $d > 0$, and $\tau = 1$ or -1; e is positive and squarefree, and every prime factor of e remains prime in K. Then there exists an element Φ of $\mathfrak{H}_n^0$ such that*

(2.18a) $\qquad \det(\sqrt{d}\,\Phi) = (-1)^{(\sigma-1)/2} e\sqrt{d}$ *and* $s(\Phi) = \sigma$ *if* $d < 0$,

(2.18b) $\qquad \det(\sqrt{d}\,\Phi) = \tau e\sqrt{d}$ *if* $d > 0$.

Moreover, every element of $\mathfrak{H}_n^0$ is of this type, and its genus is determined by (σ, e) if $d < 0$ and by (τ, e) if $d > 0$. For $d > 0$ the sets $(1, e)$ and $(-1, e)$ determine the same genus if and only if $-1 \in N_{K/\mathbf{Q}}(K^\times)$.

Proof. Given (n, σ, τ, e) as in our theorem, take $\varepsilon = (-1)^{(\sigma-1)/2}e$ if $d < 0$ and $\varepsilon = (-1)^{(n-1)/2}\tau e$ if $d > 0$. Then Theorem 2.11 with $\xi = 1_n$ and $\mathfrak{a} = e\mathbf{Z}$ guarantees an element Φ of $\mathfrak{H}_n^0$ satisfying (2.15). We can easily verify that (2.18a, b) hold. That every $\Phi \in \mathfrak{H}_n^0$ is of this type also follows from Theorem 2.11, as $d_0(\Phi)$ can be represented by e or $-e$ with a positive integer e as in our theorem. Since the last assertion is obvious, our proof is complete.

COROLLARY 2.16. *Let $K = \mathbf{Q}(\sqrt{d})$ and $\mathfrak{H}_n^1$ be as in §2.13; let $0 < n \in \mathbf{Z}$ and $\sigma \in \mathbf{Z}$.*

(i) *If $d < 0$, there exists an element Φ of $\mathfrak{H}_n^1$ such that $\det(\sqrt{d}\,\Phi) = 1$ and $s(\Phi) = \sigma$ exactly when $n \in 2\mathbf{Z}$ and $\sigma \in 4\mathbf{Z}$.*

(ii) *Suppose $d > 0$; then there exists an element Φ of $\mathfrak{H}_n^1$ such that $\det(\sqrt{d}\,\Phi) = 1$ if and only if $n \in 2\mathbf{Z}$. Moreover, there exists an element Φ' of $\mathfrak{H}_n^1$ such that $\det(\sqrt{d}\,\Phi') = -1$ if and only if $n \in 2\mathbf{Z}$ and $-1 \in N_{K/\mathbf{Q}}(K^\times)$.*

Proof. From Theorem 2.15 we see that $\det(\sqrt{d}\,\Phi) = \pm 1$ for $\Phi \in \mathfrak{H}_n^1$ cannot happen if n is odd. Take $e = 1$ in Theorem 2.14. Then $r = 0$, and we obtain our results immediately from that theorem. Notice that if $d > 0$ and $-1 \notin N_{K/\mathbf{Q}}(K^\times)$, then $\{K, -1\}$ is a division algebra, and so $-1 \notin N_{K/\mathbf{Q}}(K_p^\times)$ for some prime number p.

The above corollary is a natural analogue of a well-known fact on unimodular quadratic form over $\mathbf{Q}$.

2.17. EXAMPLES. (2) Take $n = 2$, $d = 21$, and $e_1 = 11 \cdot 13$; then the class number of K is 1 and $-1 \notin N_{K/\mathbf{Q}}(K^\times)$. For $\Phi_1 = \begin{bmatrix} 7 & 2/\sqrt{21} \\ -2/\sqrt{21} & -1 \end{bmatrix}$ we have $\det(\sqrt{21}\,\Phi_1) = -e_1$ and $\Phi_1 \in \mathfrak{H}_2^0$. But we cannot have $\det(\sqrt{21}\,\Phi) = e_1$ for $\Phi \in \mathfrak{H}_2^0$.

Next take $e_2 = 3 \cdot 7 \cdot 11 \cdot 13$. Then $\det(\sqrt{21}\,\Phi)$ for $\Phi \in \mathfrak{H}_2^0$ can be $-e_2$ but cannot be e_2. Also, $\{K,\, 11 \cdot 13\}$ is ramified at $p = 3,\, 7$, but $\{K,\, -11 \cdot 13\}$ is not. From this we can derive that $\mathrm{diag}[11,\, -13]$ is reduced, but $\mathrm{diag}[11,\, 13]$ is not.

3. HERMITIAN DIOPHANTINE EQUATIONS OVER A LOCAL FIELD

3.1. Throughout this section we fix (V, φ) in the local case, and put $n = \dim(V)$. We denote by $\mathfrak{p}$ the maximal ideal of $\mathfrak{g}$, and by t the core dimension of (V, φ). Then $t \le 2$ as observed in §2.1. For an $\mathfrak{r}$-lattice L in V we put

$$C(L) = \{\alpha \in U^\varphi(V) \mid L\alpha = L\}, \qquad C^1(L) = C(L) \cap SU^\varphi(V). \tag{3.1}$$

Define $L[q,\, \mathfrak{b}]$ and $L[q]$ by (1.10b, c). Clearly $L[q,\, \mathfrak{b}]$ and $L[q]$ are stable under right multiplication by the elements of $C(L)$, and so the four orbit sets $L[q,\, \mathfrak{b}]/C(L)$, $L[q]/C(L)$, $L[q,\, \mathfrak{b}]/C^1(L)$, and $L[q]/C^1(L)$ are meaningful. Now our principal results of this section are the following two theorems.

THEOREM 3.2. *Suppose that F is local and $n > 1$. Let L be a maximal $\mathfrak{r}$-lattice in V. Then for every $q \in F^\times$ and every $\mathfrak{r}$-ideal $\mathfrak{b}$ the following assertions hold:*

(i) $\#\{L[q,\, \mathfrak{b}]/C(L)\} \le 1$.

(ii) $\#\{L[q,\, \mathfrak{b}]/C^1(L)\} < \infty$.

(iii) $\#\{L[q,\, \mathfrak{b}]/C^1(L)\} \le 1$ *if we exclude the following two cases:* (a) $n = 2$ *and* $t = 0$; (b) $t = 1$, $q\mathfrak{r} = \mathfrak{b}\mathfrak{b}^\rho$, *and* $\mathfrak{d} \ne \mathfrak{r}$.

(iv) *If* $n = 2$ *and* $t = 0$, *then*

$$\#\{L[q,\, \mathfrak{d}^{-1}]/C^1(L)\} = \begin{cases} 1 & \text{if } q \in \mathfrak{g}^\times, \\ N(q\mathfrak{g})\bigl[1 - \{K/F\}N(\mathfrak{p})^{-1}\bigr] & \text{if } q \in \mathfrak{p}, \end{cases} \tag{3.2}$$

where $N(\mathfrak{a}) = \#(\mathfrak{g}/\mathfrak{a})$, and $\{K/F\} = 1,\, -1,$ or 0, according as $K = F \times F$, K is an unramified quadratic extension of F, or K is ramified over F.

(v) $L[q,\, \mathfrak{d}^{-1}] \ne \emptyset$ *exactly in the following cases:* (a) $t = 0$ *and* $q \in \mathfrak{g}$; (b) $t = 1$, $\mathfrak{d} = \mathfrak{r}$, $d_0(\varphi) = N_{K/F}(K^\times)$, *and* $q \in \mathfrak{g}$; (c) $t = 1$, $d_0(\varphi) \not\subset \mathfrak{g}^\times N_{K/F}(K^\times)$, *and* $q \in \mathfrak{p}^{-1}$; (d) $t = 1$, $\mathfrak{d} \ne \mathfrak{r}$, *and* $q \in \mathfrak{p}\mathfrak{d}^{-2} \cup \bigl(\mathfrak{d}^{-2} \cap d_0(\varphi)\bigr)$; (e) $n = t = 2$ *and* $\mathfrak{g} \subset q\mathfrak{g} \subset \mathfrak{p}^{-1}$ *if* $\mathfrak{d} = \mathfrak{r}$; $q\mathfrak{g} = \mathfrak{p}^{-1}$ *if* $\mathfrak{d} \ne \mathfrak{r}$; (f) $n > 2 = t$ *and* $q \in \mathfrak{p}^{-1}$.

The proof will be given in §3.6 through §3.12.

The quantity $\#\{L[q,\, \mathfrak{b}]/C^1(L)\}$ in Case (b) of (iii) is not so simple. We will discuss that case in Lemma 3.13 (ii).

For every $\alpha \in U^\varphi$ we have $C(L\alpha) = \alpha^{-1}C(L)\alpha$, $C^1(L\alpha) = \alpha^{-1}C^1(L)\alpha$, and $(L\alpha)[q, \mathfrak{b}] = L[q, \mathfrak{b}]\alpha$. Therefore, in view of Lemma 1.10 (i), it is sufficient to prove Theorem 3.2 for a special choice of L. Also, since $c \cdot L[q, \mathfrak{b}] = L[cc^\rho q, c\mathfrak{b}]$ for every $c \in K^\times$, it is sufficient to prove our theorem when $\mathfrak{b} = \mathfrak{d}^{-1}$ or $\mathfrak{b} = \mathfrak{r}$.

THEOREM 3.3. *If $n > 1$, we have $\#\{\Lambda[q]/C^1(\Lambda)\} < \infty$ for every $q \in F^\times$ and every $\mathfrak{r}$-lattice Λ in V.*

Proof. Let L be a maximal lattice in V. Given Λ, we can find $c \in F^\times$ such that $c\Lambda \subset L$. Then $c\Lambda[q] \subset L[c^2q]$. For any two open compact subgroups D and E of SU^φ we have $[D : D \cap E] < \infty$. Therefore it is sufficient to prove that $\#\{L[q]/C^1(L)\} < \infty$. Given $h \in L[q]$, put $\mathfrak{b} = \varphi(h, L)$. Then $q\mathfrak{r} \subset \mathfrak{b} \subset \mathfrak{d}^{-1}$, and hence $L[q] \subset \bigcup_\mathfrak{b} L[q, \mathfrak{b}]$, where $\mathfrak{b}$ runs over the $\mathfrak{r}$-ideals $\mathfrak{b}$ such that $q\mathfrak{r} \subset \mathfrak{b} \subset \mathfrak{d}^{-1}$. Therefore the desired fact follows from Theorem 3.2 (ii).

LEMMA 3.4. *Let L be a maximal lattice in V. Suppose* $\dim(V) > 1$ *and $q \in \mathfrak{g}^\times$; then $\#\{L[q, \mathfrak{r}]/C(L)\} \le 1$. Moreover, $\#\{L[q, \mathfrak{r}]/C^1(L)\} \le 1$ if K is a field unramified over F, or the core dimension of $(Kh)^\perp$ is not 0 for some $h \in L[q, \mathfrak{r}]$.*

Proof. Let $h, k \in L[q, \mathfrak{r}]$. We see that $L + \mathfrak{r}h$ is integral, and so $h \in L$, as L is maximal. Given $x \in L$, put $y = x - \varphi[h]^{-1}\varphi(x, h)h$. Then $y \in L \cap (Kh)^\perp$. From this we can derive that $L = \mathfrak{r}h \oplus M$ with $M = L \cap (Kh)^\perp$. Similarly $L = \mathfrak{r}k \oplus M'$ with $M' = L \cap (Kk)^\perp$. Since L is maximal, M and M' must be maximal. By (1.5) we can find an element α of $SU^\varphi(V)$ such that $k = h\alpha$. Then $M\alpha$ is a maximal lattice in $(Kk)^\perp$. By Lemma 1.10 (i) we can find an element β of $U^\varphi\big((Kk)^\perp\big)$ such that $M\alpha\beta = M'$; by Lemma 1.10 (ii) such a β can be taken from $SU^\varphi\big((Kk)^\perp\big)$ if K is a field unramified over F, or the core dimension of $(Kk)^\perp$ is not 0. Extend β to V by putting $k\beta = k$. Then $\alpha\beta \in C(L)$ and $h\alpha\beta = k$. We have $\alpha\beta \in C^1(L)$ under the said conditions on K or on Kk. This proves our lemma. Notice that the core dimension of $(Kh)^\perp$ depends only on $hU(\varphi)$.

3.5. We call an element x of $\mathfrak{r}_n^1$ *primitive* if $x\mathfrak{r}_1^n = \mathfrak{r}$. Replacing $\mathfrak{r}$ by $\mathfrak{g}$, we can similarly define the primitive elements of $\mathfrak{g}_n^1$. Given an integral $\mathfrak{r}$-lattice L in V, identify V and L with K_n^1 and $\mathfrak{r}_n^1$ with respect to an $\mathfrak{r}$-basis $\{z_i\}_{i=1}^n$ of L; also let $\varphi_0 = \big(\varphi(z_i, z_j)\big)_{i,j=1}^n$. By (1.9) we have $\delta\varphi_0 \prec \mathfrak{r}$ for any element δ of $\mathfrak{r}$ such that $\delta\mathfrak{r} = \mathfrak{d}$. Moreover, $\varphi(x, L) = \mathfrak{d}^{-1}$ for $x \in L = \mathfrak{r}_n^1$ if and only if $\delta x\varphi_0$ is primitive.

To prove Theorem 3.2, we fix a maximal lattice L in V, and hereafter write simply C and C^1 for $C(L)$ and $C^1(L)$; we always assume $n > 1$.

3.6. In this subsection we consider the case $K = F \times F$, using the notation of §1.8. We can put $L = \mathfrak{g}_n^1 \times \mathfrak{g}_n^1$.

Let $h = (a, b) \in L[q, \mathfrak{r}]$ with $a, b \in F_n^1$. Then $\varphi(h, L) = (a\mathfrak{g}_1^n, b\mathfrak{g}_1^n)$, and so both a and b are primitive. Let $\{e_i\}_{i=1}^n$ be the standard basis of F_n^1. Take $\alpha \in SL_n(\mathfrak{g})$ so that $a\alpha = e_1$ and put $c = b \cdot {}^t\alpha^{-1} = (c_i)_{i=1}^n$. Then $c_1 = q$. Suppose $q \in \mathfrak{g}^\times$; then define $\beta \in SL_n(\mathfrak{g})$ by $\beta = \begin{bmatrix} 1 & x \\ 0 & 1_{n-1} \end{bmatrix}$ with $x = (q^{-1}c_2, \ldots, q^{-1}c_n)$ and put $\gamma = \alpha \cdot {}^t\beta$. Then $a\gamma = e_1$ and $b \cdot {}^t\gamma^{-1} = c\beta^{-1} = qe_1$. Thus $L[q, \mathfrak{r}]/C^1$ is represented by (e_1, qe_1) if $q \in \mathfrak{g}^\times$.

Next suppose $q \in \mathfrak{p}$; then $(c_2, \ldots, c_n)$ is primitive, and so we can find $\beta \in GL_{n-1}(\mathfrak{g})$ such that $(c_2, \ldots, c_n)\beta = (1, 0, \ldots, 0)$. Put $\gamma = \alpha \cdot \mathrm{diag}[1, {}^t\beta^{-1}]$. Then $a\gamma = e_1$ and $b \cdot {}^t\gamma^{-1} = (q, 1, 0, \ldots, 0)$. This shows that $\#\{L[q, \mathfrak{r}]/C\} = 1$. If $n > 2$, we can take $\beta \in SL_{n-1}(\mathfrak{g})$, and hence $\#\{L[q, \mathfrak{r}]/C^1\} = 1$.

Finally suppose $n = 2$ and $q \in \mathfrak{p}$; then we have shown that $L[q, \mathfrak{r}]/C^1$ can be represented by the elements of the form $\bigl(e_1, (q, s)\bigr)$ with $s \in \mathfrak{g}^\times$. Suppose $\bigl(e_1, (q, s)\bigr)\alpha = \bigl(e_1, (q, t)\bigr)$ with $\alpha = (\gamma, {}^t\gamma^{-1}) \in C^1$, $\gamma \in SL_2(\mathfrak{g})$, and $s, t \in \mathfrak{g}^\times$. Clearly $\gamma = \begin{bmatrix} 1 & 0 \\ v & 1 \end{bmatrix}$ with $v \in \mathfrak{g}$, and so $t = s - qv$. Since the procedure is reversible, we see that $\#\{L[q, \mathfrak{r}]/C^1\} = \#(\mathfrak{g}/q\mathfrak{g})^\times$, which gives (3.2) for $K = F \times F$. This completes the proof of Theorem 3.2 in the case $K = F \times F$.

Hereafter from §3.7 through §3.12 we assume that K is a field.

3.7. Let us consider the case $n = t = 2$. Let the symbols be as in Lemma 1.3 and §1.11; we identify B with $j(B)$. In view of Lemma 1.6 we can take $\varphi[\ell] = c = 1$ in the proof of Lemma 1.3, and so we can take $\gamma = 1$ in §1.11. Thus $M = \ell\mathfrak{O}$. Since φ is anisotropic, $SU^\varphi(V) = C^1(M) = \bigl\{\alpha \in \mathfrak{O}^\times \bigm| \alpha\alpha^\iota = 1\bigr\}$. From (1.5) we see that $\#\bigl\{M[q, \mathfrak{d}^{-1}]/C^1\bigr\} = 1$ if $M[q, \mathfrak{d}^{-1}] \neq \emptyset$. Since $\mathrm{Tr}_{K/F}\bigl(\varphi(\ell\alpha, \ell\beta)\bigr) = \mathrm{Tr}_{B/F}(\alpha\beta^\iota)$ for $\alpha, \beta \in B$, we have $\varphi(\ell\alpha, M) = \mathfrak{d}^{-1}$ only if $\mathrm{Tr}_{B/F}(\alpha\mathfrak{O}) = \mathfrak{g}$, which is so if and only if $\alpha\mathfrak{O} = \mathfrak{O}$ or $\alpha\mathfrak{O} = \mathfrak{P}^{-1}$. Take such an α and assume that K is unramified over F. Then $\ell\alpha \in M[\alpha\alpha^\iota, \mathfrak{r}]$. Thus $M[q, \mathfrak{d}^{-1}] \neq \emptyset$ if and only if $q\mathfrak{g}$ is $\mathfrak{g}$ or $\mathfrak{p}^{-1}$. Next suppose that K is ramified over F. Clearly $M[q, \mathfrak{d}^{-1}] \neq \emptyset$ for some q, and $q\mathfrak{g}$ is $\mathfrak{g}$ or $\mathfrak{p}^{-1}$ for the same reason as above. Suppose $q \in \mathfrak{g}^\times$. Then we can find an element $\xi \in \mathfrak{O}^\times$ such that $\xi\xi^\iota = q$. Then $\varphi[x\xi] = q\varphi[x]$ for every $x \in V$, $M\xi = M$, and $M[1, \mathfrak{b}]\xi = M[q, \mathfrak{b}]$. Let $d_0(\varphi) = sN_{K/F}(K^\times)$ as in the proof of Lemma 1.3. We may assume that $s \in \mathfrak{g}^\times$, $s \notin N_{K/F}(\mathfrak{r}^\times)$, and $\varphi_0 = \mathrm{diag}[1, -s]$. Observe that $\mathfrak{O}$ consists of the elements $a + b\omega$ with $a, b \in K$ such that $a + a^\rho \in \mathfrak{g}$ and $aa^\rho - sbb^\rho \in \mathfrak{g}$. Now let $\ell\alpha \in M[1, \mathfrak{d}^{-1}]$ with $\alpha \in B$. Then $\alpha\alpha^\iota = 1$, and so $\ell \in M[1, \mathfrak{d}^{-1}]$. Thus $\mathfrak{d}^{-1} = \varphi(\ell, M) = \varphi(\ell, \ell\mathfrak{O})$. For $a + b\omega \in \mathfrak{O}$ as above, we have $\varphi(\ell, \ell(a + b\omega)) = a^\rho$, and so $\mathfrak{O}$ contains an element $a + b\omega$ such that $a\mathfrak{r} = \mathfrak{d}^{-1}$. Put $N_{K/F}(\mathfrak{d}) = \mathfrak{p}^\kappa$ with $0 < \kappa \in \mathbf{Z}$. Then $aa^\rho\mathfrak{g} = \mathfrak{p}^{-\kappa}$, and $bb^\rho\mathfrak{g} = \mathfrak{p}^{-\kappa}$ as $aa^\rho - sbb^\rho \in \mathfrak{g}$. Putting $a^{-1}b = c$, we obtain $scc^\rho - 1 \in \mathfrak{p}^\kappa$, a contradiction, as $s \notin N_{K/F}(\mathfrak{r}^\times)$. Thus $M[q, \mathfrak{d}^{-1}] \neq \emptyset$ only if $q\mathfrak{g} = \mathfrak{p}^{-1}$. This proves Theorem 3.2 when $n = t = 2$.

3.8. Take an element $u \in \mathfrak{r}$ so that $\mathfrak{r} = \mathfrak{g}[u]$ and put $\delta = u - u^\rho$. Then $\mathfrak{d} = \delta\mathfrak{r}$ and $\delta^\rho = -\delta$. We will often use these u and δ in our later treatment.

Take a decomposition of V as in (1.8a, b), and assume that it is a Witt decomposition; thus $t = \dim(Z)$. Put $g_i = \delta^{-1} f_i$ and

$$L = \sum_{i=1}^{r} (\mathfrak{r}e_i + \mathfrak{r}g_i) + M, \qquad M = \{y \in Z \mid \varphi[y] \in \mathfrak{g}\}. \tag{3.3}$$

Then L is a maximal lattice in V as noted in §1.7. With an $\mathfrak{r}$-basis $\{m_i\}_{i=1}^t$ of M, consider matrix representation with respect to $\{e_1, \ldots, e_r, m_1, \ldots, m_t, g_1, \ldots, g_r\}$. Then φ is represented by

$$\varphi_0 = \begin{bmatrix} 0 & 0 & -\delta^{-1}1_r \\ 0 & \zeta & 0 \\ \delta^{-1}1_r & 0 & 0 \end{bmatrix}, \tag{3.4}$$

where $\zeta = \big(\varphi(m_i, m_j)\big)_{i,j=1}^t$. Now write an element of $GL(V)$ as a matrix with 9 matrix blocks corresponding to the blocks of (3.4), and let P be the group consisting of the elements of U^φ whose lower left 3 blocks under the diagonal blocks are all 0; also let $P^1 = P \cap SU^\varphi$. Then

$$U^\varphi = PC(L) \quad \text{and} \quad SU^\varphi = P^1C^1(L); \tag{3.5}$$

see [S2, Proposition 5.16]. If $r = 1$, P consists of the matrices of the form

$$\begin{bmatrix} a & b & a^{-\rho}(s + u^\rho b\zeta b^*) \\ 0 & e & -\delta a^{-\rho} e\zeta b^* \\ 0 & 0 & a^{-\rho} \end{bmatrix}, \tag{3.6}$$

where $a \in K^\times$, $b \in K_t^1$, $e \in U^\zeta(Z)$, and $s \in F$.

3.9. The case $t = 0$. Represent the elements of V by row vectors in $K_r^1 \times K_r^1$ with respect to the basis $\{e_1, \ldots, e_r, g_1, \ldots, g_r\}$, and $GL(V)$ by $GL_{2r}(K)$, acting on the right. Then for $h = (y, z) \in K_r^1 \times K_r^1$ we have $\varphi[h] = \delta^{-1}(zy^* - yz^*)$ and $\varphi(h, L) = \delta^{-1}\sum_{i=1}^r (\mathfrak{r}y_i + \mathfrak{r}z_i)$. Now φ is represented by $\delta^{-1}\eta$, where

$$\eta = \begin{bmatrix} 0 & -1_r \\ 1_r & 0 \end{bmatrix}.$$

Therefore $\operatorname{diag}[\alpha, \widetilde{\alpha}] \in U^\varphi$ for every $\alpha \in GL_r(K)$, where $\widetilde{\alpha} = (\alpha^*)^{-1}$. Suppose $\varphi(h, L) = \mathfrak{d}^{-1}$ for $h = (y, z)$; then $\sum_{i=1}^r (\mathfrak{r}y_i + \mathfrak{r}z_i) = \mathfrak{r}$ and $\varphi[h] \in \mathfrak{g}$. Putting $k = (e_1, qug_1)$ with $q \in F$ (not necessarily $\neq 0$), let us prove

(3.7) $\varphi[h] = q$ *and* $\varphi(h, L) = \mathfrak{d}^{-1} \Longrightarrow h \in kC(L);\ h \in kC^1(L)$ *if* $r > 1$.

Since $\eta \in C^1$, changing h for $h\eta$ if necessary, we may assume that $\sum_{i=1}^r \mathfrak{r}y_i = \mathfrak{r}$. We can find an element $\alpha \in GL_r(\mathfrak{r})$ such that $y\alpha = e_1$; we can even take α from $SL_r(\mathfrak{r})$ if $r > 1$. Put $w = z\widetilde{\alpha}$. Then $(y, z)\operatorname{diag}[\alpha, \widetilde{\alpha}] = (e_1, w)$, and so $w_1 - w_1^\rho = q\delta$. Thus we can put $w_1 = p + qu$ with $p \in \mathfrak{g}$. Define an element $s = s^* \in M_r(\mathfrak{r})$ so that $s_{11} = p$ and $s_{1j} = w_j$ for $j > 1$, and put

$\beta = \begin{bmatrix} 1_r & s \\ 0 & 1_r \end{bmatrix}$. Then $\beta \in C^1$ and $k\beta = (e_1, w)$. If $r > 1$, then we see that $h \in kC^1$, and so $L[q, \mathfrak{d}^{-1}] = kC^1$. This proves (3.7) and also Theorem 3.2 (i), (iii), (v) when $t = 0$.

Suppose $r = 1$; then $y \in \mathfrak{r}^\times$. Since C^1 is a normal subgroup of C, the above argument shows that $h \in \bigcup_t (t, t^{-\rho}qu)C^1$, where t runs over $\mathfrak{r}^\times$. Define B as in Lemma 1.3. From (1.7b), (1.15), and the last equality in the proof of Lemma 1.3 we obtain $B = M_2(F)$, $SU^\varphi = SL_2(F)$, and $C^1 = SL_2(\mathfrak{g})$. Let $t \in \mathfrak{r}^\times \cap \mathfrak{g}[qu]$. Then $t = a + cqu$ with $a, c \in \mathfrak{g}$, and we can find an element $\gamma \in SL_2(\mathfrak{g})$ of the form $\gamma = \begin{bmatrix} a & b \\ c & d \end{bmatrix}$. We have $(1, qu)\gamma = (t, t^{-\rho}v)$ with $v \in \mathfrak{r}$. Then $v - v^\rho = q\delta$, so that $v = qu - s$ with $s \in \mathfrak{g}$. Since $tt^\rho \in \mathfrak{g}^\times$, we can put $s = tt^\rho s'$ with $s' \in \mathfrak{g}$. Put $\sigma = \begin{bmatrix} 1 & s' \\ 0 & 1 \end{bmatrix}$. Then $\sigma \in C^1$ and $(t, t^{-\rho}v)\sigma = (t, t^{-\rho}qu)$, which shows that $(t, t^{-\rho}qu) \in (1, qu)C^1$ if $t \in \mathfrak{g}[qu]$. Since the converse is obvious, we thus obtain $L[q, \mathfrak{d}^{-1}] = \bigsqcup_{t\in\tau}(t, t^{-\rho}qu)C^1$, where $\tau = \mathfrak{r}^\times/\mathfrak{g}[qu]^\times$. This proves (3.2) and completes the proof of Theorem 3.2 when $t = 0$.

3.10. The notation being as in (3.3), put $H_j = \sum_{i=1}^{j}(Ke_i + Kg_i)$. Let us now show that given $h \in V$, there exists an element α of C^1 such that $h\alpha \in H_1 + Z$. This is obvious if $h \in Z$ or $r = 1$. So assume that $h \notin Z$ and $r > 1$. Put $h = w + k$ with $k \in Z$ and $w = \sum_{i=1}^{r}(y_i e_i + z_i g_i) \in H_r$, y_i, $z_i \in K$. Then we can put $\sum_{i=1}^{r}(y_i\mathfrak{r} + z_i\mathfrak{r}) = d\mathfrak{r}$ with $d \in K^\times$. Taking $d^{-1}w$ as h of (3.7), we can find an element $\gamma \in C^1(L \cap H_r)$ such that $d^{-1}w\gamma \in H_1$. Extend γ to an element of $C^1(L)$ by defining $x\gamma = x$ for every $x \in Z$. Then we obtain the desired fact. This means that if $n > t > 0$, then it is sufficient to prove Theorem 3.2 when $r = 1$.

3.11. Case $t > 0$, $r = 1$. Writing simply e and g for e_1 and g_1, we have $V = Ke + Z + Kg$ and $L = \mathfrak{r}e + M + \mathfrak{r}g$ with a maximal lattice M in Z. Let $h = ye + x + zg$ with y, $z \in K$ and $x \in Z$. Then $\varphi[h] = \delta^{-1}(zy^\rho - yz^\rho) + \zeta[x]$, where ζ is the restriction of φ to Z, and $\varphi(h, L) = \delta^{-1}(\mathfrak{r}y + \mathfrak{r}z) + \zeta(x, M)$. Suppose $h \in L[q, \mathfrak{d}^{-1}]$. Then $\mathfrak{r}y + \mathfrak{r}z + \mathfrak{d}\zeta(x, M) = \mathfrak{r}$, and hence y, $z \in \mathfrak{r}$, $x \in \widehat{M}$, and $\varphi[h] - \zeta[x] \in \mathfrak{g}$. We identify an element of Z with a row vector of K_t^1 with respect to an $\mathfrak{r}$-basis of M. Then an element $ae + b + cg$ of V with a, $c \in K$ and $b \in Z$ can be identified with a row vector $[a \quad b \quad c]$ of K_{t+2}^1. If $t = 2$ and M corresponds to $\mathfrak{O}$ as in §1.11, then $\widehat{M}$ corresponds to $\mathfrak{P}^{-1}$, and so $\zeta[x] \in \mathfrak{p}^{-1}$; consequently $\varphi[h] \in \mathfrak{p}^{-1}$. Since $e + qug \in L[q, \mathfrak{d}^{-1}]$ if $q \in \mathfrak{g}$, we see that $L[q, \mathfrak{d}^{-1}] \neq \emptyset$ for every $q \in \mathfrak{g}$.

(a) First suppose $t = 2$ and $q \in \mathfrak{p}$. Suppose also that $\mathfrak{r}y + \mathfrak{r}z \neq \mathfrak{r}$. Then $yz^\rho \in \mathfrak{p}\mathfrak{r}$, and so $\delta^{-1}(zy^\rho - yz^\rho) \in \mathfrak{p}$. Thus $\zeta[x] \in \mathfrak{p}$, and hence $\zeta(x, M) \neq \mathfrak{d}^{-1}$. (In §3.7 we showed that $\zeta(x, M) = \mathfrak{d}^{-1}$ only if $\mathfrak{g} \subset \varphi[x]\mathfrak{g} \subset \mathfrak{p}^{-1}$.) Therefore

$\delta\varphi(h, L) \neq \mathfrak{r}$, a contradiction. This shows that (y, z) must be primitive. By (3.7) there is an element α of $C(\mathfrak{r}e + \mathfrak{r}g)$ such that $(ye + zg)\alpha = e + aug$ with $a \in \mathfrak{g}$. Extend α to an element γ of $SU^{\varphi}(V)$ by defining $w\gamma = w\sigma$ for $w \in Z$ with $\sigma \in U^{\varphi}(Z)$ such that $\det(\sigma) = \det(\alpha)^{-1}$. Since $M\sigma = M$, we see that $\gamma \in C^1$, and $h\gamma = [1 \quad k \quad au]$ with $k \in M$ such that $\zeta[k] = q - a$. Given another $h' \in L[q, \mathfrak{d}^{-1}]$, we can similarly find an element γ' of C^1 such that $h'\gamma' = [1 \quad k' \quad a'u]$ with $a' \in \mathfrak{g}$ and $k' \in M$ such that $\zeta[k'] = q - a'$. Put $b = k' - k$ and

$$\tau = \begin{bmatrix} 1 & b & s + u^{\rho}b\zeta b^* \\ 0 & 1 & -\delta\zeta b^* \\ 0 & 0 & 1 \end{bmatrix} \tag{3.8}$$

with some $s \in \mathfrak{g}$. This is a special case of (3.6) and belongs to $SU^{\varphi}(V)$. Since $b \prec \mathfrak{r}$ and $\delta\zeta \prec \mathfrak{r}$, we see that $\tau \in C^1$. We choose s so that $h\gamma\tau = [1 \quad k' \quad a'u]$, which is so if and only if $a'u = s + u^{\rho}b\zeta b^* - \delta k\zeta b^* + au$. This can be achieved by taking $s = \mathrm{Tr}_{K/F}\big(u\zeta(k, k')\big) - (u + u^{\rho})\zeta[k']$. Then $h'\gamma' = h\gamma\tau$, and so $h' \in hC^1$ as expected.

(b) Next suppose $t = 2$ and $q \notin \mathfrak{p}$. Then $q \in \mathfrak{g}^{\times}$ or $q\mathfrak{g} = \mathfrak{p}^{-1}$. If $\mathfrak{d} \neq \mathfrak{r}$ and $q \in \mathfrak{g}$, then $\zeta[x] \in \mathfrak{g}$, and so $\delta\zeta(x, M) \neq \mathfrak{r}$ as shown in §3.7. Consequently $\mathfrak{r}y + \mathfrak{r}z = \mathfrak{r}$ in such a case, and the argument of case (a) is applicable. Therefore we may assume that $q\mathfrak{g} = \mathfrak{p}^{-1}$ if $\mathfrak{d} \neq \mathfrak{r}$. Then as observed in §3.7, we can find an element v of $M[q, \mathfrak{d}^{-1}]$. The same can be said for both cases $q \in \mathfrak{g}^{\times}$ and $q\mathfrak{g} = \mathfrak{p}^{-1}$ if $\mathfrak{d} = \mathfrak{r}$. We identify v with the row vector $[0 \quad v \quad 0]$, which can be viewed as an element of $L[q, \mathfrak{d}^{-1}]$, and so $L[q, \mathfrak{d}^{-1}] \neq \emptyset$ in such cases. Combining (1.5) and (3.5), we have $h \in vP^1C^1$, and hence $h = v\pi\alpha$ with $\pi \in P^1$ and $\alpha \in C^1$. Write π in the form (3.6) and focus our attention on the element e of $U^{\zeta}(Z)$ there. Since $U^{\zeta}(Z) = C(M)$, we see that $ve \in M[q, \mathfrak{d}^{-1}]$, and hence $ve = v\varepsilon$ with $\varepsilon \in C^1(M) = SU^{\zeta}(Z)$ as shown in §3.7. Let $\beta = \mathrm{diag}[1, \varepsilon, 1]$ and $\pi_1 = \pi\beta^{-1}$. Then $\beta \in C^1, \pi_1 \in P^1$, and $v\pi_1 = h\alpha^{-1}\beta^{-1} \in L[q, \mathfrak{d}^{-1}]$. Our choice of ε shows that $v\pi_1 = [0 \quad v \quad p]$ with $p \in \mathfrak{r}$. Since $v \in M[q, \mathfrak{d}^{-1}]$, $\delta v\zeta$ is primitive (see §3.5), and so we can find an element b of $\mathfrak{r}_2^1$ such that $\delta v\zeta b^* = -p$. Define τ by (3.8) with this b and $s = 0$. Then $\tau \in C^1$ and $v\tau = v\pi_1 = h\alpha^{-1}\beta^{-1}$. This shows that $h \in vC^1$. Thus we obtain Theorem 3.2 when $n > t = 2$.

(c) Finally suppose $t = 1$; let M and ζ be as in (3.3) and (3.4). Then $\zeta = \varphi[m]$ and $M = \mathfrak{r}m$ with an element m. Clearly $\zeta \in \mathfrak{g}^{\times}$ or $\zeta\mathfrak{g} = \mathfrak{p}$; the latter case occurs only when $\mathfrak{d} = \mathfrak{r}$. We first treat the case where $\zeta \in \mathfrak{g}^{\times}$; the other case will be treated in §3.12. Thus $h = ye + zg + sm$ with $y, z, s \in K$ such that $y\mathfrak{r} + z\mathfrak{r} + s\mathfrak{d} = \mathfrak{r}$ and $\delta^{-1}(zy^{\rho} - yz^{\rho}) + \zeta ss^{\rho} = q$. Suppose $\mathfrak{d} = \mathfrak{r}$ and $y\mathfrak{r} + z\mathfrak{r} \neq \mathfrak{r}$. Then $s \in \mathfrak{r}^{\times}$ and we see that $q = \varphi[h] \in \mathfrak{g}^{\times}$. Then $\#\big\{L[q, \mathfrak{d}^{-1}]/C^1(L)\big\} \leq 1$ by Lemma 3.4.

(c1) Let us now prove the case in which $q \in \mathfrak{g}$ and $y\mathfrak{r} + z\mathfrak{r} = \mathfrak{r}$ for both ramified and unramified K. Putting $p=\delta^{-1}(zy^\rho-yz^\rho)$ and applying (3.7) to $ye+zg$, we find $\gamma\in C(L)$ such that $M\gamma=M$ and $(y,\,z)\gamma=(1,\,pu)$ with $p\in\mathfrak{g}$. Replacing γ by $\gamma\alpha$ with $\alpha\in C$ such that $m\alpha=\det(\gamma)^{-1}m$ and α is the identity map on $Ke+Kg$, we may assume that $\gamma\in C^1$. We have $h\gamma=[1\quad x\quad pu]$ with $x\in\mathfrak{r}$. We consider

$$(3.9)\qquad \sigma=\begin{bmatrix}1 & -x & u^\rho\zeta x x^\rho\\ 0 & 1 & \delta\zeta x^\rho\\ 0 & 0 & 1\end{bmatrix},$$

which is similar to (3.8) and belongs to C^1. We have then $h\gamma\sigma=[1\quad 0\quad s]$ with $s\in\mathfrak{g}$. Taking $(1,\,s)$ as $(y,\,z)$ above, we find $\gamma'\in C^1$ such that $h\gamma\sigma\gamma'=[1\quad 0\quad qu]$. Thus $h\in kC^1$ with $k=e+qug$; also Theorem 3.2 (v), (b) is valid.

(c2) Suppose K is ramified over F. If $q\in\mathfrak{g}$, then $s\in\mathfrak{r}$, so that $y\mathfrak{r}+z\mathfrak{r}=\mathfrak{r}$. and (c1) covers this case. Thus we assume that $q\notin\mathfrak{g}$. Then $s\notin\mathfrak{r}$ and $q\mathfrak{g}=ss^\rho\mathfrak{g}$. Put $\mathfrak{d}=\mathfrak{q}^\kappa$ with the maximal ideal $\mathfrak{q}$ of $\mathfrak{r}$ and $0<\kappa\in\mathbf{Z}$. Since $s\mathfrak{d}\subset\mathfrak{r}$, we can put $s^{-1}\mathfrak{r}=\mathfrak{q}^a$ with $0<a\le\kappa$; then $q\mathfrak{g}=\mathfrak{p}^{-a}$. Thus a is determined by q. Suppose $a<\kappa$; then $s\mathfrak{d}\ne\mathfrak{r}$, so that $y\mathfrak{r}+z\mathfrak{r}=\mathfrak{r}$. By the same technique as in (c1), we can find $\gamma\in C^1$ such that $h\gamma=[1\quad x\quad pu]$ with $p\in\mathfrak{g}$ and $x\in K$. Then $x\mathfrak{q}^a=\mathfrak{r}$. Let $k=[1\quad x_1\quad p_1u]\in L[q,\,\mathfrak{d}^{-1}]$ with $p_1\in\mathfrak{g}$ and x_1 such that $x_1\mathfrak{q}^a=\mathfrak{r}$. Our task is to show that $k\in hC$. For simplicity put $N(w)=ww^\rho$ for $w\in K^\times$. We have $\zeta N(x)+p=q=\zeta N(x_1)+p_1$, and so $N(x^{-1}x_1)-1\in\mathfrak{p}^a$. Thus $N(x^{-1}x_1)\in N(\mathfrak{r}^\times)\cap(1+\mathfrak{p}^a)=N(1+\mathfrak{q}^a)$ by [S2, Lemma 17.6 (2)]. We can therefore put $N(x^{-1}x_1)=N(d)$ with $d\in 1+\mathfrak{q}^a$. Put $\alpha=\mathrm{diag}[1_r,\,dx_1^{-1}x,\,1_r]$. Then $\alpha\in C^1$ and $k\alpha=[1\quad dx\quad p_1u]$. Put $b=dx-x$ and consider τ of (3.8) with this b and any $s\in\mathfrak{g}$. Then $\tau\in C^1$ and $h\gamma\tau=[1\quad dx\quad c]$ with $c\in\mathfrak{g}$ such that $c-c^\rho=\delta p_1$. Choosing s suitably, we obtain $c=p_1u$. Then $k\in hC^1$.

(c3) It remains to treat the case $a=\kappa$. Then $s\mathfrak{d}=\mathfrak{r}$ and $q\mathfrak{r}=\mathfrak{d}^{-2}$. Put $h_0=\delta h$ and $q_0=-\delta^2q$. Then $h_0\in L[q_0,\,\mathfrak{r}]$. Since $q_0\in\mathfrak{g}^\times$, by Lemma 3.4 we see that $\#\{L[q,\,\mathfrak{d}^{-1}]/C\}=\#\{L[q_0,\,\mathfrak{r}]/C\}\le 1$.

(c4) As for (d) of Theorem 3.2 (v), we have seen that $L[q,\,\mathfrak{d}^{-1}]\ne\emptyset$ only if $q\in\mathfrak{d}^{-2}$; also $e+qug\in L[q,\,\mathfrak{d}^{-1}]$ if $q\in\mathfrak{g}$. Thus it remains to consider the case where $\mathfrak{d}\ne\mathfrak{r}$ and $q\notin\mathfrak{g}$. Suppose $q\mathfrak{g}=\mathfrak{p}^{-a}$ with $0<a<\kappa$. We can find $c\in\mathfrak{g}^\times$ such that $c-1\in\mathfrak{p}^a$ and $c\notin N(\mathfrak{r}^\times)$. Then q or $c^{-1}q$ represents $d_0(\varphi)$. If $q\in d_0(\varphi)$, then $q=\zeta ss^\rho$ with $s\in K$, and $e+sm\in L[q,\,\mathfrak{d}^{-1}]$. If $q\in cd_0(\varphi)$, put $q=c\zeta ss^\rho$ with $s\in K$ and $p=\zeta ss^\rho(c-1)$. Then $p\in\mathfrak{g}$ and $e+pug+sm\in L[q,\,\mathfrak{d}^{-1}]$. Thus $L[q,\,\mathfrak{d}^{-1}]\ne\emptyset$ if $q\in\mathfrak{p}\mathfrak{d}^{-2}$. The case $q\mathfrak{r}=\mathfrak{d}^{-2}$ will be settled in Lemma 3.13.

3.12. Let us now treat the case where $t=1$ and $\zeta\mathfrak{g}=\mathfrak{p}$. Then $\mathfrak{d}=\mathfrak{r}$ and $\delta\in\mathfrak{r}^\times$. To avoid possible confusion, we use the letter π instead of ζ; thus $\varphi[m]=\pi$. Let $h=ae+bf+cm\in L[q,\,\mathfrak{r}]$. Then $\mathfrak{r}a+\mathfrak{r}b+\mathfrak{r}\pi c=\mathfrak{r}$ and $ab^\rho+a^\rho b+\pi cc^\rho=q$. Clearly $q\in\mathfrak{p}^{-1}$; also $c\in\mathfrak{r}$ if and only if $q\in\mathfrak{g}$. Given

$q \in \mathfrak{g}$, we can find $b \in \mathfrak{r}$ such that $b + b^\rho = q$. Then $e + bf \in L[q, \mathfrak{r}]$. Given $q \in \mathfrak{p}^{-1}$, $\notin \mathfrak{g}$, we can find $d \in \mathfrak{r}^\times$ such that $dd^\rho = \pi q$. Then $\pi^{-1}dm \in L[q, \mathfrak{r}]$. Thus $L[q, \mathfrak{r}] \neq \emptyset$ if and only if $q \in \mathfrak{p}^{-1}$.

(1) Let us first assume that $q \in \mathfrak{g}$; then $c \in \mathfrak{r}$ and $\mathfrak{r}a + \mathfrak{r}b = \mathfrak{r}$. By (3.7) there exists an element β of $C(\mathfrak{r}e + \mathfrak{r}f)$ such that $k\beta = e + sf$ with $s \in \mathfrak{r}$. Extend β to an element α of $C^1(L)$ by defining $m\alpha = \det(\beta)^{-1}m$. Then $h\alpha = e + sf + c_1 m$ with $c_1 \in \mathfrak{r}$. Now represent the elements of $U^\varphi(V)$ by matrices with respect to $\{e, m, f\}$. Then (3.5) holds with the subgroup P of U^φ consisting of the upper triangular matrices. Observe that P contains every matrix of the form

$$\begin{bmatrix} 1 & -x & y \\ 0 & 1 & \pi x^\rho \\ 0 & 0 & 1 \end{bmatrix} \tag{3.10}$$

with $x, y \in K$ such that $y + y^\rho = -\pi x x^\rho$. Since $\mathrm{Tr}_{K/F}(\mathfrak{r}) = \mathfrak{g}$, for any $x \in \mathfrak{r}$ we can take such a y from $\mathfrak{r}$. Let γ be the matrix of (3.10) with $x = c_1$ and $y \in \mathfrak{r}$. Then $\gamma \in C^1(L)$ and $h\alpha\gamma = e + zf$ with $z \in \mathfrak{r}$ such that $z + z^\rho = q$. Take $z_1 \in \mathfrak{r}$ so that $z_1 + z_1^\rho = q$. Denote by ε the matrix of (3.10) with $x = 0$ and $y = z_1 - z$. Then $\varepsilon \in C^1(L)$ and $(e + zf)\varepsilon = e + z_1 f$. This gives the desired result when $q \in \mathfrak{g}$.

(2) Next we consider the case $c \notin \mathfrak{r}$; put $d = \pi c$. Then $d \in \mathfrak{r}^\times$ and $q = ab^\rho + a^\rho b + \pi^{-1}dd^\rho$. Thus $\pi q \in \mathfrak{g}^\times$, and so we can find an element $d_0 \in \mathfrak{r}^\times$ such that $d_0 d_0^\rho = \pi q$. Put $k = \pi^{-1}d_0 m$. Then $k \in L[q, \mathfrak{r}]$. By (1.5), we have $h = k\alpha$ with $\alpha \in SU^\varphi$, and by (3.5) we can put $\alpha = \beta\gamma$ with $\beta \in P^1$ and $\gamma \in C^1$. Replacing β by $\beta\xi$ with a suitable diagonal matrix ξ belonging to C^1, we may assume that the center entry of β is 1. (Here we need Lemma 1.9 (i).) Let $[0 \ \ 1 \ \ j]$ be the second row of β. Then $[0 \ \ \pi^{-1}d_0 \ \ \pi^{-1}d_0 j] = k\beta = h\gamma^{-1} \in L[q, \mathfrak{r}]$, and so $\pi^{-1}d_0 j \in \mathfrak{r}$. Put $x = \pi^{-1}j^\rho$; then $x \in \mathfrak{r}$. Let ε be the matrix of (3.10) with this x and y such that $y + y^\rho = -\pi x x^\rho$. Then $\varepsilon \in C^1$ and $k\varepsilon = k\beta = h\gamma^{-1}$, which gives the desired fact.

It only remains to discuss $L[q, \mathfrak{d}^{-1}]/C^1$ when $\mathfrak{d} \neq \mathfrak{r}$ and $q\mathfrak{r} = \mathfrak{d}^{-2}$. (In (c3) we treated $L[q, \mathfrak{d}^{-1}]/C$.) The problem is settled by (ii) of the following Lemma.

LEMMA 3.13. *If K is a field ramified over F, then the following assertions hold:*

(i) *Let $W = (Kh)^\perp$ with $h \in L[q, \mathfrak{b}]$ and let $E_1 = \det\big(C \cap U^\varphi(W)\big)$, where we view $U^\varphi(W)$ as the subgroup of $U^\varphi(V)$ consisting of the elements ξ such that $h\xi = h$. Then $\#\big\{L[q, \mathfrak{b}]/C^1\big\} = [E_L : E_1] < \infty$, where E_L is as in Lemma 1.9.*

(ii) *If $1 < n - 1 \in 2\mathbf{Z}$ and $q\mathfrak{r} = \mathfrak{d}^{-2}$, then $\#\big\{L[q, \mathfrak{d}^{-1}]/C^1\big\} = 2$ or 0 according as q represents $d_0(\varphi)$ or not.*

Proof. Clearly $C \cap U^\varphi(W)$ is an open subgroup of $U^\varphi(W)$. Now $\det : U^\varphi(W) \to E$ is a continuous surjective map, and so it is an open map by virtue of a well know principle; see [S3, Lemma 8.0]. Thus E_1 is an open subgroup of E, and so $[E : E_1] < \infty$, as E is compact. Therefore $[E_L : E_1] < \infty$. Next, take a finite subset $B \subset C$ so that $\{\det(\beta)\}_{\beta \in B}$ gives E_L/E_1. Let $k \in L[q, \mathfrak{b}]$. By Theorem 3.2 (i), $k = h\alpha$ with $\alpha \in C$. Then $\det(\alpha\beta^{-1}) \in E_1$ for some $\beta \in B$, so that $\det(\alpha\beta^{-1}) = \det(\gamma)$ with $\gamma \in C \cap U^{\varphi}(W)$. Put $\xi = \alpha^{-1}\gamma\beta$. Then $\xi \in C^1$ and $h\beta = h\gamma\beta = k\xi$, that is, $k \in h\beta C^1$. Thus $L[q, \mathfrak{b}] = \bigcup_{\beta \in B} h\beta C^1$. We easily see that the last union is disjoint, and so we obtain (i). To prove (ii), let the notation be as in (3.3) and (3.4); let $M = \mathfrak{r}m$, $\varphi[m] = \zeta \in \mathfrak{g}^{\times}$, and $\mathfrak{d} = \mathfrak{q}^{\kappa}$ as in (c2) above. Then ζ represents $d_0(\varphi)$. Suppose $q\mathfrak{r} = \mathfrak{d}^{-2}$ and $h \in L[q, \mathfrak{d}^{-1}]$; put $h = \sum_{i=1}^{r}(y_i e_i + z_i g_i) + sm$ with $y_i, z_i, s \in K$. Then $y_i, z_i \in \mathfrak{r}$, and $q - \zeta s s^{\rho} \in \mathfrak{g}$, so that $s\mathfrak{r} = \mathfrak{d}^{-1}$ and $(\zeta s s^{\rho})^{-1} q \in 1 + \mathfrak{p}^{\kappa} \in N_{K/F}(\mathfrak{r}^{\times})$; thus q represents $d_0(\varphi)$. This shows that $L[q, \mathfrak{d}^{-1}] \neq \emptyset$ only for such a q. Taking such a q, we can put $q = \zeta x x^{\rho}$ with $x \in K^{\times}$. Let $k = xm$ and $W = (Kk)^{\perp}$. Then $k \in L[q, \mathfrak{d}^{-1}]$, $W = \sum_{i=1}^{r}(Ke_i + Kg_i)$, and $C \cap U^{\varphi}(W) = C(\Lambda)$ with $\Lambda = \sum_{i=1}^{r}(\mathfrak{r}e_i + \mathfrak{r}g_i)$; thus $E_1 = E_{\Lambda}$. By Lemma 1.9, $E_L = E$, $E_{\Lambda} = E_0$, and $[E : E_0] = 2$, which together with (i) shows that $\#\{L[q, \mathfrak{d}^{-1}]/C^1\} = 2$. This completes the proof.

LEMMA 3.14. *Let $W = (Kh)^{\perp}$ with $h \in L[q, \mathfrak{b}]$ and let $\Lambda = L \cap W$. Suppose that $0 < n - 1 \in 2\mathbf{Z}$. Then there is a unique maximal lattice in W containing Λ at least in the following two cases: (1) $q(\mathfrak{b}\mathfrak{b}^{\rho})^{-1} = \mathfrak{r}$; (2) $q(\mathfrak{b}\mathfrak{b}^{\rho})^{-1} = \mathfrak{p}\mathfrak{r}$ and $\{K, qd_0(\varphi)\}$ is a division algebra. Moreover, Λ is maximal and $C(\Lambda) = C(L) \cap U^{\varphi}(W)$ in Case (1); Λ is maximal and $[C(\Lambda) : C(L) \cap U^{\varphi}(W)] = N(\mathfrak{p}) + 1$ in Case (2) if K is unramified over F and $d_0(\varphi) = N_{K/F}(K^{\times})$. These assertions are true with C^1 instead of C.*

Proof. Changing h for ch with some $c \in K^{\times}$, we may assume that $\mathfrak{b} = \mathfrak{r}$. This does not change the ideal $q(\mathfrak{b}\mathfrak{b}^{\rho})^{-1}$ nor $\{K, qd_0(\varphi)\}$. Thus $q \in \mathfrak{g}^{\times}$ in Case (1) and $q\mathfrak{g} = \mathfrak{p}$ in Case (2). Suppose $q \in \mathfrak{g}^{\times}$. Then $L = \mathfrak{r}h \oplus \Lambda$ as shown in the proof of Lemma 3.4, and Λ is maximal as noted there; clearly $C(\Lambda) = C(L) \cap U^{\varphi}(W)$ and $C^1(\Lambda) = C^1(L) \cap U^{\varphi}(W)$. Next suppose $q\mathfrak{g} = \mathfrak{p}$ and $\{K, qd_0(\varphi)\}$ is a division algebra. Let u, δ, and $\{e_i, g_i\}$ be as in §3.8. Since $n \notin 2\mathbf{Z}$, we have $L = \mathfrak{r}m + \sum_{i=1}^{r}(\mathfrak{r}e_i + \mathfrak{r}g_i)$ with an element m such that $\varphi[m] = \zeta$. Theorem 3.2 (i) allows us to replace h by any element in $L[q, \mathfrak{r}]$. Thus we can put $h = e_1 + qug_1$. Put now $k = e_1 + qu^{\rho}g_1$, $Y = Kk + Km$, and $N = \mathfrak{r}k + \mathfrak{r}m$. Then $\varphi[k] = -q$, $W = Y \oplus \sum_{i=2}^{r}(Ke_i + Kg_i)$, $\Lambda = N + \sum_{i=2}^{r}(\mathfrak{r}e_i + \mathfrak{r}g_i)$, and $d_0(Y) = qd_0(\varphi)$. Since $\{K, qd_0(\varphi)\}$ is a division algebra, (Y, φ) is anisotropic, and so has a unique maximal lattice M as noted in §3.7. Put $\Lambda' = M + \sum_{i=2}^{r}(\mathfrak{r}e_i + \mathfrak{r}g_i)$. Then clearly Λ' is the unique maximal lattice in W containing Λ.

To prove the remaining part, we assume that $q\mathfrak{g} = \mathfrak{p}$, $\mathfrak{d} = \mathfrak{r}$, and $d_0(\varphi) = N_{K/F}(K^{\times})$. Since $N \subset M$ and $[\widehat{M} : M] = N(\mathfrak{p})^2 = [\widehat{N} : N]$, we obtain $M = N$, and so Λ is maximal. We easily see that $C(L) \cap U^{\varphi}(W) \subset C(\Lambda)$. Let $\gamma \in C(\Lambda)$.

Then $L\gamma$ is a maximal lattice in V containing Λ, so that $\sum_{i=2}^{r}(\mathfrak{r}e_i+\mathfrak{r}g_i)\subset L\gamma$. By [S2, Lemma 4.9 (i)] we have $L\gamma = J+\sum_{i=2}^{r}(\mathfrak{r}e_i+\mathfrak{r}g_i)$ with a maximal lattice J in $Km+Ke_1+Kg_1$. Since $m\in J$, from (1.9) we see that $\varphi(m,\,J)=\mathfrak{r}$, and so $J=\mathfrak{r}m+H$ with a maximal lattice H in Ke_1+Kg_1 as shown in the proof of Lemma 3.4. Now $h=h\gamma\in L\gamma$ and $k\in\Lambda=\Lambda\gamma$; thus $h,\,k\in J$. Put $H_0=\mathfrak{r}e_1+\mathfrak{r}g_1$. By Lemma 1.10 (ii), $H=H_0\alpha$ with $\alpha\in SU^{\varphi}(Ke_1+Kg_1)$. By (3.5) we can take α in the form (3.6). Identify $GL(Ke_1+Kg_1)$ with $GL_2(K)$ with respect to the basis $\{e_1,\,g_1\}$. Replacing α by an element of $C(H_0)\alpha$, we can put $\alpha=\begin{bmatrix}a^{-1} & b\\ 0 & a\end{bmatrix}$ with $a\in F^{\times}$ and $b\in F$. Put $z=a^{-1}e_1+bg_1$ and $w=ag_1$; then $H=\mathfrak{r}z+\mathfrak{r}w$. Since $\delta e_1=uk-u^{\rho}h\in H$ and $q\delta g_1=h-k\in H$, we see that $\mathfrak{p}\subset a\mathfrak{g}\subset\mathfrak{g}$, and so we may assume that $a=1$ or $a=q$. Now $\varphi(z,\,e_1)\in\mathfrak{r}$, and so $b\in\mathfrak{g}$. Thus if $a=1$, then $H=H_0$. Suppose $a=q$; put $\alpha_b=\begin{bmatrix}a^{-1} & b\\ 0 & a\end{bmatrix}$. Then $H_0\alpha_b=H_0\alpha_{b'}$ if and only if $b-b'\in\mathfrak{p}$. Thus there exist exactly $N(\mathfrak{p})+1$ maximal lattices in Ke_1+Kg_1 containing h and k, and so there exist at most $N(\mathfrak{p})+1$ lattices of the form $L\gamma$ with $\gamma\in C(\Lambda)$. This shows that

$$\text{(3.11)}\qquad \bigl[C(\Lambda):C(L)\cap U^{\varphi}(W)\bigr]\le N(\mathfrak{p})+1.$$

To show that this is actually an equality, define the symbols ℓ, j, ω, and $\mathfrak{O}$ as in Lemma 1.3, (1.6), and §3.7, with Y as the space V there. In the proof of Lemma 1.3 take $\varphi_0=\mathrm{diag}[\zeta,\,-q]$ and $s=\zeta^{-1}q$. Then the identification of V with K_2^1 in the proof of Lemma 1.3 (which is unrelated to the above identification of $GL(Ke_1+Kg_1)$ with K_2^2) identifies m with $(1,\,0)$ and k with $(0,\,1)$, so that m here equals ℓ of Lemma 1.3; also, $M=\ell j(\mathfrak{O})=mj(\mathfrak{O})$. Let β be an element of $\mathfrak{O}$ such that $\beta\beta^{\iota}=1$; define $\xi\in SU^{\varphi}(V)$ by $\xi=j(\beta)$ on Y and $x\xi=x$ for every $x\in Kh+\sum_{i=2}^{r}(Ke_i+Kg_i)$. Then $\xi\in C^1(\Lambda)$. Put $\beta=c+d\omega$ with $c,\,d\in\mathfrak{r}$. Then $\ell\xi=c\ell+dk$ and $k\xi=sd^{\rho}\ell+c^{\rho}k$. Since $\delta e_1=uk-u^{\rho}h$ and $q\delta g_1=h-k$, we have $q\delta g_1\xi=(1-c^{\rho})e_1-sd^{\rho}\ell+q(u-c^{\rho}u^{\rho})g_1$. Thus $\xi\in C^1(L)$ if and only if $c-1\in\mathfrak{p}\mathfrak{r}$, which is so if and only if $\beta-1\in\mathfrak{P}$. Since $\beta\beta^{\iota}=1$, this shows that there exist at least $N(\mathfrak{p})+1$ different $L\gamma$ with $\gamma\in C^1(\Lambda)$, and so

$$N(\mathfrak{p})+1\le\bigl[C^1(\Lambda):C^1(L)\cap U^{\varphi}(W)\bigr].$$

This combined with (3.11) proves that

$$\bigl[C^1(\Lambda):C^1(L)\cap U^{\varphi}(W)\bigr]=\bigl[C(\Lambda):C(L)\cap U^{\varphi}(W)\bigr]=N(\mathfrak{p})+1,$$

which completes the proof.

We insert here the classification of the structures $(V,\,s\varphi)$ with $s\in F^{\times}$. If $K=F\times F$, the matter is settled in §1.8.

PROPOSITION 3.15. (i) *If F is a local field and* $\dim(V)$ *is even, then* (V, φ) *is isomorphic to* $(V, s\varphi)$ *for every* $s \in F^\times$.

(ii) *Suppose F is an arbitrary field and* $\dim(V)$ *is odd; let* $s \in F^\times$; *then* (V, φ) *is isomorphic to* $(V, s\varphi)$ *if and only if* $s \in N_{K/F}(K^\times)$.

Proof. The first assertion is included in Lemma 1.6. The second assertion can be proved easily in the same manner as for [S3, Theorem 7.13] in the even-dimensional case.

4. HERMITIAN DIOPHANTINE EQUATIONS OVER A GLOBAL FIELD

4.1. Throughout this section we assume that F is an algebraic number field and K is a quadratic extension of F; we fix a hermitian space (V, φ) and use the notation of §2.1. For an $\mathfrak{r}$-lattice L in V we put

$$\Gamma(L) = \{\alpha \in U^\varphi(V) \mid L\alpha = L\}, \qquad \Gamma^1(L) = \Gamma(L) \cap SU^\varphi(V). \tag{4.1}$$

Given $h \in V$ such that $\varphi[h] \neq 0$, put $W = (Kh)^\perp$. We view $U^\varphi(W)$ as a subgroup of $U^\varphi(V)$, and $SU^\varphi(W)$ as a subgroup of $SU^\varphi(V)$ as explained in §1.1. For $\sigma \in U^\varphi(V)_\mathbf{A}$, the symbol $h\sigma$ is meaningful as an element of $V_\mathbf{A}$.

THEOREM 4.2. *If* $\dim(V) > 1$, *then* $\#\{\Lambda[q]/\Gamma^1(\Lambda)\} < \infty$ *for every* $q \in F^\times$ *and every $\mathfrak{r}$-lattice Λ in V.*

Proof. Assuming $\dim(V) > 1$ and $\Lambda[q] \neq \emptyset$, take $h \in \Lambda[q]$ and define W as above; put $G = SU^\varphi(V)$, $H = SU^\varphi(W)$, $D = \{x \in G_\mathbf{A} \mid \Lambda x = \Lambda\}$, and $D_v = G_v \cap D$ for $v \in \mathbf{h}$. Then $\Gamma^1(\Lambda) = G \cap D$ and $D_v = C^1(\Lambda_v)$. By Theorem 3.3 and (1.5) we have $\Lambda_v[q] = \bigcup_{\alpha \in X_v} h\alpha D_v$ with a finite subset X_v of G_v. By Theorem 3.2 (iii) and (iv) we can take $X_v = \{1\}$ if Λ_v is maximal, v is not ramified in K, and $q \in \mathfrak{g}_v^\times$. Thus $X_v = \{1\}$ for almost all $v \in \mathbf{h}$. Put $X = \prod_{v \in \mathbf{h}} X_v$. This is a finite subset of $G_\mathbf{h}$. For each $\xi \in X$ we can find a finite subset E_ξ of $H_\mathbf{h}$ such that $H_\mathbf{A} = \bigcup_{\varepsilon \in E_\xi} H\varepsilon(H_\mathbf{A} \cap \xi D \xi^{-1})$. Then $H_\mathbf{A}\xi D = \bigcup_{\varepsilon \in E_\xi} H\varepsilon\xi D$. For each (ε, ξ) such that $G \cap \varepsilon\xi D \neq \emptyset$ pick $\beta_{\varepsilon,\xi} \in G \cap \varepsilon\xi D$. Now let $k \in \Lambda[q]$. Then $k = h\xi\zeta$ for some $\xi \in X$ and $\zeta \in D$. On the other hand $k = h\alpha$ with $\alpha \in G$ by (1.5). Then $\alpha\zeta^{-1}\xi^{-1} \in H_\mathbf{A}$, so that $\alpha \in H_\mathbf{A}\xi D$. Thus $\alpha \in H\varepsilon\xi D$ for some $\varepsilon \in E_\xi$. Then $\alpha \in H\beta_{\varepsilon,\xi}D \cap G = H\beta_{\varepsilon,\xi}\Gamma^1(\Lambda)$, and hence $k = h\alpha \in hb_{\varepsilon,\xi}\Gamma^1(\Lambda)$. Since the $b_{\varepsilon,\xi}$ form a finite set, we obtain our theorem.

4.3. We now fix a maximal lattice L in V, and put

$$C = \{\gamma \in U^\varphi(V)_\mathbf{A} \mid L\gamma = L\}, \quad C^1 = C \cap SU^\varphi(V)_\mathbf{A}. \tag{4.2}$$

We are going to state our main theorem with respect to a pair (G, H) belonging to the following two types of objects:

Type U: $G = U^\varphi(V)$ and $H = U^\varphi(W)$;

Type SU: $G = SU^\varphi(V)$ and $H = SU^\varphi(W)$.

Here $W = (Kh)^{\perp}$ with a fixed $h \in V$. For a subset S of $U^{\varphi}(V)_{\mathbf{A}}$ the symbol hS is meaningful as a subset of $V_{\mathbf{A}}$. Therefore $V \cap hS$ is a well-defined subset of V.

THEOREM 4.4. *Suppose* $\dim(V) > 1$. *For a fixed* $h \in V$ *such that* $\varphi[h] \neq 0$ *put* $W = (Kh)^{\perp}$, *and take* (G, H) *of Type U or SU as above. Let* $D = D_0 G_{\mathbf{a}}$ *with an open compact subgroup* D_0 *of* $G_{\mathbf{h}}$. *Then the following assertions hold.*

(i) *For* $y \in G_{\mathbf{A}}$ *we have* $H_{\mathbf{A}} \cap GyD \neq \emptyset$ *if and only if* $V \cap hDy^{-1} \neq \emptyset$.

(ii) *Fixing* $y \in G_{\mathbf{A}}$, *for every* $\varepsilon \in H_{\mathbf{A}} \cap GyD$ *take* $\alpha \in G$ *so that* $\varepsilon \in \alpha yD$. *Then the map* $\varepsilon \mapsto h\alpha$ *gives a bijection of* $H\backslash(H_{\mathbf{A}} \cap GyD)/(H_{\mathbf{A}} \cap D)$ *onto* $(V \cap hDy^{-1})/\Delta_y$, *where* $\Delta_y = G \cap yDy^{-1}$.

(iii) *Take* $\{y_i\}_{i \in I} \subset G_{\mathbf{A}}$ *so that* $G_{\mathbf{A}} = \bigsqcup_{i \in I} Gy_iD$, *and put* $\Gamma_i = G \cap y_iDy_i^{-1}$. *Then*

$$\#\{H\backslash H_{\mathbf{A}}/(H_{\mathbf{A}} \cap D)\} = \sum_{i \in I} \#\{(V \cap hDy_i^{-1})/\Gamma_i\}. \tag{4.3}$$

(iv) *Let* $q = \varphi[h]$ *and* $\mathfrak{b} = \varphi(h, L)$. *Then for every* $y \in U^{\varphi}(V)_{\mathbf{A}}$, *we have*

$$V \cap hCy^{-1} = \bigl(Ly^{-1}\bigr)[q, \mathfrak{b}]. \tag{4.4}$$

(v) *Suppose moreover that* $\dim(V) > 2$ *and the following condition is satisfied:*

(4.5) *If* n *is odd, then* $q_v\mathfrak{r}_v \neq \mathfrak{b}_v\mathfrak{b}_v^{\rho}$ *for every* $v \in \mathbf{h}$ *ramified in* K.

Then for every $y \in SU^{\varphi}(V)_{\mathbf{A}}$ *we have*

$$V \cap hC^1y^{-1} = \bigl(Ly^{-1}\bigr)[q, \mathfrak{b}]. \tag{4.6}$$

Proof. Let y, ε, and α be as in (ii); then clearly $h\alpha \in V \cap hDy^{-1}$. If $\eta\varepsilon\zeta \in \beta yD$ with $\eta \in H$, $\zeta \in H_{\mathbf{A}} \cap D$, and $\beta \in G$, then $\beta^{-1}\eta\alpha \in G \cap yDy^{-1} = \Delta_y$, and hence $h\alpha = h\eta\alpha \in h\beta\Delta_y$. Thus our map is well defined. Next let $k \in V \cap hDy^{-1}$. Then $k = h\delta y^{-1}$ with $\delta \in D$, and moreover, by (1.5), $k = h\xi$ with $\xi \in G$. Then $h = h\xi y\delta^{-1}$, so that $\xi y\delta^{-1} \in H_{\mathbf{A}}$. Thus $\xi y\delta^{-1} \in H_{\mathbf{A}} \cap GyD$. This shows that k is the image of an element of $H_{\mathbf{A}} \cap GyD$. To prove that the map is injective, suppose $\varepsilon \in \alpha yD \cap H_{\mathbf{A}}$ and $\delta \in \beta yD \cap H_{\mathbf{A}}$ with $\alpha, \beta \in G$, and $h\alpha = h\beta\sigma$ with $\sigma \in \Delta_y$. Put $\omega = \beta\sigma\alpha^{-1}$. Then $h\omega = h$, so that $\omega \in H$. Since $\sigma \in yDy^{-1}$, we have $\beta yD = \beta\sigma yD = \omega\alpha yD$, and hence $\delta \in \beta yD \cap H_{\mathbf{A}} = \omega\alpha yD \cap H_{\mathbf{A}} = \omega(\alpha yD \cap H_{\mathbf{A}}) = \omega(\varepsilon D \cap H_{\mathbf{A}}) = \omega\varepsilon(D \cap H_{\mathbf{A}}) \subset H\varepsilon(D \cap H_{\mathbf{A}})$. This proves the injectivity, and completes the proof of (ii). At the same time we obtain (i).

Since $H_{\mathbf{A}} = \bigsqcup_{i \in I}(H_{\mathbf{A}} \cap Gy_iD)$, we can derive (iii) immediately from (ii).

As for (v), clearly $V \cap hC^1 \subset L[q, \mathfrak{b}]$. Conversely, if $x \in L[q, \mathfrak{b}]$, then $x \in hC^1$ by Theorem 3.2 (iii). Thus

$$V \cap hC^1 = L[q, \mathfrak{b}]. \tag{4.7}$$

If $k \in V \cap hC^1y^{-1}$, then $hC^1y^{-1} = kyC^1y^{-1}$. Now $\varphi(k, Ly^{-1}) = \varphi(h, L) = \mathfrak{b}$. Taking k, yC^1y^{-1}, and Ly^{-1} in place of h, C^1, and L in (4.7), we obtain $V \cap kyC^1y^{-1} = (Ly^{-1})[q, \mathfrak{b}]$. This proves (4.6) when $V \cap hC^1y^{-1} \neq \emptyset$. To prove the remaining case, suppose $\ell \in (Ly^{-1})[q, \mathfrak{b}]$; then $\varphi(\ell_v y_v, L_v) = \mathfrak{b}_v = \varphi(h, L)_v$ for every $v \in \mathbf{h}$, and so, by Theorem 3.2 (iii), $\ell y \in hC^1$. Taking ℓ, yC^1y^{-1}, and Ly^{-1} in place of h, C^1, and L in (4.7), we obtain $\emptyset \neq (Ly^{-1})[q, \mathfrak{b}] = V \cap \ell yC^1y^{-1} = V \cap hC^1y^{-1}$. This shows that $(Ly^{-1})[q, \mathfrak{b}] = \emptyset$ if $V \cap hC^1y^{-1} = \emptyset$, and hence (4.6) holds for every $y \in G_{\mathbf{A}}$. This proves (v). Assertion (iv) can be proved in the same way.

In view of (i) we can restrict the indices i on the right-hand side of (4.3) to those for which $H_{\mathbf{A}} \cap Gy_iD \neq \emptyset$. If I' denotes the set of all such i's, then $H_{\mathbf{A}} = \bigsqcup_{j \in I'} (H_{\mathbf{A}} \cap Gy_jD)$.

Combining (4.3) and (4.4), we obtain, for (G, H) of type U, an equality

$$\#\{H \backslash H_{\mathbf{A}}/(H_{\mathbf{A}} \cap C)\} = \sum_{i \in I} \#\{(Ly_i^{-1})[q, \mathfrak{b}]/\Gamma_i\}, \tag{4.8}$$

where $\{y_i\}$ is such that $G_{\mathbf{A}} = \bigsqcup_{i \in I} Gy_iC$ and $\Gamma_i = G \cap y_iCy_i^{-1}$. We can state a similar formula for (G, H) of type SU when $n > 2$ and (4.5) is satisfied. Formula (4.8) connects the class number of H with respect to $H_{\mathbf{A}} \cap C$ to the solutions h of the equation $\varphi[h] = q$ under the condition $\varphi(h, Ly_i^{-1}) = \mathfrak{b}$.

5. Nonscalar hermitian Diophantine equations

5.1. So far we discussed the equation $\varphi[h] = q$ with a scalar q. We can formulate a similar problem with nonscalar q, which can be stated in terms of matrices as follows. We take F to be local or global. Given $q^* = q \in GL_m(K)$ and $\varphi^* = \varphi \in GL_n(K)$, we consider the solutions $h \in K_n^m$ of the equation $h\varphi h^* = q$. Here and throughout this section we assume $n > m > 0$. More intrinsically, take (V, φ) as before and take also (X, q) with a nondegenerate hermitian form q on a free K-module X of dimension m. We consider $h \in \mathrm{Hom}(X, V)$ such that $\varphi[xh] = q[x]$ for every $x \in X$. Since q is nondegenerate, h must be injective. To simplify our notation, for every $k \in \mathrm{Hom}(X, V)$ we denote by $\varphi[k]$ the hermitian form on X defined by $\varphi[k][x] = \varphi[xk]$ for every $x \in X$. Then our problem concerns the solutions $h \in \mathrm{Hom}(X, V)$ of the equation $\varphi[h] = q$ for a fixed q. If $m = 1$ and $X = K$, then $q \in F^{\times}$, and an element h of V defines an element of $\mathrm{Hom}(K, V)$ that sends c to ch for $c \in K$, and every element of $\mathrm{Hom}(K, V)$ is of this type. Thus the problem about $\varphi[h] = q$ with $q \in F^{\times}$ is the one-dimensional special case. Let h be an element of $\mathrm{Hom}(X, V)$ such that $\mathrm{rank}(\varphi[h]) = m$. Then

$$\{x \in \mathrm{Hom}(X, V) \mid \varphi[x] = \varphi[h]\} = h \cdot SU^{\varphi}. \tag{5.1}$$

This is similar to (1.5), and follows easily from the Witt theorem in the unitary case. Though we take V to be coordinate-free, it is practical to take X to be

K_m^1, and so take q to be a hermitian element of $GL_m(K)$, and $q[x] = xqx^*$ for $x \in K_m^1$. For $h \in \mathrm{Hom}(X, V)$ and $1 \le i \le m$ we define "the i-th row" of h to be the element h_i of V determined by $ah = \sum_{i=1}^m a_i h_i$ for $a = (a_i)_{i=1}^m \in K_m^1 = X$.

We first prove a local finiteness result that generalizes Theorem 3.3.

THEOREM 5.2. *Suppose F is a local field; let Λ be an $\mathfrak{r}$-lattice in* $\mathrm{Hom}(X, V)$ *and let* $D = \{\gamma \in SU^\varphi(V) \mid \Lambda\gamma = \Lambda\}$. *Then, given* $h \in \mathrm{Hom}(X, V) \cap \Lambda$ *such that $\varphi[h]$ is nondegenerate, there exists a finite subset A of $SU^\varphi(V)$ such that*

$$\{\, x \in \Lambda \mid \varphi[x] = \varphi[h] \,\} = \textstyle\bigsqcup_{\alpha\in A} h\alpha D. \tag{5.2}$$

Moreover, suppose $K = F \times F$ or K is a field unramified over F; suppose also that $\varphi[h] \in GL_m(\mathfrak{r})$ and $\Lambda = \{\lambda \in \mathrm{Hom}(X, V) \mid \mathfrak{r}_m^1\lambda \subset L\}$ *with a maximal lattice L in V. Then we can take $A = \{1\}$.*

Proof. The first part is Theorem 3.3 if $m = 1$, and so we assume $m > 1$ and prove (5.2) by induction on m. Put $q = \varphi[h]$. Changing (h, q, Λ) for $(ch, cqc^*, c\Lambda)$ with a suitable $c \in GL_m(K)$, we may assume that $q = \mathrm{diag}[a, \eta]$ with $a \in F^\times$ and $\eta^* = \eta \in GL_{m-1}(K)$. Also we may assume that $\Lambda = \{\kappa \in \mathrm{Hom}(X, V) \mid M\kappa \subset L\}$ with $M = \mathfrak{r}_m^1$ and an $\mathfrak{r}$-lattice L in V. Then $D = \{\alpha \in SU^\varphi(V) \mid L\alpha = L\}$. If $x \in \Lambda$ and $\varphi[x] = q$, then $x_1, h_1 \in L$ and $\varphi[x_1] = \varphi[h_1] = a$, and hence by Theorem 3.3 there exists a finite subset B of L such that such an x_1 belongs to $\bigcup_{b\in B} bD$ and $\varphi[b] = a$ for every $b \in B$. Suppose $x_1 = b\gamma$ with $b \in B$ and $\gamma \in D$. Put $W_b = (Kb)^\perp$, $y = x\gamma^{-1}$ and $z = [y_i]_{i=2}^m$. Then $y_1 = b$ and $\varphi[y] = \varphi[h]$, so that $\varphi[z] = \eta$, and $y_i \in W_b$ for $i > 1$. We can view z as an element of $\mathrm{Hom}(K_{m-1}^1, W_b)$. Then $\mathfrak{r}_{m-1}^1 z \subset L \cap W_b$. Put $E = \{\varepsilon \in SU^\varphi(W_b) \mid (L \cap W_b)\varepsilon = L \cap W_b\}$. By induction there exists a finite subset U_b of $\mathrm{Hom}(K_{m-1}^1, W_b)$ such that

$$\{z \in \mathrm{Hom}(K_{m-1}^1, W_b) \mid \mathfrak{r}_{m-1}^1 z \subset L \cap W_b,\ \varphi[z] = \eta\} = \textstyle\bigsqcup_{u\in U_b} uE.$$

We can find a finite subset S of E such that $E = \bigsqcup_{\sigma\in S} \sigma(D \cap E)$. Then $y = \begin{bmatrix} b \\ z \end{bmatrix} = \begin{bmatrix} b \\ u\sigma\tau \end{bmatrix}$ with $u \in U_b$, $\sigma \in S$, and $\tau \in D \cap E$. Thus $x = \begin{bmatrix} b \\ u\sigma \end{bmatrix} \tau\gamma$, and $\tau\gamma \in D$. This shows that $x \in \bigsqcup_{k\in P} kD$ with a finite subset P of the left-hand side of (5.2), as the elements (b, u, σ) form a finite set. By (5.1), for each $k \in P$ there exists an element α of $SU^\varphi(V)$ such that $k = h\alpha$. This proves the first assertion.

Next suppose that the conditions on K, q, Λ, and φ as in the second assertion are satisfied. Take an $\mathfrak{r}$-basis of L and identify V, L, and φ with K_n^1, $\mathfrak{r}_n^1$, and a hermitian matrix with respect to that basis. Given $\ell \in L = \mathfrak{r}_n^1$, put $z = q^{-1}h\varphi\ell^*$ and $y = \ell - z^*h$. Since $\varphi \prec \mathfrak{r}$ and $q \in GL_m(\mathfrak{r})$, we see that $y \in L$, and for every $w \in X$ we have $wh\varphi y^* = 0$, so that $y \in (Xh)^\perp$. Put $M = \mathfrak{r}_m^1$ and $Y = (Xh)^\perp$. Then $V = Xh \oplus Y$ and $L = Mh \oplus (L \cap Y)$. Suppose $\varphi[k] = q$ with $k \in \Lambda$. Then similarly $L = Mk \oplus (L \cap Z)$ with $Z = (Xk)^\perp$. Since L

is maximal, Mh resp. Mk is maximal in Xh resp. Xk, and $L \cap Y$ resp. $L \cap Z$ is maximal in Y resp. Z. By (5.1) there exists an element $\gamma \in SU^\varphi(V)$ such that $h\gamma = k$. Then $Mh\gamma = Mk$, $Y\gamma = Z$, and $(L \cap Y)\gamma$ is $\mathfrak{r}$-maximal in Z, so that by Lemma 3.4 (i), $(L \cap Y)\gamma\varepsilon = L \cap Z$ with some $\varepsilon \in U^\varphi(Z)$. Define $\alpha \in GL(V)$ by $\alpha = \gamma$ on Xh and $\alpha = \gamma\varepsilon$ on Y. Then $\alpha \in U^\varphi(V)$, $L\alpha = L$, and $h\alpha = k$. Since $\det(\alpha) \in \mathfrak{r}^\times$, we can find an element ξ of $U^\varphi(Z)$ such that $(L \cap Z)\xi = L \cap Z$ and $\det(\xi) = \det(\alpha)^{-1}$. This is clear if $K = F \times F$; see §3.6. If K is a field unramified over F, then the fact is included in Lemma 1.9. Extend ξ to an element of $U^\varphi(V)$ by putting $xk\xi = xk$ for $x \in X$. Then $\alpha\xi \in SU^\varphi(V)$, $h\alpha\xi = k$, and $L\alpha\xi = L$. Clearly $\Lambda\alpha\xi = \Lambda$, and hence we obtain (5.2) with $A = \{1\}$. This completes the proof.

Next we prove a generalization of Theorem 4.2, which is a global version of the above theorem.

THEOREM 5.3. *Suppose that F is an algebraic number field; let Λ be an $\mathfrak{r}$-lattice in* $\mathrm{Hom}(X, V)$, $\Gamma = \{\gamma \in SU^\varphi(V) \mid \Lambda\gamma = \Lambda\}$, *and* $T_q = \{x \in \Lambda \mid \varphi[x] = q\}$ *with* $q^* = q \in GL_m(K)$. *Then T_q/Γ is a finite set.*

Proof. We assume the existence of $h \in T_q$. Put $W = (Xh)^\perp$, $G = SU^\varphi(V)$, $H = SU^\varphi(W)$, $M = \mathfrak{r}^1_m$, $D = \{\gamma \in G_\mathbf{A} \mid \Lambda\gamma = \Lambda\}$, and $D_v = D \cap G_v$ for $v \in \mathbf{h}$. We identify H with $\{\alpha \in G \mid h\alpha = h\}$. Fix a maximal lattice L in V. By Theorem 5.2, for each $v \in \mathbf{h}$ there exists a finite subset E_v of G_v such that

$$\{x \in \Lambda_v \mid \varphi[x] = q\} = \textstyle\bigsqcup_{\varepsilon \in E_v} h\varepsilon D_v.$$

Now for almost all $v \in \mathbf{h}$ we have $\Lambda_v = \{\gamma \in \mathrm{Hom}(X_v, V_v) \mid M_v\gamma \subset L_v\}$, L_v is maximal, v is unramified in K, and $q \in GL_m(\mathfrak{r}_v)$. Therefore, by Theorem 5.2, we can take $E_v = \{1\}$ for almost all $v \in \mathbf{h}$. Consequently we can find a finite subset E of $G_\mathbf{h}$ such that $T_q \subset \bigcup_{\eta \in E} h\eta D$. If $x \in T_q$, then $x \in hG$ by (5.1). Thus $x = h\alpha = h\eta\delta$ with $\alpha \in G$, $\eta \in E$, and $\delta \in D$. We have $\alpha\delta^{-1}\eta^{-1} \in H_\mathbf{A}$, and hence $\alpha \in H_\mathbf{A}\eta D$. For each $\eta \in E$ we can find a finite subset Z_η of $H_\mathbf{h}$ such that $H_\mathbf{A} = \bigsqcup_{\zeta \in Z_\eta} H\zeta(H_\mathbf{A} \cap \eta D\eta^{-1})$. Then $H_\mathbf{A}\eta D = \bigcup_{\zeta \in Z_\eta} H\zeta\eta D$, and hence $\alpha \in \bigcup_{\eta,\zeta}(G \cap H\zeta\eta D) = \bigcup_{\eta,\zeta} H(G \cap \zeta\eta D)$. For each (ζ, η) such that $G \cap \zeta\eta D \neq \emptyset$, pick any $\beta \in G \cap \zeta\eta D$. Then $G \cap \zeta\eta D = G \cap \beta D = \beta\Gamma$. Let B be the set of such β's chosen for each (ζ, η). Then $\alpha \in \bigcup_{\beta \in B} H\beta\Gamma$, and thus $h\alpha \in \bigcup_{\beta \in B} h\beta\Gamma$, which proves our theorem.

THEOREM 5.4. *Suppose that F is an algebraic number field. With a fixed $h \in \mathrm{Hom}(X, V)$ such that* $\mathrm{rank}(\varphi[h]) = m$, *put* $q = \varphi[h]$, $W = (Xh)^\perp$, $G = U^\varphi(V)$, $H = U^\varphi(W)$, *and* $\mathscr{V} = \mathrm{Hom}(X, V)$. *Let D be an open subgroup of $G_\mathbf{A}$ containing $G_\mathbf{a}$ such that $D \cap G_\mathbf{h}$ is compact, and let* $G_\mathbf{A} = \bigsqcup_{i \in I} Gy_iD$. *Then assertions (i), (ii), and (iii) of Theorem 4.4 are valid if we take the symbols h, G, H, and D there to be those of the present setting, and replace V there by $\mathscr{V}$. In particular we have*

$$\#\{H\backslash H_{\mathbf{A}}/(H_{\mathbf{A}}\cap D)\} = \sum_{i\in I}\#\{(\mathscr{V}\cap hDy_i^{-1})/\Gamma_i\}, \tag{5.3}$$

where $\Gamma_i = G\cap y_iDy_i^{-1}$. *The same is true with* $G = SU^\varphi(V)$ *and* $H = SU^\varphi(W)$.

Proof. We can repeat the proof of Theorem 4.4 with obvious modifications.

THEOREM 5.5. *In the setting of Theorem 5.4 with* $G = SU^\varphi(V)$ *and* $H = SU^\varphi(W)$ *suppose that* $n - m > 1$ *and neither* $G_{\mathbf{a}}$ *nor* $H_{\mathbf{a}}$ *is compact. Let* k *be an element of* $\mathrm{Hom}(X, V)$ *such that* $k = h\gamma_v$ *for every* $v \in \mathbf{h}$ *with some* $(\gamma_v)_{v\in\mathbf{h}} \in D\cap G_{\mathbf{h}}$. *Then there exists an element* $\alpha \in G\cap D$ *such that* $k = h\alpha$. *In particular, if* $m = 1$ *and (4.5) is satisfied, then* $\#\{L[q,\,\mathfrak{b}]/\Gamma^1(L)\} \le 1$ *for every maximal lattice* L *in* V.

Proof. By our assumptions, strong approximation holds on G and H, and so we have $G_{\mathbf{A}} = GD$ and $H_{\mathbf{A}} = H(H_{\mathbf{A}}\cap D)$. Thus we can take $\{y_i\}_{i\in I} = \{1\}$. Therefore (5.3) implies that $\#\{(\mathscr{V}\cap hD)/(G\cap D)\} = 1$, which gives the first assertion. This combined with (4.6) proves the second assertion.

5.6. Before proceeding further, let us recall the notion of the mass of an algebraic group G with respect to an open subgroup D of $G_{\mathbf{A}}$ containing $G_{\mathbf{a}}$ and such that $G_{\mathbf{h}}\cap D$ is compact. For simplicity here we take G to be U^φ or SU^φ and assume that $U^\varphi_{\mathbf{a}}$ is compact. For $x \in G_{\mathbf{A}}$ put $\Delta_x = G\cap xDx^{-1}$ and $\nu(\Delta_x) = [\Delta_x : 1]^{-1}$. Then the *the mass of* G *with respect to* D is defined by

$$\mathfrak{m}(G,\,D) = \sum_{b\in\mathscr{B}}\nu(\Delta_b), \quad \mathscr{B} = G\backslash G_{\mathbf{A}}/D. \tag{5.4}$$

For this the reader is referred to [S2, (10.9.4), (24.1.1), (24.1.2)]. If D' is a subgroup of $G_{\mathbf{A}}$ of the same type as D, then from [S2, Lemma 24.2] we obtain

$$[D : D\cap D']\mathfrak{m}(G,\,D) = \mathfrak{m}(G,\,D\cap D') = [D' : D\cap D']\mathfrak{m}(G,\,D'). \tag{5.5}$$

THEOREM 5.7. *In the setting of Theorem 5.4, suppose that* $G_{\mathbf{a}}$ *is compact. Then for every* $y \in G_{\mathbf{A}}$ *we have*

$$\nu(\Delta_y)\#\{\mathscr{V}\cap hDy^{-1}\} = \sum_{\varepsilon\in\mathscr{E}}\nu(\Delta_\varepsilon), \tag{5.6}$$

where $\mathscr{E} = H\backslash(H_{\mathbf{A}}\cap GyD)/(H_{\mathbf{A}}\cap D)$ *and* $\Delta_x = H\cap xDx^{-1}$. *Moreover, let* $G_{\mathbf{A}} = \bigsqcup_{i\in I} Gy_iD$ *and* $\Gamma_i = G\cap y_iDy_i^{-1}$; *then*

$$\sum_{i\in I}\nu(\Gamma_i)\#\{\mathscr{V}\cap hDy_i^{-1}\} = \mathfrak{m}(H,\,H_{\mathbf{A}}\cap D). \tag{5.7}$$

Proof. To prove (5.6), we may assume that $H_{\mathbf{A}}\cap GyD \ne \emptyset$. For $\varepsilon \in \mathscr{E}$ take $\alpha_\varepsilon \in G$ so that $\varepsilon \in \alpha_\varepsilon yD$. Then $H\cap\alpha_\varepsilon\Delta_y\alpha_\varepsilon^{-1} = H\cap\alpha_\varepsilon yDy^{-1}\alpha_\varepsilon^{-1} = H\cap\varepsilon D\varepsilon^{-1} = \Delta_\varepsilon$. Now $\mathscr{V}\cap hDy^{-1} = \bigsqcup_{\varepsilon\in\mathscr{E}} h\alpha_\varepsilon\Delta_y$ by the part of Theorem 5.4 corresponding to Theorem 4.4 (ii). For $\gamma,\,\gamma' \in \Gamma(\Lambda)$ we have $h\alpha_\varepsilon\gamma = h\alpha_\varepsilon\gamma'$ if and only if $\alpha_\varepsilon\gamma'\gamma^{-1}\alpha_\varepsilon^{-1} \in H$, that is, $\gamma'\gamma^{-1} \in \alpha_\varepsilon^{-1}H\alpha_\varepsilon\cap\Delta_y = \alpha_\varepsilon^{-1}\Delta_\varepsilon\alpha_\varepsilon$, so that

$$\#\{h\alpha_\varepsilon \Delta_y\} = [\Delta_y : \alpha_\varepsilon^{-1}\Delta_\varepsilon\alpha_\varepsilon] = \nu(\Delta_\varepsilon)/\nu(\Delta_y).$$

Therefore we obtain (5.6). Next, let $\mathscr{E}_i = H\backslash(H_{\mathbf{A}} \cap Gy_iD)/(H_{\mathbf{A}} \cap D)$. Then $H_{\mathbf{A}} = \bigsqcup_{i\in I}(H_{\mathbf{A}} \cap Gy_iD) = \bigsqcup_{i\in I}\bigsqcup_{\varepsilon\in\mathscr{E}_i} H\varepsilon(H_{\mathbf{A}} \cap D)$, and so $\mathfrak{m}(H, H_{\mathbf{A}} \cap D) = \sum_{i\in I}\sum_{\varepsilon\in\mathscr{E}_i}\nu(\Delta_\varepsilon)$, which combined with (5.6) proves (5.7).

COROLLARY 5.8. *Define C and C^1 by (4.2) with a maximal lattice L in V; take (G, H) of type U as in Theorem 4.4; suppose that $G_{\mathbf{a}}$ is compact. Let $G_{\mathbf{A}} = \bigsqcup_{i\in I} Gy_iC$, $L_i = Ly_i^{-1}$, and $\Gamma_i = \Gamma(L_i)$. Then*

$$\sum_{i\in I}\nu(\Gamma_i)\#\{L_i[q, \mathfrak{b}]\} = \mathfrak{m}(H, H_{\mathbf{A}} \cap C), \tag{5.8}$$

where $q = \varphi[h]$ and $\mathfrak{b} = \varphi(h, L)$. This is valid for (G, H) of type SU if we replace C and $\Gamma(\cdot\cdot)$ by C^1 and $\Gamma^1(\cdot\cdot)$, provided $n > 2$ and (4.5) is satisfied.

Proof. Take $m = 1$ and $D = C$ in Theorem 5.7. Combining (5.7) with (4.4), we obtain (5.8). The case of SU^φ follows similarly from (4.6).

5.9. Formulas (5.7) and (5.8) are similar to, but different from, the formula of Siegel about $\sum_i \nu(\Gamma_i)\#\{L_i[q]\}$. We already explained in [S3, §13.13] the main differences between our formulas in the orthogonal case given in that book and that of Siegel. In principle, our comments there apply to the present unitary case.

Now in [S2, Theorem 24.4] we gave an exact formula for $\mathfrak{m}(G, D)$ for $G = U^\varphi$ and a certain type of D, under the condition that if n is odd, then $d_0(\varphi)$ is represented by an element of $\mathfrak{g}^\times$. The group $H_{\mathbf{A}} \cap C$ in (5.8) does not necessarily belong to the types of D there, but we can compute $[D : H_{\mathbf{A}} \cap C]$ by means of Lemma 3.14 under some conditions on $(q, \mathfrak{b})$. Then we obtain $\mathfrak{m}(H, H_{\mathbf{A}} \cap C)$ from [S2, Theorem 24.4] by (5.5).

PROPOSITION 5.10. *In the setting of Theorem 5.4 suppose that $n - m$ is odd. Then the structure $(W, \det(q)\varphi)$ depends only on φ and the indices of q at the real archimedean primes of F ramified in K.*

Proof. Let ψ be the restriction of φ to W. Then we can easily verify that $d_0(\det(q)\psi) = \det(q)^{n-m}d_0(\psi) = (-1)^{n-1}d_0(\varphi)$ as $m - n$ is odd. This combined with Theorem 2.2 (i) proves our proposition.

This is an analogue of the fact concerning a quadratic form in even dimension with square discriminant given in [S4, Theorem 1.12].

We insert here some results about the relationship between various invariants associated with U^φ and those with SU^φ.

PROPOSITION 5.11. *Let D be an open subgroup of $U_{\mathbf{A}}^\varphi$ containing $U_{\mathbf{a}}^\varphi$ and such that $U_{\mathbf{h}}^\varphi \cap D$ is compact; put $P = \{x \in K^\times \mid xx^\rho = 1\}$ and $D^1 = D \cap SU_{\mathbf{A}}^\varphi$. Then $U^\varphi SU_{\mathbf{A}}^\varphi D$ is a normal subgroup of $U_{\mathbf{A}}^\varphi$ and*

(5.9a) $$[U^\varphi_{\mathbf{A}} : U^\varphi SU^\varphi_{\mathbf{A}} D] = [P_{\mathbf{A}} : P\det(D)],$$

(5.9b) $$\#(U^\varphi \backslash U^\varphi_{\mathbf{A}}/D) \le \sum_{x\in\Xi} \#(SU^\varphi \backslash SU^\varphi_{\mathbf{A}}/xD^1x^{-1}),$$

where $\Xi = U^\varphi_{\mathbf{A}}/U^\varphi SU^\varphi_{\mathbf{A}} D$. *Moreover, if* $U^\varphi_{\mathbf{a}}$ *is compact, then*

(5.9c) $$\mathfrak{m}(U^\varphi,\, D) \le [P_{\mathbf{A}} : P\det(D)] \cdot \mathfrak{m}(SU^\varphi,\, D^1).$$

Furthermore, if $P \cap \det(D) = \det(U^\varphi \cap yDy^{-1})$ *for every* $y \in U^\varphi_{\mathbf{A}}$, *then the equality holds in (5.9b), and*

(5.9d) $$\#\bigl(P \cap \det(D)\bigr)\mathfrak{m}(U^\varphi,\, D) = [P_{\mathbf{A}} : P\det(D)]\mathfrak{m}(SU^\varphi,\, D^1).$$

Proof. Since $P = \det(U^\varphi)$, we can easily show that

(5.10) $$U^\varphi SU^\varphi_{\mathbf{A}} Dx = \bigl\{y \in U^\varphi_{\mathbf{A}} \bigm| \det(y) \in P\det(Dx)\bigr\}$$

for every $x \in U^\varphi_{\mathbf{A}}$. This shows that $U^\varphi SU^\varphi_{\mathbf{A}} D$ is a normal subgroup of $U^\varphi_{\mathbf{A}}$ and $U^\varphi_{\mathbf{A}}/U^\varphi SU^\varphi_{\mathbf{A}} D$ is isomorphic to $P_{\mathbf{A}}/[P\det(D)]$, as $P_{\mathbf{A}} = \det(U^\varphi_{\mathbf{A}})$. Thus we obtain (5.9a); we also see that $U^\varphi SU^\varphi_{\mathbf{A}} \backslash U^\varphi_{\mathbf{A}}/D$ can be identified with $U^\varphi_{\mathbf{A}}/U^\varphi SU^\varphi_{\mathbf{A}} D$. Given $x \in U^\varphi_{\mathbf{A}}$, take $B_x \subset SU^\varphi_{\mathbf{A}}$ so that $SU^\varphi_{\mathbf{A}} = \bigsqcup_{b\in B_x} SU^\varphi bxD^1x^{-1}$. Then we have $U^\varphi SU^\varphi_{\mathbf{A}} xD = \bigcup_{b\in B_x} U^\varphi bxD$, and hence $U^\varphi_{\mathbf{A}} = \bigcup_{x\in\Xi}\bigcup_{b\in B_x} U^\varphi bxD$. From this we obtain (5.9b). To prove (5.9c), put $\Gamma_x = U^\varphi \cap xDx^{-1}$ and $\Gamma^1_x = SU^\varphi \cap xD^1x^{-1}$ for $x \in U^\varphi_{\mathbf{A}}$. Then $\mathfrak{m}(U^\varphi,\, D) \le \sum_{x\in\Xi}\sum_{b\in B_x} \nu(\Gamma_{bx}) \le \sum_{x\in\Xi}\sum_{b\in B_x} \nu(\Gamma^1_{bx}) = \sum_{x\in\Xi} \mathfrak{m}(SU^\varphi,\, xD^1x^{-1})$. Now formula (5.5) shows that $\mathfrak{m}(SU^\varphi,\, D^1)$ depends only on the measure of D^1. (If $U^\varphi_{\mathbf{a}}$ is not compact, we have to consider the measure of $D^1_{\mathbf{h}}$.) Since $\mathfrak{m}(SU^\varphi,\, xD^1x^{-1}) = \mathfrak{m}(SU^\varphi, D^1)$, we obtain (5.9c).

Suppose $P \cap \det(D) = \det(\Gamma_y)$ for every $y \in U^\varphi_{\mathbf{A}}$. Suppose also that $b'x = abxd$ for $a \in U^\varphi$, $d \in D$, and $b, b' \in B_x$. Then $\det(a) = \det(d^{-1}) \in P \cap \det(D) = \det(\Gamma_{bx})$, and so $\det(a) = \det(c)$ with $c \in \Gamma_{bx}$. Put $e = x^{-1}b^{-1}cbx$. Then $e \in D$, $\det(ed) = 1$, and $b'x = abxd = ac^{-1}bxed \in SU^\varphi bxD^1$. Thus $b' = b$. This shows that $U^\varphi SU^\varphi_{\mathbf{A}} xD = \bigsqcup_{b\in B_x} SU^\varphi bxD$, from which we obtain the equality in (5.9b). Also, $\nu(\Gamma^1_{bx})/\nu(\Gamma_{bx}) = [\Gamma_{bx} : \Gamma^1_{bx}] = \#\bigl(\det(\Gamma_{bx})\bigr) = \#\bigl(P \cap \det(D)\bigr)$, and so

$$\#\bigl(P \cap \det(D)\bigr)\mathfrak{m}(U^\varphi,\, D) = \sum_{x\in\Xi}\sum_{b\in B_x} \nu(\Gamma^1_{bx})$$
$$= \sum_{x\in\Xi} \mathfrak{m}(SU^\varphi,\, xD^1x^{-1}) = \#(\Xi)\cdot\mathfrak{m}(SU^\varphi,\, D^1),$$

which is (5.9d).

Department of Mathematics, Princeton University, Princeton, NJ

REFERENCES

[E] M. Eichler, Zur Zahlentheorie der Quaternionen-Algebren, J. reine angew. Math. 195 (1956), 127–151.

[L] W. Landherr, Äquivalenz Hermitscher Formen über einem beliebigen algebraischen Zahlkörper, Abh. Math. Sem. Hamb., 11 (1936), 245-248.

[S1] G. Shimura, Arithmetic of unitary groups, Ann. of Math., 79 (1964), 369–409 (=Collected Papers, I, 473-513).

[S2] G. Shimura, Euler Products and Eisenstein series, CBMS Regional Conference Series in Mathematics, No. 93, American Mathematical Society, 1997.

[S3] G. Shimura, Arithmetic and analytic theories of quadratic forms and Clifford groups, Mathematical Surveys and Monographs, vol. 109, American Mathematical Society, 2004.

[S4]G. Shimura, Quadratic Diophantine equations and orders in quaternion algebras, Amer. J. Math. 128 (2006), 481–518.

[S5] G. Shimura, Classification, construction, and similitudes of quadratic forms, Amer. J. Math. 128 (2006), 1521–1552.

[S6] G. Shimura, Integer-valued quadratic forms and quadratic Diophantine equations, Documenta Mathematica 11, (2006), 333-367.

[S7] G. Shimura, Quadratic Diophantine equations, the class number, and the mass formula, Bull. Amer. Math. Soc. 43 (2006), 285-304.

Goro Shimura
Department of Mathematics
Princeton University
Princeton
NJ 08544–1000
USA
goro@Math.Princeton.EDU

The Critical Values of Certain Dirichlet Series

Goro Shimura

Received: May 27, 2008

Communicated by Don Blasius

Abstract. We investigate the values of several types of Dirichlet series $D(s)$ for certain integer values of s, and give explicit formulas for the value $D(s)$ in many cases. The easiest types of D are Dirichlet L-functions and their variations; a somewhat more complex case involves elliptic functions. There is one new type that includes $\sum_{n=1}^{\infty}(n^2+1)^{-s}$ for which such values have not been studied previously.

2000 Mathematics Subject Classification: 11B68, 11M06, 30B50, 33E05.

Introduction

By a Dirichlet character modulo a positive integer d we mean as usual a $\mathbf{C}$-valued function χ on $\mathbf{Z}$ such that $\chi(x) = 0$ if x is not prime to d, and χ induces a character on $(\mathbf{Z}/d\mathbf{Z})^{\times}$. In this paper we always assume that χ is primitive and nontrivial, and so $d > 1$. For such a χ we put

$$L(s,\,\chi) = \sum_{n=1}^{\infty} \chi(n)n^{-s}. \tag{0.1}$$

It is well known that if k is a positive integer such that $\chi(-1) = (-1)^k$, then $L(k,\,\chi)$ is π^k times an algebraic number, or equivalently, $L(1-k,\,\chi)$ is an algebraic number. In fact, there is a well-known formula, first proved by Hecke in [3]:

$$kd^{1-k}L(1-k,\,\chi) = -\sum_{a=1}^{d-1} \chi(a)B_k(a/d), \tag{0.2}$$

where $B_k(t)$ is the Bernoulli polynomial of degree k. Actually Hecke gave the result in terms of $L(k,\,\chi)$, but here we state it in the above form. Hecke's proof is based on a classical formula

$$B_k(t) = -k!(2\pi i)^{-k} \sum_{0\neq h\in\mathbf{Z}} h^{-k}\mathbf{e}(ht) \qquad (0 < k \in \mathbf{Z},\, 0 < t < 1). \tag{0.3}$$

Shimura, Goro. "The critical values of certain Dirichlet series." Documenta Mathematica 13 (2008), 775–794.

There is also a well-known proof of (0.2), which is essentially the functional equation of $L(s, \chi)$ combined with a proof of (0.3). We will not discuss it in the present paper, as it is not particularly inspiring.

In [9] we gave many formulas for $L(1-k, \chi)$ different from (0.2). The primary purpose of the present paper is to give elementary proofs for some of them, as well as (0.2), and discuss similar values of a few more types of Dirichlet series. The point of our new proofs can be condensed to the following statement: *We find infinite sum expressions for* $L(s, \chi)$, *which are valid for all* $s \in \mathbf{C}$ *and so can be evaluated at* $s = 1-k$, *whereas the old proof of Hecke and our proofs in [9]* employ calculations at $s = k$ and involve the Gauss sum of $\overline{\chi}$.

To make our exposition smooth we put

$$\mathbf{e}(z) = \exp(2\pi i z) \qquad (z \in \mathbf{C}), \tag{0.4}$$

$$H = \{z \in \mathbf{C} \,|\, \mathrm{Im}(z) > 0\}. \tag{0.5}$$

The three additional types of Dirichlet series we consider naturally involve a complex variable s, and are defined as follows:

$$D^\nu(s;\, a,\, p) = \sum_{-a \neq n \in \mathbf{Z}} (n+a)^\nu |n+a|^{-\nu-s} \mathbf{e}\big(p(n+a)\big), \tag{0.6}$$

where $a \in \mathbf{R}$, $p \in \mathbf{R}$, and $\nu = 0$ or 1,

$$\mathcal{L}_k(s,\, z) = \sum_{m \in \mathbf{Z}} \mathbf{e}(mr)(z+m)^{-k}|z+m|^{-2s} \qquad (k \in \mathbf{Z},\ r \in \mathbf{Q},\ z \in H), \tag{0.7}$$

$$\varphi_\nu(u,\, s;\, L) = \sum_{\alpha \in L} (u+\alpha)^{-\nu}|u+\alpha|^{\nu-2s}, \tag{0.8}$$

where L is a lattice in $\mathbf{C}$, $0 \leq \nu \in \mathbf{Z}$, and $u \in \mathbf{C}, \notin L$. We should also note

$$\mathfrak{E}(z,\, s) = \mathrm{Im}(z)^s \sum_{(m,\, n)} (mz+n)^{-k}|mz+n|^{-2s} \qquad (0 \leq k \in \mathbf{Z},\ z \in H), \tag{0.9}$$

where (m, n) runs over the nonzero elements of $\mathbf{Z}^2$. The value $\mathfrak{E}(z, \mu)$ for an integer μ such that $1-k \leq \mu \leq 0$ was already discussed in [9], and so it is not the main object of study in this paper, but we mention it because (0.8) is a natural analogue of (0.9). We will determine in Section 3 the value $\varphi_\nu(u, \kappa/2; L)$ for an integer κ such that $2-\nu \leq \kappa \leq \nu$, which may be called a *nearly holomorphic elliptic function.* Now (0.6) is closely connected with $L(s, \chi)$. In [9, Theorem 4.2] we showed that $D^\nu(k; a, p)$ for $0 < k \in \mathbf{Z}$ is elementary factors times the value of a generalized Euler polynomial $E_{c,k-1}(t)$ at $t = p$. In Section 2 we will reformulate this in terms of $D^\nu(1-k; a, p)$. Finally, the nature of the series of (0.7) is quite different from the other types. We will show in Section 4 that $i^k\mathcal{L}_k(\beta, z)$ is a $\mathbf{Q}$-rational expression in π, $\mathbf{e}(z/N)$, and $\mathrm{Im}(z)$, if $\beta \in \mathbf{Z}$ and $-k < \beta \leq 0$, where N is the smallest positive integer such that $Nr \in \mathbf{Z}$. Similar results will also be given under other conditions on β. In the final section we will make some comments in the case where the base field is an algebraic number field.

1. $L(1-k,\, \chi)$

1.1. We start with an elementary proof of (0.2). Strange as it may sound, the main idea is the binomial theorem. We first note

$$B_n(t) = \sum_{\nu=0}^{n} \binom{n}{\nu} B_\nu t^{n-\nu} \qquad (0 \le n \in \mathbf{Z}), \tag{1.1}$$

$$B_0 = 1, \quad \zeta(0) = -1/2 = B_1, \tag{1.2a}$$

$$n\zeta(1-n) = -B_n \qquad (1 < n \in \mathbf{Z}), \tag{1.2b}$$

where B_n is the nth Bernoulli number. Formulas (1.1) and (1.2a) are well-known; (1.2b) is usually given only for even n, but actually true also for odd n, since $\zeta(-2m) = 0 = B_{2m+1}$ for $0 < m \in \mathbf{Z}$.

To prove (0.2), we first make a trivial calculation:

$$\begin{aligned} L(s,\, \chi) - \sum_{a=1}^{d-1} \chi(a) a^{-s} &= \sum_{m=1}^{\infty} \sum_{a=1}^{d-1} \chi(dm+a)(dm+a)^{-s} \\ &= \sum_{m=1}^{\infty} \sum_{a=1}^{d-1} \chi(a)(dm)^{-s} \left(1 + \frac{a}{dm}\right)^{-s}. \end{aligned}$$

Now we apply the binomial theorem to $(1+X)^{-s}$. Thus the last double sum equals

$$\begin{aligned} &\sum_{m=1}^{\infty} \sum_{a=1}^{d-1} \chi(a)(dm)^{-s} \sum_{r=0}^{\infty} \binom{-s}{r} \left(\frac{a}{dm}\right)^r \\ &= \sum_{r=0}^{\infty} \binom{-s}{r} \sum_{m=1}^{\infty} m^{-s-r} d^{-s} \sum_{a=1}^{d-1} \chi(a)(a/d)^r, \end{aligned} \tag{1.3}$$

where

$$\binom{\tau}{r} = \frac{\tau(\tau-1)\ldots(\tau-r+1)}{r!},$$

which is of course understood to be 1 if $r = 0$. So far our calculation is formal, but can be justified at least for $\mathrm{Re}(s) > 1$. Indeed, put $\mathrm{Re}(s) = \sigma$ and $|s| = \alpha$. Then

$$\left| \binom{-s}{r} \right| \le \frac{\alpha(\alpha+1)\ldots(\alpha+r-1)}{r!} = (-1)^r \binom{-\alpha}{r}. \tag{1.4}$$

Therefore the triple sum obtained from (1.3) by taking the absolute value of each term is majorized by

$$\begin{aligned} &\sum_{a=1}^{d-1} \sum_{r=0}^{\infty} \sum_{m=1}^{\infty} m^{-\sigma-r} d^{-\sigma} \binom{-\alpha}{r} \left(\frac{-a}{d}\right)^r \\ &\le \zeta(\sigma) d^{-\sigma} \sum_{a=1}^{d-1} \sum_{r=0}^{\infty} \binom{-\alpha}{r} \left(\frac{-a}{d}\right)^r = \zeta(\sigma) d^{-\sigma} \sum_{a=1}^{d-1} \left(1 - \frac{a}{d}\right)^{-\alpha} \end{aligned}$$

if $\sigma > 1$. Thus, for $\mathrm{Re}(s) > 1$, (1.3) can be justified, and so

$$L(s,\chi) - \sum_{a=1}^{d-1}\chi(a)a^{-s} = \sum_{r=0}^{\infty}\binom{-s}{r}\zeta(s+r)d^{-s}\sum_{a=1}^{d-1}\chi(a)(a/d)^r. \tag{1.5}$$

We can show that the last sum $\sum_{r=0}^{\infty}$ defines a meromorphic function in s on the whole $\mathbf{C}$. For that purpose given $s \in \mathbf{C}$, take a positive integer μ so that $\mathrm{Re}(s) > -\mu$ and decompose the sum into $\sum_{r=0}^{\mu+1}$ and $\sum_{r=\mu+2}^{\infty}$. There is no problem about the first sum, as it is finite. As for the latter, we have $|\zeta(s+r)| \le \zeta(2)$ for $r \ge \mu+2$. Putting $\varepsilon = (d-1)/d$, we have $|\sum_{a=1}^{d-1}\chi(a)(a/d)^r| \le (d-1)\varepsilon^r$. Therefore for $\mathrm{Re}(s) > -\mu$ the infinite sum $\sum_{r=\mu+2}^{\infty}$ can be majorized by

$$d^\mu(d-1)\zeta(2)\sum_{r=0}^{\infty}\binom{-\alpha}{r}(-\varepsilon)^r = d^\mu(d-1)\zeta(2)(1-\varepsilon)^{-\alpha}.$$

This proves the desired meromorphy of the right-hand side of (1.5).

Now, for $0 < k \in \mathbf{Z}$, without assuming that $\chi(-1) = (-1)^k$, we evaluate (1.5) at $s = 1-k$. We easily see that $\binom{-s}{r} = 0$ and $\zeta(s+r)$ is finite for $s = 1-k$ if $r > k$. We have to be careful about the term for $r = k$, as $\zeta(s+k)$ has a pole at $s = 1-k$. Since

$$\lim_{s\to 1-k}\binom{-s}{k}\zeta(s+k) = \lim_{s\to 1-k}\binom{-s}{k}\frac{1}{s-1+k} = \frac{-1}{k}, \tag{1.6}$$

the term for $r = k$ at $s = 1-k$ produces $-(d^{k-1}/k)\sum_{a=1}^{d-1}\chi(a)(a/d)^k$. Thus the evaluation of (1.5) at $s = 1-k$ gives

$$\begin{aligned} kd^{1-k}L(1-k,\chi) &= k\sum_{a=1}^{d-1}\chi(a)(a/d)^{k-1} \\ &- \sum_{a=1}^{d-1}\chi(a)(a/d)^k + \sum_{r=0}^{k-1}k\binom{k-1}{r}\zeta(1+r-k)\sum_{a=1}^{d-1}\chi(a)(a/d)^r. \end{aligned} \tag{1.7}$$

By (1.2b) we have, for $0 \le r < k-1$,

$$k\binom{k-1}{r}\zeta(1+r-k) = \frac{-k}{k-r}\binom{k-1}{r}B_{k-r} = -\binom{k}{r}B_{k-r}.$$

The term for $r = k-1$ produces $k\zeta(0)\sum_{a=1}^{d-1}\chi(a)(a/d)^{k-1}$, which combined with the first term on the right-hand side of (1.7) gives $-kB_1\sum_{a=1}^{d-1}\chi(a)(a/d)^{k-1}$. Thus we obtain

$$kd^{1-k}L(1-k,\chi) = -\sum_{r=0}^{k}\binom{k}{r}B_{k-r}\sum_{a=1}^{d-1}\chi(a)(a/d)^r,$$

which together with (1.1) proves (0.2). Notice that we did not assume that $\chi(-1) = (-1)^k$, and so we proved (0.2) for every positive integer k. If $\chi(-1) = (-1)^{k-1}$, we have $L(1-k,\chi) = 0$, which means that the right-hand side of (0.2) is 0 if $\chi(-1) = (-1)^{k-1}$. This can be proved more directly; see [9, (4.28)].

In the above calculation the term for $r = 0$ actually vanishes, as $\sum_{a=1}^{d-1}\chi(a) = 0$. However, we included the term for the following reason. In later subsections we will consider similar infinite sums with r ranging from 0 to ∞, of which the terms for $r = 0$ are not necessarily zero.

1.2. By the same technique as in §1.1 (that is, employing the binomial theorem) we will express $L(1-k, \chi)$ explicitly in terms of a polynomial Φ_{k-1} of degree $k-1$. Writing n for $k-1$, the polynomial is defined by

$$\Phi_n(t) = t^n - \sum_{\nu=1}^{[(n+1)/2]} \binom{n}{2\nu-1}(2^{2\nu}-1)\frac{B_{2\nu}}{\nu}t^{n+1-2\nu} \quad (0 \le n \in \mathbf{Z}), \tag{1.8}$$

where B_ν denotes the Bernoulli number as before. We understand that $\Phi_0(t) = 1$. We will eventually show that Φ_n is the classical Euler polynomial of degree n, but we prove Theorem 1.4 below with this definition of Φ_n, with no knowledge of the Euler polynomial. We first prove:

LEMMA 1.3. *Let χ be a primitive Dirichlet character of conductor $4d_0$ with $0 < d_0 \in \mathbf{Z}$. Then $\chi(a - 2d_0) = -\chi(a)$ for every $a \in \mathbf{Z}$.*

Proof. We may assume that a is prime to $2d_0$, as the desired equality is trivial otherwise. Then we can find an integer b such that $ab - 1 \in 4d_0\mathbf{Z}$, and we have $\chi(a - 2d_0) = \chi(a)\chi(1 - 2d_0b)$. Since $(1 - 2d_ob)^2 - 1 \in 4d_0\mathbf{Z}$, we have $\chi(1 - 2d_ob) = \pm 1$. Suppose $\chi(1 - 2d_ob) = 1$; let $x = 1 - 2d_0y$ with $y \in \mathbf{Z}$. Then $x^b - (1 - 2d_0b)^y \in 4d_0\mathbf{Z}$, and so $\chi(x)^b = 1$. Thus $\chi(x) = 1$, as b is odd. This shows that the conductor of χ is a divisor of $2d_0$, a contradiction. Therefore $\chi(1 - 2d_0b) = -1$, which proves the desired fact.

THEOREM 1.4. *Let χ be a nontrivial primitive Dirichlet character modulo d, and let k be a positive integer such that $\chi(-1) = (-1)^k$.*

(i) *If $d = 2q + 1$ with $0 < q \in \mathbf{Z}$, then*

$$L(1-k, \chi) = \frac{d^{k-1}}{2^k\chi(2) - 1}\sum_{b=1}^{q}(-1)^b\chi(b)\Phi_{k-1}(b/d). \tag{1.9}$$

(ii) *If $d = 4d_0$ with $1 < d_0 \in \mathbf{Z}$, then*

$$L(1-k, \chi) = (2d_0)^{k-1}\sum_{a=1}^{d_0-1}\chi(a)\Phi_{k-1}(2a/d). \tag{1.10}$$

Before proving these, we note that these formulas are *better than* (0.2) in the sense that $\Phi_{k-1}(t)$ is of degree $k-1$, whereas $B_k(t)$ is of degree k.

Proof. We first put

$$Z(s) = \sum_{n=1}^{\infty}(-1)^n n^{-s}, \qquad \Lambda(s) = \sum_{n=1}^{\infty}(-1)^n\chi(n)n^{-s}.$$

We easily see that

$$\Lambda(s) + L(s, \chi) = 2\sum_{n=1}^{\infty} \chi(2n)(2n)^{-s} = \chi(2)2^{1-s}L(s, \chi),$$

and a similar equality holds for $Z(s)$. Thus

$$Z(s) = \zeta(s)(2^{1-s} - 1), \qquad \Lambda(s) = L(s, \chi)\bigl\{\chi(2)2^{1-s} - 1\bigr\}.$$

We prove (i) by computing $\Lambda(1-k)$ for a given k in the same elementary way as we did in §1.1. With q as in (i) we observe that every positive integer m not divisible by d can be written uniquely $m = nd + a$ with $0 \le n \in \mathbf{Z}$ or $m = nd - a$ with $0 < n \in \mathbf{Z}$, where in either case a is in the range $0 < a \le q$. Therefore

$$\begin{aligned}\Lambda(s) &= \sum_{a=1}^{q}(-1)^a\chi(a)a^{-s}\\ &+ \sum_{a=1}^{q}\sum_{n=1}^{\infty}\Bigl\{(-1)^{nd+a}\chi(nd+a)(nd+a)^{-s} + (-1)^{nd-a}\chi(nd-a)(nd-a)^{-s}\Bigr\}.\end{aligned}$$

The last double sum can be written

$$\sum_{a=1}^{q}\sum_{n=1}^{\infty}(-1)^{n+a}d^{-s}n^{-s}\left\{\chi(a)\left(1+\frac{a}{nd}\right)^{-s} + \chi(-a)\left(1-\frac{a}{nd}\right)^{-s}\right\}.$$

Applying the binomial theorem to $(1 \pm X)^{-s}$, we obtain

$$\begin{aligned}\Lambda(s) &- \sum_{a=1}^{q}(-1)^a\chi(a)a^{-s}\\ &= \sum_{n=1}^{\infty}\sum_{r=0}^{\infty}(-1)^n(dn)^{-r-s}\binom{-s}{r}\{1+(-1)^{r+k}\}\sum_{a=1}^{q}(-1)^a\chi(a)a^r\\ &= \sum_{r=0}^{\infty}d^{-s}\binom{-s}{r}Z(s+r)\{1+(-1)^{r+k}\}\sum_{a=1}^{q}(-1)^a\chi(a)(a/d)^r.\end{aligned}$$

By the same technique as in §1.1, we can justify this for $\mathrm{Re}(s) > 1$. We can even show that the last sum $\sum_{r=0}^{\infty}$ is absolutely convergent for every $s \in \mathbf{C}$ as follows. We first note that Z is an entire function. Take a positive integer μ and s so that $\mathrm{Re}(s) > -\mu$. Then for $r \ge \mu+2$ we have $|Z(s+r)| \le \zeta(2)$. Put $|s| = \alpha$. We have also $|\sum_{a=1}^{q}(-1)^a\chi(a)(a/d)^r| \le 2^{-r}q$. Therefore for $\mathrm{Re}(s) > -\mu$ the infinite sum $\sum_{r=\mu+2}^{\infty}$ can be majorized by

$$2d^{\mu}\zeta(2)q\sum_{r=0}^{\infty}\binom{-\alpha}{r}(-2)^{-r} = 2d^{\mu}\zeta(2)q(1-2^{-1})^{-\alpha}.$$

This proves the desired convergence of $\sum_{r=0}^{\infty}$. Substituting $1-k$ for s in the above equality, we obtain $\Lambda(1-k)$ as an infinite sum, which is actually a finite sum, because $\binom{k-1}{r} = 0$ if $r \ge k$. (This time, the term $r = k$ causes no problem.) Also, we need only those r such that $k-r \in 2\mathbf{Z}$. Putting $k-r = 2\nu$, we find that

$$d^{1-k}L(1-k,\,\chi)\{\chi(2)2^k-1\} = d^{1-k}\Lambda(1-k)$$
$$= \sum_{a=1}^{q}(-1)^a\chi(a)(a/d)^{k-1} + 2\sum_{\nu=1}^{[k/2]}\binom{k-1}{2\nu-1}Z(1-2\nu)\sum_{a=1}^{q}(-1)^a\chi(a)(a/d)^{k-2\nu}.$$

From (1.2b) we obtain

$$(1.11)\qquad 2\nu Z(1-2\nu) = 2\nu(2^{2\nu}-1)\zeta(1-2\nu) = (1-2^{2\nu})B_{2\nu}.$$

Using this expression for $Z(1-2\nu)$, we obtain a formula for $L(1-k,\,\chi)$. Then comparison of it with our definition of $\varPhi_n$ proves (1.9).

Next, let $d = 4d_0$ with $1 < d_0 \in \mathbf{Z}$ as in (ii). Observe that the set of all positive integers greater than d_0 and not divisible by d_0 is the disjoint union of the sets

$$\{4\nu d_0 \pm a \,|\, 0 < a < d_0,\ 0 < \nu \in \mathbf{Z}\} \sqcup \{(4\nu+2)d_0 \pm a \,|\, 0 < a < d_0,\ 0 \le \nu \in \mathbf{Z}\}.$$

Clearly $\chi(4\nu d_0 \pm a) = \chi(\pm a)$; also $\chi\big((4\nu+2)d_0 \pm a\big) = -\chi(\pm a)$ by Lemma 1.3. Therefore we have

$$L(s,\,\chi) = \sum_{a=1}^{d_0-1}\chi(a)a^{-s}$$
$$+\sum_{\nu=1}^{\infty}\sum_{a=1}^{d_0-1}\{\chi(a)(4\nu d_0+a)^{-s} + \chi(-a)(4\nu d_0-a)^{-s}\}$$
$$-\sum_{\nu=0}^{\infty}\sum_{a=1}^{d_0-1}\{\chi(a)\big((4\nu+2)d_0+a\big)^{-s} + \chi(-a)\big((4\nu+2)d_0-a\big)^{-s}\}.$$

Employing the binomial theorem in the same manner as before, we have

$$L(s,\,\chi) - \sum_{a=1}^{d_0-1}\chi(a)a^{-s}$$
$$= \sum_{\nu=1}^{\infty}\sum_{r=0}^{\infty}\binom{-s}{r}(4\nu d_0)^{-s-r}\{1+(-1)^{k+r}\}\sum_{a=1}^{d_0-1}\chi(a)a^r$$
$$-\sum_{\nu=0}^{\infty}\sum_{r=0}^{\infty}\binom{-s}{r}\big((4\nu+2)d_0\big)^{-s-r}\{1+(-1)^{k+r}\}\sum_{a=1}^{d_0-1}\chi(a)a^r.$$

Notice that $\sum_{\nu=1}^{\infty}(4\nu)^{-s} - \sum_{\nu=0}^{\infty}(4\nu+2)^{-s} = 2^{-s}Z(s)$. Therefore

$$L(s,\,\chi) = \sum_{a=1}^{d_0-1}\chi(a)a^{-s} + \sum_{r=0}^{\infty}\binom{-s}{r}(2d_0)^{-s-r}Z(s+r)\{1+(-1)^{k+r}\}\sum_{a=1}^{d_0-1}\chi(a)a^r.$$

The validity of this formula for all $s \in \mathbf{C}$ can be proved in the same way as in the previous case. The last infinite sum $\sum_{r=0}^{\infty}$ evaluated at $s = 1-k$ becomes a finite sum $\sum_{r=0}^{k-1}$, which is actually extended only over those r such that $k - r = 2\nu$ with $\nu \in \mathbf{Z}$. Therefore, using (1.11), we obtain (1.10).

1.5. Let us now show that Φ_n coincides with the classical Euler polynomial. In [9, (4.2)] we defined polynomials $E_{c,\,n}(t)$ for $c = -\mathbf{e}(\alpha)$ with $\alpha \in \mathbf{R}, \notin \mathbf{Z}$, by

$$\frac{(1+c)e^{tz}}{e^z + c} = \sum_{n=0}^{\infty} \frac{E_{c,n}(t)}{n!} z^n. \tag{1.12}$$

If $c = 1$, the polynomial $E_{1,n}(t)$ is the classical Euler polynomial of degree n. Our task is to prove

$$E_{1,n} = \Phi_n. \tag{1.13}$$

We first note here some basic formulas:

$$E_{c,n}(t) = (1 + c^{-1})n!(2\pi i)^{-n-1} \sum_{h\in\mathbf{Z}} (h+\alpha)^{-n-1}\mathbf{e}\big((h+\alpha)t\big) \tag{1.14}$$

$$(c = -\mathbf{e}(\alpha),\ \alpha \in \mathbf{R}, \notin \mathbf{Z};\ 0 < t < 1 \text{ if } n = 0;\ 0 \le t \le 1 \text{ if } 0 < n \in \mathbf{Z}),$$

$$E_{c,n}(t + r) = \sum_{k=0}^{n} \binom{n}{k} E_{c,k}(r)t^{n-k} \qquad (0 \le n \in \mathbf{Z}), \tag{1.15}$$

$$E_{1,0}(0) = 1, \quad E_{1,n}(0) = 2(1 - 2^{n+1})(n+1)^{-1}B_{n+1} \quad (0 < n \in \mathbf{Z}). \tag{1.16}$$

Formula (1.14) was given in [9, (4.5)]; the sum $\sum_{h\in\mathbf{Z}}$ means $\lim_{m\to\infty} \sum_{|h|\le m}$ if $n = 0$. Replacing t in (1.12) by $t+r$ and making an obvious calculation, we obtain (1.15). We have $E_{c,0}(t) = 1$ as noted in [9, (4.3h)]. Clearly $E_{1,n}(0) = 0$ if n is even. Assuming n to be odd, take $t = 0$ and $\alpha = 1/2$ in (1.14), and recall that $2 \cdot m!(2\pi i)^{-m}\zeta(m) = -B_m$ if $0 < m \in 2\mathbf{Z}$. Then we obtain $E_{1,n}(0)$ as stated in (1.16). Taking $r = 0$ in (1.15) and using (1.16), we obtain (1.13). The value $E_{c,n}(0)$ for an arbitrary c is given in [9, (4.6)].

1.6. In [9, Theorem 4.14] we proved, for χ, d, and k as in Theorem 1.4,

$$L(1-k,\,\chi) = \frac{d^{k-1}}{2^k - \overline{\chi}(2)} \sum_{a=1}^{q} \chi(a)E_{1,k-1}(2a/d), \tag{1.17}$$

where $q = [(d-1)/2]$, and derived (i) and (ii) above, with $E_{1,k-1}$ in place of Φ_{k-1}, from (1.17). In fact, (i) and (ii) combined are equivalent to (1.17). Though this is essentially explained in [9, p. 36], here let us show that (1.17) for even d follows from (ii). With $d = 4d_0$ as before, we have $[(d-1)/2] = 2d_0 - 1$ and

$$\sum_{a=1}^{2d_0-1} \chi(a)\Phi_{k-1}(2a/d)$$
$$= \sum_{a=1}^{d_0-1} \Big\{\chi(a)\Phi_{k-1}(2a/d) + \chi(2d_0 - a)\Phi_{k-1}\big(2(2d_0 - a)/d\big)\Big\}.$$

We have $E_{1,n}(1-t) = (-1)^n E_{1,n}(t)$ as noted in [9, (4.3f)]. This combined with (1.13) shows that $\Phi_{k-1}(1-t) = (-1)^{k-1}\Phi_{k-1}(t)$. By Lemma 1.3, we have $\chi(2d_0 - a) = -\chi(-a) = (-1)^{k+1}\chi(a)$, and so the last sum equals

$$2\sum_{a=1}^{d_0-1}\chi(a)\Phi_{k-1}(2a/d).$$

Therefore (1.17) follows from (1.10) if $d = 4d_0$. Similarly we can derive(1.17) for odd d from (1.9), which, in substance, is shown in the last paragraph of [9, p. 36].

1.7. Our technique is applicable even to $\zeta(1-k)$. Instead of $\zeta(s)$ we consider $W(s) = \sum_{m=0}^{\infty}(2m+1)^{-s}$. We have clearly

$$\begin{aligned} W(s) &= 1 + \sum_{m=1}^{\infty}(2m+1)^{-s} = 1 + \sum_{m=1}^{\infty}(2m)^{-s}\left(1+\frac{1}{2m}\right)^{-s} \\ &= 1 + \sum_{m=1}^{\infty}(2m)^{-s}\sum_{r=0}^{\infty}\binom{-s}{r}(2m)^{-r} = 1 + \sum_{r=0}^{\infty}\zeta(s+r)\binom{-s}{r}2^{-s-r}. \end{aligned}$$

We evaluate this at $s = 1-k$ with $0 < k \in \mathbf{Z}$. Our calculation is similar to that of §1.1; we use (1.6) for determining the term for $r = k$, which produces $-(2k)^{-1}$. Thus

$$(1-2^{k-1})\zeta(1-k) = W(1-k) = 1 - \frac{1}{2k} + \sum_{r=0}^{k-1}\binom{k-1}{r}2^{k-1-r}\zeta(1-k+r).$$

Taking $k = 1$, we find a well-known fact $\zeta(0) = -1/2$. Also, $\zeta(1-k)$ appears on both sides. Therefore, putting $k - r = t$ and rearranging our sum, we obtain

$$(1-2^k)\zeta(1-k) = \frac{k-1}{2k} + \sum_{t=2}^{k-1}\binom{k-1}{t-1}2^{t-1}\zeta(1-t).$$

This holds for every even or odd integer $k > 1$. Recall that $\zeta(-m) = 0$ for $0 < m \in 2\mathbf{Z}$. Thus, taking $k = 2n$ with $0 < n \in \mathbf{Z}$, we obtain a formula for $\zeta(1-2n)$ as a linear combination of $\zeta(1-2\nu)$ for $1 \le \nu < n$ (which is 0 if $n = 1$) plus a constant as follows:

$$(1.18)\qquad (1-2^{2n})\zeta(1-2n) = \frac{2n-1}{4n} + \sum_{\nu=1}^{n-1}\binom{2n-1}{2\nu-1}2^{2\nu-1}\zeta(1-2\nu).$$

Similarly, taking $k = 2n+1$ and putting $t = 2\nu$, we obtain

$$(1.19)\qquad \sum_{\nu=1}^{n}\binom{2n}{2\nu-1}2^{2\nu-1}\zeta(1-2\nu) = \frac{-n}{2n+1}.$$

Either of these equalities (1.18) and (1.19) expresses $\zeta(1-2n)$ as a $\mathbf{Q}$-linear combination of $\zeta(1-2\nu)$ for $1 \le \nu < n$ plus a constant. The two expressions are different, as can easily be seen.

In [9, (11.8)] we gave a similar recurrence formula which can be written

$$(1.20)\quad 4(1-2^{n+1})\zeta(-n) = 1 + 2\sum_{k=2}^{n}\binom{n}{k-1}(2^k-1)\zeta(1-k) \qquad (0 < n \in \mathbf{Z}).$$

Taking n to be even or odd, we again obtain two different recurrence formulas for $\zeta(1-2n)$. It should be noted that the technique of using the binomial theorem is already in §68 of Landau [5], in which $(s-1)\zeta(s)$ is discussed, while we employ $W(s)$.

2. EXTENDING THE PARAMETERS c AND n IN $E_{c,n}$

2.1. The function $E_{c,n}(t)$ is a polynomial in t of degree n, and involves $c = -\mathbf{e}(\alpha)$ with $\alpha \in \mathbf{R}$. We now extend this in two ways: first, we take $\alpha \in \mathbf{C}, \notin \mathbf{Z}$; second, we consider $(h+\alpha)^{-s}$ instead of $(h+\alpha)^{-n-1}$. The first case is simpler. Since $E_{c,n}(t)$ is a polynomial in t and $(1+c)^{-1}$ as noted in [9, p. 26], we can define a function $\mathcal{E}_n(\alpha, t)$ by

$$\mathcal{E}_n(\alpha, t) = E_{c,n}(t), \quad c = -\mathbf{e}(\alpha), \ \alpha \in \mathbf{C}, \notin \mathbf{Z}, \ 0 \le n \in \mathbf{Z}. \tag{2.1}$$

This is a polynomial in t, whose coefficients are holomorphic functions in $\alpha \in \mathbf{C}, \notin \mathbf{Z}$. Now equality (1.14) can be extended to

$$\mathcal{E}_n(\alpha, t) = \bigl(1 - \mathbf{e}(-\alpha)\bigr) n! (2\pi i)^{-n-1} \sum_{h \in \mathbf{Z}} (h+\alpha)^{-n-1} \mathbf{e}\bigl((h+\alpha)t\bigr) \tag{2.2}$$

for all $\alpha \in \mathbf{C}, \notin \mathbf{Z}$, where $0 < t < 1$ if $n = 0$, and $0 \le t \le 1$ if $n > 0$. Indeed, if $n > 0$, the right-hand side is absolutely convergent, and defines a holomorphic function. Since (2.2) holds for $\alpha \in \mathbf{R}, \notin \mathbf{Z}$, we obtain (2.2) as expected. If $n = 0$, we have to consider $\lim_{m\to\infty} \sum_{|h| \le m} (h+\alpha)^{-1} \mathbf{e}\bigl((h+\alpha)t\bigr)$. Clearly

$$\sum_{h=-m}^{m} \frac{\mathbf{e}(ht)}{\alpha + h} = \frac{1}{\alpha} + \sum_{h=1}^{m} \frac{2\alpha \cdot \cos(2\pi h t)}{\alpha^2 - h^2} + 2i \sum_{h=1}^{m} \frac{h \cdot \sin(2\pi h t)}{h^2 - \alpha^2}.$$

The last sum on the right-hand side equals

$$\sum_{h=1}^{m} \frac{\sin(2\pi h t)}{h} + \sum_{h=1}^{m} \frac{\sin(2\pi h t)\alpha^2}{h(h^2 - \alpha^2)}.$$

It is well-known that the first sum tends to a finite value as $m \to \infty$. Obviously the last sum converges to a holomorphic function in $\alpha \in \mathbf{C}, \notin \mathbf{Z}$ as $m \to \infty$. Thus we can justify (2.2) for $n = 0$.

Formula (2.2) for $n = 0$ (with $-\alpha$ in place of α) can be written

$$\frac{\mathbf{e}(t\alpha)}{1 - \mathbf{e}(\alpha)} = \frac{1}{2\pi i} \sum_{h \in \mathbf{Z}} \frac{\mathbf{e}(th)}{h - \alpha} \qquad (\alpha \in \mathbf{C}, \notin \mathbf{Z}, \ 0 < t < 1). \tag{2.3}$$

This was first given by Kronecker [4].

2.2. We next ask if the power $(h+a)^{-n-1}$ in (1.14) can be replaced by $(h+a)^{-s}$ with a complex parameter s. Since $h+a$ can be negative, $(h+a)^{-s}$ is not suitable. Thus, for $s \in \mathbf{C}$, $a \in \mathbf{R}$, $p \in \mathbf{R}$, and $\nu = 0$ or 1 we put

$$D^{\nu}(s;\, a,\, p) = \sum_{-a \ne n \in \mathbf{Z}} (n+a)^{\nu} |n+a|^{-\nu-s} \mathbf{e}\bigl(p(n+a)\bigr), \tag{2.4}$$

$$T^\nu(s;\,a,\,p) = \Gamma\big((s+\nu)/2\big)\pi^{-(s+\nu)/2}D^\nu(s;\,a,\,p). \tag{2.5}$$

Clearly the infinite series of (2.4) is absolutely convergent for $\mathrm{Re}(s) > 1$, and defines a holomorphic function of s there. Notice that if $k-\nu \in 2\mathbf{Z}$, then $D^\nu(k;\,0,\,t) = \sum_{0\neq n\in\mathbf{Z}} n^{-k}\mathbf{e}(nt)$, which is the infinite sum of (0.3). Thus the Bernoulli polynomials are included in our discussion.

THEOREM 2.3. *The function $T^\nu(s;\,a,\,p)$ can be continued as a meromorphic function of s to the whole $\mathbf{C}$. It is entire if $\nu = 1$. If $\nu = 0$, then $T^0(s;\,a,\,p)$ is*

$$\frac{-2\delta(a)}{s} + \frac{2\mathbf{e}(ap)\delta(p)}{s-1}$$

plus an entire function, where $\delta(x) = 1$ if $x \in \mathbf{Z}$ and $\delta(x) = 0$ if $x \notin \mathbf{Z}$. Moreover,

$$T^\nu(1-s;\,a,\,p) = i^{-\nu}\mathbf{e}(ap)T^\nu(s;\,-p,\,a). \tag{2.6}$$

Proof. Put $\varphi(x) = x^\nu\mathbf{e}(-x^2z^{-1}/2 + px)$ for $x \in \mathbf{R}$ and $z \in H$. Denote by $\widehat{\varphi}$ the Fourier transform of φ. Then from [9, (2.25)] we easily obtain $\widehat{\varphi}(x) = i^{-\nu}(-iz)^\kappa(x-p)^\nu\mathbf{e}\big((x-p)^2z/2\big)$, where $\kappa = \nu + 1/2$. Put also

$$f(z) = \sum_{n\in\mathbf{Z}}(n+a)^\nu\mathbf{e}\big((n+a)^2z/2 + p(n+a)\big),$$

and $f^\#(z) = (-iz)^{-\kappa}f(-z^{-1})$. Then $f(-z^{-1}) = \sum_{n\in\mathbf{Z}}\varphi(n+a)$, which equals $\sum_{m\in\mathbf{Z}}\mathbf{e}(ma)\widehat{\varphi}(m)$ by virtue of the Poisson summation formula. In this way we obtain

$$f^\#(z) = i^{-\nu}\sum_{m\in\mathbf{Z}}\mathbf{e}(ma)(m-p)^\nu\mathbf{e}\big((m-p)^2z/2\big).$$

Now $T^\nu(2s-\nu;\,a,\,p)$ is the Mellin transform of $f(iy)$, and so we obtain our theorem by the general principle of Hecke, which is given as Theorem 3.2 in [9].

THEOREM 2.4. *For $\nu = 0$ or 1, $0 \le a \le 1$, and a positive integer k such that $k-\nu \in 2\mathbf{Z}$ we have*

$$D^0(0;\,a,\,p) = -\delta(a), \tag{2.7}$$

$$D^\nu(\nu-2m;\,a,\,p) = 0 \quad \text{if} \quad 0 < m \in \mathbf{Z}, \tag{2.8}$$

$$D^\nu(1-k;\,a,\,p) = 2(2\pi i)^{-k}(k-1)!\mathbf{e}(ap)D^\nu(k;\,-p,\,a), \tag{2.9}$$

$$D^\nu(1-k;\,a,\,p) = -2\mathbf{e}(ap)B_k(a)/k \quad \text{if} \quad p \in \mathbf{Z}, \tag{2.10}$$

$$D^\nu(1-k;\,a,\,p) = \frac{2\mathbf{e}(ap)}{1-\mathbf{e}(p)}E_{c,k-1}(a) \quad \text{if} \quad p \notin \mathbf{Z}, \tag{2.11}$$

where $c = -\mathbf{e}(-p)$, and we have to assume that $0 < a < 1$ in (2.10) and (2.11) if $k = 1$.

Proof. By Theorem 2.3, $\big[sT^0(s;\,a,\,p)\big]_{s=0} = -2\delta(a)$, from which we obtain (2.7). Next, let $0 < m \in \mathbf{Z}$. Since $\Gamma\big((s+\nu)/2\big)D^\nu(s;\,a,\,p)$ is finite and

$\Gamma\big((s+\nu)/2\big)$ has a pole at $s = \nu - 2m$, we obtain (2.8). We easily see that $\Gamma(1/2-m) = \pi^{1/2}(-2)^m \prod_{t=1}^{m}(2t-1)^{-1}$. Therefore from (2.6) we obtain (2.9). If $p \in \mathbf{Z}$, then $D^\nu(k;\,-p,\,a) = \sum_{0\neq n\in\mathbf{Z}} n^{-k}\mathbf{e}(an)$. The well-known classical formula, stated in [9, (4.9)] (and also as (0.3)), shows that the last sum equals $-(2\pi i)^k B_k(a)/k!$ for $0 \le a \le 1$ if $k > 1$, and for $0 < a < 1$ if $k = 1$. If $p \notin \mathbf{Z}$, then $D^\nu(k;\,-p,\,a) = \sum_{n\in\mathbf{Z}}(n-p)^{-k}\mathbf{e}\big(a(n-p)\big)$. By (1.14), this equals $(1+c^{-1})^{-1}(2\pi i)^k E_{c,k-1}(a)/(k-1)!$, where $c = -\mathbf{e}(-p)$, under the same condition on a. Combining these with (2.9), we obtain (2.10) and (2.11).

We note here a special case of (2.10):

$$D^\nu(1-k;\,0,\,0) = \begin{cases} -2B_k/k & \text{if } k > 1, \\ 0 & \text{if } k = 1. \end{cases} \tag{2.12}$$

It should be noted that $D^1(s;\,0,\,0) = 0$.

3. Nearly holomorphic elliptic functions

3.1. Let L be a lattice in $\mathbf{C}$. As an analogue of (2.4) we put

$$\varphi_\nu(u,\,s;\,L) = \sum_{\alpha\in L}(u+\alpha)^{-\nu}|u+\alpha|^{\nu-2s} \tag{3.1}$$

for $0 \le \nu \in \mathbf{Z}$, $u \in \mathbf{C}$, $\notin L$, and $s \in \mathbf{C}$. Clearly

$$\varphi_\nu(\lambda u,\,s;\,\lambda L) = \lambda^{-\nu}|\lambda|^{\nu-2s}\varphi_\nu(u,\,s;\,L) \quad \text{for every} \quad \lambda \in \mathbf{C}^\times, \tag{3.2a}$$

$$\varphi_\nu(u+\alpha,\,s;\,L) = \varphi_\nu(u,\,s;\,L) \quad \text{for every} \quad \alpha \in L. \tag{3.2b}$$

If L is a $\mathbf{Z}$-lattice in an imaginary quadratic field K and $u \in K$, (3.1) is the same as the series of [9, (7.1)]. The analytic properties of the series that we proved there can easily be extended to the case of (3.1). First of all, the right-hand side of (3.1) is absolutely convergent for $\mathrm{Re}(s) > 1$, and defines a holomorphic function of s there.

THEOREM 3.2. *Put $\Phi(u,\,s) = \pi^{-s}\Gamma(s+\nu/2)\varphi_\nu(u,\,s;\,L)$. Then $\Phi(u,\,s)$ can be continued to the whole s-plane as a meromorphic function in s, which is entire if $\nu > 0$. If $\nu = 0$, then $\Phi(u,\,s)$ is an entire function of s plus $v(L)^{-1}/(s-1)$, where $v(L) = \mathrm{vol}(\mathbf{C}/L)$. Moreover, $\Phi(u,\,s)$ is a C^∞ function in u, except when $\nu = 0$ and $s = 1$, and each derivative $(\partial/\partial u)^a(\partial/\partial\overline{u})^b\Phi(u,\,s)$ is meromorphic in s on the whole $\mathbf{C}$.*

Proof. This can be proved by the same argument as in [9, §7.2], except for the differentiability with respect to u and the last statement about the derivatives, which can be shown as follows. As shown in the proof of [9, Theorem 3.2], the product $\pi^{-s}\Gamma(s)\varphi_\nu(u,\,s-\nu/2;\,L)$ minus the pole part can be written

$$\int_p^\infty F(u,\, y)y^{s-1}dy + \int_p^\infty G(u,\, y)y^{\nu-s}dy,$$

where

$$F(u,\, y) = \sum_{\alpha\in L} (\overline{u}+\alpha)^\nu \exp\big(-\pi|u+\alpha|^2 y\big),$$

$$G(u,\, y) = A\sum_{\beta\in B} \exp\big(\pi i(\beta\overline{u}+\overline{\beta}u)\big) \sum_{\xi-\beta\in M} \xi^\nu \exp\big(-\pi|\xi|^2 y\big)$$

with a constant A, a finite subset B of $\mathbf{C}$, a positive constant p, and lattices L and M in $\mathbf{C}$. Therefore the differentiability and the last statement follow from the standard fact on differentiation under the integral sign.

3.3. Before stating the next theorem, we note a few elementary facts. Take L in the form $L = \mathbf{Z}\omega_1 + \mathbf{Z}\omega_2$ with complex numbers ω_1 and ω_2 such that $\omega_1/\omega_2 \in H$. We put then $v(\omega_1,\, \omega_2) = v(L)$. It can easily be seen that

$$v(\omega_1,\, \omega_2) = |\omega_2|^2\mathrm{Im}(\omega_1/\omega_2) = (2i)^{-1}(\omega_1\overline{\omega}_2 - \overline{\omega}_1\omega_2), \tag{3.3}$$

and in particular, $v(z,\, 1) = \mathrm{Im}(z)$. We also recall the function ζ of Weierstrass defined by

$$\zeta(u) = \zeta(u;\, \omega_1,\, \omega_2) = \frac{1}{u} + \sum_{0\neq\alpha\in L}\left\{\frac{1}{u-\alpha} + \frac{1}{\alpha} + \frac{u}{\alpha^2}\right\}, \tag{3.4}$$

where $L = \mathbf{Z}\omega_1 + \mathbf{Z}\omega_2$. It is well known that

$$\zeta(-u) = -\zeta(u), \qquad (\partial/\partial u)\zeta(u;\, \omega_1,\, \omega_2) = -\wp(u;\, \omega_1,\, \omega_2) \tag{3.5}$$

with the Weierstrass function $\wp$. We put as usual

$$\eta_\mu(\omega_1,\, \omega_2) = 2\zeta(\omega_\mu/2) \qquad (\mu = 1,\, 2). \tag{3.6a}$$

Then

$$\zeta(u+\omega_\mu) = \zeta(u) + \eta_\mu(\omega_1,\, \omega_2). \tag{3.6b}$$

We also need the classical nonholomorphic Eisenstein series E_2 of weight 2, which can be given by

$$E_2(z) = \frac{1}{8\pi y} - \frac{1}{24} + \sum_{n=1}^\infty\Bigg(\sum_{0<d|n} d\Bigg)\mathbf{e}(nz). \tag{3.7}$$

We are interested in the value of $\varphi_\nu(u,\, s;\, L)$ at $s = \nu/2$, which is meaningful for every $\nu \in \mathbf{Z}$, > 0, by Theorem 3.2. The results can be given as follows.

THEOREM 3.4. *For $L = \mathbf{Z}\omega_1 + \mathbf{Z}\omega_2$ with $\omega_1/\omega_2 \in H$ we have*

$$\varphi_\nu(u,\, \nu/2;\, L) = \frac{(-1)^\nu}{(\nu-1)!}\frac{\partial^{\nu-2}}{\partial u^{\nu-2}}\wp(u;\, \omega_1,\, \omega_2) \qquad (2<\nu\in\mathbf{Z}), \tag{3.8}$$

$$\varphi_2(u,\, 1;\, L) = \wp(u;\, \omega_1,\, \omega_2) - 8\pi^2\omega_2^{-2}E_2(\omega_1/\omega_2), \tag{3.9}$$

$$\varphi_1(u,\, 1/2;\, L) = \zeta(u) + 8\pi^2\omega_2^{-2}E_2(\omega_1/\omega_2)u - \pi v(L)^{-1}\overline{u}, \tag{3.10}$$

$$(3.11) \qquad \eta_\mu(\omega_1, \omega_2) = \pi\overline{\omega}_\mu v(L)^{-1} - 8\pi^2\omega_2^{-2}E_2(\omega_1/\omega_2)\omega_\mu \qquad (\mu = 1, 2).$$

Proof. If $\nu > 2$, then clearly $\varphi_\nu(u, \nu/2; L) = \sum_{\alpha\in L}(u+\alpha)^{-\nu}$, from which we obtain (3.8). The cases $\nu = 2$ and $\nu = 1$ are more interesting. We first note that

$$(3.12a) \qquad (\partial/\partial u)\varphi_\nu(u, s; L) = (-s - \nu/2)\varphi_{\nu+1}(u, s + 1/2; L),$$

$$(3.12b) \qquad (\partial/\partial \overline{u})\varphi_\nu(u, s; L) = (-s + \nu/2)\varphi_{\nu-1}(u, s + 1/2; L),$$

at least for sufficiently large $\mathrm{Re}(s)$. Since both sides of (3.12a, b) are meromorphic in s on the whole $\mathbf{C}$, we obtain (3.12a, b) for every s. The first formula with $\nu = 2$ produces

$$(\partial/\partial u)\varphi_2(u, 1; L) = -2\varphi_3(u, 3/2; L) = (\partial/\partial u)\wp(u; \omega_1, \omega_2),$$

from which we obtain $\varphi_2(u, 1; L) = \wp(u; \omega_1, \omega_2) + c(\overline{u})$ with an anti-holomorphic function $c(\overline{u})$. Since (3.12b) shows that $\varphi_2(u, 1; L)$ is holomorphic in u, we see that $c(\overline{u})$ does not involve u or $\overline{u}$, that is, it is a constant depending only on L. Suppose $L = \mathbf{Z}z + \mathbf{Z}$ with $z \in H$. For $0 < N \in \mathbf{Z}$ and $(p, q) \in \mathbf{Z}^2$, $\notin N\mathbf{Z}^2$ define a standard Eisenstein series $\mathfrak{E}_\nu^N(z, s; p, q)$ of level N by [9, (9.1)]. Then we easily see that

$$\varphi_\nu\big((pz+q)/N, s; L\big) = N^{2s}y^{\nu/2-s}\mathfrak{E}_\nu^N(z, s - \nu/2; p, q),$$
$$\varphi_\nu\big((pz+q)/N, \nu/2; L\big) = N^{\nu}\mathfrak{E}_\nu^N(z, 0; p, q).$$

Define F_ν and $\mathcal{F}_2$ as in [9, (10.10b, c, d)]. Taking $\nu = 2$, we obtain

$$\varphi_2\big((pz+q)/N, 1; L\big) = N^2\mathfrak{E}_2^N(z, 0; p, q) = (2\pi i)^2\mathcal{F}_2(z; p/N, q/N).$$

By [9, (10.13)], $\mathcal{F}_2(z; a, b) = (2\pi i)^{-2}\wp\big(az + b; z, 1\big) + 2E_2(z)$ with E_2 of (3.7). Therefore we can conclude that

$$(3.13) \qquad \varphi_2(u, 1; \mathbf{Z}z + \mathbf{Z}) = \wp(u; z, 1) - 8\pi^2E_2(z).$$

More generally, using (3.2a) we obtain (3.9).

We next consider the case $\nu = 1$. Since $(\partial/\partial u)\zeta(u; \omega_1, \omega_2) = -\wp(u; \omega_1, \omega_2)$, from (3.9) and (3.12a) we obtain

$$\begin{aligned}(\partial/\partial u)\varphi_1(u, 1/2; L) &= -\varphi_2(u, 1; L)\\ &= (\partial/\partial u)\zeta(u; \omega_1, \omega_2) + 8\pi^2\omega_2^{-2}E_2(\omega_1/\omega_2).\end{aligned}$$

We have also

$$(\partial/\partial \overline{u})\varphi_1(u, 1/2; L) = \lim_{\sigma\to 1}(1-\sigma)\varphi_0(u, \sigma; L) = -\pi/v(L),$$

since the residue of $\pi^{-s}\Gamma(s)\varphi_0(u, s; L)$ at $s = 1$ is $v(L)^{-1}$ as shown in Theorem 3.2. Therefore $\varphi_1(u, 1/2; L) = -\pi\overline{u}/v(L) + g(u)$ with a function g holomorphic in u. Clearly $\partial g/\partial u = (\partial/\partial u)\varphi_1(u, 1/2; L)$, and so we can conclude that

$$\varphi_1(u,\,1/2;\,L) = \zeta(u) + 8\pi^2\omega_2^{-2}E_2(\omega_1/\omega_2)u - \pi v(L)^{-1}\overline{u} + \xi(L) \tag{3.14}$$

with a constant $\xi(L)$ independent of u. From (3.2a) we obtain $\varphi_1(-u,\,s;\,L) = -\varphi_1(u,\,s;\,L)$. Also $\zeta(-u) = -\zeta(u)$. Thus $\xi(L) = 0$, and consequently we obtain (3.10). Since $\varphi_1(u,\,1/2;\,L)$ is invariant under $u \mapsto u + \omega_\mu$, we obtain (3.11) from (3.10) and (3.6b).

3.5. In [9] we discussed the value of an Eisenstein series $E(z,\,s)$ of weight k at $s = -m$ for an integer m such that $0 \le m \le k-1$, and observed that it is nearly holomorphic in the sense that it is a polynomial in y^{-1} with holomorphic functions as coefficients; for a precise statement, see [9, Theorem 9.6]. As an analogue we investigate $\varphi_\nu(u,\,\kappa/2;\,L)$ for an integer κ such that $2-\nu \le \kappa \le \nu$ and $\kappa - \nu \in 2\mathbf{Z}$. From (3.12b) we obtain, for $0 \le a \in \mathbf{Z}$,

$$(\partial/\partial\overline{u})^a\varphi_\nu(u,\,(\nu/2)-a;\,L) = a!\cdot\varphi_{\nu-a}(u,\,(\nu-a)/2;L). \tag{3.15}$$

THEOREM 3.6. *Let κ be an integer such that $2-\nu \le \kappa \le \nu$ and $\kappa-\nu \in 2\mathbf{Z}$. Then $\varphi_\nu(u,\,\kappa/2;L)$ is a polynomial in $\overline{u}$ of degree d with holomorphic functions in u as coefficients, where $d = (\nu-\kappa)/2$ if $\nu+\kappa \ge 4$ and $d = (\nu-\kappa+2)/2$ if $\nu+\kappa = 2$. The leading term is $\overline{u}^d\varphi_{(\nu+\kappa)/2}(u,\,(\nu+\kappa)/4;\,L)$ or $-\pi d^{-1}v(L)^{-1}\overline{u}^d$ according as $\nu+\kappa \ge 4$ or $\nu+\kappa = 2$.*

Proof. Given κ as in the theorem, put $a = (\nu-\kappa)/2$. Then $(\nu/2)-a = \kappa/2$ and $\nu - a = (\nu+\kappa)/2 \ge 1$. If $\nu - a \ge 2$, then by Theorem 3.4, $\varphi_{\nu-a}(u,\,(\nu-a)/2;L)$ is holomorphic in u, and so (3.15) shows that $\varphi_\nu(u,\,\kappa/2;\,L)$ is a polynomial in $\overline{u}$ of degree a with holomorphic functions in u as coefficients. If $\nu-a = 1$, the function $\varphi_1(u,\,1/2;\,L)$ is linear in $\overline{u}$ as given in (3.10). Therefore we obtain our theorem.

Thus, we may call $\varphi_\nu(u,\,\kappa/2;\,L)$ a *nearly holomorphic elliptic function.* In the higher-dimensional case it is natural to consider theta functions instead of periodic functions. For details of the basic ideas and results on this the reader is referred to [6] and [7].

4. THE SERIES WITH A PARAMETER IN H

4.1. To state the following lemma, we first define a confluent hypergeometric function $\tau(y;\,\alpha,\,\beta)$ for $y > 0$ and $(\alpha,\,\beta) \in \mathbf{C}^2$ by

$$\tau(y;\,\alpha,\,\beta) = \int_0^\infty e^{-yt}(1+t)^{\alpha-1}t^{\beta-1}dt. \tag{4.1}$$

This is convergent for $\mathrm{Re}(\beta) > 0$. It can be shown that $\Gamma(\beta)^{-1}\tau(y;\,\alpha,\,\beta)$ can be continued to a holomorphic function in $(\alpha,\,\beta)$ on the whole $\mathbf{C}^2$; see [9, Section A3], for example. Also, for $v \in \mathbf{C}^\times$ and $\alpha \in \mathbf{C}$ we define v^α by

$$v^\alpha = \exp(\alpha\,\log(v)), \qquad -\pi < \mathrm{Im}[\log(v)] \le \pi. \tag{4.2}$$

LEMMA 4.2. *For* $\alpha,\ \beta \in \mathbf{C}$ *such that* $\mathrm{Re}(\alpha+\beta) > 1,\ 0 \le r < 1$, *and* $z = x + iy \in H$ *we have*

$$\begin{aligned} i^{\alpha-\beta}(2\pi)^{-\alpha-\beta}\Gamma(\alpha)\Gamma(\beta)\sum_{m\in\mathbf{Z}}\mathbf{e}(mr)(z+m)^{-\alpha}(\overline{z}+m)^{-\beta} \\ = \sum_{n=1}^{\infty}\mathbf{e}\big((n-r)z\big)(n-r)^{\alpha+\beta-1}\tau\big(4\pi(n-r)y;\ \alpha,\ \beta\big) \\ + \sum_{n=1}^{\infty}\mathbf{e}\big(-(n+r)\overline{z}\big)(n+r)^{\alpha+\beta-1}\tau\big(4\pi(n+r)y;\ \beta,\ \alpha\big) \\ + \begin{cases} (4\pi y)^{1-\alpha-\beta}\Gamma(\alpha+\beta-1) & \text{if} \quad r = 0, \\ \mathbf{e}(-r\overline{z})r^{\alpha+\beta-1}\tau\big(4\pi ry;\ \beta,\ \alpha\big) & \text{if} \quad r \ne 0. \end{cases} \end{aligned}$$

Proof. If $r = 0$, this is Lemma A3.4 of [9]. The case with nontrivial r can be proved in the same way as follows. Define two functions $f(x)$ and $f_1(x)$ of $x \in \mathbf{R}$ by $f(x) = (x+iy)^{-\alpha}(x-iy)^{-\beta}$ with a fixed $y > 0$ and $f_1(x) = \mathbf{e}(rx)f(x)$. Then $\hat{f}_1(x) = \hat{f}(x-r)$, and so the Poisson summation formula (see [9, (2.9)]) shows that

$$\mathbf{e}(-rx)\sum_{m\in\mathbf{Z}}f_1(x+m) = \mathbf{e}(-rx)\sum_{n\in\mathbf{Z}}\hat{f}_1(n)\mathbf{e}(nx) = \sum_{n\in\mathbf{Z}}\mathbf{e}\big((n-r)x\big)\hat{f}(n-r).$$

In [9, p. 133] we determined $\hat{f}$ explicitly in terms of τ as follows:

$$i^{\alpha-\beta}(2\pi)^{-\alpha-\beta}\Gamma(\alpha)\Gamma(\beta)\hat{f}(t) = \begin{cases} \mathbf{e}(ity)t^{\alpha+\beta-1}\tau(4\pi ty;\ \alpha,\ \beta) & (t > 0), \\ \mathbf{e}(-ity)|t|^{\alpha+\beta-1}\tau(4\pi|t|y;\ \beta,\ \alpha) & (t < 0), \\ (4\pi y)^{1-\alpha-\beta}\Gamma(\alpha+\beta-1) & (t = 0). \end{cases}$$

Therefore we obtain our lemma.

4.3. We now need an elementary result:

$$\sum_{n=1}^{\infty}n^{k-1}x^n = \frac{xP_k(x)}{(1-x)^k} \qquad (1 \le k \in \mathbf{Z}). \tag{4.3}$$

Here x is an indeterminate and P_k is a polynomial. We have $P_1 = P_2 = 1$ and $P_{k+1} = (kx - x + 1)P_k - (x^2 - x)P'_k$ for $k \ge 2$. Thus P_k is of degree $k-2$ for $k \ge 2$. These are easy; see [9, p. 17]. We also need two formulas and an estimate given as (A3.11), (A3.14), and Lemma A3.2 in [9]:

$$\tau(y;\ n,\ \beta) = \sum_{\mu=0}^{n-1}\binom{n-1}{\mu}\Gamma(\beta+\mu)y^{-\mu-\beta} \qquad (0 < n \in \mathbf{Z}), \tag{4.4}$$

$$\big[\tau(y;\ \alpha,\ \beta)/\Gamma(\beta)\big]_{\beta=0} = 1, \tag{4.5}$$

(4.6) $\Gamma(\beta)^{-1}y^{\beta}\tau(y;\ \alpha,\ \beta)$ *is bounded when* $(\alpha,\ \beta)$ *belongs to a compact subset of* $\mathbf{C}^2$ *and* $y > c$ *with a positive constant* c.

Our principal aim of this section is to study the nature of the series

(4.7) $$\mathcal{L}_k(s,\, z) = \sum_{m\in\mathbf{Z}} \mathbf{e}(mr)(z+m)^{-k}|z+m|^{-2s},$$

for certain integer values of s. Here $k \in \mathbf{Z}$, $r \in \mathbf{R}$, $s \in \mathbf{C}$, and $z \in H$. The sum depends only on r modulo $\mathbf{Z}$, and so we may assume that $0 \le r < 1$. Clearly this series is absolutely convergent for $\mathrm{Re}(2s+k) > 1$, and defines a holomorphic function of s there.

THEOREM 4.4. *The function $\mathcal{L}_k(s,\, z)$ can be continued as a meromorphic function of s to the whole $\mathbf{C}$, which is entire if $r \notin \mathbf{Z}$. If $r \in \mathbf{Z}$, the locations of the poles of $\mathcal{L}_k(s,\, z)$ are the same as those of $\Gamma(2s+k-1)/\{\Gamma(s+k)\Gamma(s)\}$.*

Proof. Our function is the infinite series of Lemma 4.2 defined with $\alpha = s+k$ and $\beta = s$. Therefore our assertion can easily be verified by means of the formula of Lemma 4.2 and the estimate given by (4.6).

THEOREM 4.5. *Assuming that $r \in \mathbf{Q}$, let N be the smallest positive integer such that $Nr \in \mathbf{Z}$ and let $\beta \in \mathbf{Z}$. Then the following assertions hold:*

(i) *If $\beta > 0$ or $\beta + k > 0$, then $\mathcal{L}_k(s,\, z)$ is finite at $s = \beta$ and $i^k\mathcal{L}_k(\beta,\, z)$ is a rational function in π, $\mathbf{e}(z/N)$, $\mathbf{e}(-\overline{z}/N)$, and $\mathrm{Im}(z)$ with coefficients in $\mathbf{Q}$.*

(ii) *If $-k < \beta \le 0$, then $i^k\{1-\mathbf{e}(z)\}^{k+\beta}\mathcal{L}_k(\beta,\, z)$ is a polynomial in π, $\mathbf{e}(z/N)$, $\mathrm{Im}(z)$, and $\mathrm{Im}(z)^{-1}$ with coefficients in $\mathbf{Q}$.*

(iii) *If $0 < \beta \le -k$, then $i^k\{1-\mathbf{e}(-\overline{z})\}^{\beta}\mathcal{L}_k(\beta,\, z)$ is a polynomial in π, $\mathbf{e}(-\overline{z}/N)$, $\mathrm{Im}(z)$, and $\mathrm{Im}(z)^{-1}$ with coefficients in $\mathbf{Q}$.*

Proof. As we already said, we may assume that $0 \le r < 1$. Put $\alpha = \beta + k$. We first have to study the nature of $\Gamma(2s+k-1)/\{\Gamma(s+k)\Gamma(s)\}$ at $s = \beta$. This is clearly finite at $s = \beta$ if $\alpha + \beta > 1$. Suppose $\alpha + \beta \le 1$; then $\alpha \le 0$ if $\beta > 0$, and $\beta \le 0$ if $\alpha > 0$. In all cases the value is finite, and in fact is a rational number. We now evaluate the formula of Lemma 4.2 divided by $\Gamma(\alpha)\Gamma(\beta)$. If $\alpha > 0$, we have, by (4.4),

$$\tau\bigl(4\pi(n-r)y;\, \alpha,\, \beta\bigr)/\Gamma(\beta) = \sum_{\mu=0}^{\alpha-1}\binom{\alpha-1}{\mu}(n-r)^{-\mu-\beta}(4\pi y)^{-\mu-\beta}\prod_{\kappa=0}^{\mu-1}(\beta+\kappa).$$

Thus an infinite sum of the form $\sum_{n=1}^{\infty}\mathbf{e}\bigl((n-r)z\bigr)(n-r)^{\alpha-\mu-1}$ appears. Applying the binomial theorem to the power of $n-r$, we see that the sum is a $\mathbf{Q}$-linear combination of $\mathbf{e}(-rz)\sum_{n=1}^{\infty}\mathbf{e}(nz)n^{\nu}$ for $0 \le \nu \le \alpha-\mu-1$. We can handle $\tau\bigl(4\pi(n+r)y;\, \beta,\, \alpha\bigr)/\Gamma(\alpha)$ in a similar way if $\beta > 0$. Put $\mathbf{q} = \mathbf{e}(z)$ and $\mathbf{q}_r = \mathbf{e}(rz)$. Then, assuming that $0 < r < 1$, $\alpha > 0$, and $\beta > 0$, we have

$$\begin{aligned} i^k\mathcal{L}_k(\beta,\, z) &= \mathbf{q}_r^{-1}\sum_{\mu=0}^{\alpha-1}\sum_{\nu=0}^{\alpha-\mu-1} a_{\mu\nu}\pi^{\alpha-\mu}y^{-\mu-\beta}\sum_{n=1}^{\infty} n^{\nu}\mathbf{q}^n \\ &\quad + \overline{\mathbf{q}}_r\sum_{\mu=0}^{\beta-1}\sum_{\nu=0}^{\beta-\mu-1} b_{\mu\nu}\pi^{\beta-\mu}y^{-\mu-\alpha}\sum_{n=0}^{\infty} n^{\nu}\overline{\mathbf{q}}^n, \end{aligned}$$

where $a_{\mu\nu}$ and $b_{\mu\nu}$ are rational numbers depending on β, k, and r. Applying (4.3) to $\sum_{n=1}^{\infty} n^{\nu}X^{n}$ with $X = \mathbf{q}$ and $X = \overline{\mathbf{q}}$, we obtain (i). Suppose $\beta \leq 0$ and $\beta + k > 0$; then the sum involving $\tau(4\pi(n+r)y;\, \beta,\, \alpha)/\{\Gamma(\alpha)\Gamma(\beta)\}$ vanishes and we obtain (ii). The case in which $\beta > 0$ and $\beta + k \leq 0$ is similar and produces (iii). If $r = 0$, the constant term of $i^k\mathcal{L}_k(\beta,\, z)$ is $2\pi(2y)^{1-\alpha-\beta}\Gamma(\alpha+\beta-1)/[\Gamma(a)\Gamma(\beta)]$, which causes no problem. This completes the proof.

One special case is worthy of attention. Taking $\beta = 0$ and $1 < k = \alpha \in \mathbf{Z}$, and using (4.5), we obtain, for $0 \leq r < 1$,

$$\sum_{m\in\mathbf{Z}} \frac{\mathbf{e}\big(r(z+m)\big)}{(z+m)^k} = \frac{(2\pi i)^k}{(k-1)!}\sum_{\nu=1}^{k}\binom{k-1}{\nu-1} r^{k-\nu}\frac{\mathbf{q}P_\nu(\mathbf{q})}{(\mathbf{q}-1)^\nu}, \tag{4.8}$$

where $\mathbf{q} = \mathbf{e}(z)$. We assume $0 < r < 1$ and $\sum_{m\in\mathbf{Z}} = \lim_{h\to\infty}\sum_{|m|\leq h}$ when $k = 1$. In (4.7) we take z in H, but in (4.8) we can take $z \in \mathbf{C}$, $\notin \mathbf{Z}$, since both sides of (4.8) are meaningful for such z. If $k = 1$, the result is the same as (2.3).

We can mention another special case. Namely, take $z = ia$ with a positive rational number a. Then we see that the values

$$\sum_{m\in\mathbf{Z}} (a^2+m^2)^{-\beta} \tag{4.9}$$

for $0 < \beta \in \mathbf{Z}$ belong to the field generated by π and $e^{-2\pi a}$ over $\mathbf{Q}$, and therefore any three such values satisfy a nontrivial algebraic equation over $\mathbf{Q}$.

5. The rationality over a totally real base field

5.1. Throughout this section we fix a totally real algebraic number field F. The algebraicity of $\pi^{-k}L(k,\, \chi)$ can be generalized to the case of L-functions over F, but there is no known formulas similar to (0.2), (1.9), (1.10), except that Siegel proved some such formulas in [10] and [11] when $[F:\mathbf{Q}] = 2$. The paper [1] of Hecke may be mentioned in this connection. In this section we merely consider a generalization of (2.4) and prove an algebraicity result on its critical values, without producing explicit expressions.

We denote by $\mathfrak{g}$, D_F, and $\mathbf{a}$ the maximal order of F, the discriminant of F, and the set of archimedean primes of F. We also put $\mathrm{Tr}(x) = \mathrm{Tr}_{F/\mathbf{Q}}(x)$ for $x \in F$ and $[F:\mathbf{Q}] = g$. For $\alpha \in F$ and a fractional ideal $\mathfrak{a}$ in F we put $\alpha + \mathfrak{a} = \{\, \alpha + x \,|\, x \in \mathfrak{a}\,\}$ and $\widetilde{\mathfrak{a}} = \{\xi \in F \,|\, \mathrm{Tr}(\xi\mathfrak{a}) \subset \mathbf{Z}\}$.

Given α and $\mathfrak{a}$ as above, $\xi \in F$, $0 < \mu \in \mathbf{Z}$, and a (sufficiently small) subgroup U of $\mathfrak{g}^\times$ of finite index, we put

$$D_\mu(s;\, \xi,\, \alpha,\, \mathfrak{a}) = r_U \sum_{0\neq h\in U\backslash(\alpha+\mathfrak{a})} \mathbf{e}_{\mathbf{a}}(h\xi)h^{-\mu\mathbf{a}}|h|^{(\mu-s)\mathbf{a}}, \tag{5.1}$$

$$r_U = [\mathfrak{g}^\times : U]^{-1}, \tag{5.1a}$$

where $\mathbf{e_a}(\xi) = \mathbf{e}\big(\sum_{v\in\mathbf{a}}(\xi_v)\big)$ for $\xi \in F$ and $x^{t\mathbf{a}} = \prod_{v\in\mathbf{a}} x_v^t$ for $x \in \mathbf{C}^{\mathbf{a}}$ and $U\backslash X$ means a complete set of representatives for X modulo multiplication by the elements of U. We have to take U so small that the sum of (5.1) is meaningful. For instance, it is sufficient to take

$$U \subset \{u \in \mathfrak{g}^\times \mid u^{\mathbf{a}} = 1,\ u\xi - \xi \in \alpha^{-1}\widetilde{\mathfrak{g}} \cap \widetilde{\mathfrak{a}}\}.$$

The factor r_U makes the quantity of (5.1) independent of the choice of U. Clearly the sum is convergent for $\mathrm{Re}(s) > 1$. Now $D_\mu(s;\ \xi,\ \alpha,\ \mathfrak{a})$ is a special case of the series of [8, (18.1)], and so from Lemma 18.2 of [8] we see that it can be continued as a holomorphic function in s to the whole $\mathbf{C}$.

THEOREM 5.2. *For $0 < \mu \in \mathbf{Z}$ we have*

$$(2\pi i)^{-\mu g} D_F^{1/2} D_\mu(\mu;\ \xi,\ 0,\ \mathfrak{a}) \in \mathbf{Q}, \tag{5.2}$$

$$D_\mu(1-\mu;\ 0,\ \alpha,\ \mathfrak{a}) \in \mathbf{Q}. \tag{5.3}$$

Proof. The last formula is a restatement of Proposition 18.10(2) of [8]. To prove (5.2), let $\mathfrak{b} = \{x \in \mathfrak{a} \mid \mathbf{e_a}(x\xi) = 1\}$ and let R be a complete set of representatives for $\mathfrak{a}/\mathfrak{b}$. Then

$$D_\mu(s;\ \xi,\ 0,\ \mathfrak{a}) = \sum_{\beta\in R} \mathbf{e_a}(\beta\xi) D_\mu(s; 0,\ \beta,\ \mathfrak{b}).$$

Put $Q_\mu(\beta,\ \mathfrak{b}) = (2\pi i)^{-\mu g} D_F^{1/2} D_\mu(\mu;\ 0,\ \beta,\ \mathfrak{b})$. Then the quantity of (5.2) equals $\sum_{\beta\in R} \mathbf{e_a}(\beta\xi) Q_\mu(\beta,\ \mathfrak{b})$. Let $t \in \prod_p \mathbf{Z}_p^\times$ and let σ be the image of t under the canonical homomorphism of $\mathbf{Q}_{\mathbf{A}}^\times$ onto $\mathrm{Gal}(\mathbf{Q}_{\mathrm{ab}}/\mathbf{Q})$. Our task is to show that the last sum is invariant under σ. By [8, Proposition 18.10(1)] we have $Q_\mu(\beta,\ \mathfrak{b})^\sigma = Q_\mu(\beta_1,\ \mathfrak{b})$ with $\beta_1 \in F$ such that $(t\beta_1 - \beta)_v \in \mathfrak{b}_v$ for every nonarchimedean prime v of F. For $\beta \in R$ there is a unique $\beta_1 \in R$ with that property. Now $\mathbf{e}(c)^\sigma = \mathbf{e}(t^{-1}c)$ for every $c \in \mathbf{Q}/\mathbf{Z} = \prod_p(\mathbf{Q}_p/\mathbf{Z}_p)$; see [8, (8.2)]. Since $\mathbf{e_a}(\beta\xi) = \mathbf{e}\big(\mathrm{Tr}(\beta\xi)\big)$, we easily see that $\mathbf{e_a}(\beta\alpha)^\sigma = \mathbf{e_a}(\beta_1\alpha)$, which gives the desired fact.

REFERENCES

[1] E. Hecke, Bestimmung der Klassenzahl einer neuen Reihe von algebraischen Zahlkörpern, Nachr. der K. Gesellschaft der Wissenschaften zu Göttingen math.-phys. Klasse 1921, 1–23 (= Werke, 290–312).

[2] E. Hecke, Zur Theorie der elliptischen Modulfunktionen, Math. Ann. 97 (1926), 210-242 (= Werke, 428–460).

[3] E. Hecke, Analytische Arithmetik der positiven Quadratischen Formen, 1940, *Mathematische Werke,* 789–918.

[4] Kronecker, Bemerkungen über die Darstellung von Reihen durch Integrale, Journal für r. angew. Math. 105 (1889), 157–159, 345–354 (=Werke V, 329–342).

[5] E. Landau, Handbuch der Lehre von der Verteilung der Primzahlen, 1909.

[6] G. Shimura, On the derivatives of theta functions and modular forms, Duke Math. J. 44 (1977), 365–387.

[7] G. Shimura, Nearly holomorphic functions on hermitian symmetric spaces, Math. Ann. 278 (1987), 1–28.

[8] G. Shimura, Arithmeticity in the theory of automorphic forms, Mathematical Surveys and Monographs, vol. 82, American Mathematical Society, 2000.

[9] G. Shimura, Elementary Dirichlet Series and Modular Forms, Springer Monog. in Math., Springer, 2007.

[10] C. L. Siegel, Bernoullische Polynome und quadratische Zahlkörper, Nachr. der Akademie der Wissenschaften in Göttingen math.-phys. Klasse 1968, Nr. 2, 7–38 (=Abhandlungen IV, 9–40).

[11] C. L. Siegel, Zur Summation von L-Reihen, Nachr. der Akademie der Wissenschaften in Göttingen math.-phys. Klasse 1975, Nr. 18, 269-292 (=Abhandlungen IV, 305–328).

Goro Shimura
Department of Mathematics
Princeton University
Princeton
NJ 08544–1000
USA
goro@Math.Princeton.EDU

The Critical Values of Generalizations of the Hurwitz Zeta Function

Goro Shimura

Received: November 10, 2009

Communicated by Don Blasius

Abstract. We investigate a few types of generalizations of the Hurwitz zeta function, written $Z(s, a)$ in this abstract, where s is a complex variable and a is a parameter in the domain that depends on the type. In the easiest case we take $a \in \mathbf{R}$, and one of our main results is that $Z(-m, a)$ is a constant times $E_m(a)$ for $0 \leq m \in \mathbf{Z}$, where E_m is the generalized Euler polynomial of degree n. In another case, a is a positive definite real symmetric matrix of size n, and $Z(-m, a)$ for $0 \leq m \in \mathbf{Z}$ is a polynomial function of the entries of a of degree $\leq mn$. We will also define Z with a totally real number field as the base field, and will show that $Z(-m, a) \in \mathbf{Q}$ in a typical case.

2010 Mathematics Subject Classification: 11B68, 11M06, 30B50, 33E05.

Introduction

This paper is divided into four parts. In the first part we consider a generalization of Hurwitz zeta function given by

$$\zeta(s;\, a,\, \gamma) = \sum_{n=0}^{\infty} \gamma^n (n+a)^{-s}, \tag{0.1}$$

where $s \in \mathbf{C}$, $0 < a \in \mathbf{R}$, and $\gamma \in \mathbf{C}$, $0 < |\gamma| \leq 1$. Clearly the infinite series is convergent for $\mathrm{Re}(s) > 1$. For $\gamma = 1$ this becomes $\sum_{n=0}^{\infty}(n+a)^{-s}$, which is the classical Hurwitz zeta function usually denoted by $\zeta(s, a)$. This generalization is not new. It was considered by Lerch in [Le], a work five years after the paper [Hu] of Hurwitz in 1882. Its analytic properties can be summarized as follows.

Theorem 0.1. *For a and γ as above the product $(e^{2\pi i s}-1)\Gamma(s)\zeta(s;\, a,\, \gamma)$ can be continued to an entire function in s. In addition, there exists a holomorphic function in $(s, a, \gamma) \in \mathbf{C}^3$, defined for $\mathrm{Re}(a) > 0$ and $\gamma \notin \{x \in \mathbf{R} \mid x \geq 1\}$ with no condition on s, that coincides with the product when $\mathrm{Re}(s) > 1$, $0 < a \in \mathbf{R}$, and $0 < |\gamma| \leq 1$.*

The proof will be given in §1.1.

To state a more interesting fact, we first put

$$\mathbf{e}(z) = \exp(2\pi i z) \qquad (z \in \mathbf{C}), \tag{0.2}$$

and define a function $E_{c,n}(t)$ in t for $c \in \mathbf{C}$ and $0 < n \in \mathbf{Z}$, that is called the *nth generalized Euler polynomial*, by

$$\frac{(1+c)e^{tz}}{e^z + c} = \sum_{n=0}^{\infty} \frac{E_{c,n}(t)}{n!} z^n. \tag{0.3}$$

We assume $c = -\mathbf{e}(\alpha)$ with $\alpha \in \mathbf{R}, \notin \mathbf{Z}$. The function $E_{c,n}(t)$ was introduced in [S07]. If $c = 1$, $E_{1,n}(t)$ is the classical Euler polynomial of degree n. In [S07] we showed that $E_{c,n}(t)$ is a polynomial in t of degree n; it is also a polynomial in $(1+c)^{-1}$. Its properties are listed in [S07, pp. 25–26]. We mention here only

$$E_{c,n}(1-t) = (-1)^n E_{c^{-1},n}(t), \tag{0.3a}$$

(see [S07, (4.3f)]), which will become necessary later. Now we have

THEOREM 0.2. *For $0 < k \in \mathbf{Z}$, $\mathrm{Re}(a) > 0$, and $\gamma \notin \{x \in \mathbf{R} \,|\, x \geq 1\}$ the value $\zeta(1-k,\, a;\, \gamma)$ is a polynomial function of a and $(\gamma - 1)^{-1}$. More precisely, we have*

$$\zeta(1-k;\, a,\, \gamma) = E_{c,k-1}(a)/(1 + c^{-1}) \tag{0.4}$$

for such k, a, and γ, where $c = -\gamma^{-1}$.

This will be proven in §1.2.

As for the original Hurwitz function, there is a well known relation

$$\zeta(1-k,\, a) = -B_k(a)/k \quad \text{for} \quad 0 < k \in \mathbf{Z}, \tag{0.5}$$

where B_k is the k-th Bernoulli polynomial. This is essentially due to Hurwitz; see [Hu, p. 92]; cf. also [E, p.27, (11)] and [WW, p. 267, 13.14].

In [S07] and [S08] we investigated the critical values of the L-function $L(s,\, \chi)$ with a Dirichlet character χ, and proved especially (see [S07, Theorem 4.14] and [S08, Theorem 1.4])

THEOREM 0.3. *Let χ be a nontrivial primitive Dirichlet character modulo a positive integer d, and let k be a positive integer such that $\chi(-1) = (-1)^k$.*

(i) *If $d = 2q + 1$ with $0 < q \in \mathbf{Z}$, then*

$$L(1-k,\, \chi) = \frac{d^{k-1}}{2^k \chi(2) - 1} \sum_{b=1}^{q} (-1)^b \chi(b) E_{1,k-1}(b/d). \tag{0.6}$$

(ii) *If $d = 4d_0$ with $1 < d_0 \in \mathbf{Z}$, then*

$$L(1-k,\, \chi) = (2d_0)^{k-1} \sum_{a=1}^{d_0 - 1} \chi(a) E_{1,k-1}(2a/d). \tag{0.7}$$

In §1.4 we will give a shorter proof for these formulas by means of (0.4), and in Section 2 we will prove a functional equation for $\zeta(s;\, a, \gamma)$ by producing an expression for $\zeta(1-s;\, a,\, \gamma)$.

The second part of the paper concerns the analogue of (0.1) defined when the base field is a totally real algebraic number field F. If $F \neq \mathbf{Q}$, there are nontrivial units, which cause considerable difficulties, and for this reason we cannot give a full generalization of which (0.1) is a special case. However, taking such an F as the base field, we will present a function of a complex variable s and two parameters a and p in F, that includes as a special case at least $\zeta(s;\, a,\, \gamma)$ with $a \in \mathbf{Q}$ and γ a root of unity. We then prove in Theorem 3.4 a rationality result on its critical values.

The third part is a kind of interlude. Observing that the formula for $L(k,\, \chi)$ (not $L(1-k,\, \chi)$) involves the Gauss sum $G(\chi)$ of χ, we will give in Section 4 a formula for $G(\chi\lambda)/G(\chi)$ for certain Dirichlet characters χ and λ.

The fourth and final part of the paper, which has a potential of future development, concerns the analogue of (0.1) defined for a complex variable s, with nonnegative and positive definite symmetric matrices of size n in place of n and a. We will show in Section 5 that it is an entire function of s and also that its value at $s = -m$ for $0 \le m \in \mathbf{Z}$ is a polynomial function of the variable symmetric matrix of degree $\le mn$.

1. Proof of Theorems 0.1, 0.2, and 0.3

1.1. To prove Theorem 0.1, assuming that $0 < a \in \mathbf{R}$ and $0 < |\gamma| \le 1$, we start from an easy equality $\Gamma(s)(n+a)^{-s} = \int_0^\infty x^{s-1} e^{-(n+a)x}dx$. Therefore

$$\begin{aligned}\Gamma(s)\zeta(s;\, a,\, \gamma) &= \sum_{n=0}^{\infty} \Gamma(s)\gamma^n (n+a)^{-s} = \sum_{n=0}^{\infty} \int_0^\infty x^{s-1}\gamma^n e^{-(n+a)x}dx \\ &= \int_0^\infty \sum_{n=0}^{\infty} x^{s-1}\gamma^n e^{-(n+a)x}dx = \int_0^\infty \frac{x^{s-1}e^{-ax}}{1-\gamma e^{-x}}dx.\end{aligned}$$

Our calculation is justified for $\sigma = \mathrm{Re}(s) > 1$, since

$$\sum_{n=0}^{\infty} \int_0^\infty \left|x^{s-1}\gamma^n e^{-(n+a)x}\right|dx \le \sum_{n=0}^{\infty} \Gamma(\sigma)(n+a)^{-\sigma} < \infty.$$

Thus we obtain

$$\Gamma(s)\zeta(s;\, a,\, \gamma) = \int_0^\infty \frac{x^{s-1}e^{x(1-a)}}{e^x - \gamma}dx \quad \text{for} \quad \mathrm{Re}(s) > 1. \tag{1.1}$$

We now consider

$$\int_\infty^{0+} \frac{z^{s-1}e^{z(1-a)}}{e^z - \gamma}dz$$

with the standard symbol $\int_\infty^{0+}$ of contour integration. The integral is the sum of three integrals: $\int_\infty^\delta$, $\oint$ on the circle $|z| = \delta$, and $\int_\delta^\infty$, where $0 < \delta \in \mathbf{R}$; we naturally take the limit as δ tends to 0. We take $z^{s-1} = \exp\bigl((s-1)\log z\bigr)$ for the first integral $\int_\infty^\delta$ with $\log z \in \mathbf{R}$ for $0 < z \in \mathbf{R}$; for the evaluation of the other integrals we continue z^{s-1} analytically without passing through the positive real axis. Then the first and third integrals produce

$$(\mathbf{e}(s) - 1)\int_\delta^\infty \frac{x^{s-1}e^{x(1-a)}}{e^x - \gamma}dx,$$

which is meaningful for every $s \in \mathbf{C}$ and every $(a,\, \gamma) \in \mathbf{C}^2$ such that

(1.1a) $\qquad \mathrm{Re}(a) > 0 \quad$ and $\quad \gamma \notin \{x \in \mathbf{R} \mid x > 1\}$.

As for $\oint$, we first observe that given $\gamma \in \mathbf{C}$, we can find a small $\delta_0 \in \mathbf{R}, > 0$, such that $e^z \neq \gamma$ for $0 < |z| \le \delta_0$, since the map $z \mapsto w = e^z$ sends the punctured disc $0 < |z| \le \delta_0$ into a punctured disc $0 < |w-1| < \varepsilon$ that does not contain γ. (This is clearly so even for $\gamma = 1$.) Therefore $\oint$ is meaningful for every $s \in \mathbf{C}$ and sufficiently small δ in both cases $\gamma \neq 1$ and $\gamma = 1$; the integral is independent of δ because of Cauchy's theorem. Now put $z = \delta e^{i\theta}$ with δ such that $0 < \delta < \delta_0$ and $0 \le \theta < 2\pi$. Then $z^{s-1} = \exp\{(s-1)(\log \delta + i\theta)\}$, and so for $s = \sigma + i\tau$ with real σ and τ, we have $|z^{s-1}| = \delta^{\sigma-1}|e^{-\theta\tau}| \le \delta^{\sigma-1}e^{2\pi|\tau|}$. If $\gamma \neq 1$, we see that $\mathrm{Min}_{|z|\le\delta_0}|e^z - \gamma| > 0$, and so $|e^{z(1-a)}/(e^z-\gamma)|$ is bounded for $|z| \le \delta_0$. If $\gamma = 1$, the function $e^{z(1-a)}/(e^z-\gamma)$ is $1/z$ plus a holomorphic function at $z = 0$. Thus for $0 < \delta \le \delta_0$ we see that $\left|\oint\right| \le M\delta^\sigma$ if $\gamma \neq 1$ and $\left|\oint\right| \le M\delta^{\sigma-1}$ if $\gamma = 1$ with a constant M that depends on a, γ, and δ_0, and so $\oint$ tends to 0 as $\delta \to 0$ if $\mathrm{Re}(s) > 1$, and we obtain

$$(\mathbf{e}(s) - 1)\Gamma(s)\zeta(s;\, a,\, \gamma) = \int_\infty^{0+} \frac{z^{s-1}e^{z(1-a)}}{e^z - \gamma}\,dz \tag{1.2}$$

for $0 < |\gamma| \le 1$, $0 < a \in \mathbf{R}$, and $\mathrm{Re}(s) > 1$. (If $\gamma \neq 1$, the condition $\mathrm{Re}(s) > 0$ instead of $\mathrm{Re}(s) > 1$ is sufficient.) Now the right-hand side of (1.2) is meaningful for every $(s, a, \gamma) \in \mathbf{C}^3$ under condition (1.1a) and so it establishes the product on the left-hand side as an entire function of s. If $\gamma \neq 1$, the contour integral can define the product for the variables s, a, γ as described in Theorem 0.1. The integrand as a function of (z, s, a, γ) is not finite in a domain including $\gamma = 1$, and so the result for $\gamma = 1$ must be stated separately.

Remark. If $0 < |\gamma| \le 1$ and $\gamma \neq 1$, then the series of (0.1) is convergent for $\mathrm{Re}(s) > 0$ and defines a holomorphic function of s there. This is clear if $|\gamma| < 1$. For $|\gamma| = 1$ and $\gamma \neq 1$, this follows from [S07, Lemma 4.3].

We are interested in the value of $\zeta(s;\, a,\, \gamma)$ at $s = -m$ with $0 \le m \in \mathbf{Z}$. We first note

$$\big\{(\mathbf{e}(s) - 1)\Gamma(s)\big\}_{s=-m} = 2\pi i(-1)^m/m!.$$

For $s = -m$ the function z^{s-1} in the integrand is a one-valued function, and so $\int_\infty^\delta + \int_\delta^\infty = 0$. Therefore

$$\int_\infty^{0+} = \oint = 2\pi i \cdot \mathrm{Res}_{z=0}\, \frac{z^{-m-1}e^{(1-a)z}}{e^z - \gamma}. \tag{1.3}$$

By (0.3), for $\gamma \neq 1$ we have

$$\frac{z^{-m-1}e^{(1-a)z}}{e^z - \gamma} = \frac{1}{1-\gamma}\sum_{n=0}^\infty \frac{E_{-\gamma,n}(1-a)}{n!}z^{n-m-1},$$

and so the residue in question is $(1-\gamma)^{-1}E_{-\gamma,m}(1-a)/m!$. Combining this with (0.3a) and (1.2), we obtain

$$\zeta(-m;\, a,\, \gamma) = E_{c,m}(a)/(1 + c^{-1}) \quad \text{for} \quad 0 \le m \in \mathbf{Z}, \tag{1.4}$$

where $c = -\gamma^{-1}$. This proves (0.4). In [S07, p. 26] we showed that $E_{c,n}(t)$ is a polynomial in t and $(1+c)^{-1}$. For $c = -\gamma^{-1}$ we have $(1+c)^{-1} = 1+(\gamma-1)^{-1}$, and so we obtain Theorem 0.2.

For the reader's convenience, we give a proof of (0.5) here. If $\gamma = 1$, instead of (0.2) we use $ze^{tz}/(e^z-1) = \sum_{n=0}^{\infty} B_n(t)z^n/n!$. Then we obtain (0.5) in the same manner as in the case $\gamma \neq 1$.

1.3. Let us insert here a historical remark. Fixing a positive integer m and an integer a such that $0 \le a \le m$, Hurwitz considered in [Hu] an infinite series

$$f(s,\, a) = \sum_{n=0}^{\infty}(mn+a)^{-s}. \tag{1.5}$$

Since this depends on m, he also denoted it by $f(s,\, a|m)$. He proved analytic continuation of these functions and stated a functional equation for $f(1-s,\, a)$, basically following Riemann's methods for the investigation of $\zeta(s)$ in [R]. At that time not much was known about Dirichlet's L-function beyond his formulas for the class number of a binary quadratic form and his theorem about prime numbers in an arithmetic progression. Employing the results on $f(s,\, a)$, Hurwitz was able to prove that the L-function for a quadratic character has analytic continuation and satisfies a functional equation. Using the standard notation $\zeta(s,\, a) = \sum_{n=0}^{\infty}(n+a)^{-s}$ employed at present, we have $f(s,\, a|m) = m^{-s}\zeta(s,\, a/m)$, and so he considered $\zeta(s,\, a)$ only for $a \in \mathbf{Q}$. It is noticeable however that he proved essentially (0.5) as we already said.

As noted at the beginning of the paper, Lerch investigated the series of (0.1); one can also find an exposition of this topic in [E, p. 27, §1.11]. The paper [Li] of Lipschitz may be mentioned in this connection.

1.4. Let the symbols be as in Theorem 0.3(i). Put

$$\Lambda(s) = \sum_{n=1}^{\infty}(-1)^n\chi(n)n^{-s}.$$

Then $\Lambda(s) + L(s,\, \chi) = 2\sum_{n=1}^{\infty}\chi(2n)(2n)^{-s} = \chi(2)2^{1-s}L(s,\, \chi)$, and so

$$\Lambda(s) = L(s,\chi)\bigl\{\chi(2)2^{1-s} - 1\bigr\}. \tag{1.6}$$

Since $\bigl\{a \bigm| 1 \le a < d\bigr\} = \bigl\{a \bigm| 1 \le a \le q\bigr\} \sqcup \bigl\{d-a \bigm| 1 \le a \le q\bigr\}$, we have

$$\Lambda(s) = \sum_{a=1}^{q}\sum_{n=0}^{\infty}(-1)^{nd+a}\chi(a)(nd+a)^{-s} + \sum_{a=1}^{q}\sum_{n=0}^{\infty}(-1)^{nd+d-a}\chi(-a)(nd+d-a)^{-s},$$

and so $d^s\Lambda(s)$ equals

$$\sum_{a=1}^{q}(-1)^a\chi(a)\left\{\sum_{n=0}^{\infty}(-1)^n\left(n+\frac{a}{d}\right)^{-s} - \chi(-1)\sum_{n=0}^{\infty}(-1)^n\left(n+\frac{d-a}{d}\right)^{-s}\right\}$$

$$= \sum_{a=1}^{q}(-1)^a\chi(a)\bigl\{\zeta(s;\, a/d,\, -1) - (-1)^k\zeta(s;\, 1-a/d,\, -1)\bigr\}.$$

Putting $s = 1-k$ and employing (0.4) and (0.3a), we obtain (0.6).

Our next task is to prove (0.7). Let $d = 4d_0$ with $1 < d_0 \in \mathbf{Z}$ as in Theorem 1.3(ii). We note an easy fact(see [S08, Lemma 1.3]):

(1.7) $$\chi(2d_0+a) = -\chi(a) \quad \text{for every} \quad a \in \mathbf{Z}.$$

Observe that $\{x \in \mathbf{Z} \mid x > 0,\ d_0 \nmid x\}$ is a disjoint union

$$(*)\quad \{nd+a \mid 0<a<d_0,\ 0 \le n \in \mathbf{Z}\} \sqcup \{nd+2d_0+a \mid 0<a<d_0,\ 0 \le n \in \mathbf{Z}\}$$
$$\sqcup \{nd-a \mid 0<a<d_0,\ 0<n \in \mathbf{Z}\} \sqcup \{nd+2d_0-a \mid 0<a<d_0,\ 0 \le n \in \mathbf{Z}\}.$$

The sum of $\sum \chi(x)x^{-s}$ for x belonging to the first two sets equals

$$\sum_{a=1}^{d_0-1} \left\{ \sum_{\nu=0}^{\infty} \chi(a)(4\nu d_0+a)^{-s} + \sum_{\nu=0}^{\infty} \chi(2d_0+a)\bigl(2(2\nu+1)d_0+a\bigr)^{-s} \right\}.$$

Employing (1.7), we see that this equals

$$\sum_{a=1}^{d_0-1} \sum_{m=1}^{\infty} (-1)^m \chi(a)(2md_0+a)^{-s} = (2d_0)^{-s} \sum_{a=1}^{d_0-1} \chi(a)\zeta(s;\ 2a/d, -1). \tag{1.8}$$

Similarly, from the last two sets of $(*)$ we obtain

$$\sum_{a=1}^{d_0-1} \left\{ \sum_{\nu=1}^{\infty} \chi(-a)(4\nu d_0-a)^{-s} + \sum_{\nu=0}^{\infty} \chi(2d_0-a)\bigl(2(2\nu+1)d_0-a\bigr)^{-s} \right\}$$
$$= -\sum_{a=1}^{d_0-1} \sum_{m=0}^{\infty} \chi(-a)(-1)^m(2md_0+2d_0-a)^{-s}$$
$$= -(2d_0)^{-s} \sum_{a=1}^{d_0-1} \chi(-a)\zeta(s;\ 1-2a/d,\ -1)$$

by (1.7). Thus, adding (1.8) to this and putting $s = 1-k$, from (0.4) we obtain

$$(2d_0)^{1-k}L(1-k,\ \chi) = 2^{-1} \sum_{a=1}^{d_0-1} \chi(a)\bigl\{E_{1,k-1}(2a/d) - \chi(-1)E_{1,k-1}(1-2a/d)\bigr\}.$$

Suppose $\chi(-1) = (-1)^k$; then applying (0.3a) to $E_{1,k-1}(1-2a/d)$, we obtain (0.7). The proof of Theorem 0.3 is now complete

1.5. The case $d = 4$ is excluded in Theorem 0.3(ii). In this case, however, the matter is simpler. Indeed, for $\mu_4(n) = \left(\dfrac{-1}{n}\right)$ we have

$$L(s,\ \mu_4) = \sum_{m=0}^{\infty} (-1)^m(2m+1)^{-s}$$
$$= 2^{-s} \sum_{m=0}^{\infty} (-1)^m(m+1/2)^{-s} = 2^{-s}\zeta(s;\ 1/2,\ -1),$$

and so by (0.4) we obtain

$$L(1-k,\ \mu_4) = 2^{k-2}E_{1,k-1}(1/2) = E_{1,k-1}/2 \tag{1.9}$$

for every odd positive integer k, where $E_{1,n}$ denotes the nth Euler number. This is classical, except that the result is usually given in terms of $L(k,\ \mu_4)$ instead of $L(1-k,\ \mu_4)$.

1.6. Let us now show that a special case of Theorem 0.3(ii) can be given in a somewhat different way. Let ψ be a primitive character whose conductor d is odd, μ_4 the primitive character modulo 4 as above, and k a positive integer such that $\psi(-1) = (-1)^{k+1}$; put $m = (d-1)/2$. Then

$$L(1-k,\, \psi\mu_4) = (-1)^m (2d)^{k-1} \sum_{j=1}^{m} (-1)^j \psi(2j) E_{1,k-1}\bigl(\tfrac{1}{2} + \tfrac{j}{d}\bigr). \tag{1.10}$$

This was given in [S07, (6.2)], if in terms of $L(k,\, \psi\mu_4)$, but we can derive it also from Theorem 0.3(ii) as follows. Take $\chi = \psi\mu_4$ in (0.7). Then the sum on the right-hand side of (0.7) is $\sum_{a=1}^{2m} \chi(a) E_{1,k-1}(a/2d)$, which equals $\sum_{b=1}^{2m} \chi(d-b) E_{1,k-1}\bigl((d-b)/2d\bigr)$. Since the bth term is nonvanishing only for even b, employing (0.3a), we see that the last sum equals $\psi(-1)$ times $\sum_{j=1}^{m} (\psi\mu_4)(d-2j) E_{1,k-1}\bigl((d+2j)/2d\bigr)$. Since $d - 2j = 2(m-j)+1$, we have $\mu_4(d-2j) = (-1)^{m-j}$, and so we obtain (1.10).

1.7. We can show that $\zeta(-m;\, a,\, \gamma)$ for $0 \le m \in \mathbf{Z}$ is a polynomial in a by a formal calculation as follows. Assuming that $|\gamma| < 1$, we have

$$\zeta(-m;\, a,\, \gamma) = a^m + \sum_{n=1}^{\infty} \gamma^n (n+a)^m = a^m + \sum_{\nu=0}^{m} \binom{m}{\nu} a^{m-\nu} c_\nu$$

with $c_\nu = \sum_{n=1}^{\infty} n^\nu \gamma^n$, and so $\zeta(-m;\, a,\, \gamma)$ is a polynomial in a at least for $|\gamma| < 1$, and so Theorem 0.1 guarantees the same in a larger domain as described in that theorem

We have $c_{\nu-1} = \gamma(1-\gamma)^{-\nu} P_\nu(\gamma)$ for $1 \le \nu \in \mathbf{Z}$ with a polynomial P_ν introduced in [S07, (2.16)]; see also [S08, (4.3)]. We showed that $P_{\nu+1}(\gamma) = (\gamma - 1)^\nu E_{-\gamma,\nu}(0)$ for $\nu > 0$ in [S07, (4.6)] and that $\gamma^{n-2} P_n(\gamma^{-1}) = P_n(\gamma)$ in [S07, (2.19)]; also, $E_{c,n}(t) = \sum_{k=0}^{n} \binom{n}{k} E_{c,k}(0) t^{n-k}$ by [S08, (1.15)]. Combining these together, we obtain (0.4).

Though clearly this is not the best way to prove (0.4), at least it explains an elementary aspect of the nature of the problem. In Section 5, we will return to this idea in our discussion in the higher-dimensional case.

2. The functional equation for $\zeta(s;\, a,\, \gamma)$

2.1. For $\mathrm{Re}(s) > 1$ Hurwitz proved (see [Hu, p. 93, 1)] and [WW, p. 269])

$$\begin{aligned} &\zeta(1-s,\, a) \\ &\quad = \frac{2\Gamma(s)}{(2\pi)^s}\left\{ \cos(\pi s/2) \sum_{n=1}^{\infty} \frac{\cos(2\pi n a)}{n^s} + \sin(\pi s/2) \sum_{n=1}^{\infty} \frac{\sin(2\pi n a)}{n^s} \right\}. \end{aligned} \tag{2.1}$$

Therefore it is natural to ask if there is a meaningful formula for $\zeta(1-s,\, a;\, \gamma)$. Though this was essentially done by Lerch in [Le], here we take a different approach. For $s \in \mathbf{C}$, $a \in \mathbf{R}$, $p \in \mathbf{R}$, and $\nu = 0$ or 1 we put

$$D^\nu(s;\, a,\, p) = \sum_{-a \ne n \in \mathbf{Z}} (n+a)^\nu |n+a|^{-\nu-s} \mathbf{e}\bigl(p(n+a)\bigr), \tag{2.2}$$

$$(2.3)\qquad T^\nu(s;\,a,\,p) = g_\nu(s)D^\nu(s;\,a,\,p),\quad g_\nu(s) = \pi^{-(s+\nu)/2}\Gamma\big((s+\nu)/2\big).$$

These were introduced in [S08]. In particular, we proved

$$(2.4)\qquad T^\nu(1-s;\,a,\,p) = i^{-\nu}\mathbf{e}(ap)T^\nu(s;\,-p,\,a).$$

Let $\gamma = \mathbf{e}(p)$ and $b = 1-a$ with $0 < a < 1$. Then it is easy to verify that

$$(2.5a)\qquad D^0(s;\,a,\,p) = \mathbf{e}(ap)\zeta(s;\,a,\,\gamma) + \mathbf{e}(-bp)\zeta(s;\,b,\,\gamma^{-1}),$$

$$(2.5b)\qquad D^1(s;\,a,\,p) = \mathbf{e}(ap)\zeta(s;\,a,\,\gamma) - \mathbf{e}(-bp)\zeta(s;\,b,\,\gamma^{-1}),$$

and so

$$(2.6)\qquad 2\mathbf{e}(ap)\zeta(s;\,a,\,\gamma) = D^0(s;\,a,\,p) + D^1(s;\,a,\,p).$$

Employing (2.3) and (2.4), we obtain

$$(2.7)\qquad 2\zeta(1-s;\,a,\,\gamma) = \frac{g_0(s)}{g_0(1-s)}D^0(s;\,-p,\,a) - \frac{ig_1(s)}{g_1(1-s)}D^1(s;\,-p,\,a).$$

From (2.5a, b) we see that $D^\nu(s;\,-p,\,a)$ is a linear combination of $\zeta(s;\,-p,\,\delta)$ and $\zeta(s;\,1+p,\,\delta^{-1})$, where $\delta = \mathbf{e}(a)$. Thus, for $-1 < p < 0$ we have

$$2\zeta(1-s;\,a,\,\gamma) = \mathbf{e}(-ap)A\zeta(s;\,-p,\,\delta) + \mathbf{e}(-a-ap)B\zeta(s;\,1+p,\,\delta^{-1})$$

with

$$(2.8)\qquad A = \frac{g_0(s)}{g_0(1-s)} - \frac{ig_1(s)}{g_1(1-s)},\qquad B = \frac{g_0(s)}{g_0(1-s)} + \frac{ig_1(s)}{g_1(1-s)}.$$

Recalling that $\Gamma(s/2)\Gamma((s+1)/2) = 2^{1-s}\pi^{1/2}\Gamma(s)$ and $\Gamma(s)\Gamma(1-s) = \pi/\sin(\pi s)$, we find that

$$(2.9)\qquad A = 2^{1-s}\pi^{-s}\mathbf{e}(-s/4)\Gamma(s),\qquad B = 2^{1-s}\pi^{-s}\mathbf{e}(s/4)\Gamma(s),$$

and so we obtain

$$(2.10)\qquad \zeta(1-s;\,a,\,\gamma) = \frac{\mathbf{e}(-ap)\Gamma(s)}{(2\pi)^s}\Big\{\mathbf{e}(-s/4)\zeta(s;\,-p,\,\delta) + \mathbf{e}(s/4-a)\zeta(s;\,p+1,\,\delta^{-1})\Big\}.$$

at least when $-1 < p < 0$ and $0 < a < 1$, where $\gamma = \mathbf{e}(p)$ and $\delta = \mathbf{e}(a)$.

This does not apply to the case $\gamma = 1$. In this case we have

$$2\zeta(s;\,a,\,1) = D^0(s;\,a,\,0) + D^1(s;\,a,\,0)\quad \text{for}\quad 0 < a \le 1,$$
$$D^\nu(s;\,0,\,a) = \delta\zeta(s;\,1,\,\delta) + (-1)^\nu\delta^{-1}\zeta(s;\,1,\,\delta^{-1}).$$

Repeating the same argument as in the case $\gamma \ne 1$, we find that

$$2\zeta(1-s;\,a,\,1) = \delta A\zeta(s;\,1,\,\delta) + \delta^{-1}B\zeta(s;\,1,\,\delta^{-1})$$

with the same A and B as in (2.8) and (2.9), and so

$$(2.11)\qquad \zeta(1-s;\,a,\,1) = \frac{\Gamma(s)}{(2\pi)^s}\left\{\mathbf{e}\Big(a-\frac{s}{4}\Big)\zeta(s;\,1,\,\delta) + \mathbf{e}\Big(\frac{s}{4}-a\Big)\zeta(s;\,1,\,\delta^{-1})\right\}.$$

where $\delta = \mathbf{e}(a)$, $0 < a \le 1$. This gives (2.1). Indeed, we have

$$\mathbf{e}(\pm a)\zeta(s;\,1,\,\delta^{\pm 1}) = \sum_{m=1}^{\infty} \frac{\delta^{\pm m}}{m^s} = \sum_{m=1}^{\infty} \frac{\cos(2\pi ma) \pm i\sin(2\pi ma)}{m^s},$$

and so a simple calculation transforms (2.11) into (2.1).

Returning to (2.10) in which we assumed $-1 < p < 0$, put $q = -p$. Then $0 < q < 1$ and

$$\text{(2.12)}\qquad \begin{aligned}&\frac{(2\pi)^s}{\Gamma(s)}\zeta\big(1-s;\,a,\,\mathbf{e}(q)^{-1}\big)\\ &= \mathbf{e}(-s/4)\sum_{h=0}^{\infty}\frac{\mathbf{e}\big(a(h+q)\big)}{(h+q)^s} + \mathbf{e}(s/4)\sum_{h=0}^{\infty}\frac{\mathbf{e}\big(a(q-h-1)\big)}{(h+1-q)^s}.\end{aligned}$$

Then employing (0.4), for $0 < s = k \in \mathbf{Z}$ we obtain

$$\text{(2.13)}\qquad \frac{(2\pi i)^k}{(k-1)!(1+c^{-1})}E_{c,k-1}(a) = \sum_{h\in\mathbf{Z}}\frac{\mathbf{e}\big(a(h+q)\big)}{(h+q)^k},$$

where $c = -\mathbf{e}(q)$. This formula was given in [S07, (4.5)]. Thus we have given a proof of (2.13) different from that of [S07]. (The case $k = 1$ must be handled carefully; see [S07, pp. 26–27].) In [S08] we asked the question whether the parameter k in (2.13) can be extended to a complex variable, and presented $D^\nu(s;\,a,\,p)$ as an answer. Since (2.13) is a special case of (2.12), we can now say that (2.12) is another answer to that question.

3. The case of a totally real number field

3.1. Let F be a totally real algebraic number field. We ask whether we can define a function similar to $\zeta(s;\,a,\,\gamma)$ by taking the totally positive integers in F in place of n in (0.1). We are going to give a partially affirmative answer to this question. We let $\mathfrak{g}$ denote the maximal order of F, $\mathfrak{d}$ the different of F relative to $\mathbf{Q}$, and $\mathbf{a}$ the set of all archimedean primes of F. For each $v \in \mathbf{a}$ we denote by F_v the v-completion of F, identified with $\mathbf{R}$. In other words, v defines an injection of F into $\mathbf{R}$, and for $\xi \in F$ we denote by ξ_v the image of ξ under this injection. We put $[F:\mathbf{Q}] = g$, $F_{\mathbf{a}} = \prod_{v\in\mathbf{a}} F_v$, and $F_{\mathbf{a}}^\times = \prod_{v\in\mathbf{a}} F_v^\times$. Then $F_{\mathbf{a}}$ can be identified with $\mathbf{R}^g$, and for $\xi \in F$ the map $\xi \mapsto (\xi_v)_{v\in\mathbf{a}}$ defines an injection of F into $F_{\mathbf{a}}$. We then put

$$\mathbf{e}_{\mathbf{a}}(\xi) = \mathbf{e}\big(\textstyle\sum_{v\in\mathbf{a}}\xi_v\big) \qquad (\xi \in F_{\mathbf{a}}),$$

$$\xi^k = \prod_{v\in\mathbf{a}}\xi_v^{k_v}, \qquad \xi^{\mathbf{a}} = \prod_{v\in\mathbf{a}}\xi_v, \qquad (k = (k_v)_{v\in\mathbf{a}} \in \mathbf{Z}^{\mathbf{a}},\ \xi \in F_{\mathbf{a}}^\times).$$

We also put $N(\xi) = N_{F/\mathbf{Q}}(\xi)$ for $\xi \in F$. Then $N(\xi\mathfrak{g}) = |N(\xi)| = |\xi^{\mathbf{a}}|$ for $\xi \in F^\times$. We write $\xi \gg 0$ if $\xi_v > 0$ for every $v \in \mathbf{a}$ and put $\mathfrak{g}_+^\times = \big\{u \in \mathfrak{g}^\times \,\big|\, u \gg 0\big\}$.

Now for $s \in \mathbf{C}$, a fractional ideal $\mathfrak{b}$, and $a, p \in F$ we put

$$\text{(3.1)}\qquad \zeta(s;\,\mathfrak{b},\,a,\,p) = [\mathfrak{g}_+^\times : U]^{-1}\sum_{0\ll\xi\in\mathfrak{b}+a \pmod{U}} |N(\xi)|^{-s}\mathbf{e}_{\mathbf{a}}(p\xi).$$

Here the sum is taken over $F^\times/U$ under the condition that $\xi \gg 0$ and $\xi - a \in \mathfrak{b}$. We take a subgroup U of $\mathfrak{g}_+^\times$ of finite index such that $u - 1 \in a^{-1}\mathfrak{b} \cap$

$p^{-1}(\mathfrak{b}^{-1}\mathfrak{d}^{-1} \cap a^{-1}\mathfrak{d}^{-1})$ for every $u \in U$. Clearly such a U exists, the sum of (3.1) is meaningful, and $\zeta(s;\, \mathfrak{b},\, a,\, p)$ is well defined independently of the choice of U.

This is an analogue of (0.1), but it should be remembered that here both a and p belong to F. Thus (1.5) is a special case of (3.1).

Next, let $k \in \mathbf{Z}^{\mathbf{a}}$. With s, $\mathfrak{b}$, a, and p as above, we put

$$D_k(s;\, \mathfrak{b},\, a,\, p) = [\mathfrak{g}_+^\times : U]^{-1} \sum_{0 \neq \xi \in \mathfrak{b}+a \ (\mathrm{mod}\ U)} \xi^{-k}|\xi|^{k-s\mathbf{a}}\mathbf{e}_{\mathbf{a}}(p\xi), \tag{3.2}$$

where the summation is the same as in (3.1) except that this time we do not impose the condition $\xi \gg 0$. This is well defined and both (3.1) and (3.2) are convergent for $\mathrm{Re}(s) > 1$.

LEMMA 3.2. *Let $k \in \mathbf{Z}^{\mathbf{a}}$ with $k_v = 0$ or 1 for every $v \in \mathbf{a}$. Then the product*

$$D_k(s;\, \mathfrak{b},\, a,\, p)\prod_{v\in\mathbf{a}} \Gamma\big((s+k_v)/2\big)$$

can be continued to a meromorphic function of s on the whole s-plane that is holomorphic except for possible simple poles at $s = 0$ and 1, which occur only when $k = 0$. The pole at $s = 0$ occurs if and only if $k = 0$ and $a \in \mathfrak{b}$.

This is included in [S00, Lemma 18.2], as $D_k(s;\, \mathfrak{b},\, a,\, p)$ is a special case of the series $D_k(s,\, \kappa)$ of [S00, (18.1)]. Notice that $D_k(s;\, \mathfrak{b},\, a,\, p)$ is finite at $s = 0$ for every k.

LEMMA 3.3. *For $0 < \mu \in \mathbf{Z}$ and $k \in \mathbf{Z}^{\mathbf{a}}$ with $k_v = 0$ or 1 the following assertions hold:*

(i) *$D_k(1-\mu;\, \mathfrak{b},\, a,\, p) = 0$ if $k_v - \mu \notin 2\mathbf{Z}$ for some $v \in \mathbf{a}$.*

(ii) *Suppose $k_v - \mu \in 2\mathbf{Z}$ for every $v \in \mathbf{a}$; then $D_k(1-\mu;\, \mathfrak{b},\, a,\, p) \in \mathbf{Q}_{\mathrm{ab}}$, where $\mathbf{Q}_{\mathrm{ab}}$ denotes the maximal abelian extension of $\mathbf{Q}$; in particular, $D_k(1-\mu;\, \mathfrak{b},\, a,\, 0) \in \mathbf{Q}$.*

Proof. Suppose $k_v - \mu \notin 2\mathbf{Z}$ for one particular v. Then $\Gamma\big((s+k_v)/2\big)$ has a pole at $s = 1-\mu$, and so (i) follows from Lemma 3.2. As for (ii), that $D_k(1-\mu;\, \mathfrak{b},\, a,\, 0) \in \mathbf{Q}$ is given in Proposition 18.10(ii) of [S00]. For any $p \in F$, we see that $D_k(s;\, \mathfrak{b},\, a,\, p)$ is a finite $\mathbf{Q}_{\mathrm{ab}}$-linear combination of $D_k(s;\, \mathfrak{b}',\, a',\, 0)$ with several $(\mathfrak{b}',\, a')$, and so $D_k(1-\mu;\, \mathfrak{b},\, a,\, p) \in \mathbf{Q}_{\mathrm{ab}}$.

Now our principal result on $\zeta(s;\, \mathfrak{b},\, a,\, p)$ can be stated as follows.

THEOREM 3.4. (i) *$\zeta(s;\, \mathfrak{b},\, a,\, p)$ can be continued to a meromorphic function of s on the whole s-plane, which is holomorphic except for a possible simple pole at $s = 1$.*

(ii) *Let $0 < \mu \in \mathbf{Z}$. Then $\zeta(1-\mu;\, \mathfrak{b},\, a,\, p) \in \mathbf{Q}_{\mathrm{ab}}$, and in particular, $\zeta(1-\mu;\, \mathfrak{b},\, a,\, 0) \in \mathbf{Q}$.*

Proof. For $\xi \in F^\times$ we note an easy fact

$$\sum_{k\in\mathbf{Z}^{\mathbf{a}}/2\mathbf{Z}^{\mathbf{a}}} \xi^{-k}|\xi|^k = \begin{cases} 2^g & \text{if } \ \xi \gg 0, \\ 0 & \text{otherwise,} \end{cases}$$

where $g = [F : \mathbf{Q}]$. Therefore we obtain

$$2^g\zeta(s;\,\mathfrak{b},\,a,\,p) = \sum_{k\in\mathbf{Z}^{\mathbf{a}}/2\mathbf{Z}^{\mathbf{a}}} D_k(s;\,\mathfrak{b},\,a,\,p), \tag{3.3}$$

and so assertion (i) follows immediately from Lemma 3.2. Take μ as in (ii). By Lemma 3.3(i), the terms on the right-hand side of (3.3) vanish except for the term with k such that $k_v - \mu \in 2\mathbf{Z}$ for every v. Therefore we obtain the desired result from Lemma 3.3(ii), and our proof is complete.

4. Some explicit expressions for Gauss sums

4.1. There are two kinds of formulas for the critical values of $L(s,\,\chi)$: one is for $L(k,\,\chi)$ and the other for $L(1-k,\,\chi)$. The former involves π and the Gauss sum of χ, whereas the latter does not. In a sense $L(1-k,\,\chi)$ is conceptually more natural than $L(k,\,\chi)$, but there is an interesting aspect in the computation of $L(k,\,\chi)$, since it allows us to find an explicit expression for a certain Gauss sum. This can be achieved by computing $L(k,\,\chi)$ in two different ways, which involve two *different* Gauss sums. Let us begin with the definition of a Gauss sum and an easy lemma.

Given a primitive or an imprimitive Dirichlet character χ' modulo a positive integer, we take the primitive character χ associated with χ', and define the Gauss sum $G(\chi')$ to be the same as the Gauss sum $G(\chi)$ of χ, given by

$$G(\chi) = \sum_{a=1}^{d} \chi(a)\mathbf{e}(a/d), \tag{4.1}$$

where d is the conductor of χ. Now we have an elementary

LEMMA 4.2. (i) *Let $\chi_1,\,\ldots,\,\chi_m$ be Dirichlet characters. Then the number $G(\chi_1)\cdots G(\chi_m)/G(\chi_1\cdots\chi_m)$ belongs to the field generated by the values of $\chi_1,\,\ldots,\,\chi_m$ over $\mathbf{Q}$.*

(ii) *Let ψ and χ be primitive characters of conductor c and d, respectively. If c and d are relatively prime, then*

$$G(\psi\chi) = \psi(d)\chi(c)G(\psi)G(\chi). \tag{4.2}$$

Proof. For the proof of (i) see [S78, Proposition 4.12], which generalizes [S76, Lemma 8]. In the setting of (ii) take $r,\,s \in \mathbf{Z}$ so that $cr + ds = 1$. Then for $x,\,y \in \mathbf{Z}$ the map $(x,\,y) \mapsto xds + ycr$ gives a bijection of $(\mathbf{Z}/c\mathbf{Z}) \times (\mathbf{Z}/d\mathbf{Z})$ onto $\mathbf{Z}/cd\mathbf{Z}$, and so

$$\begin{aligned} G(\psi\chi) &= \sum_{x=1}^{c}\sum_{y=1}^{d} \mathbf{e}\big((xds + ycr)/cd\big)\psi(xds)\chi(ycr) \\ &= \sum_{x=1}^{c} \psi(xds)\mathbf{e}(xs/c) \sum_{y=1}^{d} \chi(ycr)\mathbf{e}(yr/d), \end{aligned}$$

which proves (4.2).

THEOREM 4.3. *Let χ be a primitive character of conductor d such that $\chi(-1) = -1$, and let $\lambda(m) = \left(\frac{3}{m}\right)$. Suppose d is odd and $0 < d/9 \in \mathbf{Z}$. Then*

$$(4.3)\qquad \frac{G(\chi\lambda)}{G(\chi)} = \Bigl\{2\sqrt{3}\chi(2)\sum_{a=1}^{d-1}(\chi\lambda)(a)\Bigr\}\Big/\Bigl\{\sum_{j=1-g}^{d-g}(-1)^j\chi(j)\Bigr\},$$

where $g = [d/6] + 1$.

Proof. In [S07, Theorem 6.3(v)] we gave a formula for $L(k, \chi\lambda)$. Taking $k = 1$ and $\bar{\chi}$ in place of χ, we obtain

$$2\sqrt{3}\chi(2)(\pi i)^{-1}G(\chi)L(1, \bar{\chi}\lambda) = \sum_{j=1-g}^{d-g}(-1)^j\chi(j),$$

since $E_{1,0}(t) = 1$. Now $\bar{\chi}\lambda$ is primitive and has conductor $4d$, and so the formula [S07, (4.34)] applied to $\bar{\chi}\lambda$ produces

$$(\pi i)^{-1}G(\chi\lambda)L(1, \bar{\chi}\lambda) = \sum_{a=1}^{d-1}(\chi\lambda)(a).$$

Taking the quotient of these two formulas, we obtain (4.3).

In [S07, Theorem 6.3] we gave eight formulas for $L(k, \chi\lambda)$, where λ is a "constant" character and χ is a "variable" character. In the above theorem we employed only one of those formulas. We can actually state results about $G(\chi\lambda)/G(\chi)$ in the other seven cases, but they are not so interesting, since we can apply (4.2) to $\chi\lambda$ in those cases, and the case we employed in the above theorem is the only case to which (4.2) is not applicable. Even in that case, the significance of (4.3) is rather obscure. Still, the formula is clear-cut and nontrivial, and we state it here with the hope that future researchers will be able to clarify its nature in a better perspective.

We end our discussion of this subject by showing the quantity of (4.3) can be determined in a different way. We begin with some preliminary results.

LEMMA 4.4. *Let* χ *and* ψ *be primitive characters of conductor* p^m *and* p^n, *respectively, where* p *is a prime number and* m, n *are positive integers. Suppose* $m \ge n$ *and* $\chi\psi$ *has conductor* p^m. *Then*

$$(4.4)\qquad G(\chi)G(\psi) = G(\chi\psi)\sum_{a=1}^{p^n}\chi(1 - p^{m-n}a)\psi(a).$$

This was given in [S76, (4.2)].

LEMMA 4.5. *Let* χ *be a primitive character of conductor* c, *where* $c = 3^m$ *with* $m > 1$ *or* $c = 2^m$ *with* $m > 3$, *and let* μ_3 *and* μ_4 *denote the primitive characters of conductor* 3 *and* 4, *respectively. Let* $\chi' = \chi\mu_3$ *if* $c = 3^m$ *and* $\chi' = \chi\mu_4$ *if* $c = 2^m$. *Then* $G(\chi') = \varepsilon G(\chi)$ *with* $\varepsilon = \pm 1$ *determined by* $\chi(1 - 3^{m-1}) = \mathbf{e}(\varepsilon/3)$ *if* $c = 3^m$ *and* $\chi(1 - 2^{m-2}) = \mathbf{e}(\varepsilon/4)$ *if* $c = 2^m$.

Proof. We first consider the case $c = 2^m$. By (4.4), $G(\chi)G(\mu_4)/G(\chi') = \beta - \gamma$ with $\beta = \chi(1-2^{m-2})$ and $\gamma = \chi(1+2^{m-2})$. Since $m > 3$, we have $(1-2^{m-2})^2 \equiv 1-2^{m-1} \pmod{2^m}$, $(1-2^{m-2})(1+2^{m-2}) \equiv 1 \pmod{2^m}$, and $(1-2^{m-2})^4 \equiv 1 \pmod{2^m}$, and so $\beta^4 = \beta\gamma = 1$. Suppose $\beta = \pm 1$. Then $\chi(1 - 2^{m-1}) = 1$, and so $\chi(1 - 2^{m-1}a) = \chi(1 - 2^{m-1})^a = 1$ for every $a \in \mathbf{Z}$, which means that χ

has conductor $\le 2^{m-1}$, a contradiction. Thus $\beta \neq \pm 1$, and so $\beta = \varepsilon i$ with $\varepsilon = \pm 1$. Since $G(\mu_4) = 2i$, we have $G(\chi)/G(\chi') = (\beta - \beta^{-1})/2i = \varepsilon$, which proves the case $c = 2^m$.

If $c = 3^m$, we have similarly $G(\chi)G(\mu_3)/G(\chi') = \beta - \gamma$ with $\beta = \chi(1-3^{m-1})$ and $\gamma = \chi(1+3^{m-1})$. We easily see that $\beta\gamma = \beta^3 = 1$ and $\beta \neq 1$. Since $G(\mu_3) = \sqrt{3}i$, we have $G(\chi)/G(\chi') = (\beta - \beta^{-1})/\sqrt{3}i = \varepsilon$, and ε is determined by $\beta = \mathbf{e}(\varepsilon/3)$. This completes the proof.

4.6. Returning to the setting of Theorem 4.3, put $d = 3^m f$ with $m \in \mathbf{Z}$ and $0 < f \in \mathbf{Z}$, $3 \nmid f$; put also $\chi = \chi_0\chi_1$ with characters χ_0 and χ_1 of conductor 3^m and f, respectively. Since $\lambda = \mu_3\mu_4$, we have $\chi\lambda = \chi_0\mu_3\chi_1\mu_4$. By (4.2) we have $G(\chi_1\mu_4) = 2i\chi_1(4)\mu_4(f)G(\chi_1)$, and so

$$\begin{aligned} G(\chi\lambda) &= (\chi_1\mu_4)(3^m)(\chi_0\mu_3)(4f)G(\chi_0\mu_3)G(\chi_1\mu_4) \\ &= 2i(-1)^m\chi_0(f)\chi(4)\chi_1(3^m)\lambda(f)G(\chi_0\mu_3)G(\chi_1). \end{aligned}$$

Also, $G(\chi) = \chi_0(f)\chi_1(3^m)G(\chi_0)G(\chi_1)$. Therefore

$$G(\chi\lambda)/G(\chi) = 2i(-1)^m\lambda(f)\chi(4)G(\chi_0\mu_3)/G(\chi_0).$$

Applying Lemma 4.5 to $G(\chi_0\mu_3)/G(\chi_0)$, we thus obtain

$$G(\chi\lambda)/G(\chi) = 2i(-1)^m\lambda(f)\chi(4)\varepsilon, \tag{4.5}$$

where $\varepsilon = \pm 1$ is determined by $\chi_0(1-3^{m-1}) = \mathbf{e}(\varepsilon/3)$.

5. THE CASE OF A DOMAIN OF POSITIVITY

5.1. There is a natural analogue of $\zeta(s;\, a,\, \gamma)$ defined on the space of symmetric matrices. To be explicit, with a positive integer n we denote by V the set of all real symmetric matrices of size n, and write $h > 0$ resp. $h \ge 0$ for $h \in V$ when h is positive definite resp. nonnegative. We put $V_{\mathbf{C}} = V \otimes_{\mathbf{R}} \mathbf{C}$,

$$\kappa = (n+1)/2, \quad P = \{h \in V \mid h > 0\}, \quad \text{and}$$
$$\mathfrak{H} = \{x + iy \in V_{\mathbf{C}} \mid x \in V,\, y \in P\}.$$

For $h \in V$ we denote by $\lambda(h)$ and $\mu(h)$ the maximum and minimum absolute value of eigenvalues of h, respectively. We easily see that V is a normed space with $\lambda(h)$ as the norm of h, that is, $\lambda(ch) = |c|\lambda(h)$ for $c \in \mathbf{R}$ and $\lambda(h+k) \le \lambda(h) + \lambda(k)$. For $0 \le d \in \mathbf{Z}$ we denote by S_d the space of all $\mathbf{C}$-valued homogeneous polynomial functions on V of degree d. Here are two easy facts:

$$\mathrm{tr}(gh) \ge \lambda(h)\mu(g) \quad \text{if} \quad g,\, h \in P. \tag{5.1}$$

(5.2) $|\xi(h)| \le c_\xi\lambda(h)^d$ for every $\xi \in S_d$ and $h \in V$ with a positive constant c_ξ that depends only on ξ.

To prove (5.1), we may assume that g is diagonal. For $g = \mathrm{diag}[\mu_1, \ldots, \mu_n]$ we have $\mathrm{tr}(gh) = \sum_{i=1}^n \mu_i h_{ii} \ge \mu(g)\mathrm{tr}(h) \ge \mu(g)\lambda(h)$, since $\mathrm{tr}(h)$ is the sum of all eigenvalues of h. As for (5.2), given $h \in V$, take an orthogonal matrix p so that $h = {}^t p \cdot \mathrm{diag}[\kappa_1, \ldots, \kappa_n] p$ with $\kappa_i \in \mathbf{R}$. Since $|p_{ij}| \le 1$ and $|\kappa_i| \le \lambda(h)$, we see that $|h_{ij}| \le n\lambda(h)$, from which we obtain (5.2).

We now consider three types of infinite series:

$$\Phi(s;\, L,\, a,\, z) = \sum_{h \in L} \det\bigl(2\pi i(h - z)\bigr)^{-s} \mathbf{e}\bigl(\mathrm{tr}(a(h - z))\bigr), \tag{5.3}$$

$$F(s;\, L,\, a,\, z) = \sum_{h \in L,\, h+a>0} \det(h + a)^{-s} \mathbf{e}\bigl(\mathrm{tr}(hz)\bigr), \tag{5.4}$$

$$F_0(s;\, L,\, a,\, z) = \sum_{0 \le h \in L} \det(h + a)^{-s} \mathbf{e}\bigl(\mathrm{tr}(hz)\bigr). \tag{5.5}$$

Here $s \in \mathbf{C}$, L is a lattice in V, $a \in V$, $z \in \mathfrak{H}$, and the sum of (5.4) is extended over all $h \in L$ such that $h + a \in P$; the sum of (5.3) is simply over all $h \in L$; in (5.5) we assume that $a > 0$ and the sum is extended over all nonnegative h in L. For $z \in \mathfrak{H}$ and $s \in \mathbf{C}$ we define $\det(-2\pi i z)^s$ so that it coincides with $\det(2\pi p)^s$ if $z = ip$ with $p \in P$.

LEMMA 5.2. *For every $\xi \in S_d$ the infinite series*

$$\sum_{h \in L,\, h+a>0} \det(h + a)^s \xi(h + a) \mathbf{e}\bigl(\mathrm{tr}(hz)\bigr) \tag{5.6}$$

converges absolutely and locally uniformly for $(s,\, a,\, z) \in \mathbf{C} \times V \times \mathfrak{H}$.

Proof. For a fixed positive number α the number of $h \in L$ such that $\lambda(h) \le \alpha$ is finite, as $\{h \in V \mid \lambda(h) \le \alpha\}$ is compact and L is discrete in V. Thus, to prove the convergence of (5.6), we can restrict h to those that satisfy $\lambda(h) > \lambda(a) + 1$ for every $a \in A$, where A is a fixed compact subset of V. For such an h we have $1 \le \lambda(h + a) < 2\lambda(h)$ and $\det(h + a) \le \lambda(h + a)^n \le 2^n \lambda(h)^n$. Also, by (5.2), $|\xi(h + a)| \le c_\xi \lambda(h + a)^d \le c_\xi 2^d \lambda(h)^d$. Thus for $\mathrm{Re}(s) = \sigma$ we have $|\det(h + a)^s \xi(h + a)| \le 2^{n\sigma+d} c_\xi \lambda(h)^{n\sigma+d}$. On the other hand, for $0 < N \in \mathbf{Z}$ the number of $h \in L$ such that $N \le \lambda(h) < N + 1$ is less than CN^b with positive constants C and b. Put $g = 2\pi \mathrm{Im}(z)$. Then $|\mathbf{e}(\mathrm{tr}(hz))| = e^{-\mathrm{tr}(gh)}$. Since $-\mathrm{tr}(gh) \le -\lambda(h)\mu(g)$ by (5.1), our partial sum can be majorized by $2^{n\sigma+d} c_\xi C \sum_{N=1}^{\infty} (N + 1)^{n\sigma+d+b} e^{-N\mu(g)}$, which proves our lemma.

THEOREM 5.3. (i) *The infinite series of (5.3)* converges absolutely and locally uniformly in $(s,\, a,\, z) \in \{s \in \mathbf{C} \mid \mathrm{Re}(s) > n\} \times V \times \mathfrak{H}$. Thus it defines a holomorphic function in s for $\mathrm{Re}(s) > n$.

(ii) *The infinite series of (5.4)* and *(5.5) converge absolutely and locally uniformly in $(s,\, a,\, z) \in \mathbf{C} \times V \times \mathfrak{H}$, and so they define entire functions of s.*

(iii) *For* $\mathrm{Re}(s) > n$ *we have*

$$\mathrm{vol}(V/L) F(\kappa - s;\, L,\, a,\, z) = \Gamma_n(s) \Phi(s;\, L',\, a,\, z), \tag{5.7}$$

where $L' = \{x \in V \mid \mathrm{tr}(xL) \subset \mathbf{Z}\}$ and

$$\Gamma_n(s) = \pi^{n(n-1)/4} \prod_{k=0}^{n-1} \Gamma\bigl(s - (k/2)\bigr).$$

(iv) *For* $0 \leq m \in \mathbf{Z}$ *the values* $F(-m;\, L,\, a,\, z)$, $F_0(-m;\, L,\, a,\, z)$, *and* $\Phi(\kappa + m;\, L,\, a,\, z)$ *are polynomial functions of* a *of degree* $\leq mn$, *whose coefficients depend on* L *and* z.

Statement (iv) for F_0 can be taken literally, but the cases of F and Φ require some clarifications, which will be given in the proof.

Proof. Assertion (i) is included in [S82, Lemma 1.3]; (ii) follows from Lemma 5.2. To prove (iii), given $s \in \mathbf{C}$ and $p \in P$, we consider a function g on V defined by

$$g(u) = \begin{cases} e^{-\mathrm{tr}(up)} \det(u)^{s-\kappa} & (u \in P), \\ 0 & (u \notin P), \end{cases}$$

and define its Fourier transform $\hat{g}$ by

$$\hat{g}(t) = \int_V \mathbf{e}\big(-\mathrm{tr}(ut)\big) g(u) du \qquad (t \in V),$$

where $du = \prod_{i \leq j} du_{ij}$. As shown in [S82, (1.22)], $\hat{g}(t) = \Gamma_n(s) \det(p + 2\pi i t)^{-s}$, provided $\mathrm{Re}(s) > \kappa - 1$. Then the Poisson summation formula establishes equality (5.7) with both sides multiplied by $\mathbf{e}\big(\mathrm{tr}(az)\big)$ for $z = (-2\pi i)^{-1} p$ when the series of (5.3) and (5.4) are convergent, which is the case at least when $\mathrm{Re}(s) > n$. Since both sides of (5.7) are holomorphic in z if $\mathrm{Re}(s) > n$, we obtain (iii) as stated.

To prove (iv), take a $\mathbf{C}$-basis B of $\sum_{d=0}^{mn} S_d$. For $0 \leq m \in \mathbf{Z}$ we have

$$F_0(-m;\, L,\, a,\, z) = \sum_{0 \leq h \in L} \det(h + a)^m \mathbf{e}\big(\mathrm{tr}(hz)\big). \tag{5.8}$$

We can put $\det(h+a)^m = \sum_{\beta \in B} \beta(a) f_\beta(h)$ with polynomial functions f_β, and so

$$F_0(-m;\, L,\, a,\, z) = \sum_{\beta \in B} \beta(a) G_\beta(z) \quad \text{with} \tag{5.8a}$$

$$G_\beta(z) = \sum_{0 \leq h \in L} f_\beta(h) \mathbf{e}\big(\mathrm{tr}(hz)\big). \tag{5.8b}$$

Thus $F_0(-m;\, L,\, a,\, z)$ is a polynomial in a as stated in (iv). This argument is basically valid for F in place of F_0, but the functions corresponding to G_β in that case may depend on a. To avoid that difficulty, we first take a compact subset A of V and restrict a to A. As shown in the proof of Lemma 5.2, we can find a subset M of L independent of a such that

$$\big\{h \in L \,\big|\, h > 0,\, h + a > 0\big\} = M \sqcup K_a,$$

with a finite set K_a for each $a \in A$. Then, taking F in place of F_0, we obtain, for $a \in A$,

$$F(-m;\, L,\, a,\, z) = \sum_{\beta \in B} \beta(a) H_\beta(z) \quad \text{with} \tag{5.9a}$$

$$H_\beta(z) = \sum_{h \in K_a} f_\beta(h) \mathbf{e}\big(\mathrm{tr}(hz)\big) + \sum_{h \in M} f_\beta(h) \mathbf{e}\big(\mathrm{tr}(hz)\big). \tag{5.9b}$$

Thus the statement about $F(-m;\, L,\, a,\, z)$ in (iv) must be understood in the sense of (5.9a, b). It is a polynomial in a whose coefficients $H_\beta(z)$ is the sum of a "principal part" that is independent of a and a finite sum depending on a.

As for the value of Φ, from (5.7) we see that Φ is an entire functions of s. Also $\Gamma_n(s)^{-1}$ is nonzero for $(n-1)/2 < s \in \mathbf{R}$. Thus $\Phi(\kappa + m;\, L',\, a,\, z)$ is a nonzero constant times $F(-m;\, L,\, a,\, z)$, and so is a polynomial in a in the sense explained above.

5.4. Let us add some remarks. Clearly F_0 of (5.5) is a natural generalization of (0.1), but we introduced F and Φ as in (5.4) and (5.3), as we think they are natural objects of study closely related to (5.5). It must be remembered, however, that (5.5) includes (0.1) as a special case only if $|\gamma| < 1$. To define something like (5.5) that includes (0.1) with $|\gamma| = 1$ is one of the open problems in this area.

Next, there are four classical types of domains of positivity associated with tube domains discussed in [S82]. Our V, P, and $\mathfrak{H}$ in this section belong to the easiest type. We can in fact define the analogues of (5.3), (5.4), and (5.5) for all three other types of domains, and prove the results similar to Theorem 5.3 in those cases.

As to the nature of the polynomials in the variable a obtained in Theorem 5.3(iv), we do not have their description as explicit as what we know about $E_{c,m}(t)$. Still, we can show that they are of a rather special kind. For that purpose we need a matrix of differential operators $\partial_a = (\partial_{ij})_{i,j=1}^n$ on V as folows. Taking a variable symmetric matrix $a = (a_{ij})$ on V, we put $\partial_{ii} = \partial/\partial a_{ii}$ and $\partial_{ij} = 2^{-1}\partial/\partial a_{ij}$ for $i \neq j$. Then for every $\varphi \in S_d$ we can define a diffrential operator $\varphi(\partial_a)$. In particular, taking $\varphi(a) = \det(a)$, we put

$$\Delta_a = \det(\partial_a) = \sum_\sigma \operatorname{sgn}(\sigma)\partial_{1\sigma(1)} \cdots \partial_{n\sigma(n)}, \tag{5.10}$$

where σ runs over all permutations of $\{1,\, \ldots\, ,\, n\}$. It is well known that

$$\Delta_a\bigl(\det(a)^s\bigr) = \prod_{k=0}^{n-1}(s + k/2)\cdot \det(a)^{s-1}. \tag{5.11}$$

This is a special case of a general formula on $\varphi(\partial_a)\det(z)^s$ for $\varphi \in S_d$ given in [S84].

Fixing L and z, for $0 \le m \in \mathbf{Z}$ put

$$\mathscr{E}_m(a) = F_0(-m;\, L,\, a,\, z). \tag{5.12}$$

We have shown that $\mathscr{E}_m$ is a polynomial of degree $\le mn$. Notice that $\mathscr{E}_0(a) = \sum_{0 \le h \in L} \mathbf{e}\bigl(\operatorname{tr}(hz)\bigr)$. From (5.8) and (5.11) we obtain

$$\Delta_a \mathscr{E}_m(a) = \prod_{k=0}^{n-1}(m + k/2)\cdot \mathscr{E}_{m-1}(a). \tag{5.13}$$

This is a generalization of the formula $(d/dt)E_{c,m}(t) = mE_{c,m-1}(t)$, noted in [S07, (4.3c)].

References

[E] A. Erdélyi (ed.), Higher Transcendental Functions I, McGraw-Hill, New York, 1953.

[Hu] A. Hurwitz, Einige Eigenschaften der Dirichlet'schen Functionen $F(s) = \sum \left(\frac{D}{n}\right) \cdot 1/n^s$, die bei der Bestimmung der Classenanzahlen binärer quadratischer Formen auftreten, Zeitschrift für Math. und Phys. xxvii (1882), 86–101 (= Mathematische Werke I, 72–88).

[Le] M. Lerch, Note sur la fonction $\mathfrak{K}(w,\, x,\, s) = \sum_{k=0}^{\infty} e^{2k\pi ix}/(w+k)^s$, Acta mathematica, 11 (1887), 19–24.

[Li] R. Lipschitz, Untersuchung der Eigenschaften einer Gattung von unendlichen Reihen, Journ. für die reine u. angew. Mathematik, 105 (1889), 127–156.

[R] B. Riemann, Über die Anzahl der Primzahlen unter einer gegebenen Grösse, Monatsberichten der Berliner Akademie 1859 (= Gesammelte Mathematische Werke, 145–153).

[S76] G. Shimura, The special values of the zeta functions associated with cusp forms, Communications on pure and applied Mathematics, 29 (1976), 783–804 (=Collected Papers II, 740–761).

[S78] G. Shimura, The special values of the zeta functions associated with Hilbert modular forms, Duke Mathematical Journal 45 (1978), 637–679. (Corrections to this paper were given in Duke Mathematical Journal 48 (1981), 697. A better revised version of the paper can be found in Collected Papers III, 75–114.)

[S82] G. Shimura, Confluent hypergeometric functions on tube domains, Mathematische Annalen, 260 (1982), 269–302 (=Collected Papers III, 388–421).

[S84] G. Shimura, On differential operators attached to certain representations of classical groups, Inventiones mathematicae, 77 (1984), 463–488 (=Collected Papers III, 584–609).

[S00] G. Shimura, Arithmeticity in the theory of automorphic forms, Mathematical Surveys and Monographs, vol. 82, American Mathematical Society, 2000.

[S07] G. Shimura, Elementary Dirichlet series and modular forms, Springer Monographs in Mathematics, Springer, 2007.

[S08] G. Shimura, The critical values of certain Dirichlet series, Documenta Mathematika 13, (2008), 775–794.

[WW] E. T. Whittaker and G. N. Watson, A Course of Modern Analysis, 4th ed., Cambridge, 1927.

Goro Shimura
Department of Mathematics
Princeton University
Princeton
NJ 08544–1000
USA
goro@Math.Princeton.EDU

Some remarks and open problems on the Woods Hole fixed point formula

Goro Shimura

I gave a series of lectures titled "The zeta function of an algebraic variety and automorphic functions" at the Summer Institute on Algebraic Geometry held at Woods Hole, Massachusetts, July 6-31, 1964. This is reproduced as [64e] in my Collected Papers, Vol. I. Though the lecture is not directly related to the formula in the title of this article, I mention it, since the type of Hecke operators discussed there and its higher-dimensional generalizations were among the themes in which I was interested at that time.

In [E57] Eichler proved a trace formula for an algebraic correspondence on an algebraic curve over the complex number field, including such Hecke operators as special cases. In the Spring of 1964 I came up with the idea of a new formula that would generalize Eichler's formula. It concerns an algebraic correspondence between two complex algebraic varieties of the same dimension, or rather, an algebraic correspondence between the holomorphic bundles on such varieties, consistent with the correspondence on the base varieties. When the bundles are trivial, the formula is just the Lefschetz fixed point formula for complex algebraic varieties.

I told the idea to Mikio Ise who was at the Institute for Advanced Study, and also to André Weil, but otherwise I kept it to myself. During the Woods Hole conference, I recalled this conjectural formula, and told it first to John Tate, who I knew since 1958. Then Tate told it to Atiyah and Bott. Eventually I discussed the matter with them. I met them for the first time then. The idea was completely new to them; in fact nobody there, except me, was aware of Eichler's result. Practically all participants of the conference became interested in this topic. I remember that a large room was filled by many people who with their pencils and notebooks tried to do something on the question, but only those two came up with concrete results. Later Atiyah and Bott said there were some who did not believe in the formula, but I was not aware of any such opinion during the conference.

Atiyah and Bott announced their results in the Bulletin of the American Mathematical Society, and later published the full version in the Annals of Mathematics. Initially they acknowledged that the idea was due to me, but later they tried to minimize my contribution, in their Annals article, for example. In fact in the April 2001 Notices of the American Mathematical Society Bott did not mention my involvement in the matter, and so I wrote a letter to the editor of the Notices. See Notices, vol. 48, No. 7 (August 2001, p. 678). Bott then conceded (in another Notices article) that he was wrong and apologized to me. However, Atiyah noted somewhere in his complete works that they learned it from me. (I already noted these in my book [S08], p. 131.)

Here are some comments. One is relatively minor. Bott, in his response to my letter to the editor of the Notices, still stated that he and Atiyah had something before hearing my idea. However, I don't see how that can be, as he never said so in their first article on this topic, and Atiyah never said so. I complained about this to the editor of the Notices, but he and I were tired of the matter and we decided not to pursue the issue.

There is a far more meaningful point. Atiyah and Bott proved a formula for a map, not for a correspondence. Thus their formula does not include Eichler's formula of 1957, and I believe, my conjectural formula is still an open problem. It should be noted that all the algebraic correspondences on an algebraic curve form a ring, which is an important object of study, and the classical Lefschetz fixed point formula is applicable to a corresspondence, not merely to a map. This is one of the main reasons why I am writing the present article.

Let me mention two more important points, one concerned with the case involving modular forms of half-integral weight, and the other with the possibility of an algebraic approach. As for forms of half-integral weight, we have to deal with certain coverings of the base algebraic varieties. I discussed the problem in the one-dimensional case in [S74]. Actually Kappus [K62] and Eichler [E63] reformulated Eichler's formula in [E57] more algebraically, and I employed some of their ideas in my paper. At the time of the Woods Hole conference I was aware of [K62] and [E63] as Eichler had sent me a complimentary copy of [E63], but I didn't mention them at the conference, since I thought [E57] was enough. The formula I gave in [S74] is practical at least to the extent that one can use it effectively for numerical calculations of the traces of Hecke operators in the case of half-integral weight.

More importantly, the methods of [K62], [E63], and [S74] seem to suggest the possibility of a purely or partly algebraic approach to my conjectural formula. I hope the problems raised in the last few paragraphs will be taken up by future researchers.

References

[E57] M. Eichler, Eine Verallgemeinerung der Abelschen Integrale, Math. Z. 67 (1957), 267–298.

[E63] M. Eichler, Einführung in die Theorie der algebraischen Zahlen und Funktionen, Birkhäuser, 1963 (English edition, Academic Press, 1966).

[K62] H. Kappus, Darstellung von Korrespondenzen algebraisher Funktionenkörper und ihre Spuren, J. Reine u. Angew. Math. 21 (1962), 123–140.

[S74] G. Shimura, On the trace formula for Hecke operators, Acta mathematica, 132 (1974), 245–281 (= Collected Papers, vol. II, [74]).

[S08] G. Shimura, The Map of My Life, Springer, 2008.

Printed in the United States of America